Time Series Modelling

Time Series Modelling

Editor

Christian H. Weiss

MDPI • Basel • Beijing • Wuhan • Barcelona • Belgrade • Manchester • Tokyo • Cluj • Tianjin

Editor
Christian H. Weiss
Helmut-Schmidt-University
Germany

Editorial Office
MDPI
St. Alban-Anlage 66
4052 Basel, Switzerland

This is a reprint of articles from the Special Issue published online in the open access journal *Entropy* (ISSN 1099-4300) (available at: https://www.mdpi.com/journal/entropy/special issues/Time Series Model).

For citation purposes, cite each article independently as indicated on the article page online and as indicated below:

LastName, A.A.; LastName, B.B.; LastName, C.C. Article Title. *Journal Name* **Year**, *Volume Number*, Page Range.

ISBN 978-3-0365-2121-3 (Hbk)
ISBN 978-3-0365-2122-0 (PDF)

© 2021 by the authors. Articles in this book are Open Access and distributed under the Creative Commons Attribution (CC BY) license, which allows users to download, copy and build upon published articles, as long as the author and publisher are properly credited, which ensures maximum dissemination and a wider impact of our publications.

The book as a whole is distributed by MDPI under the terms and conditions of the Creative Commons license CC BY-NC-ND.

Contents

About the Editor vii

Christian H. Weiß
Time Series Modelling
Reprinted from: *Entropy* **2021**, *23*, 1163, doi:10.3390/e23091163 1

Ayumu Nono, Yusuke Uchiyama and Kei Nakagawa
Entropy Based Student's *t*-Process Dynamical Model
Reprinted from: *Entropy* **2021**, *23*, 560, doi:10.3390/e23050560 5

Adriana AnaMaria Davidescu, Simona-Andreea Apostu and Andreea Paul
Comparative Analysis of Different Univariate Forecasting Methods in Modelling and Predicting the Romanian Unemployment Rate for the Period 2021–2022
Reprinted from: *Entropy* **2021**, *23*, 325, doi:10.3390/e23030325 17

Michael R. Lindstrom, Hyuntae Jung and Denis Larocque
Functional Kernel Density Estimation: Point and Fourier Approaches to Time Series Anomaly Detection
Reprinted from: *Entropy* **2020**, *22*, 1363, doi:10.3390/e22121363 49

Eliana Vivas, Héctor Allende-Cid, Rodrigo Salas and Lelys Bravo
A Systematic Review of Statistical and Machine Learning Methods for Electrical Power Forecasting with Reported MAPE Score
Reprinted from: *Entropy* **2020**, *22*, 1412, doi:10.3390/e22121412 65

Raanju R. Sundararajan, Ron Frostig and Hernando Ombao
Modeling Spectral Properties in Stationary Processes of Varying Dimensions with Applications to Brain Local Field Potential Signals
Reprinted from: *Entropy* **2020**, *22*, 1375, doi:10.3390/e22121375 89

Dietmar Bauer and Rainer Buschmeier
Asymptotic Properties of Estimators for Seasonally Cointegrated State Space Models Obtained Using the CVA Subspace Method
Reprinted from: *Entropy* **2021**, *23*, 436, doi:10.3390/e23040436 113

Ines Nüßgen and Alexander Schnurr
Ordinal Pattern Dependence in the Context of Long-Range Dependence
Reprinted from: *Entropy* **2021**, *23*, 670, doi:10.3390/e23060670 155

Jie Huang and Fukang Zhu
A New First-Order Integer-Valued Autoregressive Model with Bell Innovations
Reprinted from: *Entropy* **2021**, *23*, 713, doi:10.3390/e23060713 193

Zhengwei Liu and Fukang Zhu
A New Extension of Thinning-Based Integer-Valued Autoregressive Models for Count Data
Reprinted from: *Entropy* **2021**, *23*, 62, doi:10.3390/e23010062 211

Kaizhi Yu, Huiqiao Wang
A New Overdispersed Integer-Valued Moving Average Model with Dependent Counting Series
Reprinted from: *Entropy* **2021**, *23*, 706, doi:10.3390/e23060706 229

Congmin Liu, Jianhua Cheng and Dehui Wang
Statistical Inference for Periodic Self-Exciting Threshold Integer-Valued Autoregressive Processes
Reprinted from: *Entropy* **2021**, *23*, 765, doi:10.3390/e23060765 . . . . **243**

Cong Li, Shuai Cui and Dehui Wang
Monitoring the Zero-Inflated Time Series Model of Counts with Random Coefficient
Reprinted from: *Entropy* **2021**, *23*, 372, doi:10.3390/e23030372 . . . . **275**

Byungsoo Kim, Sangyeol Lee and Dongwon Kim
Robust Estimation for Bivariate Poisson INGARCH Models
Reprinted from: *Entropy* **2021**, *23*, 367, doi:10.3390/e23030367 . . . . **291**

Yuliya Shapovalova, Nalan Bastürk and Michael Eichler
Multivariate Count Data Models for Time Series Forecasting
Reprinted from: *Entropy* **2021**, *23*, 718, doi:10.3390/e23060718 . . . . **317**

Manuel Stapper
Count Data Time Series Modelling in Julia—The CountTimeSeries.jl Package and Applications
Reprinted from: *Entropy* **2021**, *23*, 666, doi:10.3390/e23060666 . . . . **341**

About the Editor

Christian H. Weiß is a Professor in the Department of Mathematics and Statistics at the Helmut Schmidt University in Hamburg, Germany. He received his doctoral degree in mathematical statistics from the University of Würzburg, Germany. His research areas include time series analysis, statistical quality control, and computational statistics. He is an author of several textbooks and has published his work in international scientific journals such as Bernoulli, Entropy, Journal of the American Statistical Association, Journal of Multivariate Analysis, Journal of Quality Technology, Journal of the Royal Statistical Society, and Journal of Time Series Analysis.

Editorial
Time Series Modelling

Christian H. Weiß

Department of Mathematics and Statistics, Helmut Schmidt University, 22043 Hamburg, Germany; weissc@hsu-hh.de

Keywords: time series; models

Time series consist of data observed sequentially in time, and they are assumed to stem from an underlying stochastic process. The scope of time series approaches thus covers models for stochastic processes as well as inferential procedures for model fitting, model diagnostics, forecasting, and various other applications. While time series data have been collected for a relatively long time in history (one may recall the famous time series on sunspot numbers), the development of methods and stochastic models for such time series is more recent. Indeed, one of the motivations for announcing the Special Issue in 2020 was the fact that this year can be considered a twofold 'anniversary year' of time series modeling. On the one hand, the correlogram, the autoregressive (AR), and the moving-average (MA) models for time series, all of which are nowadays part of any course on time series analysis and covered by any statistical software, date back to the 1920s (mainly driven by G. U. Yule, G. T. Walker, and E. E. Slutzky; see Nie and Wu [1] for a detailed discussion). On the other hand, the first comprehensive textbook on time series was published by Box and Jenkins [2] in 1970, so 2020 allowed the celebration of both the semi-centennial and centennial anniversary at the same time. In keeping with this anniversary, it was indeed possible to collect articles on a wide range of topics in this Special Issue: stochastic models for time series as well as methods for their analysis, univariate and multivariate time series, real-valued and discrete-valued time series, applications of time series methods to forecasting and statistical process control, and software implementations of methods and models for time series. The remainder of this editorial provides a brief summary of the contributions to this Special Issue, grouping the articles thematically.

Roughly one-half of the contributed articles deal with real-valued time series (thus having a continuous range). In Nono et al. [3], an entropy-based Student's t-process dynamical model is proposed for dealing with non-Gaussian and non-linear univariate time series, whose relevance is demonstrated by an application to financial time series. The paper by Davidescu et al. [4] is centered around the time series of Romanian unemployment rates, which serves as the base for comparing the forecast performance of several well-established time series models. Not a single time series, but a large collection of univariate time series is considered by Lindstrom et al. [5], who use functional kernel density estimation for uncovering anomalous time series within such a collection. They apply their approaches to time series on aviation safety events as provided by the International Air Transport Association. Another data-intensive application area is electrical power forecasting, where both statistical and machine-learning methods are used. Vivas et al. [6] provide a systematic review of both types of methods (as well as of hybrid models) regarding forecast performance. Multiple time series are also considered by Sundararajan et al. [7], but now with a focus on multivariate time series having unequal dimensions. They propose and investigate a frequency-specific spectral ratio statistic, which is used to uncover differences in the spread of spectral information in a pair of such time series, and which is applied to data from stroke experiments. Another article on multivariate time series, following types of integrated vector ARMA models, is the one by Bauer and Buschmeier [8], who investigate estimators resulting from canonical variate analysis as well as novel cointegration tests. For illustration, they present an application to hourly electricity consumption data.

The final article presented in the group of real-valued time series also constitutes a bridge to the next group of articles—namely, to those on discrete-valued time series. Nüßgen and Schnurr [9] consider a multivariate long-range dependent Gaussian time series, but they analyze its dependence structure based on discrete ordinal patterns derived thereof. The estimators of ordinal pattern dependence are analyzed asymptotically and within a simulation study.

The second half of contributed articles deals with time series having a discrete range, or more precisely, with count time series, where the observations are count values from the set of non-negative integers. A common approach to adapt the ARMA model known for real-valued time series to the case of count time series consists of substituting the multiplications within the ARMA recursive model by types of thinning operators; see Chapters 2–3 in Weiß [10] for detailed background. The resulting integer-valued counterparts to the ordinary AR and MA models are then referred to as INAR and INMA models, respectively. In Huang and Zhu [11], a new type of the classical INAR model using binomial thinning is proposed, where the innovations follow the one-parameter Bell distribution. Stochastic properties and estimation approaches are investigated, and applications to time series consisting of crime counts and strike counts are presented. Liu and Zhu [12], by contrast, develop an extension of the INAR model, where a new type of thinning operator is used, relying on the extended binomial distribution. The resulting model is able to flexibly adapt to different types of dispersion behavior, which is also demonstrated by several real-data examples. Furthermore, Yu and Wang [13] consider an extension of the binomial thinning operator, achieved by allowing for a dependent counting series, and this time, the operator is used within the class of INMA models. Properties of, and estimation for, this new type of INMA model are investigated, and they are illustrated by an application to a crime-counts time series. While the three aforementioned articles consider stationary and linear count time series, the contribution by Liu et al. [14] deals with non-stationary and non-linear time series as obtained from the periodic self-exciting threshold INAR model. Properties and estimation are discussed, and an application to monthly counts of claimants is presented. In Li et al. [15], again an INAR model is considered (using a randomized binomial thinning operator); however, now the main focus is not on the model itself, but on an approach for statistical process control. The authors use a cumulative sum chart for process monitoring, discuss its performance evaluation, and apply it to a crime-counts time series. Contrary to the aforementioned papers, the articles by Kim et al. [16] and Shapovalova et al. [17] refer to multivariate count time series. For a bivariate count time series following an integer-valued generalized AR conditional heteroscedastic (INGARCH) model, Kim et al. [16] propose a minimum density power divergence estimator being robust against outliers. The asymptotics of the estimator are investigated, and an application to bivariate crime counts is presented. Shapovalova et al. [17] consider two types of models for multivariate count time series: a log-linear multivariate INGARCH model and a non-linear state-space model. These models serve as a base for a forecast performance comparison. As real-world applications, count time series about bank failures and transactions are used. Last but not least, Stapper [18] developed a comprehensive software package (in the Julia language) for count time series modeling. The package fits different types of INARMA and INGARCH models, and it offers functions for model diagnostics, forecasting, etc. In his paper, Stapper [18] illustrates the application and the potential of "CountTimeSeries.jl" with several real-data examples and simulation experiments.

Acknowledgments: The Guest Editor is grateful to all authors for their contributions to this Special Issue, to the anonymous peer-reviewers for carefully reading the submissions as well as for their constructive feedback.

Conflicts of Interest: The author declares no conflict of interest.

References

1. Nie, S.Y.; Wu, X.Q. A historical study about the developing process of the classical linear time series models. In Proceedings of the Eighth International Conference on Bio-Inspired Computing: Theories and Applications (BIC-TA), Huangshan, China, 12–14 July 2016; Yin, Z., Pan, L., Fang, X., Eds.; Springer: Berlin/Heidelberg, Germany, 2013; Volume 212, pp. 425–433.
2. Box, G.E.P.; Jenkins, G.M. *Time Series Analysis: Forecasting and Control*, 1st ed.; Holden-Day: San Francisco, CA, USA, 1970.
3. Nono, A.; Uchiyama, Y.; Nakagawa, K. Entropy Based Student's t-Process Dynamical Model. *Entropy* **2021**, *23*, 560. [CrossRef] [PubMed]
4. Davidescu, A.A.; Apostu, S.A.; Paul, A. Comparative Analysis of Different Univariate Forecasting Methods in Modelling and Predicting the Romanian Unemployment Rate for the Period 2021–2022. *Entropy* **2021**, *23*, 325. [CrossRef] [PubMed]
5. Lindstrom, M.R.; Jung, H.; Larocque, D. Functional Kernel Density Estimation: Point and Fourier Approaches to Time Series Anomaly Detection. *Entropy* **2020**, *22*, 1363. [CrossRef] [PubMed]
6. Vivas, E.; Allende-Cid, H.; Salas, R. A Systematic Review of Statistical and Machine Learning Methods for Electrical Power Forecasting with Reported MAPE Score. *Entropy* **2020**, *22*, 1412. [CrossRef] [PubMed]
7. Sundararajan, R.R.; Frostig, R.; Ombao, H. Modeling Spectral Properties in Stationary Processes of Varying Dimensions with Applications to Brain Local Field Potential Signals. *Entropy* **2020**, *22*, 1375. [CrossRef]
8. Bauer, D.; Buschmeier, R. Asymptotic Properties of Estimators for Seasonally Cointegrated State Space Models Obtained Using the CVA Subspace Method. *Entropy* **2021**, *23*, 436. [CrossRef]
9. Nüßgen, I.; Schnurr, A. Ordinal Pattern Dependence in the Context of Long-Range Dependence. *Entropy* **2021**, *23*, 670. [CrossRef] [PubMed]
10. Weiß, C.H. *An Introduction to Discrete-Valued Time Series*, 1st ed.; John Wiley & Sons, Inc.: Chichester, UK, 2018.
11. Huang, J.; Zhu, F. A New First-Order Integer-Valued Autoregressive Model with Bell Innovations. *Entropy* **2021**, *23*, 713. [CrossRef] [PubMed]
12. Liu, Z.; Zhu, F. A New Extension of Thinning-Based Integer-Valued Autoregressive Models for Count Data. *Entropy* **2021**, *23*, 62. [CrossRef]
13. Yu, K.; Wang, H. A New Overdispersed Integer-Valued Moving Average Model with Dependent Counting Series. *Entropy* **2021**, *23*, 706. [CrossRef] [PubMed]
14. Liu, C.; Cheng, J.; Wang, D. Statistical Inference for Periodic Self-Exciting Threshold Integer-Valued Autoregressive Processes. *Entropy* **2021**, *23*, 765. [CrossRef] [PubMed]
15. Li, C.; Cui, S.; Wang, D. Monitoring the Zero-Inflated Time Series Model of Counts with Random Coefficient. *Entropy* **2021**, *23*, 372. [CrossRef] [PubMed]
16. Kim, B.; Lee, S.; Kim, D. Robust Estimation for Bivariate Poisson INGARCH Models. *Entropy* **2021**, *23*, 367. [CrossRef] [PubMed]
17. Shapovalova, Y.; Baştürk, N.; Eichler, M. Multivariate Count Data Models for Time Series Forecasting. *Entropy* **2021**, *23*, 718. [CrossRef] [PubMed]
18. Stapper, M. Count Data Time Series Modelling in Julia—The CountTimeSeries.jl Package and Applications. *Entropy* **2021**, *23*, 666. [CrossRef] [PubMed]

Article

Entropy Based Student's *t*-Process Dynamical Model

Ayumu Nono [1,*], Yusuke Uchiyama [2] and Kei Nakagawa [3]

1. Graduated School of Engineering, The University of Tokyo, 7-3-1 Hongo, Bunkyo-ku, Tokyo 113-8656, Japan
2. MAZIN Inc., 3-29-14 Nishi-Asakusa, Tito City, Tokyo 111-0035, Japan; uchiyama@mazin.tech
3. NOMURA Asset Management Co. Ltd., 2-2-1 Toyosu, Koto-ku, Tokyo 135-0061, Japan; k-nakagawa@nomura-am.co.jp
* Correspondence: nono-ayumu303@g.ecc.u-tokyo.ac.jp

Abstract: Volatility, which represents the magnitude of fluctuating asset prices or returns, is used in the problems of finance to design optimal asset allocations and to calculate the price of derivatives. Since volatility is unobservable, it is identified and estimated by latent variable models known as volatility fluctuation models. Almost all conventional volatility fluctuation models are linear time-series models and thus are difficult to capture nonlinear and/or non-Gaussian properties of volatility dynamics. In this study, we propose an entropy based Student's *t*-process Dynamical model (ETPDM) as a volatility fluctuation model combined with both nonlinear dynamics and non-Gaussian noise. The ETPDM estimates its latent variables and intrinsic parameters by a robust particle filtering based on a generalized H-theorem for a relative entropy. To test the performance of the ETPDM, we implement numerical experiments for financial time-series and confirm the robustness for a small number of particles by comparing with the conventional particle filtering.

Keywords: finance; volatility fluctuation; Student's *t*-process; entropy based particle filter; relative entropy

Citation: Nono, A.; Uchiyama, Y.; Nakagawa, K. Entropy Based Student's *t*-Process Dynamical Model. *Entropy* **2021**, *23*, 560. https://doi.org/10.3390/e23050560

Academic Editor: Christian H. Weiss

Received: 30 March 2021
Accepted: 27 April 2021
Published: 30 April 2021

Publisher's Note: MDPI stays neutral with regard to jurisdictional claims in published maps and institutional affiliations.

Copyright: © 2021 by the authors. Licensee MDPI, Basel, Switzerland. This article is an open access article distributed under the terms and conditions of the Creative Commons Attribution (CC BY) license (https://creativecommons.org/licenses/by/4.0/).

1. Introduction

Asset allocation and pricing derivatives have been studied in both academia and industry as significant problems in financial engineering and quantitative finance. For these problems, various methodologies have been developed based on the variation of asset returns. In an idealized situation, the variation of returns has been assumed to follow the Gaussian distribution [1]. However, it is known that the variation of returns follows non-Gaussian distributions with fat tails [2]. To explain this observed fact, volatility, which quantifies the magnitude of fluctuating returns, has been introduced and utilized. Volatility, in particular, is often used as an indicator for constructing asset allocations that focus on macroeconomic fundamentals, and there are many studies related to them. Both researchers and investors have begun to attend to develop mathematical models of volatility fluctuations. For example, Yuhuang et al. investigated the impact of fundamental data on oil price volatility by focusing on time-varying skewness and kurtosis [3]. Hou et al. studied volatility spillovers between the Chinese fuel oil futures market and the stock index futures market, taking into account the time-varying characteristics of the markets [4].

In general, volatility is defined as the variance of a conditional Gaussian distribution for the variation of returns, namely, given as a latent variable in the literature of Bayesian statistical modeling. Based on this idea, various time-series models for the dynamics of asset returns have been developed and proposed. Such time-series models are generally called volatility fluctuation models, on which forecasting, state estimation and smoothing can be implemented.

In recent years, volatility fluctuation models with a machine learning technique have been proposed [5]. Since volatility is a latent variable, it is necessary for machine learning models to incorporate latent variables into their own methodology. The Gaussian process is a candidate, such as a Bayesian learning model [6], and its applications for several problems in finance have been reported [7,8]. The Student's *t*-process is an extension of the Gaussian, for non-Gaussian distributed data such as asset returns. It has been proposed [9] and applied to the analysis of financial time-series and asset allocations, and it is confirmed for this model to estimate robustly [10].

This study extends the Student's *t*-process latent variable model to a dynamic latent variable model incorporating the structure of time-series. To estimate dynamic latent variables, we used the particle filter method [11]. The particle filter is used to estimate the latent variables. Conventional particle filters have problems called weight bias and the particle impoverishment problem (PIP), directly affecting the estimation accuracy [12]. Then, the merging particle method [13] and Monte Carlo filter particle filter [14] have been proposed. However, these methods are computationally expensive because they need a large number of particles. Therefore, we used an Entropy-based particle filter (EBPF), which constructs a parametric prior distribution on the generalized H-theorem for relative entropy [15]. It is expected to prevent the bias of particle weights and the loss of particle diversity while reducing the computational cost. Using EBPF in this experiment, and comparing it with conventional methods, we confirmed that it is effective for finance problems.

In summary, to estimate robustly and avoid the particle filter's problem, we combined *t*-process dynamical model and EBPF. We call the proposed model an entropy based Student's *t*-process dynamical model (ESTDM), in the following. We will verify this model's usefulness. The remains of this paper are summarized as follows—Section 2 introduces related statistical and machine learning models. In Section 3, we derive and propose ESTDM with its filtering method. In Section 4, we show the performance of volatility estimation using the proposed method and discuss the results. Section 5 is devoted to our conclusions and future perspectives.

2. Related Work

2.1. Volatility Fluctuation Models

One of the most basic and utilized volatility fluctuation models is the GARCH model [16] given as follows:

$$x_t \sim \mathcal{N}(0, \sigma_t^2), \tag{1}$$

$$\sigma_t^2 = \alpha_0 + \sum_{j=1}^{q} \alpha_j \sigma_{t-j}^2 + \sum_{i=1}^{p} \beta_i x_{t-i}^2, \tag{2}$$

where x_t is a time-dependent random variable sampled from a Gaussian distribution with mean 0 and variance σ_t^2, and the time evolution of the variance is given by Equation (2). The parameters α_j and β_i take positive values, which can be estimated by observed data. Positive integers p and q are the order of the regression, respectively. Then this model is known as the GARCH(p,q) model. For the sake of simplicity, the order parameters are often fixed as $p = q = 1$. Various families of GARCH model have been developed and proposed in the area of econometrics and quantitative finance [17]. For instance, asymmetric effect has been introduced into a multivariate GARCH model [18,19].

2.2. Gaussian Process

For any finite number of vectors $\{x_1, x_2, \cdots, x_n\}$ and a stochastic process $f(\cdot)$, if the joint probability density function $\{f(x_1), f(x_2), \cdots, f(x_n)\}$ is a Gaussian distribution, $f(\cdot)$ is called a Gaussian process [6]. Since the Gaussian process samples an infinite-dimensional vector, the mean value function $m(\cdot)$ and the covariance function $K(\cdot, \cdot)$ are introduced as follows:

$$m(x) = \mathbb{E}[f(x)], \tag{3}$$

$$K(x, x') = \mathbb{E}[(f(x) - m(x))(f(x') - m(x'))^{\mathrm{T}}]. \tag{4}$$

Then, given a matrix $X = [x_1, x_2, \cdots, x_n]^T$, $p(f|X) = \mathcal{N}(m(X), K(X,X))$ is the probability density function of the Gaussian process. When we explicitly state that the stochastic process f is sampled from the Gaussian process, we write $f \sim \mathcal{GP}(m, K)$. Without loss of generality, the mean function of the Gaussian process is often assumed to be zero. The covariance function is represented by the kernel function $k(\cdot, \cdot)$, which is a positive symmetric bi-variate function, satisfying

$$K(x, x') = k(x, x'). \quad (5)$$

Hence, $K(X, X)$ is a positive definite symmetric matrix. As a kernel function, for example, the radial basis function

$$k_{\text{RBF}}(x, x') = \alpha \exp(-l^{-2}||x - x'||^2) \quad (6)$$

is often used. Here, α and l are hyper parameters.

For a pair of observed data $\mathcal{D} = \{(x_1, y_1), (x_2, y_2), \cdots, (x_n, y_n)\}$, let $X = [x_1, x_2, \cdots, x_n]^T$, $Y = [y_1, y_2, \cdots, y_n]^T$. The hyper parameters of the Gaussian process can be estimated by gradient and Monte Carlo methods on $\mathcal{D}$. From the trained Gaussian process, the prediction $Y^* = [y_1^*, y_2^*, \cdots, y_m^*]^T$ for unknown input $X^* = [x_1^*, x_2^*, \cdots, x_m^*]^T$ is sampled from the conditional Gaussian distribution $\mathcal{N}(f^*, K^*)$. The mean function f^* and the covariance function K^* of the conditional Gaussian distribution are given by

$$f^* = m_X + K_{X^*, X} K_{X, X}^{-1} Y, \quad (7)$$

$$K^* = K_{X^*, X^*} - K_{X^*, X} K_{X, X}^{-1} K_{X, X^*}. \quad (8)$$

It is seen that the mean and covariance functions of the Gaussian process propagate the information of previously observed data to predicted values.

2.3. Student's t-Process

In the Gaussian process, it is assumed for the probability density function to be the Gaussian distribution. Thus, when we apply the Gaussian process to data following a probability distribution with fat tails, such as financial time-series, it is impossible to perform an accurate estimation. A model that extends the Gaussian process to such data is the Student's t-process [9]. The Student's t-process is a stochastic process $f(\cdot)$ with ν degrees of freedom and a Student's t-distribution defined as follows:

$$\mathcal{T}(m, K, \nu) = \frac{\Gamma\left(\frac{\nu+n}{2}\right)}{[(\nu-2)\pi]^{\frac{n}{2}} \Gamma\left(\frac{\nu}{2}\right) |K_{X,X}|^{\frac{1}{2}}} \left[1 + \frac{1}{\nu-2}(Y - m_X)^T K_{X,X}^{-1}(Y - m_X)\right]^{-\frac{\nu+n}{2}}. \quad (9)$$

Here, $m(\cdot)$ and $K(\cdot, \cdot)$ are the mean and covariance functions, respectively, and $\Gamma(\cdot)$ is the gamma function. When the stochastic process $f(\cdot)$ is a Student's t-process, it is denoted by $f \sim \mathcal{TP}(m, K; \nu)$. As with the Gaussian process, the mean function of the Student's t-process is often assumed to be zero without loss of generality.

The predictive distribution of the Student's t-process is also the Student's t-distribution $\mathcal{T}(m^*, K^*, \nu^*)$, where degrees of freedom, mean and covariance functions are then updated as follows:

$$\nu^* = \nu + n, \quad (10)$$

$$m^* = m_X + K_{X^*, X} K_{X, X}^{-1} (Y - m_X) \quad (11)$$

$$K^* = \frac{\nu - \beta - 2}{\nu - n - 2} \left[K_{X^*, X^*} - K_{X^*, X} K_{X, X}^{-1} K_{X, X^*} \right]$$

$$\beta = (Y - m_X)^T K_{X, X}^{-1} (Y - m_X). \quad (12)$$

Unlike the Gaussian process, in the Student's t-process, we can confirm that the effect of the number of data is reflected in the update equations of the degrees of freedom and the covariance function.

2.4. Student's t-Process Latent Variable Model

In the Student's t-process latent variable model, the input matrix X is given as a latent variable. Assume that the observed data $y \in \mathbb{R}^D$ and the latent variable $x \in \mathbb{R}^Q$ are related as $y = f(x)$ by the Student's t-process $f \sim \mathcal{TP}(m, K; \nu)$. When we let $Y \in \mathbb{R}^{D \times N}$ be the sequence of N observed data, and $X \in \mathbb{R}^{Q \times N}$ be the sequence of N latent variables, we can define the following model as Student's t-process latent variable model [10]:

$$p(Y|X) = \frac{\Gamma\left(\frac{\nu+D}{2}\right)}{[(\nu-2)\pi]^{\frac{D}{2}} \Gamma\left(\frac{\nu}{2}\right) |K_{X,X}|^{\frac{1}{2}}} \left[1 + \frac{1}{\nu-2}(Y - m_X)^T K_{X,X}^{-1} (Y - m_X)\right]^{-\frac{\nu+D}{2}}. \quad (13)$$

Since the Student's t-distribution converges to the Gaussian distribution in the limit of $\nu \to \infty$, we can see that the Student's t-process latent variable model embraces the Gaussian process latent variable model [20].

3. Proposed Model

3.1. Student's t-Process Dynamical Model

Since volatility fluctuations cannot be observed directly, they are modeled as dynamic latent variables, such as the family of GARCH models, most of which are given by linear time-series models [18]. To describe nonlinear dynamics of volatility fluctuations, we extend the Student's t-process latent variable model to dynamic latent variables, namely, Student's t-process dynamical model (TPDM), which is expected to be robust for both observable and unobservable with outliers.

Suppose p_t represents an asset price at time t, the log-return is given by $r_t = \log(p_t/p_{t-1})$. Let σ_t^2 denote the volatility of r_t. Here, for an observable r_t and a latent variable σ_t^2, we provide a volatility fluctuation model by a TPDM as follows:

$$r_t \sim \mathcal{T}(0, \sigma_t^2; \nu), \quad (14)$$

$$v_t \equiv \log \sigma_t^2, \quad (15)$$

$$v_t = f(v_{t-1}, r_{t-1}; \nu) + \epsilon_t \quad (16)$$

$$\epsilon_t \sim \mathcal{N}(0, \sigma_n^2), \quad (17)$$

where the observable r_t as centered at 0 and following a Student's t-distribution with ν degrees of freedom, whose parameter is given by σ_t^2. The dynamic latent variable v_t is defined by Equation (15) to take the whole real number as its range of value. The time evolution of the dynamic latent variable v_t is given by Equation (16) with a Gaussian white noise whose variance is σ_n. The stochastic process f on the right-hand side of Equation (16) follows a Student's t-process given by

$$f \sim \mathcal{TP}(m, K; \nu), \quad (18)$$

$$m(\xi_{t-1}) = a v_{t-1} + b x_{t-1}, \quad (19)$$

$$k(\xi_{t-1}, \xi'_{t-1}) = \gamma \exp\left(-l^{-2} ||\xi_{t-1} - \xi'_{t-1}||^2\right), \quad (20)$$

where $\xi_t = (x_t, v_t)$, and the hyper parameters are $\theta = (\nu, \sigma_n, a, b, \gamma, l)$. Given a series of observed data $r_{1:T} = \{r_1, r_2, \cdots, r_T\}$, it is possible to obtain the volatility fluctuations by estimating a series of dynamic latent variables $v_{1:T} = \{v_1, v_2, \cdots, v_T\}$.

3.2. Particle Filter

Particle filter is a method of state estimation by Monte Carlo sampling, where a large number of particles approximates posterior distributions. Hence, it can be applied to nonlinear systems, where posterior distributions are intractable [21]. For N particles, let $\{v_{1:t-1}^i\}_{i=1}^N$ and W_{t-1}^i ($i = 1, 2, \cdots, N$) be the realizations of the dynamic latent variables and their associated weights up to time $t-1$, respectively. The weights are normalized to $\sum_{i=1}^N W_{t-1}^i = 1$. With these values, the posterior distribution $p(v_{1:t-1}|x_{1:t-1})$ at time $t-1$ can be approximated as follows [22]:

$$\hat{p}(v_{1:t-1}|x_{1:t-1}) = \sum_{i=1}^N W_{t-1}^i \delta(v_{1:t-1}), \quad (21)$$

where $\delta(\cdot)$ is the Dirac's delta function. In other words, the posterior distribution is approximated by a mixture of the delta functions.

It is however known that an insufficient number of particles fails to approximate the posterior distribution by the degeneracy of ensemble. Indeed, each particle's weights become unbalanced and biased toward a tiny number of particles as the time step progresses [11,12]. To overcome this problem, a huge amount of particles is used for filtering processes in practice.

3.3. Entropy-Based Particle Filter

In the use of the conventional particle filter, it is necessary to sample a large number of particles for accuracy. That leads to the growth of estimation time. In the case of online estimation, reducing run time is desired. For this purpose, we introduce a robust particle filter for a small number of particles.

Let us reconsider approximating the probability density function for the dynamic latent variable, $Q(v,t)$, called a background distribution. In the conventional particle filter, the background distribution is approximated by the mixture of delta functions. This approximation works well only when the background distribution exhibits an extensively sharp peak. Nevertheless, the delta function has no width, and the distribution peaks only at a single point.

To improve the accuracy for the approximation of the background distribution, we replace the mixture of the delta functions with that of Gaussian distributions as

$$\hat{Q}(v,t) = \sum_{i=1}^M W_t^i \mathcal{N}(\mu_t^i, \sigma_t^{2,i}), \quad (22)$$

where $\mu_t^i, \sigma_t^{2,i}$ ($1 \leq i \leq M$) are the mean and variance of the Gaussian distributions at t. Unlike the delta function, the Gaussian distribution has a certain width in its distribution. Hence, the mixture of the Gaussian distributions is capable of fitting properly to data with large variance and fat tails.

With the use of finite samples from the background distribution $Q(v,t)$, the posterior/filter distribution $P(v,t)$ is inferred by the minimum principal for relative entropy, which is known as an entropy based particle filter [15]. The relative entropy (Kullback-Liebler divergence) between the filter distribution $P(v,t)$ and the background distribution $Q(v,t)$ are defined as follows [23–25]:

$$H[P|Q] = \int_{\Omega_v} P(v,t) \log\left(\frac{P(v,t)}{Q(v,t)}\right) dv, \quad (23)$$

where Ω_v is the domain of the dynamic latent variable v_t. On the properties of the relative entropy as a quasi-distance for probability density functions, the filter distribution is obtained as the minimizer for the relative entropy in Equation (23). Combined with the entropy based particle filter, the state estimation of the ESTDM is implemented. An overview of its algorithm is explained in the following Algorithm 1.

Algorithm 1 Entropy Based Student's t-Process Dynamical Model (ETPDM)

Require: Initial particles $X_0 = [X_0^0, ..., X_0^M]$, Initial particles' weights $W_0^i = 1/M$
Ensure: $\sum_{i=1}^{M} W_t^i = 1.0$ at any time t
1: **while** There are observations to be assimilated **do**
2:   Compute importance weights proportional to the likelihood with observation x_t

$$W_t^i \propto p(y_t|X_t) \quad (24)$$

According to weights $\{W_t^i\}$, resample M particles $\{X_t^j\}_{j=1}^M$.
Then we can compute filter distribution $Q'(x)$ at time t

$$Q_t' = \sum_{i=1}^{M} W_t^j \mathcal{N}(X_t^j). \quad (25)$$

At the same time, we're also able to estimate expected status v_t, extracting any finite number of samples $\{v^k\}$ from background density Q_t

$$v_t = \mathbb{E}(v^k). \quad (26)$$

With stochastic process $f \sim \mathcal{T}(m, k; \nu)$, generate new particles

$$\{X_{t+1}^i\} = f(X_t^j, x_t) \quad (27)$$

Then we can predict distribution $\hat{Q}(x)$ at time $t+1$

$$\hat{Q}_{t+1} = \sum_{i=1}^{M} W_t^i \mathcal{N}(X_{t+1}^i). \quad (28)$$

3:   **return** Log likelihood for estimation $p(y_t|v_t)$.
4: **end while**

4. Numerical Experiments

In this section, we implement numerical experiments to validate the ETPDM for the time-series of a foreign exchange rate. As a dataset, we use USD/JPY exchange rate in 2010—every 1-min sampled, 30-min sampled and 1-h sampled. Figures 1–3 show the time-series of the log-return of the USD/JPY exchange rate r_t and volatility fluctuations estimated by respective the ETPDM, the conventional particle filter for the GARCH model (cp-GARCH) and the conventional particle filter for the Student's t-process dynamical model (cp-TPDM). Warm up period of the estimations is $0 \leq t \leq 20$, where the values of volatility show zeros. In Figure 1a, intermittent jumps are observed, which are evidence of the non-Gaussian behavior of r_t. Indeed, the estimated volatility fluctuations show higher peaks at the same time point of the intermittent jumps in Figure 1b–d. That means all of the models capture the nature of volatility fluctuations of the USD/JPY exchange rate effectively. Besides, the same can be said for other types of data sets—30-min and 1-h—in Figure 2 or Figure 3, which means that these models can be applied to data of any sampling rate. Previous volatility estimation studies used the GARCH model with various estimation methods. A typical example is the particle filter [26], or the Markov chain Monte Carlo simulation [27]. In all of these studies, including this experiment, the GARCH model has been implemented well.

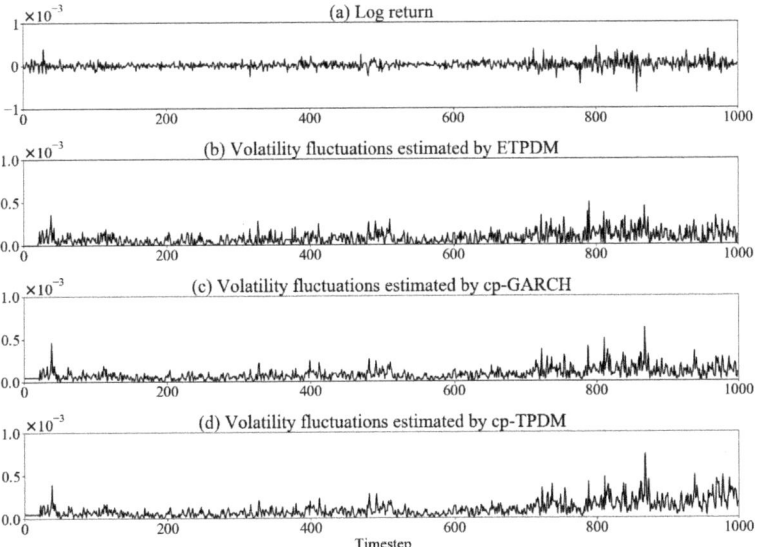

Figure 1. Overview of estimation results in 1-min chart.

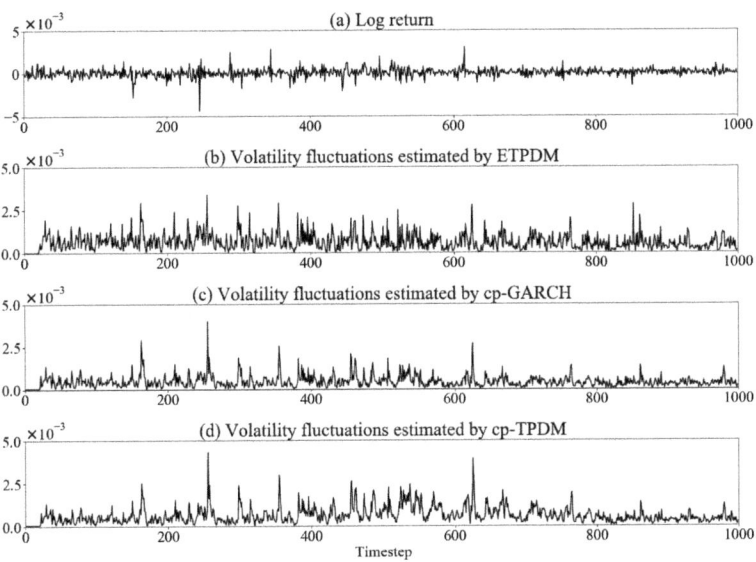

Figure 2. Overview of estimation results in 30-min chart.

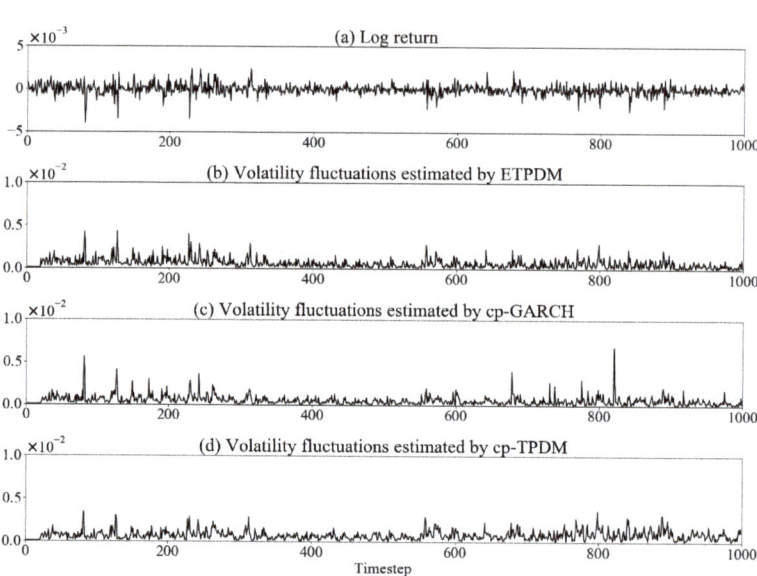

Figure 3. Overview of estimation results in 1-h chart.

Figure 4 show the estimated log-likelihoods of the ETPDM, cp-GARCH and the cp-TPDM. The likelihood tends to be higher for the ETPDM, the cp-TPDM and cp-GARCH in that order. As mentioned in Section 2, TPDM is a superordinate model that encompasses Gaussian process dynamical model, the GPDM, and the fact that the likelihood of the GARCH was at a lower level than that of TPDM is consistent with the results of previous studies comparing the GARCH and the GPDM [26]. Likelihood is the most reliable indicator to quantify the model performance, and the EPTDM had the best performance of three models. Besides, in the case of cp-TPDM, the log-likelihoods scatter around −0.7 in the range of particle numbers from 10 to 500 without convergence. This means the performance of the conventional particle filter is insufficient for the given data. On the other hand, the log-likelihoods of the ETPDM exhibit good convergence for the particle numbers larger than 100 in Figure 4b, which indicates the ETPDM is expected to be robust for fewer sampling.

To investigate the effectiveness of particle filtering, we introduce an effective particles rate

$$R_{eff} = \frac{1}{N \sum_{i=1}^{N} (W_t^i)^2} \qquad (29)$$

as a measure for evaluating the bias of sampled particles. This value gives the maximum value $R_{eff} = 1$ when the weights are uniformly distributed as $W^i = 1/N$ ($i = 1, 2, ..., N$). In Figure 5a, the effective particle rates for the cp-TPDM scatter for whole particle numbers from 10 to 500. This kind of worse performance for the effective particle rates stems from the weight bias problem of the particle filtering. In other words, the conventional particle filter is hard to overcome the particle impoverishment problem, even if, by increasing particle numbers. On the contrary, for the case of the ETPDM as shown in Figure 5b, we can see that the effective particle rate converges beyond 50%. This is the expected advantage of the ETPDM, which stems from the finite band of each Gaussian distribution as a component of the prior distribution. Thus, the ETPDM serves as an accurate estimation for lower particle numbers and then would contribute to effective online estimation. Focusing on another comparison of the ETPDM, the cp-GARCH, also looks good in the view of the effective particles rate. However, when we also consider the likelihood in Figure 4a, we can say that the practical ensembles didn't affect to performance because cp-GARCH was less suitable for this problem than the ETPDM. This is another evidence that the ETPDM have better potential.

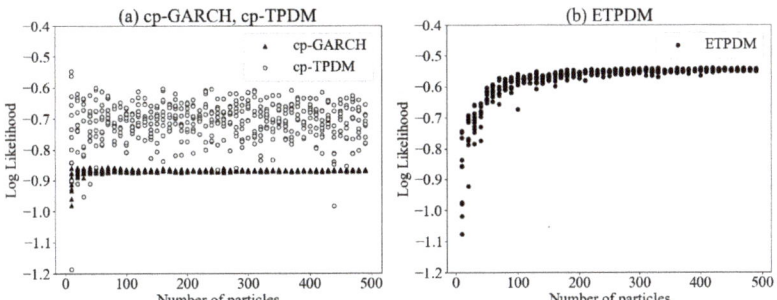

Figure 4. Estimated log-likelihoods of (**a**) the cp-GARCH, the cp-TPDM and (**b**) the ETPDM.

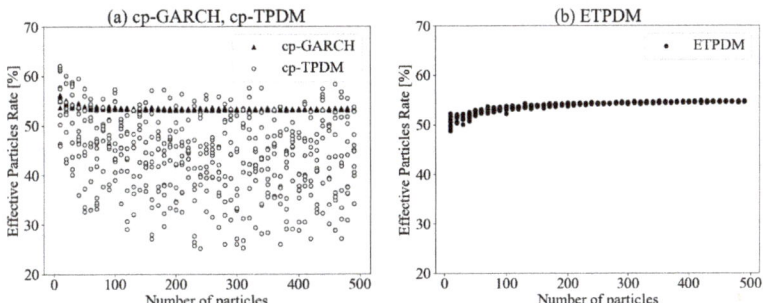

Figure 5. Effective particle rates of (**a**) cp-GARCH, the cp-TPDM and (**b**) the ETPDM.

In order to validate the robustness for estimating intermittent return dynamics, we investigate the degree of freedom ν of the Student's t-process dynamical models. For this purpose, we split time window of return fluctuations; one is low volatility window ($50 \leq t \leq 250$) and the other is high volatility counterpart ($360 \leq t \leq 560$). The descriptive statistics of the return fluctuations in the two windows are shown in Table 1. As is seen in the table, kurtoses in both time windows are larger than 3, namely, corresponding return fluctuations follow non-Gaussian statistics. Prior research has confirmed that when the data set follows a Gaussian distribution, the strengths of models that excel at robust estimation do not come into play [28]. Therefore, such a data set that follows a non-Gaussian distribution is appropriate for the purpose of this experiment. Figure 6 exhibit the log-likelihoods of the cp-TPDM and the ETPDM in (a) low volatility window and (b) high volatility one. In these figures, the log-likelihoods of the ETPDM in both time windows have maxima in $6 \leq \nu \leq 7$ though the estimations by the the cp-TPDM are unstable. This result evidences the robustness of the ETPDM.

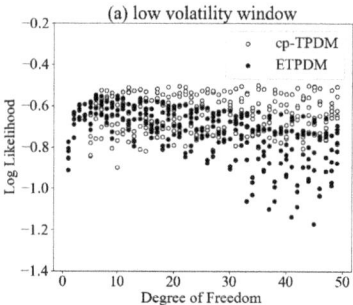 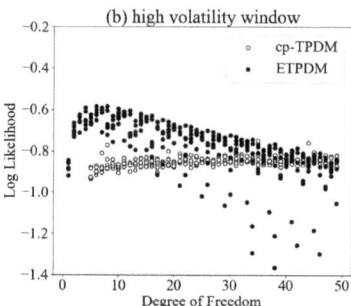

Figure 6. Log-likelihoods of the cp-TPDM and the ETPDM in (**a**) low volatility window and (**b**) high volatility one.

Table 1. Two types of window.

Window Type	Mean	Variance	Skewness	Kurtosis
high volatility window	-1.0×10^{-6}	8.31×10^{-4}	-0.974	5.22
low volatility window	1.0×10^{-6}	4.41×10^{-4}	-0.531	3.56

5. Conclusions

In this study, we proposed the ETPDM to implement robust estimation for dynamical latent variables of nonlinear and non-Gaussian fluctuations. In estimating the dynamic latent variables and hyper parameters, the entropy based particle filter with the Gaussian mixture distribution was adopted. To validate the performance of the ETPDM, we carried out the numerical experiment for the return fluctuations of a foreign exchange rate compared with the cp-GARCH and the cp-TPDM. As a result, we confirmed the advantages of the ETPDM; (i) good convergence property, (ii) high effective particle rate and (iii) robustness for a small number of particles.

Based on its advantages, the ETPDM is applicable for online volatility estimation for the problem of asset allocation and derivative pricing in a short time span. As a basis distribution for background distribution, we employed the Gaussian distribution in our numerical experiments. Nevertheless, the framework of the entropy based particle filter is able to be extended to other probability density functions. Additionally, we can adapt this research to any other time-series data, not just asset data. It has the potential to be applied to control engineering, such as the self-positioning estimation problem. These are our future works.

Author Contributions: Conceptualization, Y.U.; methodology, A.N.; software, A.N.; validation, A.N.; formal analysis, A.N.; investigation, A.N.; resources, A.N.; data curation, K.N.; writing—original draft preparation, A.N.; writing—review and editing, Y.U., and K.N.; visualization, A.N.; project administration, Y.U.; All authors have read and agreed to the published version of the manuscript.

Funding: This research received no external funding.

Data Availability Statement: The data that support the findings of this study are available from the corresponding author, Nono, A., upon reasonable request.

Conflicts of Interest: The authors declare no conflict of interest.

References

1. Cochrane, J.H. *Asset Pricing: Revised Edition*; Princeton University Press: Princeton, NJ, USA, 2009.
2. Mandelbrot, B.B. The variation of certain speculative prices. In *Fractals and Scaling in Finance*; Springer: Berlin/Heidelberg, Germany, 1997; pp. 371–418.
3. Shang, Y.; Dong, Q. Oil volatility forecasting and risk allocation: Evidence from an extended mixed-frequency volatility model. In *Applied Economics*; Routledge: London, UK, 2021; pp. 1127–1142.

4. Hou, Y.; Li, S.; Wen, F. Time-varying volatility spillover between Chinese fuel oil and stock index futures markets based on a DCC-GARCH model with a semi-nonparametric approach. *Energy Econ.* **2019**, *83*, 119–143. [CrossRef]
5. Horvath, B.; Muguruza, A.; Tomas, M. Deep Learning Volatility. *arXiv* **2019**, arXiv:1901.09647.
6. Rasmussen, C.; Williams, C. *Gaussian Processes for Machine Learning*; Adaptative computation and machine learning series; The MIT Press: Cambridge, MA, USA, 2006.
7. Gonzalvez, J.; Lezmi, E.; Roncalli, T.; Xu, J. Financial Applications of Gaussian Processes and Bayesian Optimization. *arXiv* **2019**, arXiv:1903.04841.
8. Nirwan, R.S.; Bertschinger, N. Applications of Gaussian process latent variable models in finance. In *Advances in Intelligent Systems and Computing*; Springer: Berlin/Heidelberg, Germany, 2019; pp. 1209–1221.
9. Shah, A.; Wilson, A.; Ghahramani, Z. Student-t processes as alternatives to Gaussian processes. In *Artificial Intelligence and Statistics*; 2014; pp. 877–885. Available online: http://proceedings.mlr.press/v33/shah14.html (accessed on 30 March 2021)
10. Uchiyama, Y.; Nakagawa, K. TPLVM: Portfolio Construction by Student's t-Process Latent Variable Model. *Mathematics* **2020**, *8*, 449. [CrossRef]
11. Sarkka, S. *Baysian Filtering and Smoothing*; Cambridge University Press: Cambridge, UK, 2013.
12. Murata, M.; Hiramatsu, K. On Ensemble kalman Filter, Particle Filter, and Gaussian Particle Filter. *Trans. Inst. Syst. Control Inf. Eng.* **2016**, *29*, 448–462.
13. Nakano, S.; Ueno, G.; Haguchi, T. Merging particle filter for sequential data assiimilation. *Nonlinear Process. Geophys.* **2007**, *14*, 395–408. [CrossRef]
14. Murata, M.; Nagano, H.; Kashino, K. Monte Carlo filter particle filter. In Proceedings of the 2015 European Control Conference (ECC), Linz, Austria, 15–17 July 2015; Volume 14, pp. 2836–2841.
15. Eyink, G.L.; Kim, S. A Maximum Entropy Method for Particle Filtering. *J. Stat. Phys.* **2006**, *123*, 1071–1128. [CrossRef]
16. Bollerslev, T. Generalized autoregressive conditional heteroskedasticity. *J. Econ.* **1986**, *31*, 307–327. [CrossRef]
17. Bollerslev, T. Glossary to arch (garch). *CREATES Res. Pap.* **2008**, *49*, 1–46. [CrossRef]
18. Teräsvirta, T. An introduction to univariate GARCH models. In *Handbook of Financial Time Series*; Springer: Berlin/Heidelberg, Germany, 2009; pp. 17–42.
19. Nakagawa, K.; Uchiyama, Y. GO-GJRSK Model with Application to Higher Order Risk-Based Portfolio. *Mathematics* **2020**, *8*, 1990. [CrossRef]
20. Lawrence, N.D. Gaussian process latent variable models for visualisation of high dimensional data. *Nips* **2003**, *2*, 5.
21. Kato, T. Introduction of Particle Filter and Its Implementation. *IPSJ SIG Tech. Rep.* **2007**, *157*, 161.
22. Wu, Y.; Hernández-Lobato, J.M.; Ghahramani, Z. Gaussian Process Volatility Model. In *Advances in Neural Information Processing Systems 27*; Ghahramani, Z., Welling, M., Cortes, C., Lawrence, N.D., Weinberger, K.Q., Eds.; Curran Associates, Inc.: Red Hook, NY, USA, 2014; pp. 1044–1052.
23. Vințe, C.; Ausloos, M.; Furtună, T.F. A Volatility Estimator of Stock Market Indices Based on the Intrinsic Entropy Model. *Entropy* **2021**, *23*, 484. [CrossRef]
24. Cover, T.M.; Thomas, J.A. *Elements of Information Theory*; John Wiley & Sons: Hoboken, NJ, USA, 1991.
25. Kullback, S.; Leibler, R.A. On information and sufficiency. *Ann. Math. Stat.* **1951**, *22*, 79–86. [CrossRef]
26. Wu, Y.; Hernndez-Lobato, J.M.; Ghahramani, Z. Gaussian process volatility model. In *Advances in Neural Information Processing Systems*; The MIT Press: Cambridge, MA, USA, 2014.
27. Fleming, J.; Kirby, C. A closer look at the Relation between GARCH and Stochastic Autoregressive volatility. *J. Financ. Econ.* **2003**, *1*, 365–419.
28. Jacquier, E.; Polson, N.G.; Rossi, P.E. Bayesian analysis of stochastic volatility models withfat-tails and correlated errors. *J. Econ.* **2004**, *122*, 185–212. [CrossRef]

Article

Comparative Analysis of Different Univariate Forecasting Methods in Modelling and Predicting the Romanian Unemployment Rate for the Period 2021–2022

Adriana AnaMaria Davidescu [1,2,*], Simona-Andreea Apostu [1,3] and Andreea Paul [4,5]

1 Department of Statistics and Econometrics, Bucharest University of Economic Studies, 010552 Bucharest, Romania; simona.apostu@csie.ase.ro
2 National Scientific Research Institute for Labor and Social Protection, 010643 Bucharest, Romania
3 National Institute of Economy, 050711 Bucharest, Romania
4 Department of International Economic Relations, Bucharest University of Economic Studies, 15–17 Dorobanti St., Sector 1, 010404 Bucharest, Romania; andreea.paul@inaco.ro or office@inaco.ro
5 President of Think-Tank INACO—The Initiative for Competitiveness, 030237 Bucharest, Romania
* Correspondence: adriana.alexandru@csie.ase.ro

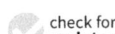

Citation: Davidescu, A.A.; Apostu, S.-A.; Paul, A. Comparative Analysis of Different Univariate Forecasting Methods in Modelling and Predicting the Romanian Unemployment Rate for the Period 2021–2022. *Entropy* **2021**, *23*, 325. https://doi.org/10.3390/e23030325

Academic Editor: Christian H. Weiss

Received: 29 January 2021
Accepted: 4 March 2021
Published: 9 March 2021

Publisher's Note: MDPI stays neutral with regard to jurisdictional claims in published maps and institutional affiliations.

Copyright: © 2021 by the authors. Licensee MDPI, Basel, Switzerland. This article is an open access article distributed under the terms and conditions of the Creative Commons Attribution (CC BY) license (https://creativecommons.org/licenses/by/4.0/).

Abstract: Unemployment has risen as the economy has shrunk. The coronavirus crisis has affected many sectors in Romania, some companies diminishing or even ceasing their activity. Making forecasts of the unemployment rate has a fundamental impact and importance on future social policy strategies. The aim of the paper is to comparatively analyze the forecast performances of different univariate time series methods with the purpose of providing future predictions of unemployment rate. In order to do that, several forecasting models (seasonal model autoregressive integrated moving average (SARIMA), self-exciting threshold autoregressive (SETAR), Holt–Winters, ETS (error, trend, seasonal), and NNAR (neural network autoregression)) have been applied, and their forecast performances have been evaluated on both the in-sample data covering the period January 2000–December 2017 used for the model identification and estimation and the out-of-sample data covering the last three years, 2018–2020. The forecast of unemployment rate relies on the next two years, 2021–2022. Based on the in-sample forecast assessment of different methods, the forecast measures root mean squared error (RMSE), mean absolute error (MAE), and mean absolute percent error (MAPE) suggested that the multiplicative Holt–Winters model outperforms the other models. For the out-of-sample forecasting performance of models, RMSE and MAE values revealed that the NNAR model has better forecasting performance, while according to MAPE, the SARIMA model registers higher forecast accuracy. The empirical results of the Diebold–Mariano test at one forecast horizon for out-of-sample methods revealed differences in the forecasting performance between SARIMA and NNAR, of which the best model of modeling and forecasting unemployment rate was considered to be the NNAR model.

Keywords: unemployment rate; SARIMA; SETAR; Holt–Winters; ETS; neural network autoregression; Romania

1. Introduction

Unemployment is a socio-economic problem facing all countries of the world, affecting both the standard of living of the people and the socio-economic status of the nations. Unemployment represents the result of a poor demand in the economy; a low demand implies a lower need for labor, which will lead either to reduced working hours or redundancies. Although unemployment is a consequence of a fundamental change in an economy, its frictional, structural, and cyclical behavior contributes to its existence.

The pandemic led to a large number of unemployed in Romania; in March, the unemployment rate rose to 4.6% compared to 3.9% in February 2020. The provisions of the military ordinances on stopping the spread of the new coronavirus have led many companies to partially or completely cease operations, which has led to the highest unemployment rate in the last two years.

Given the pandemic context, the unemployment rate in March was still low, due to the fact that in March, the effects of the pandemic were not entirely felt; companies waited until the last moment to see what measures the state would take to support technical unemployment. The first measures taken by companies were rest leave and other types of leave to be granted to employees.

The projection was that the unemployment rate in 2020 will increase, the month of March being only the beginning of the health crisis in Romania. According to the Ministry of Labor and Social Protection, more than 276,000 people were in a position where their employment contract was terminated on 30 April 2020. The industries with the most terminated employment contracts were wholesale and retail trade, manufacturing, and construction.

Although the effects of the coronavirus crisis have been seen in the economy since the measures taken in March 2020, forecasts indicated that the highest level of unemployment will reach 3.98% in the second quarter of 2020. Even the most pessimistic forecasts indicated that the unemployment rate in 2020 will not exceed 7%. The explanations for these low values compared to real figures were given by the fact that the persons returned to the country and the persons with terminated employment contracts are not included in the number of unemployed; at the end of March 111,340 terminated employment contracts had been registered and 250,000 people returned to Romania from abroad. Another explanation is the fact that the labor market was not growing at all during the crisis; therefore, people were not searching for a job, which is an essential condition to be declared unemployed.

The crisis caused by the coronavirus affected activities in many sectors and the number of unemployed increased, but this has not been reflected by the unemployment rate, as the real number of those unemployed was not included in the reporting base. Therefore, unemployment was lower, but this was not real, as the unemployed were not included in the statistics but rather in structural unemployment: the employment rate was reduced.

In this context, it becomes even more important to be able to provide future predictions of unemployment rate, and in order to do that, different univariate forecasting models (seasonal model autoregressive integrated moving average (SARIMA), self-exciting threshold autoregressive (SETAR), Holt–Winters, ETS (error, trend, seasonal), neural network autoregression (NNAR)) have been applied in order to identify the most appropriate model and to forecast the future values of unemployment rate. In order to do that, the period January 2000 to December 2020 has been used and divided into two sub-samples: the in-sample data or the training dataset covering the period January 2000–December 2017 used in the model identification and estimation and the test dataset or the out-of-sample data covering the last three years, 2018–2020. The forecast of unemployment rate relies on the next two years, 2021–2022.

Analyzing the patterns of unemployment rate, the research aims to respond to the following questions: Does the unemployment rate exhibits a non-stationary nonlinear pattern? Does the unemployment rate exhibit a seasonal pattern? Do the more sophisticated methods such as SETAR, NNAR, or SARIMA performs better than simple methods (HW or ETS)? What is the univariate forecasting method that performs best within in-sample data? What is the univariate forecasting method that has the best performance for the out-of-sample dataset? Which is the method that best captures the pandemic shock?

What is the combination of methods that could offer reliable future values for the Romanian unemployment rate?

Relying on these questions, the following three main hypotheses can be formulated:

Hypotheses 1 (H1). *The Romanian unemployment rate exhibits a non-stationary nonlinear and seasonal pattern over the period January 2000 to December 2017.*

Hypotheses 2 (H2). *NNAR and SARIMA models registered the best out-of-sample forecast performance from the all four methods applied.*

Hypotheses 3 (H3). *The combination of NNAR and SARIMA models offers the best approach in forecasting the unemployment rate for 2021–2022.*

The paper is organized as follows. The literature review presents an overview of the most important studies regarding this topic of forecasting unemployment rate, while Section 3 is dedicated to the presentation of five different forecasting models (SARIMA, NNAR, SETAR, Holt Winters, ETS). Section 4 incorporates information related to the data used in the analysis and the main empirical results of all five forecasting methods. The last part of this section ends with the comparison of models forecasting performance both for in sample and out-of-sample datasets. The final section of the paper presents the main conclusions about the relevance of this research.

2. Literature Review

The phenomenon of unemployment is the result of the dysfunctions of the economy, in the field of employment, being present both in the period of market economy transition and in the period of economic growth [1]. Unemployment is a very important labor market issue, being a mismatch between the labor demand and supply. This indicator has major social and economic implications, being one of the factors to be examined in macroeconomic growth and very important in comparing the country's economic performance from a work perspective [2], affecting people's living standard and the nation's socio-economic status.

In this context, unemployment represents one of the biggest social problems of the world, being present in each country, the intensity of the phenomenon differing according to the economic development of a society. Population growth implies an increase regarding workforce, the jobs being insufficient in the short term [3]. The adjustment of the economic structure, the education system, and the establishment of the specialty does not satisfy the needs of economic restructuring; the professional skills of the rural labor force cannot satisfy the demand for jobs, aggravating the severity of unemployment. One of the solutions to this problem is the establishment of an early unemployment warning system, the forecast being absolutely necessary [4].

Forecasting the unemployment rate is very important for many economic decisions, especially setting relative policies by the government. The unemployment rate is correlated to the economic development of a society; therefore, different forecasting techniques are used for its forecast, from the simple OLS (ordinary least squares) method to the GARCH (generalized autoregressive conditional heteroskedasticity) models and neural networks. The econometric models are often related to stationary time series, seasonality, and trend analysis, and exponential smoothening to the simple OLS technique including ARIMA (autoregressive integrated moving average) models [5].

The ARMA and GARCH models were used by Chiros [6] to predict the unemployment rate in the UK; Parker and Rothman [7] modeled quarterly unemployment rates using the AR model (2), Power and Gasser [8] highlighted that the ARIMA (1,1,0) model has better forecasting performance for unemployment rates in Canada. Etuk et al. [9] indicated that the ARIMA (1,2,1) model is suitable for forecasting the unemployment rate in Nigeria.

Rothman [10] used six nonlinear models for out-of-sample forecasting, Koop and Potter [11] used the autoregressive threshold (ART) for modeling and forecasting the monthly unemployment rate, and Proietti [12] used seven forecasting models (linear and nonlinear). Johnes [13] used autoregressive models, GARCH, SETAR (Self-Exciting Threshold AutoRegressive) and neural networks in order to predict the monthly unemployment rate in the United Kingdom, the SETAR model registering the best results. Peel and Speight [14] also concluded that the SETAR model is better, in terms of root mean squared error (RMSE), compared to AR models.

As an alternative to ARMA models, Gil-Alana [15] used an exponential Bloomfield spectral model to model unemployment in the UK, the results indicating that this model is suitable for forecasting this phenomenon.

Forecasting the unemployment rate in Italy, Naccarato et al. [16] used both official data and the Google Trends query rate, estimating two different models: ARIMA and VAR (vector-autoregressive models), the VAR model registering a lower forecast error.

The autoregressive integrated moving average (ARIMA) models were introduced by Box and Jenkins [17], also developing the practical process to select the most suitable ARIMA model. ARIMA models are more secure in case of short-term forecasts compared to long-term forecasts [18]. For seasonal and non-seasonal data, the SARIMA (seasonal model autoregressive integrated moving average) is used. The SARIMA model is an extension of the simple ARIMA models, being used for inflation forecasting [19–21], for exchange rate forecasting [22,23], for tourist arrivals and income forecasting [24,25], as well as for unemployment forecasting. The literature includes a lot of studies on forecasting using ARIMA models, respectively the Box–Jenkins methodology, which is widely used by many researchers to highlight future unemployment rates [26].

Among them, Wong et al. [27] developed autoregressive integrated moving average (ARIMA) models in order to analyze and forecast important indicators in the Hong Kong construction labor market: employment level, productivity, unemployment rate, underemployment rate, and real wage. Ashenfelter and Card [28] analyzed unemployment, nominal wages, consumer prices, and the nominal interest rate, using the autoregressive moving average model. Kurita [29] forecasted the unemployment rate using autoregressively integrated fractional moving average, the model being much better than naive predictions.

Predictions of unemployment rate in the world using the ARIMA model were made by Chih-Chou and Chao-Ton [30], Etuk et al. [22] and Nkwatoh [31] in Nigeria using the ARIMA and ARCH model, Kanlapat et al. [32] in Thailand, Nlandu et al. [33] in Barbados, using the seasonal integrated autoregressive moving average model (SARIMA), Dritsakis and Klazoglou [34] in the USA using SARIMA and GARCH models, and Didiharyono and Syukri [35] in South Sulawesi using the ARIMA model.

In the European Union, the unemployment rate is forecasted using Box–Jenkins and TRAMO/SEATS methods [36,37]. In European countries, the unemployment rate was predicted using the Box–Jenkins methodology in Germany using the ARIMA and VAR models [38], in the Czech Republic using the SARIMA model [39,40], in the German regions using a model spatial GVAR [41], in Greece, both as a dynamic process and as a static process using SARIMA models [42,43], and in Slovakia using ARIMA and GARCH models [44].

Unemployment predictions using VAR were realized also by Kishor and Koenig [45], taking into account that data are subject to revisions. The accuracy of forecasts based on VAR models can be measured using the trace of the mean-squared forecasts error matrix, generalized forecasts error second moment [46], transfer functions [47], and combined forecasts based on VAR models are a good strategy for improving predictions' accuracy [48].

Wang et al. [49] used back propagation neural networks (BPNN) and the Elman neural network to predict unemployment rate. Neural networks are also used by Peláez [50] to forecast the unemployment rate, together with econometric models.

As the asymmetric behavior of unemployment rate can be modeled using a nonlinear time series model, Skalin and Terasvirta [51] proposed STAR. Peel and Speight [14] forecasted the unemployment rate in the UK using self-exciting threshold autoregressive (SETAR) models and an autoregressive model, in terms of RMSE, SETAR models registering better forecasting performance. Koop and Potter [11] used threshold autoregressive (TAR) in order to forecast the US unemployment rate, Johns [13] forecasted the unemployment rate using AR(4), AR(4)-GARCH(1,1), SETAR(3,4,4), and neural network, highlighting that SETAR is the best model.

According to the international definition [52], the unemployed are people aged between 15 and 74 who at the same time satisfy three conditions: they do not have a job, are available to start work in the next two weeks, and have been actively looking for a job anytime in the last four weeks. The unemployment rate represents the share of the unemployed in the active population, the active population in a country including all persons who provide labor available for the production of goods and services during the reference period, including employees and the unemployed.

Unemployment was first introduced in Romania in 1991, and the first study to assess unemployment according to ILO standards was conducted in 1994 [1]. Specific to a country in transition, unemployment in Romania was the result of the enterprise restructuring and the contraction of production [53].

In the first period after 1990, although many occupations appeared in Romania, the number of unemployed increased; 1994 had the highest registered unemployment rate [54]. In the period 1995–1996, the number of unemployed decreased by 46.28% and then increased significantly until 1999 due to socio-economic imbalances that arose from the closure of other productive structures. After 1999, the economic activities were restructured and privatized, especially in the case of large companies, leading to large layoffs, but also to the emergence of new jobs, the result being the unemployment reduction. Since 2000, employment in Romania has registered a continuous increase, with small fluctuations, leading to a reduction in unemployment [55].

In order to substantiate the macroeconomic policies in Romania, it is important and topical to forecast the labor supply, employment, and unemployment. In Romania, as in other European countries, unemployment is monitored and assessed very seriously. The most common method used in order to predict the unemployment in Romania involves ARIMA models.

Son et al. [56] analyzed the unemployment rate in EU-27 countries, focusing on Romania, concluding that the unemployment rate can be modeled by using a linear autoregressive model. Others studies using ARIMA models in order to predict the unemployment rate in Romania were realized by Madaras [57], Bratu [58], and Simionescu [59], while Dobre and Alexandru used the VARMA and VAR models [60], and at the level of two Romanian counties (Brasov and Harghita), studies used the Box–Jenkins methodology and NAR model based on the artificial neural network. Comparing the forecasted values with the officially recorded unemployment rate from the same period, we noticed that by the end of the period, the differences between the real and the predicted values became larger in the NAR model than in the ARMA model forecast, medium-term forecasts, forecasts based on the ARMA model being more accurate.

Other forecasts of the unemployment rate in Romania were realized by Bratu and Marin [61] using several techniques: econometric, exponential modeling, smoothing technique, and moving average method; of these, predictions based on the exponential smoothing technique recording the highest degree of accuracy. Voinegu et al. [62] predicted the unemployment rate using Holt's improved model, the monthly series being constructed and disseminated in three forms: adjusted, seasonally adjusted, and trend adjusted. Other predictions used the Kalman approach, the Kalman filter being appropriate for calculating the natural unemployment rate [63]. In the short term, Zamfir [64] modeled the unemployment rate using stochastic models.

Simionescu [65] predicted the unemployment rate in Romanian counties using Internet data and official data as well as a methodology consisting of different types of models with panel data. In the case of the quarterly unemployment rate, updated vector-autoregressive models (VAR models) and a Bayesian VAR model were used, but the VAR model exceeded the Bayesian approach in terms of predicted accuracy [66].

In order to analyze the dynamics of the unemployment rate in Eastern Europe, including Romania, Lukianenko et al. [67] constructed econometric regression models with nonlinearities due to discrete changes in modes. Using the Markov switching model, regularities were captured by modeling the asymmetry in the unemployment rate during contractionary states, revealing the specifics of the labor market for each country and the differences in the flexibility of reactions to changes in the economic environment.

3. Data and Methodology

In order to determine the best model to forecast the Romanian unemployment rate, we have investigated the monthly unemployment rate covering the period 2000M01 to 2020M12. The data were provided by Eurostat (European Union labour force survey, EU-LFS).

When choosing models, it is common practice to split the available data into two portions, training and test data, where the training data are used to estimate any parameters of a forecasting method and the test data are used to evaluate its accuracy. Therefore, the training set or "in-sample data" was set to the period 2000M01–M2017M12, and the test set or the "out-of-sample data" was set to the period 2018M01-2020M12. The forecast of unemployment rate will rely on the next two years of the period 2021–2022.

The main objective of the paper is to compare the forecasting potential of five models: exponential smoothing models (additive and multiplicative Holt–Winters (HW) models, and ETS model), the SARIMA model, the neural network autoregression (NNAR) model, and the SETAR model, and to predict future values of unemployment rate beyond the period under consideration.

Therefore, with the study, the forecasting performance was derived from the five models in view of identifying the best suited forecasting procedure for the monthly unemployment rate, taking into account the following steps:

1. Fit the Holt–Winters models (additive and multiplicative) on the training dataset (January 2000 to December 2017)
2. Fit the ETS model on the training dataset
3. Fit the NNAR model on the training dataset
4. Fit the SARIMA model on the training dataset
5. Fit the SETAR model on the training dataset
6. Compare the in-sample forecast accuracy measures for the all models
7. Compare the out-of-sample forecast accuracy measures for the models over the period January 2018 to December 2020
8. Compare the forecast projections of unemployment rate for all models over the period January 2021 to December 2022.

3.1. Holt–Winters Method and ETS Models

We will start our technical demarche by introducing the class of exponential smoothing methods as widely used forecasting procedures referring particularly to the Holt–Winters (HW) method, which is a commonly used forecasting method in time series analysis incorporating both trend and seasonal components, irrespective of whether they are additive or multiplicative in nature. The additive method is preferred when the seasonal variations are roughly constant through the series, while the multiplicative method is preferred when the seasonal variations are changing proportional to the level of the series.

The Holt–Winters' additive method can be written as follows:

$$L_t = \alpha(y_t - S_{t-s}) + (1-\alpha)(L_{t-1} + b_{t-1}) \qquad (1)$$

$$b_t = \gamma(L_t - L_{t-1}) + (1-\gamma)b_{t-1} \qquad (2)$$

$$S_t = \delta(y_t - L_t) + (1-\delta)S_{t-1}. \qquad (3)$$

The Holt–Winters' multiplicative method can be written as follows:

$$L_t = \alpha\frac{y_t}{S_{t-s}} + (1-\alpha)(L_{t-1} + b_{t-1}) \qquad (4)$$

$$b_1 = \gamma(L_t - L_{t-1}) + (1-\beta)b_{t-1} \qquad (5)$$

$$S_t = \delta\frac{y_t}{L_t} + (1-\delta)S_{t-1} \qquad (6)$$

where $t = 1, \ldots, n$, s represents the length of seasonality (months), L_t represents the level of the series, and b_t denotes the trend and S_t seasonal component [22]. The constants used for this model are α (level smoothing constant), γ (trend smoothing constant), and δ (seasonal smoothing constant). In order to choose the most adequate smoothing constants, we tested different values of the smoothing constants. The model is selected according to the certain forecast accuracy such as MAPE (the mean absolute percentage error), the best model being the model who register the minimum value for MAPE.

The ETS (error, trend, seasonal) model represents time series models that support the exponential smoothing methods, consisting of a trend component (T), a seasonal component (S), and an error term (E). These are based on error–trend–season probabilities of Hyndman, being defined an extended class of ES methods using probability calculations based on the state space, with support for model selection and the calculation of standard forecast errors [68].

The long-term movement is characterized by the trend term, the pattern with known periodicity is reflected by the seasonal term, and the error term represents the irregular, unpredictable component of the series.

ETS models generate both point forecasts and prediction intervals (or forecast). If the same values of the smoothing parameters are used, the point forecasts are identical but will generate different prediction intervals.

The individual components of an ETS specification may be specified as being of the following form: N = none, A = additive, M = multiplicative:

E: A, M
T: N, A, M
S: N, A, M.

An ETS (A,A,A) decomposition is a Holt–Winters method with an additive seasonal component, and an ETS (M,A,M) represents a Holt–Winters method with a multiplicative seasonal component.

The automatic selection of the model is based on the ETS smoothing. For each ETS model, the probability and the forecast error can be calculated by comparing the information criterion based on probability or an out-of-sample AMSE (The average mean square error estimator finds the parameter values and initial state values that minimize the average mean square error of the step forecasts of the specified ETS model) in order to determine the model that best fits the most accurate data or forecasts. Automatic selection for unemployment rate forecasting using the ETS framework will be done using Akaike Information Criterion corrected (AICc) minimization.

3.2. The Neural Network Autoregression Model

Artificial neural networks are used to model complex nonlinear relationships between input variables and output variables. An autoregression model of the neural network (NNAR) has delayed values of a time series as input in the model, and it predicted values of the time series as output. The major difference of the NNAR method compared to the HW method is the non-existence of the restriction of stationary parameters. Considering the seasonality of the monthly unemployment rate, the specification of the neural network will be NNAR(p,P,k)m, and the graphical representation from Figure 1. By adding an intermediate layer with hidden neurons, the neural network becomes nonlinear, and without the hidden layer, NNA(p,P,0)m becomes SARIMA(p,0,0) (P,0,0)m.

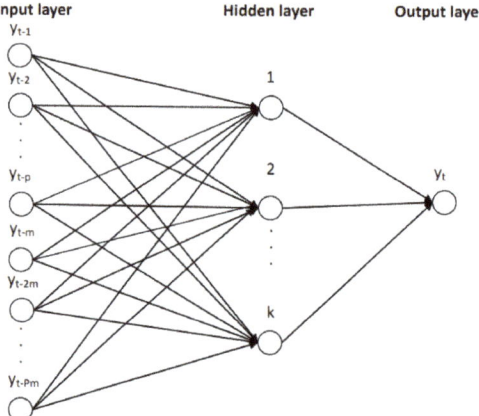

Figure 1. A diagrammatic representation of the NNAR(p,P,k)m model. Source: Touplan [69]. NNAR: neural network autoregression.

The NNAR model represents a feedforward neural network, involving a linear combination function and an activation function. The linear combination function has the following form [70,71]:

$$net_j = \sum_i w_{ij} y_{ij}. \tag{7}$$

The hidden layer has a nonlinear sigmoid function in order to issue the input for the next layer:

$$s(z) = \frac{1}{1+e^{-z}}. \tag{8}$$

In the case of NNAR(p,k) with p delayed entries and k nodes in the hidden layer, the model involves delayed time series values as entries in a neural network, considering a feed-forward network with a single hidden layer. The seasonal component is present in the data (m = 12), so the last observed values from the same season will be added as inputs, NNAR becoming NNAR(p,P,k)12.

The forecasting procedure is iterative; the one-step ahead forecast uses historical inputs; and the two-steps ahead forecast uses the one-step ahead forecast and the historical data.

3.3. Seasonal Autoregressive Integrated Moving Average Model (SARIMA) Model

Taking into account the seasonal pattern exhibited by the monthly unemployment rate, a seasonal process may be considered; therefore, the ARIMA model will become a SARIMA model. The seasonal autoregressive integrated moving average (SARIMA) model is a generalized form of an ARIMA model that accounts for both seasonal and non-seasonal data. The SARIMA model is denoted as ARIMA(p,d,q) (P,D,Q)S and has the following specification based on the backshift operator [72,73]:

$$\phi(B)\phi(B^s)(1-B)^d(1-B^s)^D Y_t = (B)(B^s)\varepsilon_t \tag{9}$$

$$\phi(B) = 1 - \phi_1 B - \phi_2 B^2 - \ldots - \phi_p B^p \tag{10}$$

$$\phi(B^s) = 1 - \phi_1 B^s - \phi_2 B^{2s} - \ldots - \phi_p B^{2Ps} \tag{11}$$

$$(B) = 1 +_1 B +_2 B^2 + \ldots +_q B^q \tag{12}$$

$$(B^s) = 1 +_1 B^s +_2 B^{2s} + \ldots +_Q B^{Qs} \tag{13}$$

where Y_t represents the time series data at period t, B denotes the backshift operator, ε_t is a sequence of i.i.d. variables (mean zero and variance σ^2), s is the seasonal order, ϕ_i and ϕ_j are the non-seasonal and seasonal AR parameters, θi and θj are respectively non-seasonal and seasonal MA parameters, p, d, and q denote the non-seasonal AR, I, and MA orders, respectively, and P, D, and Q respectively represent the seasonal AR, I, and MA orders.

Similar to the Box–Jenkins methodology, also, the SARIMA model follows a five-step iterative procedure: identification, estimation, selection, diagnostics, and forecasting [34,60,69].

Before fitting a particular model to time series data, the stationarity of a series must be checked [74]. In order to identify if the time series in stationary, the graphical representation of the series together with the correlogram of the series in level, Bartlett test, and Ljung–Box test can be applied. In order to test if the series has a unit root, the Augmented Dickey–Fuller and Philips–Perron tests can be used. To obtain a stationary time series, the corresponding value of d is estimated, in the case of a non-stationary series in mean, the series is differentiated, and in the case of a non-stationary series in variance, the series is logarithmized.

In addition, the series needs to be tested against the presence of a structural break using the Zivot–Andrews test. The Zivot and Andrews endogenous structural break test is a sequential test that uses the full sample and a different dummy variable for each possible break date. The break date is selected where the t-statistics of a unit root ADF (Augmented Dickey Fuller) test is at a minimum (most negative). Consequently, a break date will be chosen when the null hypothesis of a unit root will be rejected. The Zivot–Andrews test uses three scenarios: a structural break in the level of the series, a one-time change in the slope of the trend, and a structural break in the level and slope of the trend function of the series. Therefore, under the test, the null hypothesis assumes that the series yt contains a unit root without any structural break, against the alternative that the series is a trend-stationary process with a one-time break occurring at an unknown time point.

Another important feature that needs to be investigated for a series exhibiting a seasonal pattern under the stationarity condition is to test for the presence of a seasonal unit root using the HEGY test [75]. The HEGY test is used in case of a seasonal and non-seasonal unit root in a time series. A time series y_t is considered as an integrated seasonal process if it has a seasonal unit root as well as a peak at any seasonal frequency in its spectrum other than the zero frequency.

The test distinguishes between deterministic seasonality—which can be removed by seasonal adjustment—and stochastic seasonality—which refers to unit roots at the seasonal frequencies [76].

Once the stationarity has been achieved, the *identification stage* involves determining the proper values of p, P, and q, Q based on the correlogram of the stationary series (ACF and PACF plot). Checking the ACF and PACF plots, we should both look at the seasonal and nonseasonal lags. Usually, the ACF and the PACF have spikes at lag k and cut off after lag k at the non-seasonal level. The ACF and the PACF also have spikes at lag ks and cut off after lag ks at the seasonal level. The number of significant spikes suggests the order of the model [74].

An SAR signature usually occurs when the autocorrelation at the seasonal period is positive, whereas an SMA signature usually occurs when the seasonal autocorrelation is negative.

In the model selection stage, we need to decide on the optimal model from several alternative estimated models in the situation in which two or more models compete in the selection of the best model for the study.

In order to be able to make a decision, we can rely on the penalty information criteria (Akaike Information Criterion (AIC), the Akaike Information Criterion corrected (The AICc includes a penalty that discourages overfitting, and increasing the number of parameters improves the goodness of fit [72]) (AICc), and the Bayesian Information Criterion (BIC), choosing as an optimal model the model with the smallest values of AIC, AICc, and BIC.

In the model estimation stage, the parameters of the chosen model are estimated using the method of maximum likelihood estimation (MLE).

The diagnostic checking stage is the next stage investigating if the estimated model or models are firstly validated in accordance with the classical tests: t-test for the statistical significance of the parameters and F-test for the statistical validity of the model.

Secondly, the main hypotheses on the model residuals need to be tested, showing that they are white noise, homoscedastic, and do not exhibit autocorrelation. The normality of the residuals has been checked using Jarque–Bera test, while for non-autocorrelation, the Ljung–Box test has been applied. When the variance of the residuals is not constant, the issue of conditional heteroscedasticity is one of the key problems that is likely to encounter when fitting models. For checking autoregressive conditional heteroskedasticity (ARCH) in the residuals, the squared residuals correlograms and the ARCH-LM test can be used. In case there is no ARCH in the residuals, the autocorrelations and partial autocorrelations should be zero; regardless, the lags and the Q-statistics should be insignificant.

If at the level of this stage, one of the hypotheses is invalidated, we need to return to the first stage of the model and rebuild a better model. Otherwise, if the model passes this stage, the forecasting process can be implemented to predict future time series based on the most reliable model validated in the previous stages.

The final stage is forecasting in order to design future time series values, using the most convenient model according to previous stages [43].

3.4. SETAR Model

The SETAR model is part of the more general class of threshold autoregressive models (TAR) and represents an extension of autoregressive models, bringing as its main advantage in modeling a time series and a higher flexibility in parameters due to a regime-switching behavior. Thus, this particular type of model allows for the prediction of future values of unemployment rate, assuming that the behavior of the time series changes when the series switch the regime, and this switching is dependent on the past values of the series. The model relies on an autoregressive model of lags p, on each regime, and it is denoted to be SETAR(k,p), where k is the number of thresholds (k + 1 regime assumed in the model) and p is the order of an AR(p).

Even if the process is assumed to be linear in each regime, the switching from one regime to another transforms the process into a nonlinear one.

The general specification of a two-regime SETAR(2,p,d) of the following regime to the others proves the entire regime as nonlinear [66,67,73]. The two-regime version of the SETAR model of order p is given by:

$$y_t = \phi_0^{(1)} + \sum_{i=1}^{p(1)} \phi_i^{(1)} y_{t-i} + \varepsilon_t^{(1)}, \text{ if } y_{t-d} \leq \tau \tag{14}$$

$$y_t = \phi_0^{(2)} + \sum_{i=1}^{p(2)} \phi_i^{(2)} y_{t-i} + \varepsilon_t^{(2)}, \text{ if } y_{t-d} > \tau \tag{15}$$

where $\phi_i^{(1)}$ and $\phi_i^{(2)}$ are the coefficient in the lower and higher regime, respectively, which needs to be estimated; τ is the threshold value; $p(1)$ and $p(2)$ are the order of the linear AR model in the low and high regime, respectively. y_{t-d} is the threshold variable governing the transition between the two regimes, d being the delay parameter, which is a positive integer (d < p); $\varepsilon_t^{(1)}$ and $\varepsilon_t^{(2)}$ are a sequence of independently and identically distributed random variables with zero mran and constant variance [77].

The main phases for setting a SETAR model are the order selection of the model based on AR(p) order identification together with the test for threshold nonlinearity, model identification requiring the selection of the delay parameter d together with the location of the threshold value, model estimation and evaluation, and the last stage forecasting the future values of unemployment rate.

Thus, the first stage in applying the SETAR model is to analyze the existence of a nonlinearity behavior, and for that, it is important to first determine the appropriate lag length of the autoregressive model AR(p) for the analyzed time series, and the choice could rely on the minimum value of AIC. Secondly, we will test the existence of nonlinearity using the Tsay F test, the null hypothesis of linearity being rejected if the p-value of the test is smaller than the significance level assumed.

Proving that there is nonlinearity in the time series, we can pass to the second stage—model identification—and we will consider a two-regime SETAR model with the order p of autoregressive parts equal in both regimes, SETAR(2,p,d).

In the third stage, the selection of delay parameter together with the location of the threshold value is realized, taking into account that the possible value d is less than order. Therefore, several SETAR models with different delay parameters and threshold values can be identified, and based on a grid search method, the best model is selected to be the model with the smallest value for the residual sum of squares.

The model is estimated using the MLE, and then, the adequacy of the selected model is evaluated based on diagnostics tests on residuals. The ARCH-LM test is used for testing the hypothesis of constant variance and Breusch–Godfrey is used for testing for higher-order serial correlation in the residuals.

3.5. Forecasting Performance Comparison

In order to provide predictions of the future values of unemployment rate based on past and present data and analysis of trends, it is important to use both in-sample and out-of-sample forecasting performance methods, even if the out-of-sample is known to offer more reliable results. Therefore, a model with good performance in the out-of-sample forecasting performance is picked as the best model. The forecasting performance of models was evaluated on two sub-samples: in-sample data, 2000M01–2017M12, which is used to estimate and identify the model and also to provide in-sample forecasting performance, and out-of-sample data, 2018M01–2020M12, which is used for analyzing the forecasting performance.

Forecasting accuracy offers valuable information about the goodness fit of the forecasting model and shows the capacity of the model to predict future values of unemployment rate. Three criteria have been used to evaluate the performance of models both on in-sample data and out-of-sample data: the root mean squared error (RMSE), the mean absolute error (MAE), and the mean absolute percent error (MAPE). The better forecast performance of the model is that with the smaller error statistics.

Another test used to check the existence of differences between the forecast accuracy of two models was the Diebold–Mariano test [78], which assumes in the null hypothesis the absence of such a difference against the alternative of the existence of a statistical difference between the forecast accuracy of the models.

4. Data and Empirical Results

We have used in the empirical analysis the ILO unemployment rate for Romania covering the period 2000M01–2020M12, summing up a total of 252 monthly observations. The data source is the Employment and Unemployment database of Eurostat. We used for the model estimation and identification the estimation period 2000M1–2017M12 as training data and the period 2018M01–2020M12 as test data, while the forecast projections have been made for the next two years, 2021–2023.

The evolution of unemployment rate revealed an oscillating trend, from peaks (8.1% in January 2001 and January–March 2002) to minimum levels (5% in September 2008). The winter months of the years 2000, 2001, and 2002 registered increases in the unemployment due to lack of jobs, the year 2002 recording the highest rate of the monthly unemployment rate (144%). A potential explanation could be the dismissals that took place as a result of the implementation of restructuring and privatization programs of different sectors of activity. The impasse in the general economic and social development of Romania, the low living standard, and the lack of future perspectives from the period 1998–2000 reactivated the migration phenomenon, causing many Romanians to look for a job in more developed countries. However, after 1998, illegal migration predominated, which was mainly directed to Italy and Spain.

Compared to previous years, in 2004, the unemployment rate decreased; the number of persons entering unemployment was lower than the previous year by 92,442 persons. The 278,080 unemployed related to 2004 came from the redundancies that took place as a result of restructuring and privatization programs of different sectors of activity, and of these, only 67,042 people came from collective redundancies; the remaining 211,038 people came from current redundancies personal.

Young people represent the best professionally trained age group in Romania, but also the most exposed to unemployment, highlighting the brain-drain phenomenon. The decrease in the unemployment rate in the period 2002–2006 is due both to legal and illegal departures of persons to work abroad. Thus, in 2006, according to the figures offered by Eurostat, it was estimated that over two million Romanians work in the countries of Western Europe or other developed countries. The economic crisis from 2008 created another peak in the evolution of unemployment rate, registering in the first three months of 2010 the values of 7.7%, 7.7%, and 7.9% and oscillating around this value until the first three months of 2015 (7.5%, 7.4%, and 7.2%).

The unemployment rate in 2008 decreased compared to the previous year (6.4%), but during the economic crisis of 2008–2009, there was a substantial increase in the unemployment rate. Although the number of jobs in the economy is constantly decreasing, the unemployment rate is decreasing, the explanation of this paradox being given by the following:

1. Working abroad: according to official estimates, in the first nine months of 2010, the number of those who went to work abroad exceeded 380,000, of which 140,000 went on their own, 140,000 went through recruitment agencies, and 102,000 went through the NAE (National Agency for Employment)
2. Retirement of some of the employees. Quarterly, 70,000–80,000 people retire; therefore 200–300,000 employees must be replaced annually. It is very likely that companies will no longer replace some of the people who have retired, so that the number of employees can decrease without increasing the number of unemployed.
3. Undeclared work. In second quarter of 2010, the number of undeclared workers increased by almost 100,000.

For the last years, the trend for unemployment rate was continuously downward, with a minimum point in the first month of 2020 (3.8%), and since February 2020, the unemployment rate registered an ascendant trend. The reversed trend was due to the high unemployment rate (18.5%) among young people (15–24 years) and seasonality in the construction and tourism sectors.

In 2019, the unemployment rate decreased to 3.9%, compared to 4.2% in 2018, affecting to a greater extent the graduates of lower and secondary education, for which the rate was 6.3% and 4%, respectively, according to data from the National Institute of Statistics (NIS). On the other hand, the unemployment rate for people with higher education was much lower, 1.6% in 2019.

In 2020, in the context of the coronavirus crisis, the unemployment rate started to increase since February, with the taking of safety measures, reaching 5.2% in May, which was the highest level since 2017. According to the NIS, the number of unemployed people exceeded 460,000, with over 110,000 more people than the same period last year.

In August, the unemployment rate decreased by 0.1 points compared to the previous month, but it increased by 1.5 points compared to the same month last year. Thus, August was the first month since the beginning of the COVID-19 pandemic on the Romanian territory when the unemployment rate registered a decrease. In March, the unemployment rate was 4.6%.

In autumn, in October 2020, the unemployment rate increased by 0.2 points compared to the previous month (5.1%), unemployment among men being higher than among women by 0.5 percentage points, according to the NIS. Unfortunately, youth unemployment (18–24 years) is approaching 20%. As for the number of unemployed, Romanians looking for a job were 477,000, with over 100,000 more than in October of the previous year.

In January–October 2020, the medium unemployment rate stood at 4.9%, which was up 1.1 points year/year, an evolution determined by the incidence of the health crisis (and the consequences of this unprecedented shock), partially offset by the implementation of an unprecedented relaxed mix of economic policies.

Figure 2 revealed that the Romanian unemployment rate exhibited seasonal fluctuations over the period 2000–2020, with peaks in the last and the first months of the year. Figure 2 depicts the evolution of the monthly unemployment rate, revealing a clear seasonal component in the data, which was confirmed also by the autocorrelation plot (Figure 3).

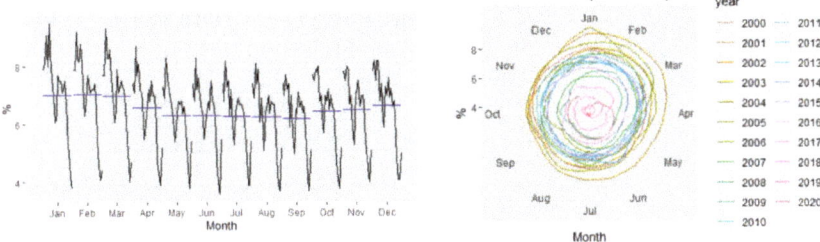

Figure 2. The seasonal pattern in the monthly ILO unemployment rate.

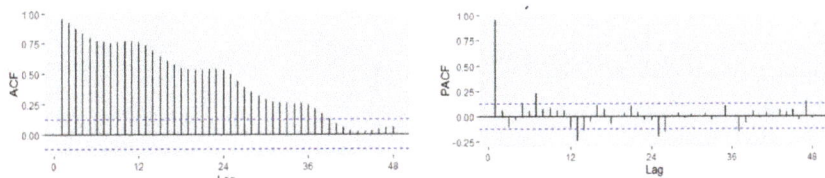

Figure 3. Autocorrelation and partial correlation plot of Romania's monthly unemployment rate for the horizon 2000–2020.

4.1. Holt–Winters Results

The empirical results of Holt–Winters additive and multiplicative models revealed that because both models have exactly the same number of parameters to estimate, the training RMSE from both models can be compared, revealing that the method with multiplicative seasonality fits the data best. In addition, based on the informational criteria (AIC, AICc, or BIC), the optimal model is also the multiplicative version of HW. Table 1 gives the results of the both in-sample and out-of-sample forecasting accuracy measures of the Holt–Winters methods for the unemployment rate.

According to the RMSE measure, the multiplicative model performs better than the additive one, while based on the other forecast accuracy measures (MAPE, MASE, or MAE), the optimal model is the additive one, for which they registered the minimum values (Table 2).

Analyzing the evolution of monthly unemployment rate for the period 2021–2022, it can be highlighted the fact that the forecast projections tend to under evaluate the actual series, not capturing the impact of the pandemics, and revealing a downward trend in both cases, which is more accentuated in the case of the multiplicative model (Figure 4).

Table 1. The empirical results of HW for the forecast of unemployment rate.

Model 1: Holt–Winters' Multiplicative Method	Model 2: Holt–Winters' Additive Method
Smoothing parameters: Alpha (level) = 0.6928 Beta (trend) = 0.0001 Gamma (seasonal) = 0.0001	Smoothing parameters: Alpha (level) = 0.7503 Beta (trend) = 0.0001 Gamma (seasonal) = 0.0001
AIC = 630.187 AICc = 633.278 BIC = 687.566	AIC = 645.789 AICc = 648.8807 BIC = 703.169

Table 2. Forecasting performance of Holt–Winters.

	Holt–Winters' Multiplicative Method		Holt–Winters' Additive Method	
	Training Dataset	Testing Dataset	Training Dataset	Testing Dataset
ME	−0.0124	−0.2670	0.0006	−0.0371
RMSE	0.2771	0.6906	0.2804	0.7480
MAE	0.2086	0.6524	0.2109	0.6273
MPE	−0.3191	−7.8322	−0.1259	−2.6101
MAPE	3.0368	15.1393	3.0699	13.8268
MASE	0.3317	1.0374	0.3353	0.9974

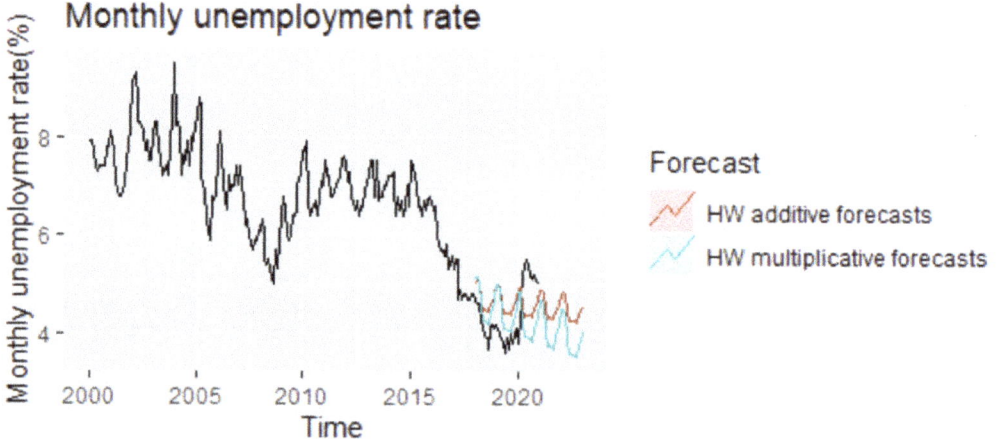

Figure 4. The forecast of unemployment rate based on Holt–Winters (HW) models for the period 2021–2022.

4.2. ETS Models Results

In the process of obtaining a reliable forecast of the monthly unemployment rate, the ETS automatic selection framework, based on minimizing the AICc, revealed the optimal model to be an ETS(M,N,M) with multiplicative error, no trend, and multiplicative season. The empirical results highlighted that on the training dataset, the ETS model produces better results in comparison with HW additive or multiplicative methods (Table 3). The ETS(M,N,M) model will provide different point forecasts to the multiplicative Holt–Winters' method, because the parameters have been estimated differently, the default estimation method being maximum likelihood rather than minimum sum of squares (Table 4).

Table 3. The empirical results of ETS (error, trend, seasonal) models for the forecast of unemployment rate.

ETS(M,N,M) Model: Multiplicative Error, No Trend, Multiplicative Season
Smoothing parameters:
Alpha(level) = 0.7914
Gamma(seasonal) = 0.0001
AIC = 627.799
AICc = 630.199
BIC = 678.428

Table 4. Forecasting performance of ETS model.

	ETS Model
	Training dataset
ME	−0.0166
RMSE	0.2788
MAE	0.2097
MPE	−0.3682
MAPE	3.0569
MASE	0.3335

The plot of ETS(M,N,M) components displays the states over time, while Figure 3 shows point forecasts and prediction intervals generated from the model. The empirical results of the model pointed out an under evaluation of the real values during the period of the test dataset from 2018 to 2020, highlighting an oscillating evolution characterized by a strong seasonal pattern also for the forecast projections period, 2021–2022 (Figure 5).

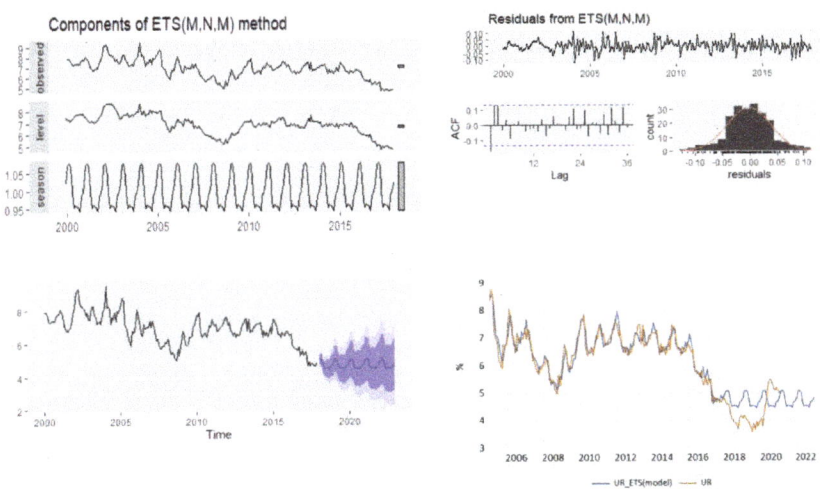

Figure 5. The forecast of unemployment rate based on the results of ETS(M,N,M).

4.3. NNAR Model

In order to fit the NNAR model, the series of unemployment rate has been explored on the training dataset in the process of identifying the order of an AR term present in the data, using the correlogram of the series. Based on the ACF and PACF plots, a pure AR(1) process can be highlighted for the non-seasonal part. Analyzing the ACF plot, the decaying spikes at every 12-month interval indicate a seasonal component present in the data (Figure 6). As the autocorrelation at the seasonal period (ACF at lag 12) is positive, an autoregressive model for the seasonal part should be considered; therefore, the order P was set to 1. Therefore, a NNAR(1,1,k)$_{12}$ model is fitted, and the in-sample and out-sample root mean square error (RMSE), mean absolute error (MAE), mean absolute scale error (MASE), and mean absolute percentage error (MAPE) are provided in Table 5 where k = 1, ..., 14.

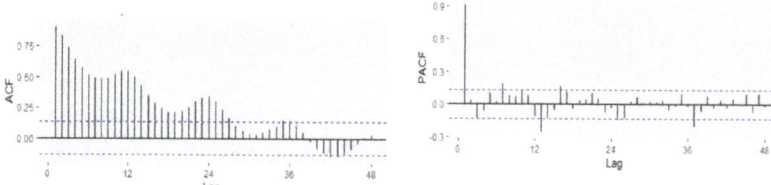

Figure 6. Autocorrelation and partial correlation plot of Romania's monthly unemployment rate.

Table 5. Forecasting performance of NNAR(1,1,k)$_{12}$.

	Training Dataset				Test Dataset			
k	RMSE	MAE	MAPE	MASE	RMSE	MAE	MAPE	MASE
1	0.3570	0.2734	3.9654	0.4348	0.6792	0.6399	16.2143	1.0174
2	0.3477	0.2662	3.8562	0.4233	0.9019	0.8542	21.6274	1.3582
3	0.3402	0.2604	3.7626	0.4141	0.8510	0.8044	20.3754	1.2790
4	0.3329	0.2553	3.6772	0.4059	2.0452	1.8630	47.2547	2.9622
5	0.3297	0.2524	3.6264	0.4013	1.6242	1.4196	36.1478	2.2572
6	0.3228	0.2464	3.5341	0.3918	0.7710	0.7208	18.2993	1.1461
7	0.3195	0.2443	3.5057	0.3884	0.7739	0.7221	18.3387	1.1482
8	0.3173	0.2421	3.4737	0.3850	0.8042	0.7518	19.0849	1.1954
9	0.3167	0.2421	3.4681	0.3850	0.7873	0.7356	18.6744	1.1696
10	0.3150	0.2411	3.4513	0.3834	0.5979	0.5508	14.0168	0.8758
11	0.3087	0.2362	3.3860	0.3757	0.6936	0.6450	16.3913	1.0256
12	0.3033	0.2329	3.3456	0.3704	0.6220	0.5747	14.6184	0.9139
13	0.3058	0.2339	3.3533	0.3719	0.7008	0.6510	16.5462	1.0351
14	0.3064	0.2357	3.3779	0.3749	0.6944	0.6452	16.4001	1.0260

The selection of the best model relied on the lowest values of all the forecast accuracy measures (RMSE, MAE, MAPE, and MASE), but especially on the values of MAPE and MASE, which are scale independent and used to compare forecast accuracy across series on different scales and seen as an appropriate measure when the out-of-sample data are not of the same length as the in-sample data. Based on the results of Table 5, MASE and MAPE are lower for the training dataset with 12 nodes in the hidden layer, whereas the out-of-sample MASE and MAPE are lower for 10 nodes in the hidden layer. Therefore, we can consider as the best choice the model NNAR(1,1,10)$_{12}$. The forecast of the unemployment rate based on the NNAR(1,1,10)$_{12}$ model results revealed a downward trend with a peak in September 2018 (4.43%) and with a forecasting value for 2021–2022 oscillating around the value of 4.35% (Figure 7).

4.4. SARIMA Model

For fitting a SARIMA model, we used data covering the period January 2000 to December 2017. The descriptive statistics values of the unemployment rate for the training dataset are displayed in Figure 8. The series exhibited a strong seasonal pattern over the horizon 2000–2017.

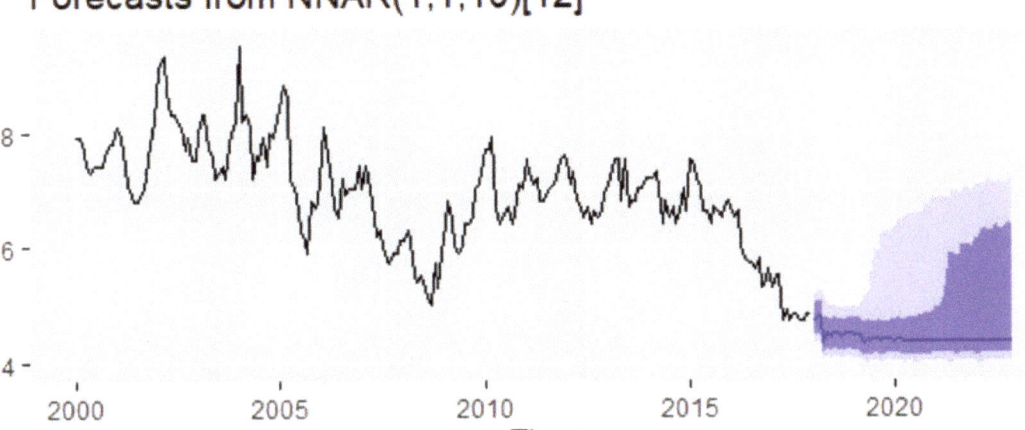

Figure 7. Forecasts from a neural network with one seasonal and non-seasonal lagged input and one hidden layer containing ten neurons.

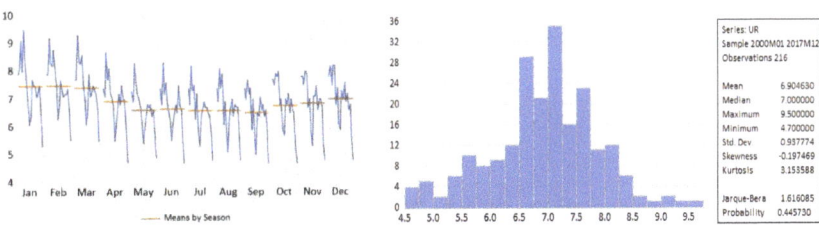

Figure 8. Descriptive statistics of unemployment rate for the horizon 2000–2017.

4.4.1. Testing for Non-Stationarity

In order to fit a suitable time series model, the stationarity need to be investigated based on Augmented Dickey–Fuller and Philips–Perron tests. The graphical inspection of the autocorrelation and partial correlation plot of Romania's quarterly unemployment rate (Figure 9) revealed that the values of autocorrelation coefficients decrease slowly, pointing out a nonstationary and relatively stable seasonal pattern of our time series.

The time-series plot of the first difference of the series highlighted that the unemployment rate is a non-stationary mean time series. The information is also confirmed by the empirical results of Bartlett and Ljung–Box tests.

The time-series plot of the first difference of the series highlighted that the first difference of the unemployment rate seems to be a stationary mean time series. Therefore, the original quarterly series is a non-stationary time series.

Diagram (b) from Figure 9 indicates that a possible stationarity exists in first differences. Alternately, we investigated the presence of unit roots by applying the Augmented Dickey–Fuller and Phillips–Peron tests initially to the series in level and then to the series in first differences. The empirical results on unemployment rate are displayed in Table 6, indicating that the series of unemployment rate is stationary in first differences, being integrated of order 1.

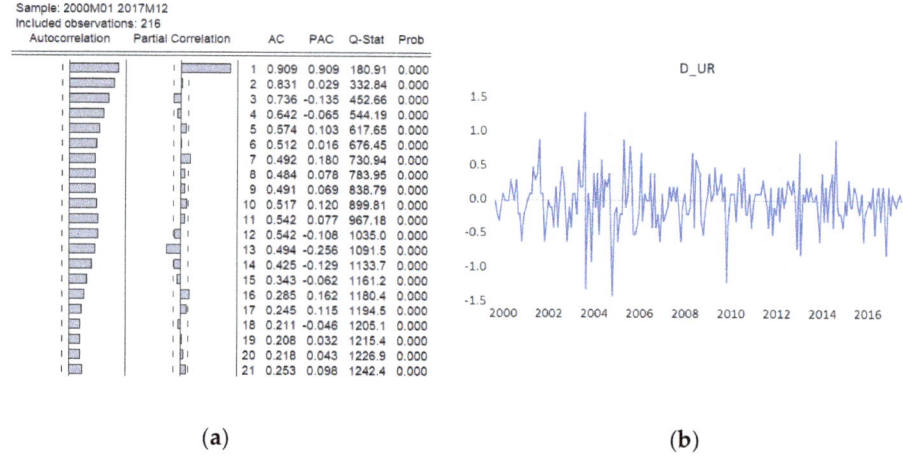

Figure 9. Autocorrelation and partial correlation plot of Romania's monthly unemployment rate (**a**) and first difference of the original time series (**b**).

Table 6. Unit root analysis of the Romanian unemployment rate.

Variable	Unit Root [Transf.]		Level		First Difference	
			ADF	PP	ADF	PP
Unemployment rate	I(1) [ΔUR]	T&C	−3.56 **	−3.52 **	−15.87 ***	−16.20 ***
		C	−2.58 *	−2.72 *	−15.90 ***	−16.01 ***
		None	−0.90	−0.98	−15.91 ***	−16.01 ***

Note: ***, **, and * means stationary at 1%, 5%, and 10%; T&C represents the most general model with a constant and trend; C is the model with a constant and without trend; None is the most restricted model without a drift and trend. For the ADF test, the number of lags was determined using SCH criterion for a maximum of 14 lags to remove serial correlation in the residuals. For both PP tests, the value of the test was computed using Newey–West Bandwith (as determined by Bartlett kernel). Tests for unit roots have been carried out in E-VIEWS 11.

The next step was to test the presence of a structural break around 2009 (from Figure 10), taking into account that the presence of a structural break will invalidate the results of unit root tests. Therefore, the Zivot–Andrews test has been used, the empirical result revealing that there is not enough evidence to reject both the null hypothesis that unemployment has a unit root with structural break in trend, and in both intercept and trend (Table 7).

Thus, the empirical results proved that the unemployment rate is non-stationary and integrated of order 1, I(1).

However, because the series of unemployment exhibits a seasonal pattern over the training period, the study will use a seasonal ARIMA model instead of non-seasonal models; therefore, it is necessary to check whether the seasonality is needed to be differenced or not, testing if the stochastic seasonality is present within the data, the empirical results of Hegy test revealing the rejection of seasonal unit root and the acceptance of only a non-seasonal unit root. Therefore, seasonal difference is not needed.

Therefore, we can conclude that the unemployment rate is a non-stationary series, without stochastic seasonality and integrated of order 1. Thus, the rate of unemployment will be modeled at the first difference of the series within the SARIMA model and self-exciting threshold autoregressive (SETAR) model.

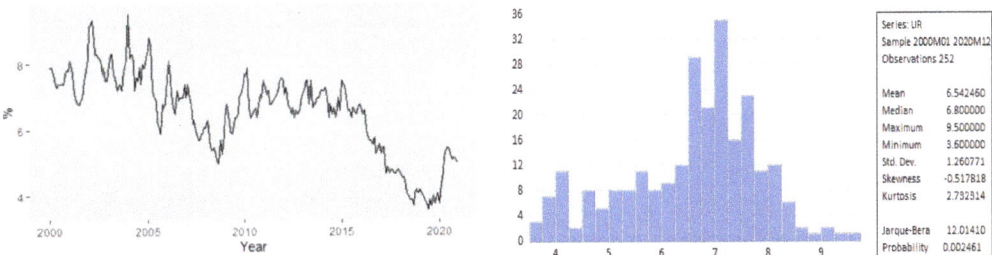

Figure 10. The Romanian ILO unemployment rate for the period 2000M1–2020M12.

Table 7. Zivot–Andrews unit root test having a structural break for unemployment rate.

Series (Trend Specification: Trend and Intercept)		Allowing for Break in Trend	Allowing for Break in Both Intercept and Trend
Unemployment Rate	Minimum t-stat (Lag length has been established using SBC criterion for maximum 14 lags) p-value	−4.139 (0.13)	−4.484 (0.243)
	Critical values		
	1%	−5.067	−5.719
	5%	−4.524	−5.175
	10%	−4.261	−4.893
		Potential break point at 2015M09	Potential break point at 2009M06

4.4.2. Identification of the Model

For the first difference of the UR, the model identification implies the identification of proper values of p, P, q, and Q using the ACF and PACF plot. The seasonal part of an AR or MA model will be seen in the seasonal lags. The ACF plot has a spike at lags 4 and 6 and an exponential decay starting from seasonal lag 12, suggesting a potential non-seasonal MA component-MA(4) or MA(6) (Table 8).

Table 8. HEGY test of seasonality for level of unemployment series.

Null	Simulated p-Value *	The Presence of Non-Seasonal Unit Root **	The Presence of Seasonal Unit Root **
Unemployment rate			
Non-seasonal unit root (zero frequency)	0.736310		
Seasonal unit root (2 months per cycle)	0.005643		
Seasonal unit root (4 months per cycle)	0.000000	Yes	No
Seasonal unit root (2.4 months per cycle)	0.000177		
Seasonal unit root (12 months per cycle)	0.000177		
Seasonal unit root (3 months per cycle)	0.000000		
Seasonal unit root (6 months per cycle)	0.000000		

Note: The HEGY test was applied taking into account intercept and trend and seasonal dummies; the maximal number of lags was eight following Schwarz criterion and a number of 1000 simulations. * If the probability is higher than 0.10, then the presence of the non-seasonal unit root cannot be rejected. ** If the probability is higher than 0.10, then the presence of a seasonal unit root cannot be rejected.

The PACF plot shows that lags 4, 6, and 12 are significant, capturing also potential non-seasonal AR components together with a seasonal AR(1) (Figure 11). In our case, because the autocorrelation at the seasonal lags (12, 24) is positive, a combination of seasonal and non-seasonal autoregressive models can be identified. Thus, several models have been specified, and based on AIC and BIC together with the goodness of fit measures, the best model has been identified.

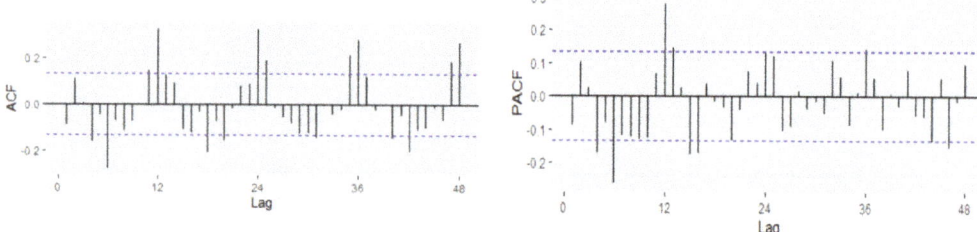

Figure 11. Autocorrelation and partial correlation plot of the first difference of the unemployment rate.

Thus, several models have been specified, and based on AIC and BIC together with the goodness of fit measures, the best model has been identified, taking into account the lowest values of AIC and SBC. The best model has been an ARIMA$(0,1,6)(1,0,1)_{12}$ considered based on the minimum value of AIC and SBC (Table 9).

Table 9. AIC and SBC for the suggested ARIMA models.

Model	AIC	AICc	BIC
ARIMA$(4,1,4)(1,0,0)_{12}$	133.2	134.28	166.9
ARIMA$(4,1,4)(2,0,0)_{12}$	129.99	131.29	167.07
ARIMA$(4,1,4)(3,0,0)_{12}$	124.03	125.58	164.48
ARIMA$(4,1,4)(3,0,1)_{12}$	116.73	118.54	160.55
ARIMA$(0,1,4)(3,0,0)_{12}$	148.39	149.09	175.36
ARIMA$(4,1,4)(0,0,3)_{12}$	130.87	132.41	171.31
ARIMA$(4,1,4)(0,0,1)_{12}$	136.36	137.43	170.06
ARIMA$(6,1,0)(1,0,0)_{12}$	148.51	149.21	175.48
ARIMA$(6,1,0)(2,0,0)_{12}$	132.34	133.22	162.68
ARIMA$(6,1,0)(3,0,0)_{12}$	124.33	125.41	158.04
ARIMA$(6,1,6)(3,0,0)_{12}$	128.28	131.03	182.21
ARIMA$(0,1,6)(1,0,0)_{12}$	146.22	146.92	173.19
ARIMA$(0,1,6)(2,0,0)_{12}$	131.15	132.03	161.49
ARIMA$(0,1,6)(3,0,0)_{12}$	124.17	125.25	157.87
ARIMA$(0,1,6)(1,0,1)_{12}$	108.42	109.3	138.76
ARIMA$(0,1,6)(2,0,1)_{12}$	109.83	110.91	143.54

4.4.3. Model Estimation

Based on the model identified in the previous stage, we can proceed to the phase of model estimation using maximum likelihood method (ML), the empirical results being presented in Table 10. All coefficients statistically are significant at the 10% significance level.

Table 10. Estimates of parameters for SARIMA(0,1,6)(1,0,1)$_{12}$.

	Estimate S	td. Error	z Value	Pr(>|z|)
ma6	−0.12316	0.069532	−1.7712	0.07653 *
sar1	0.983605	0.015399	63.8766	2.2×10^{-16} ***
sma1	−0.8462	0.066935	−12.6421	2.2×10^{-16} ***

Note: *** and * means stationary at 1% and 10%.

4.4.4. Diagnostic Checking of the Model

Apart from classical tests, the t-test for the statistical significance of the parameters, and the F-test for the validity of the model, the selection of the best model depends also on the performance of residuals. For that, the series of residuals has been investigated to follow a white noise. The empirical results of the Ljung–Box test show that the p-values of the test statistic exceed the 5% level of significance for all lag orders, which implies that there is no significant autocorrelation in residuals (Figure 12).

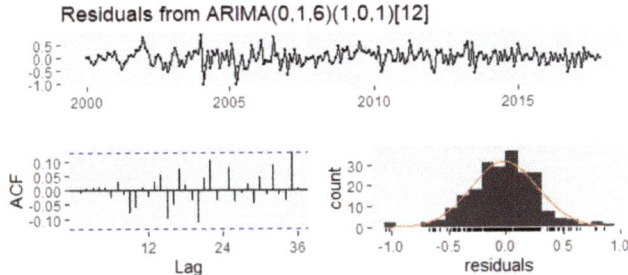

Figure 12. Diagnostic plot of SARIMA(0,1,6)(1,0,1)$_{12}$.

For checking autoregressive conditional heteroskedasticity (ARCH) in the residuals, the ARCH-LM test has been used, and the empirical results confirmed that there is no autoregressive conditional heteroscedasticity (ARCH) in the residuals (Table 11). Therefore, we can conclude that residuals are not autocorrelated and do not form ARCH models, the SARIMA(0,1,6)(1,0,1)$_{12}$ model being reliable for forecasting (Table 12).

Table 11. Empirical results of JB test and autoregressive conditional heteroskedasticity (ARCH)-LM test for model residuals.

	Ljung–Box Test	p-Value	ARCH-LM Test	p-Value
12	2.9459	0.5669	9.1184	0.6928
24	15.123	0.5157	44.267	0.2345
36	25.531	0.5988	51.336	0.1878
48	40.434	0.4511	58.159	0.1495

Table 12. Forecasting performance of SARIMA(0,1,6)(1,0,1)$_{12}$.

	Training Dataset	Testing Dataset
RMSE	0.28861	0.764092
MAE	0.22163	0.615342
MAPE	3.20478	13.37031
MASE	0.35240	0.97840

The forecast of the unemployment rate based on the ARIMA(0,1,6)(1,0,1)$_{12}$ model results revealed a downward trend with a forecasting value for 2021–2022 oscillating around the value of 3–4% (Figure 13).

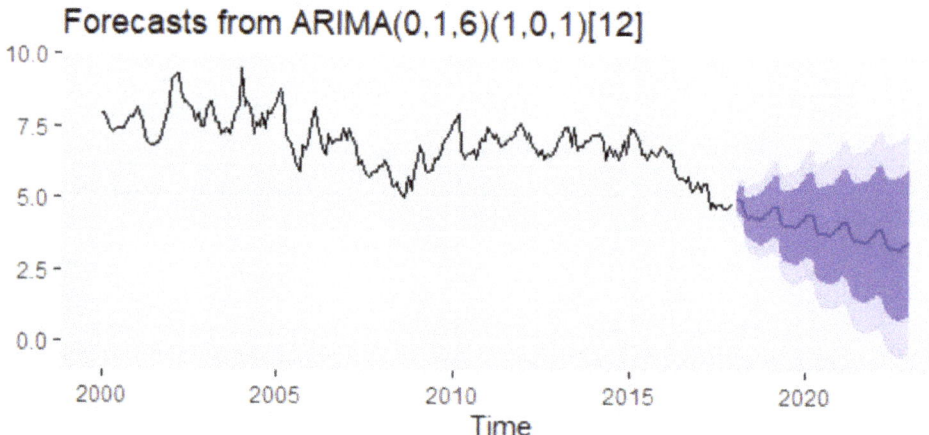

Figure 13. Forecasts of unemployment rate based on the results of ARIMA(0,1,6)(1,0,1)$_{12}$.

4.5. Self-Exciting Threshold Autoregressive (SETAR Model)

In fitting a SETAR model for the Romanian unemployment rate, the first stages require the identification of the autoregressive order and testing the existence of nonlinear thresholds. The autoregressive order has been identified based on the PACF plot. Following Desaling [74], we explored the unemployment rate in level for identifying the lag autoregressive order, since the non-stationarity in UR does not cause the non-stationarity of nonlinear thresholds in the SETAR model, even if the existence of a unit root in one regime can occur. Significant spikes can be observed at lags 1, 7, and 13 (Figure 14).

Autocorrelation	Partial Correlation		AC	PAC	Q-Stat	Prob
		1	0.956	0.956	233.14	0.000
		2	0.919	0.053	449.26	0.000
		3	0.873	-0.110	645.25	0.000
		4	0.827	-0.040	821.93	0.000
		5	0.794	0.134	985.45	0.000
		6	0.765	0.049	1137.8	0.000
		7	0.757	0.221	1287.6	0.000
		8	0.754	0.071	1436.6	0.000
		9	0.758	0.079	1587.8	0.000
		10	0.764	0.052	1742.2	0.000
		11	0.769	0.055	1899.1	0.000
		12	0.760	-0.119	2053.4	0.000
		13	0.729	-0.244	2195.6	0.000
		14	0.688	-0.140	2322.7	0.000
		15	0.640	-0.054	2433.4	0.000
		16	0.603	0.108	2531.8	0.000
		17	0.573	0.069	2621.1	0.000
		18	0.546	-0.076	2702.7	0.000
		19	0.533	-0.008	2780.7	0.000
		20	0.526	0.033	2857.1	0.000

Figure 14. Partial autocorrelation plot of unemployment series.

At these lags, we have tested the presence of nonlinear thresholds applying the Tsay test of threshold nonlinearity, the empirical results being presented in Table 13, revealing that there is enough evidence to reject the null hypothesis of no nonlinear threshold in autoregressive order 1, 7, 8, 9, 10, 11, 12, and 13, the *p*-value being mostly less than 1%. Therefore, at these lags, the SETAR model is better than the simple AR model.

Table 13. The empirical results of the Tsay test.

Order	F-Statistics	*p*-Value	AIC
AR(1)	4.798	0.029 **	0.734
AR(2)	1.935	0.125	-
AR(3)	1.363	0.231	-
AR(4)	1.097	0.366	-
AR(5)	1.119	0.341	-
AR(6)	1.267	0.202	-
AR(7)	1.744	0.016 **	0.689
AR(8)	1.994	0.001 ***	0.693
AR(9)	2.116	0.001 ***	0.697
AR(10)	1.989	0.001 ***	0.696
AR(11)	2.151	0.001 ***	0.698
AR(12)	2.257	0.001 ***	0.702
AR(13)	2.034	0.003 ***	0.628

Note: ***, ** means statistical significance at 1%, 5%.

For the lags exhibiting a nonlinear threshold, we have used the lowest values of AIC to select the optimal model for which we will design the SETAR model. Thus, an AR(13) with possible values of delay parameter d = 1 ... 12 < *p* has been used in setting the SETAR model. Since the number of potential regimes in the autoregressive model depends on the number of threshold values, a grid search method has been performed to determine the regimes and estimate the thresholds value under the condition of one threshold in AR based on the smallest value of sum of squared residuals. Thus, the delay parameter d = 10 registered the smallest value for residuals sum of squares; therefore, a SETAR model with two regimes of order 13 and threshold decay 1, a SETAR(2,13,1) model with a threshold variable could be appropriate to explain the nonlinearity in the data (Figure 15).

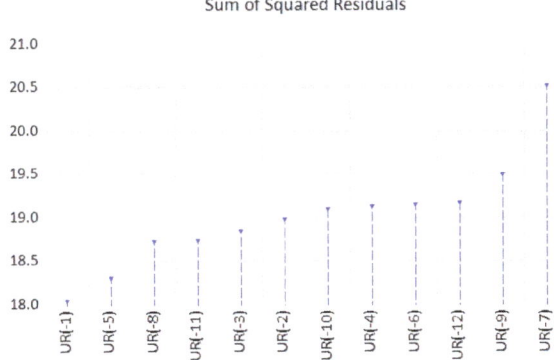

Figure 15. Grid search method estimation of one threshold value.

Table 14 displays the estimated parameters of the SETAR(2,13,1) with the threshold of 7.79, the model having the following specification:

$$y_t = \begin{cases} 0.13 + 0.82 y_{t-1} + \ldots - 0.307 y_{t-13}, \; if \; y_{t-1} < 7.799 \\ 2.344 + 0.539 y_{t-1} + \ldots - 0.019 y_{t-13}, \; if \; y_{t-1} > 7.799 \end{cases}$$

Table 14. Estimates of parameters for SETAR(2,13,1).

Variable	Coefficient	Std. Error	Prob.
	UR(−1) < 7.7999999–171 obs		
C	0.130	0.264	0.623
UR(−1)	0.820	0.079	0.000 ***
UR(−2)	0.254	0.099	0.011 ***
UR(−3)	−0.055	0.099	0.579
UR(−4)	−0.136	0.098	0.168
UR(−5)	0.063	0.092	0.499
UR(−6)	−0.126	0.092	0.172
UR(−7)	0.078	0.092	0.401
UR(−8)	0.069	0.097	0.475
UR(−9)	0.010	0.093	0.917
UR(−10)	−0.030	0.099	0.762
UR(−11)	0.103	0.101	0.310
UR(−12)	0.237	0.099	0.018 **
UR(−13)	−0.307	0.076	0.000 ***
	7.7999999 <= UR(−1)–32 obs		
C	2.344	2.092	0.264
UR(−1)	0.539	0.202	0.008 ***
UR(−2)	0.214	0.184	0.247
UR(−3)	0.022	0.194	0.909
UR(−4)	−0.643	0.207	0.002 ***
UR(−5)	0.682	0.335	0.043
UR(−6)	0.062	0.264	0.814
UR(−7)	0.141	0.299	0.637
UR(−8)	−0.639	0.267	0.018 **
UR(−9)	−0.526	0.301	0.083
UR(−10)	0.479	0.218	0.030 **
UR(−11)	0.833	0.181	0.000 ***
UR(−12)	−0.440	0.257	0.089
UR(−13)	−0.019	0.198	0.923

Note: ***, ** means statistical significance at 1%, 5%.

After the estimation stage, the residuals of the model have been checked for best fit, verifying them for the information of serial autocorrelation, constant variance, and zero mean based on ARCH-LM and Breusch–Godgrey tests. Having the p-values greater than a 1% significance level, we can conclude that the residuals are not autocorrelated and with constant variance (Table 15).

Table 15. Residuals diagnostic test for SETAR(2,13,1).

	BG Test (F-Stat)	*p*-Value	ARCH-LM Test	*p*-Value
12	1.180	0.301	0.722	0.728
24	0.99	0.473	0.738	0.805
36	1.179	0.247	0.991	0.493
48	1.197	0.213	1.068	0.381

The forecast of unemployment rate based on the results of the SETAR(2,13,1) model (Table 16) revealed an upward trend, over evaluating the phenomenon (Figure 16).

Table 16. Forecasting performance of SETAR(2,13,1).

	Training Data Set	Testing Data Set
RMSE	0.931	0.834
MAE	0.803	0.715
MAPE	11.598	17.742
MASE	12.022	15.770

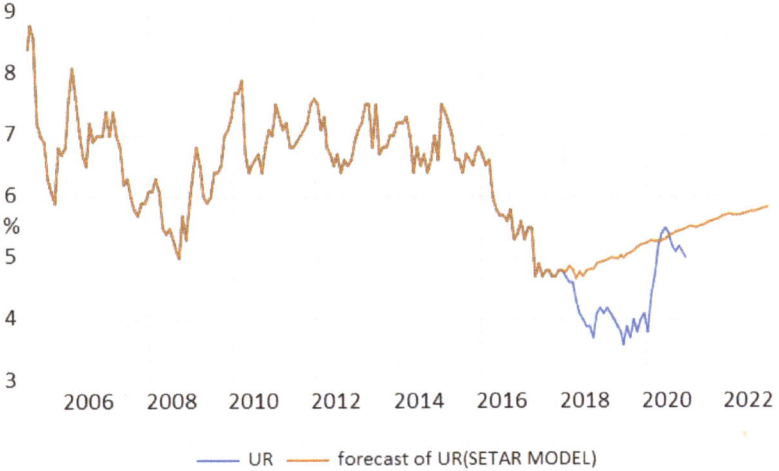

Figure 16. Forecasts of unemployment rate based on the results of the SETAR(2,13,1) model.

4.6. Comparison of Models Forecasting Performance

Analyzing the forecasting performance of all models for the in-sample dataset based on RMSE, MAE, and MAPE as well as on the results of the Diebold and Marino test, it can observed that all three criteria suggested that multiplicative HW registered better forecast performance for the training dataset. The p-value of the Diebold and Marino test highlighted the existence of differences in forecast accuracy between almost all models, with the exception of multiplicative HW and ETS, for which the probability being higher than 10% does not provide enough evidence to reject the null hypothesis (Table 17).

The out-of-sample forecasting performance of models has performed with a one-step ahead recursive method. Based on RMSE and MAE values, the NNAR model has better forecasting performance, while MAPE stipulates the SARIMA model to register higher performance. For the out-of-sample data, the empirical results of the DM test pointed out differences in the predictive power for almost all models, with the exception of multiplicative HW and NNAR, for which the p-value is greater than the 10%, so the null hypothesis can not be rejected (Table 18).

Analyzing comparatively the forecast performance of all methods during the period 2018–2022 and taking into account the presence of the pandemic shock, it is worth mentioning that ETS and Multiplicative HW are the methods that best capture the pandemic shock from 2020, offering forecast values relatively close to the real values of unemployment rate from the pandemics (Figure 17).

Table 17. In-sample forecasting performance of models.

Measures	Model					
	Holt–Winters Additive	Holt–Winters Multiplicative	ETS	NNAR	SARIMA	SETAR
RMSE	0.2804	0.2771	0.2788	0.315	0.28861	0.931
MAE	0.2109	0.2086	0.2097	0.2411	0.22163	0.803
MAPE	3.0699	3.0368	3.0569	3.4513	3.20478	11.598
DM Test for in Sample at h = 1						
Models	DM Test Statistics			p-Value		
HW Additive vs. HW Multiplicative	7.7819			0		
HW Additive vs. ETS	−7.7841			0		
HW Additive vs. NNAR	4.2089			0		
HW Additive vs. SARIMA	1.6588			0.0986		
HW Additive vs. SETAR	55.592			0		
HW Multiplicative vs. ETS	0.3324			0.7399		
HW Multiplicative vs. NNAR	8.1815			0		
HW Multiplicative vs. SARIMA	8.0321			0		
HW Multiplicative vs. SETAR	55.568			0		
ETS vs. NNAR	8.1791			0		
ETS vs. SARIMA	8.0342			0		
ETS vs. SETAR	55.568			0		
NNAR vs. SARIMA	−3.3088			0.001		
NNAR vs. SETAR	54.421			0		
SARIMA vs. SETAR	−55.615			0		

Table 18. Out-of-sample forecasting performance of models.

Measures	Model					
	Holt–Winters Additive	Holt–Winters Multiplicative	ETS	NNAR	SARIMA	SETAR
RMSE	0.748	0.6906		0.5979	0.764092	0.834
MAE	0.6273	0.6524		0.5508	0.615342	0.715
MAPE	13.8268	15.1393		14.0168	13.37031	17.742
DM Test for Out of Sample at h = 1						
Models	DM Test Statistics			p-Value		
HW Additive vs. HW Multiplicative	−13.541			0		
HW Additive vs. ETS	14.388			0		
HW Additive vs. NNAR	7.4791			0		
HW Additive vs. SARIMA	16.703			0		
HW Additive vs. SETAR	−11.61			0		
HW Multiplicative vs. ETS	13.616			0		
HW Multiplicative vs. NNAR	1.4745			0.1457		
HW Multiplicative vs. SARIMA	17.175			0		
HW Multiplicative vs. SETAR	−16.362			0		
ETS vs. NNAR	−3.2896			0.0016		
ETS vs. SARIMA	−15.773			0		
ETS vs. SETAR	17.254			0		
NNAR vs. SARIMA	−12.841			0		
NNAR vs. SETAR	18.072			0		
SARIMA vs. SETAR	−17.303			0		

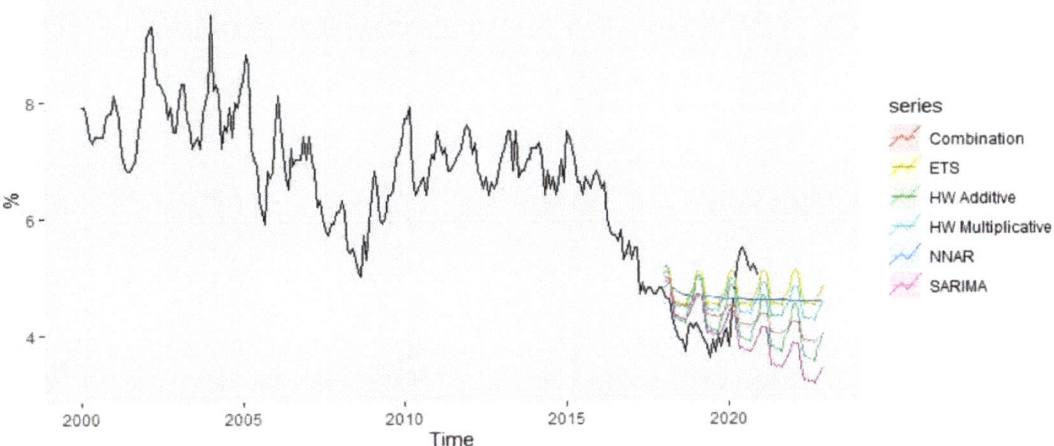

Figure 17. Forecast combination of the Romanian unemployment rate.

Based on the methods offering the best results for out-of-sample forecasting, NNAR and SARIMA, the forecasted values of unemployment rate for the period 2021–2022 have been examined, revealing the existence of a slight difference (Figure 18).

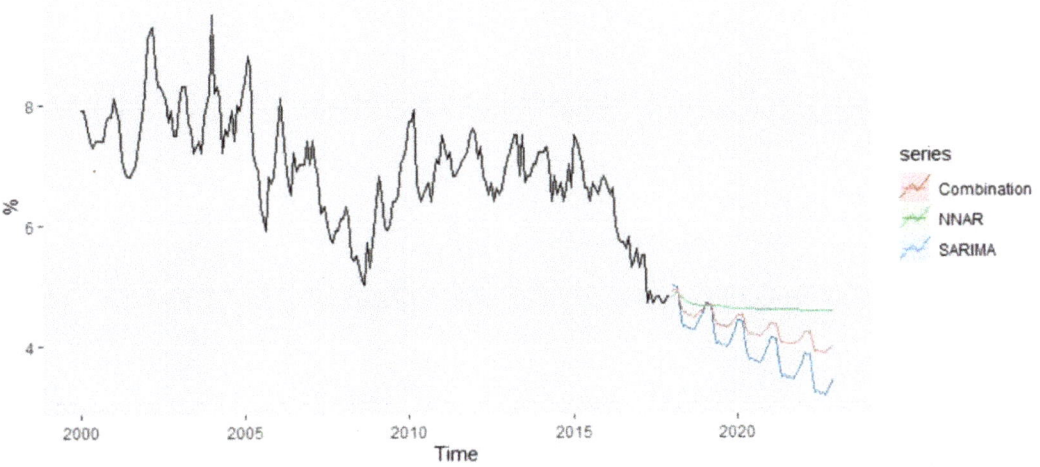

Figure 18. The forecasts of unemployment rate for the period 2021–2022.

According to NNAR, the predicted value of unemployment rate for January 2021 is estimated to be 4.35% compared with 5% in December 2020, and over the whole period, the forecast values oscillate around 4.35%. On the other hand, the forecast values based on the SARIMA model revealed a predicted value of 4.22% for the unemployment rate of January 2021 and highlighted a descending trend over the horizon 2021–2022, with a predicted value of 3.54% in December 2022.

An alternative to improving the forecast accuracy is to average the resulting forecasts based on these two methods, which are considered to be suitable for the modeling and forecasting of unemployment rate.

5. Conclusions

Making predictions about unemployment rate, one of the core indicators of the Romanian labor market with fundamental impact on the government future social policy strategies, is of great importance, mostly in this period of a major shock in the economy caused by the pandemic.

In this context, the aim of the research has been to evaluate the forecasting performance of several models and to build future values of unemployment rate for the period 2021–2022 using the most suitable results. In order to do that, we have employed exponential smoothing models, both additive and multiplicative Holt–Winters (HW) models together with an ETS model, the SARIMA model, the neural network autoregression (NNAR) model, and the SETAR model, which allow taking into account a nonlinear behavior and a switching regime on the time series and predicting future values of unemployment rate beyond the period under consideration.

The empirical results revealed for unemployment rate a non-stationary nonlinear and seasonal pattern in data. The out-of-sample forecasting accuracy of the models based on the performance measures RMSE and MAE pointed out the NNAR model as performing better, while MAPE indicated SARIMA to have the best performance. The empirical results of the Diebold–Mariano test at one forecast horizon for out-of-sample methods revealed differences in the forecasting performance between SARIMA and NNAR; of these, the best model of modeling and forecasting unemployment rate was considered to be the NNAR model.

Author Contributions: Conceptualization, A.A.D., S.-A.A. and A.P.; methodology, A.A.D., S.-A.A. and A.P.; software, A.A.D., S.-A.A. and A.P.; validation, A.A.D., S.-A.A. and A.P.; formal analysis, A.A.D., S.-A.A. and A.P.; investigation, A.A.D., S.-A.A. and A.P.; resources, A.A.D., S.-A.A. and A.P; data curation, A.A.D., S.-A.A. and A.P.; writing—original draft preparation, A.A.D., S.-A.A. and A.P.; writing—review and editing, A.A.D., S.-A.A. and A.P.; visualization, A.A.D., S.-A.A. and A.P.; supervision, A.A.D., S.-A.A. and A.P.; project administration, A.A.D., S.-A.A. and A.P.; funding acquisition, A.A.D., S.-A.A. and A.P. All authors have read and agreed to the published version of the manuscript.

Funding: This research received no external funding.

Institutional Review Board Statement: Not applicable.

Informed Consent Statement: Informed consent was obtained from all subjects involved in the study.

Data Availability Statement: Publicly available datasets were analyzed in this study. This data can be found here: https://ec.europa.eu/eurostat/en/web/products-datasets/-/UNE_RT_M (accessed on 15 January 2021).

Conflicts of Interest: The authors declare no conflict of interest.

References

1. Bădulescu, A. Unemployment in Romania. A retrospective study. *Theor. Appl. Econ.* **2006**, *2*, 71–76.
2. Kavkler, A.; Dncic, D.E.; Babucea, A.G.; Biani, I.; Bohm, B.; Tevdoski, D.; Tosevska, K.; Borsi, D. Cox regression models for unemployment duration in Romania, Austria, Slovenia, Croatia and Macedonia. *Rom. J. Econ. Forecast.* **2009**, *2*, 81–104.
3. Cai, F.; Wang, M. Growth and structural changes in employment in transition China. *J. Comp. Econ.* **2010**, *38*, 71–81. [CrossRef]
4. Bussiere, M.; Fratzscher, M. Towards a new early warning system of financial crises. *J. Int. Money Financ.* **2006**, *25*, 953–973. [CrossRef]
5. Sune, K.; Farrukh, J. Modeling and Forecasting Unemployment Rate in Sweden Using Various Econometric Measures. Available online: https://www.diva-portal.org/smash/get/diva2:949512/FULLTEXT01.pdf (accessed on 14 October 2020).
6. Chiros, F. Forecasting the UK unemployment rate: Model comparisons. *Int. J. Appl. Econom. Quant. Stud.* **2005**, *2*, 57–72.
7. Parker, R.E.; Rothman, P. The current depth-of-recession and unemployment-rate forecasts. *Stud. Nonlinear Dyn. Econ.* **1998**, *2*, 151–158. [CrossRef]
8. Power, B.; Gasser, K. *Forecasting Future Unemployment Rates*; ECON 452: Kingston, OT, USA, 2012.
9. Etuk, E.B.; Ifeduba, A.V.; Chiaka, U.E.; Ifeanyi, I.; Okeudo, N.J.; Esonu, B.O.; Udedibie, A.B.I.; Moreki, J.C. Nutrient composition and feeding value of sorghum for livestock and poultry: A review. *J. Anim. Sci. Adf.* **2012**, *2*, 510–524.

10. Rothman, P. Forecasting asymmetric unemployment rates. *Rev. Econ. Stat.* **1998**, *80*, 164–168. [CrossRef]
11. Koop, G.; Potter, S.M. Dynamic asymmetries in U.S. unemployment. *J. Bus. Econ. Stat.* **1999**, *17*, 298–312.
12. Proietti, T. Forecasting the US unemployment rate. *Comput. Stat. Data Anal.* **2003**, *42*, 451–476. [CrossRef]
13. Johnes, G. Forecasting unemployment. *Appl. Econ. Lett.* **1999**, *6*, 605–607. [CrossRef]
14. Peel, D.A.; Speight, A.E.H. Threshold nonlinearities in unemployment rates: Further evidence for the UK and G3 economies. *Appl. Econ.* **2000**, *32*, 705–715. [CrossRef]
15. Gil-Alana, L. A fractionally integrated exponential model for UK unemployment. *J. Forecast.* **2001**, *20*, 329–340. [CrossRef]
16. Naccarato, A.; Falorsi, S.; Loriga, S.; Pierini, A. Combining official and Google Trends data to forecast the Italian youth unemployment rate. *Technol. Forecast. Soc. Chang.* **2018**, *130*, 114–122. [CrossRef]
17. Box, G.E.P.; Jenkins, G.M. *Time Series Analysis, Forecasting and Control*; Holden-Day: San Francisco, CA, USA, 1976.
18. Pankratz, A. *Forecasting with Univariate Box-Jenkins Models: Concepts and Cases*; John Wiley & Sons, Inc.: Hoboken, NJ, USA, 1983.
19. Fannoh, R.; Otieno, O.G.; Mungatu, G.J.K. Modeling the inflation rates in Liberia SARIMA approach. *Int. J. Sci. Res.* **2012**, *3*, 1360–1367.
20. Gikungu, S.W.; Waititu, A.G.; Kihoro, J.M. Forecasting inflation rate in Kenya Using SARIMA Model. *Am. J. Theor. Appl. Stat.* **2015**, *4*, 15–18. [CrossRef]
21. Nasiru, S.; Sarpong, S. Empirical approach to modelling and forecasting inflation in Ghana. *Curr. Res. J. Econ. Theory* **2012**, *4*, 83–87.
22. Etuk, E.H.; Uchendu, B.; Edema, U.V. ARIMA fit to Nigerian unemployment data. *J. Basic Appl. Sci. Res.* **2012**, *2*, 5964–5970.
23. Etuk, E.H. An additive SARIMA model for daily exchange rates of the Malaysian ringgit (MYR) and Nigerian Naira (NGN). *Int. J. Empir. Financ.* **2014**, *2*, 193–201.
24. Brida, J.G.; Garrido, N. Tourism forecasting using SARIMA Models in chilenean regions. *Int. J. Leis. Tour. Mark.* **2009**, *2*, 176–190. [CrossRef]
25. Chang, Y.-W.; Liao, M.-Y. A Seasonal ARIMA model of tourism forecasting: The case of Taiwan. *Asia Pac. J. Tour. Res.* **2010**, *15*, 215–221. [CrossRef]
26. Bedowska-Sojka, B. Unemployment rate forecasts: Evidence from the Baltic States. *East. Eur. Econ.* **2015**, *53*, 57–67. [CrossRef]
27. Wong, J.M.W.; Chan, A.P.C.; Chiang, Y.H. Time series forecasts of the construction labour market in Hong Kong: The Box-Jenkins approach. *Constr. Manag. Econ.* **2005**, *23*, 979–991. [CrossRef]
28. Ashenfelter, O.; Card, D. Time series representations of economic variables and alternative models of the labour market. *Rev. Econ. Stud.* **1982**, *49*, 761–782. [CrossRef]
29. Kurita, T. A forecasting model for Japan's unemployment rate. *Eurasian J. Bus. Econ.* **2010**, *3*, 127–134.
30. Chiu, C.C.; Chao-Ton, S. A novel neural network model using Box-Jenkins technique and response surface methodology to predict unemployment rate. In Proceedings of the Tenth IEEE International Conference on Tools with Artificial Intelligence (Cat. No.98CH36294), Taipei, Taiwan, 10–12 November 1998; pp. 74–80.
31. Nkwatoh, L.S. Forecasting unemployment rates in Nigeria using univariate time series models. *Int. J. Bus. Commer.* **2012**, *1*, 33–46.
32. Kanlapat, M.; Nipaporn, C.; Bungon, K. A forecasting model for thailand's unemployment rate. *Mod. Appl. Sci.* **2013**, *7*, 10–16.
33. Nlandu, M.; Williams, D.; Rudolph, B. *Modelling and Forecasting the Unemployment Rate in Barbados*; Central Bank of Barbados: Bridgetown, Barbados, 2014; Available online: http://www.centralbank.org.bb/news/article/7306/modelling-and-forecasting-the-unemployment-rate-in-barbados (accessed on 15 August 2020).
34. Dritsakis, N.; Klazoglou, P. Forecasting unemployment rates in USA using box-jenkins methodology. *Int. J. Econ. Financ.* **2018**, *8*, 9–20.
35. Didiharyono, D.; Muhammad, S. Forecasting with ARIMA model in anticipating open unemployment rates in South Sulawesi. *Int. J. Sci. Technol. Res.* **2020**, *9*, 3838–3841.
36. Gagea, M.; Balan, C.B. Prognosis of monthly unemployment rate in the European Union through methods based on econometric models. *Ann. Fac. Econ.* **2008**, *2*, 848–853.
37. Mladenovic, J.; Ilic, I.; Kostic, Z. Modeling the unemployment rate at the EU level by using box-jenkins methodology. *KnE Soc. Sci.* **2017**. [CrossRef]
38. Funke, M. Time-series forecasting of the German unemployment rate. *J. Forecast.* **1992**, *11*, 111–125. [CrossRef]
39. Stoklasová, R. Model of the unemployment rate in the Czech Republic. In Proceedings of the 30th International Conference on Mathematical Methods in Economics, Karviná, Czech Republic, 11–13 September 2012; pp. 836–841.
40. Jeřábková, V. Unemployment in the Czech Republic and its predictions based on Box-Jenkins methodology. In Proceedings of the 12th International Scientific Conference Applications of Mathematics and Statistics in Economy, Uherske Hradiste, Czech Republic, 26–28 August 2009.
41. Schanne, N.; Wapler, R.; Weyh, A. Regional unemployment forecasts with spatial interdependencies. *Int. J. Forecast.* **2010**, *26*, 908–926. [CrossRef]
42. Dritsaki, C. Forecast of SARIMA models: An application to unemployment rates of Greece. *Am. J. Appl. Math. Stat.* **2016**, *4*, 136–148.
43. Dritsakis, N.; Athianos, S.; Stylianou, T.; Samaras, I. Forecasting unemployment rates in Greece. *Int. J. Sci. Basic Appl. Res.* **2018**, *37*, 43–55.

44. Rublikova, E.; Lubyova, M. Estimating ARIMA-ARCH model rate of unemployment in Slovakia. *Forecast. Pap. Progn. Pr.* **2013**, *5*, 275–289.
45. Kishor, N.K.; Koenig, E.F. VAR estimation and forecasting when data are subject to revision. *J. Bus. Econ. Stat.* **2012**, *30*, 181–190. [CrossRef]
46. Clements, M.P.; Hendry, D.F. On the limitations of comparing mean squared forecast errors (with discussion). *J. Forecast.* **2003**, *12*, 617–639. [CrossRef]
47. Robinson, W. Forecasting Inflation Using VAR Analysis. Bank of Jamaica. Available online: http://boj.org.jm/uploads/pdf/papers_pamphlets/papers_pamphlets_forecasting_inflation_using_var_analysis.pdf (accessed on 11 October 2020).
48. Lack, C. *Forecasting Swiss Inflation Using VAR Models*; Swiss National Bank Economic Studies: Bern, Switzerland, 2006.
49. Wang, L.; Zeng, Y.; Chen, T. Back propagation neural network with adaptive differential evolution algorithm for time series forecasting. *Expert Syst. Appl.* **2015**, *42*, 855–863. [CrossRef]
50. Peláez, R.F. Using neural nets to forecast the unemployment rate. *Bus. Econ.* **2006**, *41*, 37–44. [CrossRef]
51. Skalin, J.; Teräsvirta, T. Modeling asymmetries and moving equilibria in unemployment rates. *Macroecon. Dyn.* **2002**, *6*, 202–241. [CrossRef]
52. International Labor Organisation. Global Employment Trends for Youth. 2013. Available online: https://www.ilo.org/wcmsp5/groups/public/---dgreports/---dcomm/documents/publication/wcms_212423.pdf (accessed on 12 August 2020).
53. Kotzeva, M.; Pauna, B. Labour Market Dynamics and Characteristics in Bulgaria and Romania—Challenges for a Successful Integration in the European Union. Global Development Network Southeast Europe. 2006. Available online: www.wiiw.ac.at/balkan/files/KOTZEVA%20PAUNA.pdf (accessed on 3 August 2020).
54. National Institute of Statistics. Labor force in Romania. Available online: https://insse.ro/cms/sites/default/files/field/publicatii/labour_force_in_romania_2017.pdf (accessed on 3 August 2020).
55. Pivodă, R.M. Sequences of Unemployment in Romania. *Ovidius Univ. Ann.* **2012**, *12*, 367–372.
56. Son, L.; Cariga, G.G.; Ciuca, V.; Paşnicu, D. An Autoregressive short-run forecasting model for unemployment rates in Romania and the European Union. *Recent Adv. Math. Comput. Bus. Econ. Biol. Chem.* **2016**, *2769*, 193–198.
57. Madaras, S. The impact of the economic crisis on the development of unemployment at the national and county level in Romania. *Econ. Forum* **2014**, *17*, 136–149.
58. Bratu, S.M. Some empirical strategies for improving the accuracy of unemployment rate forecasts in Romania. *Ann. Spiru Haret Univ. Econ. Ser.* **2012**, *4*, 671–677.
59. Simionescu, M. The accuracy assessment of macroeconomic forecasts based on econometric models for Romania. *Procedia Econ. Financ.* **2014**, *8*, 671–677. [CrossRef]
60. Dobre, I.; Alexandru, A.A. Modelling unemployment rate using Box-Jenkins procedure. *J. Appl. Quant. Methods* **2008**, *3*, 156–166.
61. Simionescu, M.; Marin, E. Short run and alternative macroeconomic forecasts for Romania and strategies to improve their accuracy. *J. Int. Stud.* **2012**, *5*, 30–46.
62. Voineagu, V.; Pisica, S.; Caragea, N. Forecasting monthly unemployment by econometric smoothing techniques. *J. Econ. Comput. Econ. Cybern. Stud. Res.* **2002**, *46*, 255–267.
63. Simionescu, M. The kalman filter approach for estimating the natural unemployment rate in Romania. *Acta Univ. Danub.* **2014**, *10*, 148–159.
64. Zamfir, C.G. Unemployment prospective in Romania. *Econ. Appl. Inform.* **2018**, *24*, 79–86.
65. Simionescu, M. Improving unemployment rate forecasts at regional level in Romania using Google Trends. *Technol. Forecast. Soc. Chang.* **2020**, *155*, 120026. Available online: https://www.sciencedirect.com/science/article/pii/S004016251930455X (accessed on 12 October 2020).
66. Simionescu, M. The assessment of some point and forecast intervals for unemployment rate in Romania. *Int. J. Econ. Pract. Theor.* **2015**, *5*, 88–94.
67. Lukianenko, M.O.; Bazhenova, O. Regime switching modeling of unemployment rate in Eastern Europe. *Ekon. Cas.* **2020**, *68*, 380–408.
68. Hyndman, R.J.; Koehler, A.B.; Snyder, R.D.; Grose, S. A state space framework for automatic forecasting using exponential smoothing methods. *Int. J.* **2002**, *18*, 439–454. [CrossRef]
69. Alonso, A.M.; Peña, D.; Romo, J. Forecasting time series with sieve bootstrap. *J. Stat. Plan. Inference* **2002**, *100*, 1–11. [CrossRef]
70. Thoplan, R. Simple v/s sophisticated methods of forecasting for Mauritius monthly tourist arrival data. *Int. J. Stat. Appl.* **2014**, *4*, 217–223.
71. Hyndman, R.J.; Athanasopoulos, G. *Forecasting: Principles and Practice*; OTexts: Melbourne, VIC, Australia, 2012.
72. Badawi, M. Comparative Analysis of SARIMA and SETAR Models in Predicting Pneumonia Cases in the Northern Region of Ghana. Ph.D. Thesis, University for Development Studies, Tamale, Ghana, 2016.
73. Desaling, M. Modeling and Forecasting Unemployment Rate in Sweden Using Various Econometric Measures. Ph.D. Thesis, Örebro University, Örebro, Sweden, 2016.
74. Aidoo, E. Modelling and Forecasting Inflation Rate in Ghana: An Application of SARIMA Models. Ph.D. Thesis, School of Technology and Business Studies, Hogskolen, Dalarma, 2010, unpublished work.
75. Hylleberg, S.; Engle, R.; Granger, C.; Yoo, B. Seasonal integration and cointegration. *J. Econ.* **1990**, *44*, 215–238. [CrossRef]

76. Dell'Anno, R.; Davidescu, A.A. Estimating shadow economy and tax evasion in Romania. A comparison by different estimation approaches. *Econ. Anal. Policy* **2019**, *63*, 130–149. [CrossRef]
77. Boero, G.; Marrocu, E. The performance of SETAR models: A regime conditional evaluation of point, interval and density forecasts. *Int. J. Forecast.* **2004**, *20*, 305–320. [CrossRef]
78. Diebold, F.X.; Mariano, R.S. Comparing predictive accuracy. *J. Bus. Econ. Stat.* **1995**, *13*, 253–263.

Article

Functional Kernel Density Estimation: Point and Fourier Approaches to Time Series Anomaly Detection

Michael R. Lindstrom [1,*], Hyuntae Jung [2] and Denis Larocque [3]

1. Department of Mathematics, University of California, Los Angeles, CA 90024, USA
2. Global Aviation Data Management, International Air Transport Association (IATA), Montréal, QC H2Y 1C6, Canada; jungh@iata.org
3. Department of Decision Sciences, HEC Montréal, Montréal, QC H2Y 1C6, Canada; denis.larocque@hec.ca
* Correspondence: mikel@math.ucla.edu

Received: 16 November 2020; Accepted: 27 November 2020; Published: 30 November 2020

Abstract: We present an unsupervised method to detect anomalous time series among a collection of time series. To do so, we extend traditional Kernel Density Estimation for estimating probability distributions in Euclidean space to Hilbert spaces. The estimated probability densities we derive can be obtained formally through treating each series as a point in a Hilbert space, placing a kernel at those points, and summing the kernels (a "point approach"), or through using Kernel Density Estimation to approximate the distributions of Fourier mode coefficients to infer a probability density (a "Fourier approach"). We refer to these approaches as Functional Kernel Density Estimation for Anomaly Detection as they both yield functionals that can score a time series for how anomalous it is. Both methods naturally handle missing data and apply to a variety of settings, performing well when compared with an outlyingness score derived from a boxplot method for functional data, with a Principal Component Analysis approach for functional data, and with the Functional Isolation Forest method. We illustrate the use of the proposed methods with aviation safety report data from the International Air Transport Association (IATA).

Keywords: time series; anomaly detection; unsupervised learning; kernel density estimation; missing data

1. Introduction

Being able to detect anomalies has many applications, including in the fields of medicine and healthcare management [1,2]; in data acquisition, such as filtering out anomalous readings [3]; in computer security [4]; in media monitoring [5]; and many in the realm of public safety such as identifying thermal anomalies that may precede earthquakes [6], identifying potential safety issues in bridges over time [7], detecting anomalous conditions for trains [8], system level anomaly detection among different air fleets [9], and identifying which conditions pose increased risk in aviation [10]. Given a dataset, anomaly detection is about identifying individual data that are quantitatively different from the majority of other members of the dataset. Anomalous data can come in a variety of forms such as an abnormal sequence of medical events [11] and finding aberrant trajectories of pantograph-caternary systems [12]. In our context, we look for time series of aviation safety incident frequencies for fleets of aircrafts that differ substantially from the rest. By identifying the aircraft types or airports that have significant different patterns of frequencies of specific incidents, our model can provide insights on the potential risk profile for each aircraft type or airport and highlight areas of focus for human analysts to perform further investigations.

Identifying anomalous time series can be divided into different types of anomalous behaviour [13] such as: point anomalies (a single reading is off), collective anomalies (a portion of a time series that reflects an abnormality), or contextual anomalies (when a time series behaves very differently from most others). Identifying anomalous time series from a collection of time series, as in our problem, can be done through dimensionality reduction (choosing representative statistics of the series, applying PCA, and identifying points that are distant from the rest) and through studying dissimilarity between curves (a variant of classical clustering like kmeans) [14]. After reducing the dimension, some authors have used entropy-based methods, instead, to detect anomalies [15]. Archetypoid analysis [16] is another method, which selects time series as archetypeoids for the dataset and identifies anomalies as those not well represented by the archetypeoids. Very recently, authors have used a generalization of Isolation Forests to identify anomalies [17] and have examined the Fourier spectrum of time series and looked at shifting frequency anomalies [18]. Our approach, like Functional Isolation Forest, is geometric in flavor and we employ Kernel Density Estimation and analysis of Fourier modes to detect anomalies.

In this manuscript, we present two alternative means of anomaly detection based on Kernel Density Estimation (KDE) [19]. We use two approaches: the first and simplest considers each time series as element of a Hilbert space $\mathcal{H}$ and employs KDE, treating each time series in $\mathcal{H}$ as if it were a point in one-dimensional Euclidean space, placing a Gaussian kernel at each curve with scale parameter $\zeta > 0$. We refer to this as the **point approach** to Functional KDE Anomaly Detection, because each curve in $\mathcal{H}$ is treated as a point. This approach then *formally* generates a proxy for the "probability density" over $\mathcal{H}$. Anomalous series are associated with smaller values of this density. This is distinct from considering a single time series as collection of points sampled from a distribution and using KDE upon points in the time series as has been done before [20]. This is a very simple, and seemingly effective method, with ζ chosen as a hyper-parameter. We also present a **Fourier approach**, which approximates a probability density over $\mathcal{H}$ through estimating empirical distributions for each Fourier mode with KDE. This allows us to estimate the likelihood of a given curve. Curves with lower likelihoods are more anomalous. Both methods naturally handle missing data, without interpolating. In real flight operations, sometimes it is not possible to capture and record complete information because incident data is documented from voluntary reporting, which may result in incomplete datasets. Therefore, model robustness to the impact of missing data is crucial to derive the correct understanding, which may save human lives and prevent damaged aircrafts.

The rest of our paper is organized as follows: in Section 2, we present the details and implementation of our methods; in Section 3, we conduct some experiments to investigate the strengths and weaknesses of the approaches and compare them with three other methods (Functional Isolation Forest available in Python and the PCA and functional boxplot methods available in R); following this, we apply our techniques to data from the International Air Transport Association (IATA); finally, in Section 4, we discuss our results and present some recommendations.

2. Functional Kernel Density Estimation

2.1. Review of Kernel Density Estimation

We first recall KDE over $\mathbb{R}^d$, $d \in \mathbb{N}$. Given a sample $\mathcal{S} \subset \mathbb{R}^d$ of n points from a distribution with probability density function (pdf) $f : \mathbb{R}^d \to [0, \infty)$ with $\int_{\mathbb{R}^d} f(x)\mathrm{d}x = 1$, KDE provides an empirical estimate for the probability density given by [19]

$$\tilde{f}(x) = \frac{1}{n} \sum_{y \in \mathcal{S}} |\Xi|^{-1/2} K(\Xi^{-1/2}(x-y)) \tag{1}$$

where Ξ is a symmetric, positive definite matrix known as the *bandwidth matrix* and K is a Kernel function. We choose the form of a multivariate Gaussian function so

$$\tilde{f}(x) = \frac{1}{n} \sum_{y \in \mathcal{S}} \frac{e^{-\frac{1}{2}(x-y)^T \Xi^{-1}(x-y)}}{(2\pi)^{d/2} |\Xi|^{1/2}}, \qquad x \in \mathbb{R}^d \qquad (2)$$

and we choose [19]

$$\Xi = \alpha \operatorname{diag}(\tilde{\sigma}_1, \tilde{\sigma}_2, ..., \tilde{\sigma}_d) \qquad (3)$$

where $\tilde{\sigma}_i$ is the sample standard deviation of the ith-coordinate of the sample points in $\mathcal{S}$ and

$$\alpha = \left(\frac{4}{(d+2)n} \right)^{1/(d+4)}. \qquad (4)$$

We used tilde (~) rather than hat (^) to denote estimators as later on we use the hats for Fourier Transform modes and wish to avoid ambiguities. In general tildes will be used for estimates derived from samples.

2.2. Setup, Assumptions, and Notation

The proposed methods are applicable to situations where we look for anomalous time-series relative to the sample we have. We study time series, which we consider more abstractly as being discrete samples from curves of form $x(t)$ where $x : [0, T] \to \mathbb{R}$ for some $T > 0$. The space of all such curves is quite general and we limit the scope to Hilbert spaces on $[0, T]$. For example, we may consider spaces $\mathcal{H} = L^2([0, T])$ or $H^1([0, T])$, the space of square integrable functions or the space of square integrable functions whose derivative is also square integrable, respectively. Within our Hilbert space, $\mathcal{H}$, there is an inner product $(\cdot, \cdot) : \mathcal{H}^2 \to \mathbb{C}$ and an induced norm, $||\cdot|| : \mathcal{H} \to [0, \infty)$ where $||x|| = (x, x)^{1/2}$. With this norm, we can define distances between elements of $\mathcal{H}$.

Observations are made at p different times, $t_0, t_1, ..., t_{p-1}$ where $t_i = i\Delta$ with $\Delta = T/p$ and $i = 0, 1, ..., p - 1$. We also have $t_p = T$, but this time is not included in the data. Although observations are made at these times, some time series could have missing values. When a value is missing, we will say its "value" is Not-a-Number (NaN). While the set of observation points are uniformly spaced, the times at which a given time series has non-NaN values may not be.

We denote by n the number of time series observed, given to us as a sample of form $\mathcal{X} = \{\{(t_j^{(k)}, x_j^{(k)})\}_{j=0}^{P_k-1}\}_{k=1}^{n}$, where $k = 1, ..., n$ indexes the time series, P_k is the number of available (i.e., non-NaN) points for time series k, $0 \leq t_0^{(k)} < t_1^{(k)} < ... < t_{P_k-1}^{(k)} < T$ are the times for series k, with corresponding non-NaN values $x_0^{(k)}, x_1^{(k)}, ..., x_{P_k-1}^{(k)} \in \mathbb{R}$.

2.3. Preprocessing

The methods often performed better if we normalized the data by a standard centering and rescaling. At each fixed observation time, the values of the time series were shifted to have mean zero and then rescaled to have unit variance. When the variance was already zero, the values were mapped to 0. Further remarks are given in Section 4.

Even though our methods do not assume stationary or other similar properties, applying transformations to the data before applying them can be done. For example, we may wish to make the series stationary, or to extract some characteristics (e.g., the cyclical part, or the seasonal part). This can be useful if we want to focus on finding specific types of anomalies.

2.4. Point Approach to Functional KDE Anomaly Detection

Our first method can be summarized as follows: treat each $x \in \mathcal{H}$ as a point in one dimension. Select a value for the KDE scale hyper-parameter $\xi > 0$, and define a score functional over $\mathcal{H}$ by

$$S_P[a] = \sum_{x \in \mathcal{X}} e^{-\frac{||x-a||^2}{2\xi^2}}, \quad a \in \mathcal{H}, \tag{5}$$

which, at least formally, can be thought of as a proxy to a "probability density" functional. More rigorously, one should consider measures on Hilbert spaces [21]. Assuming anomalous curves are truly rare, they should be very distant from the majority of curves and $S_P[\cdot]$ should be smaller at such curves. See Figure 1 for a conceptual illustration. We find that choosing ξ to be the mean of $\{||a||\}_{a \in \mathcal{X}}$ to work well; another natural choice would be the median. These choices are natural because they represent a natural size/scale for the series. This approach can also be interpreted from a Fourier perspective which we remark on in Appendix A.

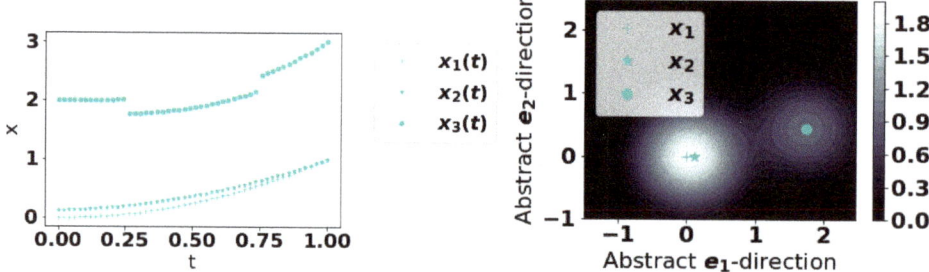

Figure 1. A visual depiction of the Point method. The curves are time series in a Hilbert space $\mathcal{H}$ but after applying KDE, there is a score associated to each point in $\mathcal{H}$. In the cartoon, curves 1 and 2 are similar and curve 3 is anomalous. (**Left**): the time series. (**Right**): a representation of them with associated scores in the color scale. In reality, the space is infinite dimensional and this is only a conceptual illustration.

This method can be implemented with the following steps:

1. Choose $\xi > 0$.
2. For each $x \in \mathcal{X}$, compute its score from (5) where, for example, in the case of $\mathcal{H} = L^2([0,T])$,

$$||x - a||^2 = \int_0^T |(x(t) - a(t)|^2 dt. \tag{6}$$

3. Identify anomalies as curves with the lowest score.

The integral in (6), even with some data points missing, can be computed as below:

1. To compute $I = \int_0^T |(x(t) - a(t)|^2 dt$, determine all t-values where both x and a are not NaN. Call these $t_0^*, t_1^*, ..., t_{r-1}^*$.
2. Define $t_r^* = T - t_{r-1}^* + t_0^*$, $x_r^* = x_0^*$ and $y_r^* = y_0^*$.
3. Estimate the integral as

$$I \approx \frac{1}{2} \sum_{m=0}^{r-1} (t_{m+1}^* - t_m^*)(|x(t_m^*) - y(t_m^*)|^2 + |x(t_{m+1}^*) - y(t_{m+1}^*)|^2).$$

This is a second-order accurate (trapezoidal) approximation to I where we have extended the signal periodically at the endpoint. This ensures that in a pathological case such as there being only a single point of observation for the integrand with value v, then the inner product evaluates to Tv.

2.5. Fourier Approach to Functional KDE Anomaly Detection

We first observe that most Hilbert spaces of interest such as $L^2([0, T])$ have a countable, orthogonal basis $\mathcal{B} = \{\exp(2\pi i k t/T) | k \in \mathbb{Z}\}$. By considering time series as being represented by these basis vectors, we can more accurately consider a true probability density over $\mathcal{H}$. In practice, we pick $L \in \mathbb{N}$ large and represent $a \in \mathcal{H}$ by

$$a(t) \approx \sum_{j=-L}^{L} \hat{a}_k e^{2\pi i k t/T}.$$

Then, up to a Fourier mode of size L, we can define a probability density at $a \in \mathcal{H}$ by $\prod_{k=-L}^{L} \zeta_k(\hat{a}_k)$ where ζ_k is a pdf over $\mathbb{C}$ for mode k.

Our time series are discrete with finitely many points so we consider a Non-Uniform Discrete Fourier Transform (NUDFT). To estimate the probability density over $\mathcal{H}$ at a, we:

1. Compute $p^* = \min\{P_1, P_2, ..., P_n\}$.
2. Compute the Discrete Fourier coefficients

$$\hat{x}_j^{(k)} = \frac{1}{P_k} \sum_{m=0}^{P_k - 1} \exp(-2\pi i j t_m/T) x^{(k)}(t_m)$$

for each $k = 1, ..., n$ and for $j = 0, 1, ..., p^* - 1$.

3. For each $0 \leq j \leq p^* - 1$, use KDE to estimate the pdf over $\mathbb{C}$ for $\hat{x}_j$, by using KDE (Equations (2)–(4)) for $\mathbb{R}$ or $\mathbb{R}^2$ when the coefficients are all purely real/imaginary or contain a mix of real and imaginary components, respectively. Call the empirical distribution $\tilde{\zeta}_j$ for each j.
4. For any $a \in \mathcal{H}$ define an estimated pdf via

$$\rho_F[a] = \prod_{j=0}^{p^*-1} \zeta_j(\hat{a}_j). \quad (7)$$

5. Let the score of $a \in \mathcal{H}$ be

$$S_F[a] = \log \rho_F[a]. \quad (8)$$

6. Identify anomalies in $\mathcal{X}$ as those whose scores given by (8) are smallest.

Due to missing data, this method does lose some information since the higher Fourier modes necessary to fully reconstruct a given time series may be discarded. Additionally, as the missing data may result in non-uniform sampling, the typical aliasing of the Discrete Fourier Transform does not take effect. In general for one of the series $x^{(k)}$, we will not have $\hat{x}_{P_k - j}^{(k)} = \overline{\hat{x}_j^{(k)}}$, where the bar denotes complex conjugation. See the remark on aliasing in Appendix B.

In multiplying the pdfs in each mode to estimate the probability density at a point in the Hilbert space, we have implicitly assumed the modes are independent. It may seem intuitive to decouple the modes by applying a Mahalanobis transformation upon the modes prior to KDE, but this results in poor outcomes. Thus, this implicit independence seems to work well in practice, without adjustments.

A Discrete Fourier transform of a signal $x_0, x_1, ..., x_{P_S - 1}$ measured at times $\tilde{t}_0, \tilde{t}_1, ..., \tilde{t}_{P_S - 1}$ is a representation in a new basis $\{e^{(k)}\}_{k=0}^{P_S - 1}$ where $e_j^{(k)} = e^{2\pi i k \tilde{t}_j/T}$ for $j = 0, ..., P_S - 1$. In general, such a basis of vectors for a NUDFT will not be orthogonal [22]. However, if $m = p - P_S \ll p$ and the $\tilde{t}$'s are a subset of a uniformly spaced set of times, we can show that the vectors are *almost* orthogonal with a cosine similarity of size $O(m/p)$. Details appear in Appendix C. This orthogonality is not strictly necessary to run the method, but doing so allows a deeper justification of multiplying the pdfs in each mode if the Fourier modes are truly independent because the Discrete Fourier Transform is then approximately a projection onto an orthogonal basis of modes, each of which are independent.

3. Method Performance

We begin by illustrating the performance of our methods for some synthetic data and compare Functional KDE to other methods. The first one is the Functional Boxplot (FB) [23]. The `fbplot` function in the R package `fda` is used to obtain a center outward ordering of the time series based on the band depth concept which is a generalization to functional data of the univariate data depth concept [24]. The idea is that anomalous curves will be the ones with the largest ranks, that is, the ones that are farther away from the center. The second method is the recently proposed Functional Isolation Forest (FIF) [17], which is also depth-based and assigns a score to a curve, with higher values indicating that it is more anomalous. We used the code provided for FIF directly on GitHub [25] with the default settings given. The third is the method proposed in [26] and implemented in the R package `anomalousACM` [27]. This method works in three steps: (i) extract features (e.g., mean, variance, trend) from the time-series; (ii) use Principal Component Analysis (PCA) to identify patterns; (iii) Use a two dimensional outlier detection algorithm with the first two principal components as inputs. It will be referenced as the PCA method in what follows After testing them on synthetic data, we apply our techniques to real data to identify anomalies in time series for aviation events.

The methods against which we compare our methods did not have standard means of managing missing entries. For these methods, we replace missing data (NaN) in a series using Python's default interpolation scheme. For the methods proposed in this paper, we do not have to use imputation.

3.1. Synthetic Data

We apply the Point and Fourier Approaches to Functional KDE, Functional Boxplot, and Functional Isolation Forest to the two scenarios described below.

Scenario 1: we define a base curve

$$x_0(t) = a_0(1 + \tanh(b_0(t - t_0))) + c_0 \sin(\omega_0 t / T),$$

with $a_0 = 5$, $b_0 = 2$, $T = 50$, $t_0 = T/2 = 25$, and $\omega_0 = 2\pi$. Ordinary curves are generated via

$$x_0(t) + \epsilon(t),$$

where $\epsilon(t)$ represents Gaussian white noise at every t with mean $\mu_g = 0$ and standard deviation $\sigma_g = 0.05$. We then consider a series of 7 anomalous curves:

- $C_1(t) = x_0(t)\left(1 + r_1 \frac{(t-t^*)^2}{1+(t-t_0)^2} \Theta(t - t_0)\right) + \epsilon(t)$, where $r_1 = 0.05$ and Θ denotes the Heaviside function. Thus, the function is scaled up after t_0.
- $C_2(t) = x_0(t) + \left(1 + r_2 \Theta(t - t_0)\right)\epsilon(t)$, where $r_2 = 3$. Thus, the noise is larger after t_0.
- $C_3(t) = x_0(t) - r_3(t - t_0)\Theta(t - t_0) + \epsilon(t)$, where $r_3 = 0.05$. Thus, there is a decreasing component added to the function after t_0.
- $C_4(t) = 2a_0 \Theta(t - t_0) + c_0 \sin(\omega_0 t / T) + \epsilon(t)$, i.e., the tanh is replaced by a discontinuous function.
- $C_5(t) = x_0(t) + \mathcal{E}(t)$, where $\mathcal{E}(t)$ represents an exponential random variable at every t with mean 0.05.
- $C_6(t) = a_0(1 + \tanh(2b_0(t - t_0))) + c_0 \sin(\omega_0 t / T) + \epsilon(t)$, which has a slightly steeper transition rate than the base curve.
- $C_7(t) = a_0(1 + \tanh(b_0(t - t_0)) + c_0 \sin((1 + r_7 t/T)\omega_0 t / T) + \epsilon(t)$, where $r_7 = 0.1$ so the frequency increases with time.

Over 50 trials, we generate 70 time series, 63 normal curves, and 7 anomalous curves with each of C_1 through C_7 being used once. See Figure 2 for an illustration. We used a uniform mesh with 50 points, $0, 1, ..., 49$. Since we used a 9 : 1 ratio of regular to anomalous series, successful methods, after ranking curves in ascending order of "regular," should rank anomalous curves as among the bottom 10%.

We can also determine the 95th percentile for the percentile rank of each curve, to give an estimate for how much of the data would need to be re-examined to capture such anomalies. These trials can also be done by dropping data points independently at random with a fixed probability to simulate missing data. We ran sets of trials with 0% and 10% of drop probabilities. Results for the mean percentile rank and 95th percentile of the percentile ranks are presented in Tables 1 and 2.

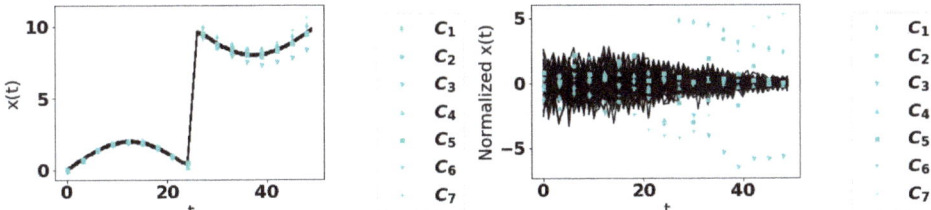

Figure 2. Plot of 63 normal curves and the 7 anomalous curves $C_i(t)$, $i = 1, ..., 7$. Left: un-normalized. Right: normalized.

Table 1. Mean percentiles (out of 100) for curves C_1–C_7 in Scenario 1. A correct classification is a percentile less than or equal to 10 (in bold in the table). The -N suffix denotes the data were normalized by the pre-processing described in Section 2.3; the -U suffixed denotes the data were un-normalized. Note that method FB is not affected by the normalization.

Method	Lost	C_1	C_2	C_3	C_4	C_5	C_6	C_7
Point-N	0%	**4.3**	5.8	**1.4**	21	12	24	**2.9**
Point-U	0%	**4.3**	5.7	**1.4**	14	**8.8**	16	**2.9**
Fourier-N	0%	**4.7**	**1.9**	**2.8**	28	51	29	**8.5**
Fourier-U	0%	5.8	**4.0**	**2.8**	38	43	35	**1.7**
FIF-N	0%	51	24	72	56	79	58	13
FIF-U	0%	19	**2.2**	**4.8**	18	53	19	**5.3**
PCA	0%	**7.5**	20.	**4.3**	**7.5**	53	**9.4**	11
FB	0%	**4.5**	5.6	**1.8**	36	21	37	**2.5**
Point-N	10%	**4.3**	6.0	**1.4**	23	18	29	**2.9**
Point-U	10%	**4.3**	5.7	**1.4**	20.	14	23	**2.9**
Fourier-N	10%	**4.3**	**4.5**	**2.5**	28	43	36	**4.0**
Fourier-U	10%	45	59	50	46	49	53	49
FIF-N	10%	50.	21	75	48	74	51	13
FIF-U	10%	45	15	29	42	48	44	32
PCA	10%	32	20.	**6.1**	36	47	35	**7.7**
FB	10%	**7.5**	**8.7**	**4.2**	47	24	49	**5.1**

Scenario 2: we utilized the testing examples of Staerman et al. [17]. The data consist of 105 time series over $[0, 1]$ with 100 time points. There are 100 regular curves defined by $x(t) = 30(1-t)^q t^q$ where q is equi-spaced in $[1, 1.4]$–thus there is a large family of normal curves. Then, there are 5 anomalous curves:

- $D_1(t) = 30(1-t)^{1.2} t^{1.2} + \beta \chi_{[0.2, 0.8]}$, where β is chosen from a Normal distribution with mean 0 and standard deviation 0.3 and χ_I is the characteristic function of I (there is a jump discontinuity at 0.2 and 0.8).
- $D_2(t) = 30(1-t)^{1.6} t^{1.6}$, being anomalous in its magnitude.
- $D_3(t) = 30(1-t)^{1.2} t^{1.2} + \sin(2\pi t)$.
- $D_4(t) = 30(1-t)^{1.2} t^{1.2} + 2\chi_{\{\tau\}}$, where $\tau = 0.7$ is a single point.
- $D_5(t) = 30(1-t)^{1.2} t^{1.2} + \frac{1}{2}\sin(10\pi t)$.

Each curve was sampled uniformly at 100 points. We did not drop any data points and, owing to the limited randomness, we only present the results of one trial. The results are presented in Table 3.

Table 2. The 95th percentile of the percentile ranks (out of 100) for curves C_1–C_7 in Scenario 1. See Table 1 caption for -N vs. -U distinction.

Method	Lost	C_1	C_2	C_3	C_4	C_5	C_6	C_7
Point-N	0%	4.3	6.5	1.4	47	42	62	2.9
Point-U	0%	4.3	5.7	1.4	30.	12	46	2.9
Fourier-N	0%	8.6	3.6	4.3	67	99	74	13
Fourier-U	0%	5.7	4.3	4.2	84	94	77	2.9
FIF-N	0%	84	69	92	90	100	96	30
FIF-U	0%	57	7.5	11	55	97	53	11
PCA	0%	25	67	7.1	17	96	27	47
FB	0%	5.7	5.7	2.9	75	74	75	2.9
Point-N	10%	4.3	7.1	1.4	52	51	72	2.9
Point-U	10%	4.3	5.7	1.4	46	43	59	2.9
Fourier-N	10%	6.5	10.	5.1	61	91	89	5.7
Fourier-U	10%	84	97	92	93	93	92	89
FIF-N	10%	81	67	97	91	99	95	39
FIF-U	10%	84	27	56	88	95	94	56
PCA	10%	82	60.	31	73	94	73	19
FB	10%	11	13	10.	75	75	75	9.4

Table 3. Percentiles (out of 100) for curves D_1–D_5 in Scenario 2. A correct classification is a percentile less than or equal to 4.8 (in bold in the table) since $5/105 = 4.8\%$. See Table 1 caption for -N vs. -U distinction.

Method	D_1	D_2	D_3	D_4	D_5
Point-N	74	**0.95**	6.7	73	**1.9**
Point-U	83	**0.95**	44	85	71
Fourier-N	**3.8**	**4.8**	**1.9**	2.9	**0.95**
Fourier-U	**1.9**	8.6	30	**0.95**	2.9
FIF-N	**1.9**	28	**3.8**	10	**0.95**
FIF-U	**1.9**	2.9	**3.8**	**4.8**	**0.95**
PCA	**4.8**	2.9	**3.8**	**1.9**	**0.95**
FB	75	**0.95**	21	75	75

Our method is unsupervised and thus the distinction as to what constitutes an anomaly requires considering a curve's score relative to the others and making a decision based upon this. This can involve human judgment. However, since our method returns a scalar score, we can also use a univariate outlier test on the score to formally test the hypothesis H_0: There are no anomalies. The Rosner test [28] is such a test and is available in the R package EnvStats [29]. In Appendix D, by considering scenarios where anomalies are present or absent, we show the validity of this approach.

3.2. Aviation Safety Reports

We now consider how our methods behave in identifying anomalous time series for aviation safety events. A discussion on method performance is deferred to Section 4.

We were provided IATA data for safety-related events of different types on a month-by-month basis from 2018–2020 for different aircraft types and airports. Aircraft types were given IDs from 1 to 64 (not every ID in the range was included). We were also given separate data pertaining to flight frequency in order to normalize and obtain event rates (cases per 1000 flights). Events of interest could include phenomena occurring during a flight such as turbulence or striking birds, or physical problems such as a damaged engine. We study two events: A and B. Event Type A is a contributing factor for one specific type of accident; Event Type B is the aircraft defense against that type of accident. To illustrate our method while preserving the confidentiality of the data, we do not state what A and B represent.

We plot histograms for the scores of Type A and B Events in Figure 3. These histograms suggest that, for events A and B, anomalous curves could be those with scores below 10 for the Point approach.

Then, we consider curves anomalous by the Fourier method if they have scores below −60 for event A and −30 for event B. As the method is unsupervised, the notion of where to draw the line of being anomalous is somewhat subjective. The idea here is to raise a flag so that experts can investigate the anomalous cases more closely. The aircraft types identified as anomalous for both methods are presented in Table 4. It appears for these data, the curves deemed anomalous by the Fourier method are a subset of the curves deemed anomalous by the Point method. In Figure 4, we plot the anomalous curves (with markers) along with the normal curves (dotted lines) for the fleet IDs that were common to both approaches.

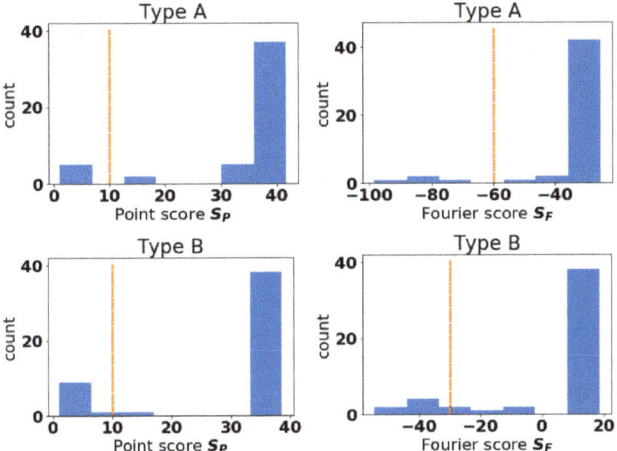

Figure 3. Histogram of scores for Point and Fourier methods for Type A Point (**top-left**), Type A Fourier(**top-right**), Type B Point (**bottom-left**) and Type B Fourier (**bottom-right**). The dashed vertical line represents the division we chose between anomalous (left of line) and normal (right of line). The Sturges estimate was used to set bin widths [30].

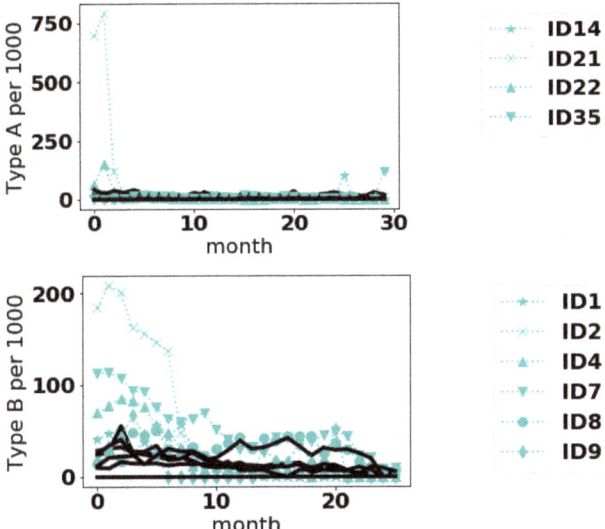

Figure 4. Plots of the time series for Type A and Type B events. Anomalous are dotted curves with markers in the legend; normal curves are solid black curves.

Table 4. IDs of anomalous flights for events A and B. Columnwise, the bolded IDs are common to both methods for a given event type.

Method	Type A	Type B
Point	**14, 21, 22,** 23, **35**	**1, 2, 4,** 5, 6, **7, 8, 9,** 25, 34
Fourier	**14, 21, 22, 35**	**1, 2, 4, 7, 8, 9**

4. Discussion and Conclusions

4.1. Method Performance

From Tables 1 and 2 with regards to Scenario 1, the Point method and FB are superior. They correctly classify C_1–C_3 and C_7 as anomalous. The Point and PCA methods significantly outperform the other methods in the more difficult C_4–C_6 curves. With these data, the Point method generally performs better without normalization. Generally all methods failed to identify the replacement of Gaussian white noise with exponential noise (C_5) as anomalous, although the un-normalized Point approach succeeded. Additionally, all methods considered, except PCA, had difficulty identifying the discontinuous replacement of the hyperbolic tangent (C_4) and a slightly steeper hyperbolic tangent (C_6). The PCA method was not as good as the others for identifying the noise increases (C_2) and the frequency increase with time (C_7). This suggests that not all methods are effective at detecting the same types of anomalies and that they may be complementary.

From Table 3 for Scenario 2, the Fourier approach with normalized data, FIF on un-normalized data, and PCA classify equally correctly. Data can always be normalized and this is therefore not a problem for the Fourier method. In this example, the Point approach fares better with normalization. However, this method and the FB method are not as effective as the FIF, PCA, and Fourier methods.

From our experiments, when there was a large family of curves as with Scenario 2, the Fourier method performed better at detecting anomalies, especially when provided normalized data. But when the family of curves were all close to the same, except for noise, the Point method was better, with or without normalization. Providing more theoretical understanding as to whether these are general phenomena is left for future work.

4.2. Aviation Safety Data

From Figure 4, it appears the methods can detect different sorts of anomalies. In the case of Type A events, the anomalous curves appear to have anomalously large values at an isolated point or over small range of values. The anomalies in Type B events are more interesting and subtle. Even some of the normal curves have sizeable event frequencies, sometimes even exceeding the anomalous curves. But on the average it seems the anomalous curves are higher. In the case of curve 9, the reason it is deemed anomalous is not immediately intuitive. Whether such differences are of a concern to safety would require follow-up from safety inspectors.

To prevent aviation accidents, identifying the potential hazards and risks before they evolve into accidents is the key to proactive safety management. While collecting and analyzing data manually is a time-consuming process, especially on a global scale, the risk identification process may remain reactive process if there is not an automated process. The application of the anomaly detection will enable proactive data-driven risk identification in global aviation safety, by continuously monitoring aviation safety data across multiple criteria (e.g., airport, aircraft type and date), then automatically raising a flag when the model detects any anomalous patterns.

The proposed model shows potential value in automatically detecting potential risk areas with robustness from missing data; however, the interpretation of the model still requires future study. As safety risk is an outcome of complex interactions between multiple factors, including human, machine, environment, and other hidden factors, understanding the full context of such risk requires in-depth investigation and validation from multiple experts. While the model can identify some anomalous patterns, this does not take into consideration the interactions. For example, some aircraft fleets fly more frequently over certain pathways than others. Thus, some differences identified as anomalous due to aircraft type may actually stem from location. Therefore, there will always be a human layer between the model and the interpretation of the model.

4.3. Comments on the Models

There are various degrees of freedom the proposed methods allow for, which are worth noting. Firstly, the point method could be generalized to compute $H^1([0,T])$, and higher Sobolev norms too, but that could lead to additional hyper-parameters in how heavily to weigh the derivative terms. With the Fourier approach, it may seem more appropriate to replace the NUDFT with a weighted combination of terms that more accurately reflects the non-uniform spacing, i.e., a Riemann Sum. Interestingly such an approach tends to make the results slightly worse, hence our choice to use the standard NUDFT.

We anticipate these methods perform well when the time series are sampled at regular intervals and a small portion of entries are missing. If the number of missing entries is very large, this makes inner products computed with the Point method less accurate (without additional interpolations) and the preprocessing of shifting and rescaling could result in poorer outcomes due to a limited sample size upon which to base the normalizations. For many applications, however, most data are present.

4.4. Future Work

We note that our proposed methods aim to identify anomalous time series relative to the sample taken. In general, even if all time series are sampled from the same distribution, due to low probability events, some time series could still be anomalous relative to the sample given. As such, our work has mostly been an empirical investigation of the methods; however, by adding further assumptions on the underlying distribution of time series, it could be possible to obtain a more theoretical basis for the method performance. This would be worth investigating, but is beyond our current work. Going hand-in-hand with this theory it would be interesting to investigate the optimal choice of ζ in the point approach, to understand how the Fourier modes being treated as independent works as effectively as it does, or to more rigorously establish classes of problems when the Point or Fourier approaches are superior.

In conclusion, we have presented two approaches to detecting anomalous time series using KDE to generate functionals to score a series for its degree of anomalousness. The methods handle missing data and perform well in comparison to other methods.

Author Contributions: M.R.L. contributed to conceptualization, methodology, software, formal analysis, investigation, visualization, writing–original draft preparation, writing–review and editing. H.J. contributed to data curation, writing–original draft preparation, and writing–review and editing. D.L. contributed to methodology, funding acquisition, writing–original draft preparation, and writing–review and editing. All authors have read and agreed to the published version of the manuscript.

Funding: This research was supported by the Natural Sciences and Engineering Research Council of Canada (NSERC) and by Fondation HEC Montréal.

Acknowledgments: The authors wish to thank the Associate Editor and two reviewers for their interesting and constructive comments that helped prepare an improved version of the paper. The authors would like to express gratitude to Odile Marcotte for her seamless organization of the Tenth Montréal Industrial Problem Solving Workshop held virtually in 2020. The IATA workshop problem led to these ideas. The authors also thank IATA for providing data.

Conflicts of Interest: The authors declare no conflict of interest.

Appendix A. Fourier Perspective of the Point Method

From Parseval's identity [31], we can also write the terms of (5) as

$$e^{-\frac{||x-a||^2}{2\zeta^2}} = e^{-\frac{\sum_{k \in \mathbb{Z}} |\hat{x}_k - \hat{a}_k|^2}{2T\zeta^2}},$$

i.e., each kernel can be thought of as a Gaussian in $\mathbb{C}^\infty$ with constant variance in all directions. Unfortunately this thinking can only be true in spirit because such a series would not be in $L^2([0, T])$ as $\sum_{k \in \mathbb{Z}} |\hat{a}_k|^2$ would almost surely diverge.

Appendix B. Aliasing

Observe that if the *real* time series $x_0, x_1, ..., x_{N-1}$ is observed at p equally spaced points $t_j = j\Delta$, $\Delta = T/p, j = 0, ..., p - 1$ then for $j = 1, ..., p - 1$:

$$\hat{x}_{p-j} = \sum_{m=0}^{p-1} e^{-2\pi i t_{p-j} m \Delta / T} x_m$$

$$= \sum_{m=0}^{p-1} e^{-2\pi i (p-j) m \Delta / T} x_m$$

$$= \sum_{m=0}^{p-1} e^{2\pi i j m \Delta / T} x_m$$

$$= \overline{\hat{x}_j}$$

where in getting from the first to second line we used $\exp(-2\pi i p m \Delta / T) = \exp(-2\pi i m) = 1$. On the other hand, if data are only observed at $\tilde{t}_0 < \tilde{t}_1 < ... < \tilde{t}_{P_S}$, a subset of the times $t_0, ..., t_p$ with $P_S < p$ then $\tilde{t}_j$ is not, in general $j\Delta$ and the identity does not hold.

Appendix C. Approximate Orthogonality

Before our approximate orthogonality result, we first define the standard inner product for vectors over $\mathbb{C}^N$:

$$(x, y) = \sum_{j=1}^{N} \bar{x}_i y_i.$$

Theorem A1 (Approximate Orthogonality). *Let $t_j = j\Delta$ for $j = 0, 1, ..., p - 1$ where $\Delta = T/p$ for $T > 0$. Let $\{\tilde{t}_j\}_{j=0}^{P_S-1} \subset \{t_0, t_1, ..., t_{p-1}\}$. Define $m = p - P_S$ and define the basis vectors $\{e^{(k)} = e^{2\pi i k \tilde{t}_j / T}, j = 0, ..., P_S | k = 0, ..., P_S\}$. Then*

$$\frac{(e^{(k)}, e^{(k')})}{|e^{(k)}||e^{(k')}|} = \begin{cases} 1, & k = k' \\ O(m/p), & k \neq k' \end{cases}.$$

In other words the cosine similarity of the two vectors is either 1 or $O(m/p)$.

Proof. We trivially note that $|e^{(k)}| = \sqrt{P_S}$ for any k. Next, if $k = k'$ then

$$\frac{(e^{(k)}, e^{(k')})}{|e^{(k)}||e^{(k')}|} = \frac{1}{P_S} \sum_{j=0}^{P_S-1} 1$$

$$= 1.$$

Let us define the set $B = \{j | t_j \notin \{\tilde{t}_k | k = 0, ..., P_S\}$ for $j = 0, ..., p\}$, i.e., it is a listing of all regular time values that have been lost in only observing at the $\tilde{t}$'s. Note that $|B| = m$. Also let $q = k' - k$ so that when $k \neq k'$:

$$\frac{\langle e^{(k)}, e^{(k')} \rangle}{|e^{(k)}||e^{(k')}|} = \frac{1}{P_S} \sum_{j=0}^{N_S-1} \exp(2\pi i q \tilde{t}_j / T)$$

$$= \frac{1}{P_S} \Big(\sum_{j=0}^{p-1} \exp(2\pi i q t_j / T) - \sum_{j \in B} \exp(2\pi i q t_j / T) \Big)$$

$$= \frac{-1}{P_S} \sum_{j \in B} \exp(2\pi i q t_j / T)$$

where in arriving at the final equality we used that the sum of $e^{2\pi i q t_j}$ over $j = 0, ..., p-1$ is 0 ($1 + \eta + \eta^2 + ... + \eta^{p-1} = 0$ if $\eta^p = 1$ and $\eta \neq 1$). As each term in the remaining sum is bounded by 1 and $|B| = m$, we have that

$$\frac{\langle e^{(k)}, e^{(k')} \rangle}{|e^{(k)}||e^{(k')}|} = O(m/P_S)$$

$$= O(m/p).$$

□

Appendix D. Rosner Test on the Scores

We briefly explore the capacity of our methods to test the hypothesis H_0: There are no anomalies. The idea is to compute the scores from our methods, which are scalars, and use a univariate outlier test on them. In this experiment, we use Rosner test [28] with the R package EnvStats [29]. Specifically, we test H_0 for each of the 50 trial datasets of Scenario 1 and report the proportion of time the null hypothesis is rejected at the $\alpha = 0.05$ level. In this case, the proportion of rejection measures the power of the test and we wish to have the highest values possible. However, to verify the validity of the test, we also run the tests on the samples containing only the 63 normal curves. This time, we want the proportion of rejection to be close to the level $\alpha = 0.05$. The results, presented in Table A1, show that this method is working. Setting aside the un-normalized Fourier approach (Fourier-U) with 10% of missing data, the proportion of rejection varies between 0.02 and 0.08 when the data contain no anomalies, showing that the test is able to maintain its prescribed level. When the data contain anomalies, the power ranges between 0.98 and 1, showing that the anomalies are detected in almost all cases. The only exception is the Fourier-U method with 10% of missing data, which never rejects H_0 whether or not the data contain anomalies. But this is consistent with the fact that this method had a very poor performance in this case and was not able to detect the anomalies as seen in Table 1.

Table A1. Proportions of time the null hypothesis H_0 is rejected at the $\alpha = 0.05$ level. See Table 1 caption of main manuscript for -N vs -U distinction.

Method	Lost	Anomalies Present	H_0 Reject Fraction
Point-N	0%	No	0.04
Point-N	0%	Yes	1
Point-U	0%	No	0.08
Point-N	0%	Yes	1
Fourier-N	0%	No	0.08
Fourier-N	0%	Yes	0.98
Fourier-U	0%	No	0.02

Table A1. *Cont.*

Method	Lost	Anomalies Present	H_0 Reject Fraction
Fourier-U	0%	Yes	1
Point-N	10%	No	0.04
Point-N	10%	Yes	1
Point-U	10%	No	0.08
Point-N	10%	Yes	1
Fourier-N	10%	No	0.06
Fourier-N	10%	Yes	1
Fourier-U	10%	No	0
Fourier-U	10%	Yes	0

References

1. Cleophas, T.J.; Zwinderman, A.H.; Cleophas-Allers, H.I. *Machine Learning in Medicine*; Springer: Berlin/Heidelberg, Germany, 2013; Volume 9.
2. Howedi, A.; Lotfi, A.; Pourabdollah, A. An Entropy-Based Approach for Anomaly Detection in Activities of Daily Living in the Presence of a Visitor. *Entropy* **2020**, *22*, 845. [CrossRef]
3. Chen, X.; Li, Z.; Wang, Y.; Tang, J.; Zhu, W.; Shi, C.; Wu, H. Anomaly detection and cleaning of highway elevation data from google earth using ensemble empirical mode decomposition. *J. Transp. Eng. Part Syst.* **2018**, *144*, 04018015. [CrossRef]
4. Siboni, S.; Cohen, A. Anomaly Detection for Individual Sequences with Applications in Identifying Malicious Tools. *Entropy* **2020**, *22*, 649. [CrossRef]
5. Bernikova, O.; Granichin, O.; Lemberg, D.; Redkin, O.; Volkovich, Z. Entropy-Based Approach for the Detection of Changes in Arabic Newspapers' Content. *Entropy* **2020**, *22*, 441. [CrossRef]
6. Saradjian, M.; Akhoondzadeh, M. Thermal anomalies detection before strong earthquakes (M > 6.0) using interquartile, wavelet and Kalman filter methods. *Nat. Hazards Earth Syst. Sci.* **2011**, *11*, 1099. [CrossRef]
7. Xu, X.; Ren, Y.; Huang, Q.; Fan, Z.Y.; Tong, Z.J.; Chang, W.J.; Liu, B. Anomaly detection for large span bridges during operational phase using structural health monitoring data. *Smart Mater. Struct.* **2020**, *29*, 045029. [CrossRef]
8. Holst, A.; Bohlin, M.; Ekman, J.; Sellin, O.; Lindström, B.; Larsen, S. Statistical anomaly detection for train fleets. *AI Mag.* **2013**, *34*, 33. [CrossRef]
9. Das, S.; Matthews, B.L.; Lawrence, R. Fleet level anomaly detection of aviation safety data. In Proceedings of the 2011 IEEE Conference on Prognostics and Health Management, Montreal, QC, Canada, 20–23 June 2011; pp. 1–10.
10. Ketabdari, M.; Giustozzi, F.; Crispino, M. Sensitivity analysis of influencing factors in probabilistic risk assessment for airports. *Saf. Sci.* **2018**, *107*, 173–187. [CrossRef]
11. Huang, Z.; Lu, X.; Duan, H. Anomaly detection in clinical processes. In *AMIA Annual Symposium Proceedings*; American Medical Informatics Association: Bethesda, MD, USA, 2012; Volume 2012, p. 370.
12. Aydin, I.; Karakose, M.; Akin, E. Anomaly detection using a modified kernel-based tracking in the pantograph–catenary system. *Expert Syst. Appl.* **2015**, *42*, 938–948. [CrossRef]
13. Braei, M.; Wagner, S. Anomaly Detection in Univariate Time-series: A Survey on the State-of-the-Art. *arXiv* **2020**, arXiv:2004.00433.
14. Blázquez-García, A.; Conde, A.; Mori, U.; Lozano, J.A. A review on outlier/anomaly detection in time series data. *arXiv* **2020**, arXiv:2002.04236.
15. Martos, G.; Hernández, N.; Muñoz, A.; Moguerza, J.M. Entropy measures for stochastic processes with applications in functional anomaly detection. *Entropy* **2018**, *20*, 33. [CrossRef]
16. Vinue, G.; Epifanio, I. Robust archetypoids for anomaly detection in big functional data. In *Advances in Data Analysis and Classification*; Springer: Berlin/Heidelberg, Germany, 2020; pp. 1–26.
17. Staerman, G.; Mozharovskyi, P.; Clémençon, S.; d'Alché-Buc, F. Functional Isolation Forest. *Proc. Mach. Learn. Res.* **2019**, *101*, 332–347.

18. Jackson, A.C.; Lacey, S. Seasonality and Anomaly Detection in Rare Data Using the Discrete Fourier Transformation. In Proceedings of the 2019 First International Conference on Digital Data Processing (DDP), London, UK, 15–17 November 2019; pp. 13–17.
19. Hyndman, R.L.; Zhang, X.; King, M.L. Bandwidth selection for multivariate kernel density estimation using mcmc. In Proceedings of the Econometric Society 2004 Australasian Meetings, Melbourne, Australia, 7–9 July 2004; number 120.
20. Guo, Y.; Xu, Q.; Li, P.; Sbert, M.; Yang, Y. Trajectory shape analysis and anomaly detection utilizing information theory tools. *Entropy* **2017**, *19*, 323. [CrossRef]
21. Maniglia, S.; Rhandi, A. Gaussian measures on separable Hilbert spaces and applications. *Quaderni di Matematica* **2004**. Available online: http://siba-ese.unisalento.it/index.php/quadmat/issue/view/775 (accessed on 30 November 2020).
22. Mobli, M.; Hoch, J.C. Nonuniform sampling and non-Fourier signal processing methods in multidimensional NMR. *Prog. Nucl. Magn. Reson. Spectrosc.* **2014**, *83*, 21–41. [CrossRef] [PubMed]
23. Sun, Y.; Genton, M.G. Functional boxplots. *J. Comput. Graph. Stat.* **2011**, *20*, 316–334. [CrossRef]
24. Ramsay, J.O.; Graves, S.; Hooker, G. fda: Functional Data Analysis; R Package Version 5.1.5.1. Available online: https://cran.r-project.org/web/packages/fda/index.html (accessed on 30 November 2020).
25. Staerman, G.; Mozharovskyi, P.; Clémençon, S.; d'Alché-Buc, F. FIF: Functional Isolation Forest. Available online: https://github.com/GuillaumeStaermanML/FIF (accessed on 15 September 2020).
26. Hyndman, R.J.; Wang, E.; Laptev, N. Large-scale unusual time series detection. In Proceedings of the 2015 IEEE international conference on data mining workshop (ICDMW), Atlantic City, NJ, USA, 14–17 November 2015; pp. 1616–1619.
27. Hyndman, R.J.; Wang, E.; Laptev, N. anomalousACM: Unusual Time Series Detection; R Package Version 0.1.0. Available online: https://github.com/robjhyndman/anomalous-acm (accessed on 30 November 2020).
28. Rosner, B. Percentage points for a generalized ESD many-outlier procedure. *Technometrics* **1983**, *25*, 165–172. [CrossRef]
29. Millard, S.P. *EnvStats: An R Package for Environmental Statistics*; Springer: New York, NY, USA, 2013.
30. Sturges, H.A. The choice of a class interval. *J. Am. Stat. Assoc.* **1926**, *21*, 65–66. [CrossRef]
31. Folland, G.B. *Real Analysis: Modern Techniques and Their Applications*; John Wiley & Sons: Hoboken, NJ, USA, 1999; Volume 40.

Publisher's Note: MDPI stays neutral with regard to jurisdictional claims in published maps and institutional affiliations.

© 2020 by the authors. Licensee MDPI, Basel, Switzerland. This article is an open access article distributed under the terms and conditions of the Creative Commons Attribution (CC BY) license (http://creativecommons.org/licenses/by/4.0/).

Review

A Systematic Review of Statistical and Machine Learning Methods for Electrical Power Forecasting with Reported MAPE Score

Eliana Vivas [1,*], Héctor Allende-Cid [1] and Rodrigo Salas [2]

1. Escuela de Ingeniería Informática, Pontificia Universidad Católica de Valparaíso, Brasil 2950, Valparaíso, Chile; hector.allende@pucv.cl
2. Escuela de Ingeniería C. Biomédica, Universidad de Valparaíso, Chacabuco 2092-2220, Valparaíso, Chile; rodrigo.salas@uv.cl
* Correspondence: eliana.vivas.r@mail.pucv.cl; Tel.: +56-9-870-25259

Received: 29 October 2020; Accepted: 10 December 2020; Published: 15 December 2020

Abstract: Electric power forecasting plays a substantial role in the administration and balance of current power systems. For this reason, accurate predictions of service demands are needed to develop better programming for the generation and distribution of power and to reduce the risk of vulnerabilities in the integration of an electric power system. For the purposes of the current study, a systematic literature review was applied to identify the type of model that has the highest propensity to show precision in the context of electric power forecasting. The state-of-the-art model in accurate electric power forecasting was determined from the results reported in 257 accuracy tests from five geographic regions. Two classes of forecasting models were compared: classical statistical or mathematical (MSC) and machine learning (ML) models. Furthermore, the use of hybrid models that have made significant contributions to electric power forecasting is identified, and a case of study is applied to demonstrate its good performance when compared with traditional models. Among our main findings, we conclude that forecasting errors are minimized by reducing the time horizon, that ML models that consider various sources of exogenous variability tend to have better forecast accuracy, and finally, that the accuracy of the forecasting models has significantly increased over the last five years.

Keywords: electric power; forecasting accuracy; machine learning

1. Introduction

Electric power forecasting plays a substantial role in the administration and balance of current power systems. The load forecasts help to identify strategies to optimize the operating mechanisms in a determined period and thus ensure the demand even in situations adverse to the system [1]. Accompanying the rapid advances in forecasting theory [2,3] and machine learning [4–6], the technology in the energy forecasting research area has also developed rapidly [7]. Additionally, the popular prediction methods for the generation and demand of energy can be divided into two categories. The first category is statistical or mathematical methods, and the second category is modern statistical-learning-based methods (also known as machine learning). In addition, hybrid methods can be found that apply not only statistical tools but also other elements, such as mathematical optimization or signal processing [8,9]. Additionally, other authors [10] consider hybrid approaches that focus on a series of individual methods, such as noise reduction, seasonal adjustment and clustering, to process the data in advance, whereas combined methods use weight coefficients. With respect to the techniques implemented to forecast energy in recent years, in the international context, we can find a wide diversity; e.g., the application of kernel-based multitask learning methodologies [11], energy load forecasting methodologies based on deep neural networks as in [12,13], methodologies based on the classic time series approach as in [14–16], and mathematical representations as in [17–19]. Developing a model that achieves the highest forecasting precision in the context of electric power has been the object of study in recent years. Additionally, the determination of the appropriate input variables in load forecasting constitutes an important part of the forecasting procedure [20]. Due to the importance of the area, several review papers have appeared that present insights into current applications and future challenges and opportunities [21,22]. However, existing review papers examine the applications of a single model, e.g., an ANN [23], or cover only one energy domain, e.g., solar radiation prediction [24,25], and do not perform comparisons among specific metrics, such as MAPE, for multiple applications. Therefore, a systematic review to identify the type of model that has the highest propensity to show precision in the forecast is the main objective of this paper.

Motivation and Scope of the Review

The number of papers published on the topic of electric power forecasting has been growing at an exponential rate throughout the last decade, as Figure 1 shows. The order of magnitude of the increase in the number of scientific publications on the subject revolves around 61.59%, between 2016–2020, with respect to 2011–2015. Generally, the studies are site-specific, and the results strongly depend on the nature of the model and the time horizon of the forecast, along with a large number of other characteristics pertaining to the data and models. This is a major limitation, which makes a generalization of the results difficult. A test of a given model over all different mentioned factors is needed to measure the average effect of the model [25]. Consequently, the contribution of this paper is to present the state-of-the-art of models in electric power systems and discuss their likely future trends, considering:

- (I) The models that tend to provide precision in electric power forecasts according to the literature.
- (II) Exogenous sources that tend to lead to accurate forecasting of electrical energy according to the literature.
- (III) Relationships between the times of forecasting and the accuracy of existing models.

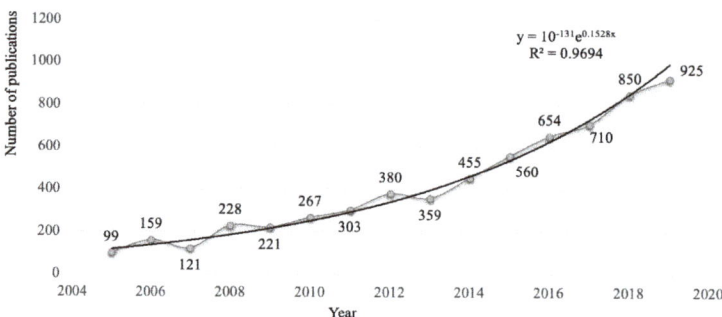

Figure 1. Number of articles published per year. The line represents an exponential fit, highlighting the yearly growth trend. The publications from 2020 are excluded since only partial data are available.

The rest of this paper is organized as follows. In Section 2, the methodology of the research is presented. In Section 3, a description of the data set is presented. In Section 4, a performance analysis of the forecasting models is presented, and finally, the overall discussion and conclusions are presented.

2. Theoretical and Referential Framework

This chapter presents an analysis of the documents found in the literature during the last 15 years on the subject of electric power forecasting.

2.1. Selection Criteria

The number of documents published on the topic of electric power forecasting has been growing at an exponential rate throughout the last 15 years, as Figure 1 shows. We analyze in the review the documents published for electric power forecasting contained in SCOPUS, Web of Sciences, Science Direct and IEEE (Figure 2), according to the criteria shown in Figure 3 and following the steps of the PRISMA (Preferred Reporting items for Systematic Review and Meta-Analyses) methodology.

A large number of papers published between January 2005 and March 2020 were analyzed. The qualitative and quantitative synthesis of the analysis was collected from 164 documents selected based on the criteria shown in Figure 3; the documents that only forecast electric power in buildings, universities, homes, and rooms were excluded; likewise, if the time horizons are not mentioned in the Abstract, the article was also skipped in our research. Similarly, if in an article, MAPE was not used as a criterion for accuracy, it was not considered in our review. When considering the accuracy of the results reported by the selected papers in terms of the MAPE Equation (1), we can compare samples of different magnitudes, thus ensuring a common basis for intercomparison analyses.

It is important to highlight that under our filtering criteria a significant volume of valuable references may have been excluded; in this sense, our searches may not be specific. If the readers are interested exclusively in consulting documents related to forecasting under machine learning methods, then they could consult [21], or if they are interested in specific documents on solar energy, they could consult [25]; in the case of documents associated with the forecast under classical statistical techniques, there are specific documents that can be consulted, such as [26].

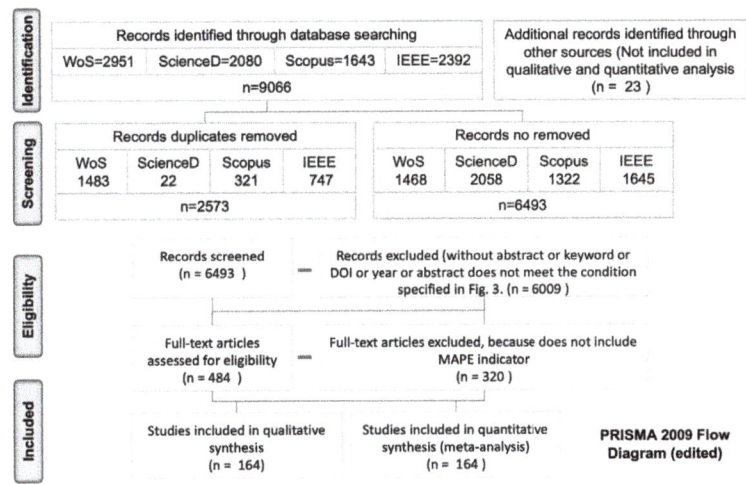

Figure 2. PRISMA Flow Diagram.

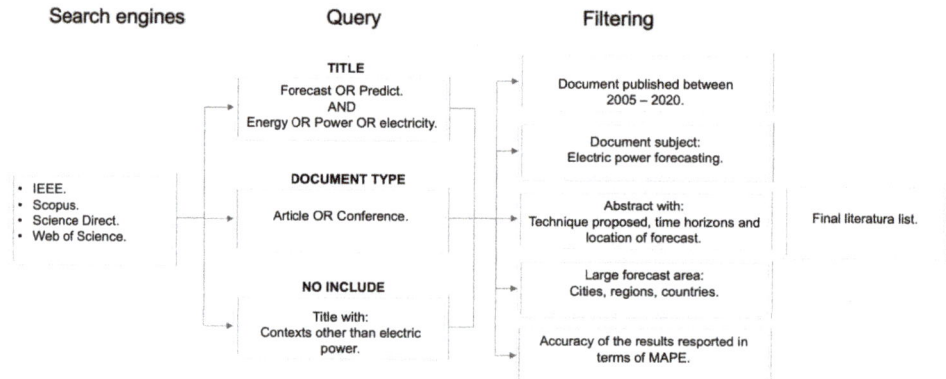

Figure 3. Search methodology for finding relevant literature.

2.2. Statistical Indicators of Accuracy in Electric Power Forecasting

The number of papers from the publications we studied that are eligible according to the criteria was 164. We collected the mean absolute percentage error (MAPE), a statistical indicator of accuracy. This index indicates an average of the absolute percentage errors (Equation (1)); the lower the MAPE, the higher is the accuracy [27].

$$MAPE = \frac{1}{m_k} \sum_{k=1}^{m} \left| \frac{t_k - y_k}{t_k} \right| * 100 \qquad (1)$$

where t_k is the actual value of electric power, y_k is the forecasting value produced by the model, and m is the total number of observations. The final quality-controlled database from the 164 documents contained 4883 entries (MAPE, type of MAPE, country, date, input variables, model, type of model, latitude, longitude, and size of sample), and we saved 257 entries associated with a MAPE value linked to the best model proposed in the cases of study with data from 33 countries. The locations are represented on the world map in Figure 4. We can see that the studied publications cover all continents. Occupying the first positions in the list of the countries with the most electrical energy forecast documents, under the criteria used, are Australia, China, Iran, and Turkey.

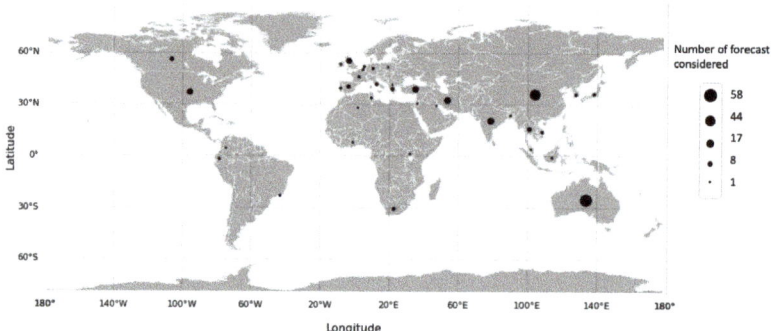

Figure 4. Number of forecasts by country considered in the review.

3. Description of the Dataset

The analysis was performed from five perspectives: the class of a forecasting model (MSC or ML), the type of model (hybrid or not), the time horizon, and the input variables and performance trend over time (MAPE). The dataset analyzed in this paper contains 257 entries associated with a MAPE value linked to the best model proposed in the document.

The MAPE value was classified according to the criteria drawn up by [28], which contain typical MAPE values for business and industrial data and their interpretation in four evaluation criteria (in our case, four prediction capabilities); this table was used in [29–32], and can be seen in Table 1.

Table 1. MAPE qualitative criteria.

MAPE (%)	Prediction Capability
<10	Highly accurate prediction (HAP)
10–20	Good prediction (GPR)
20–50	Reasonable prediction (RP)
>50	Inaccurate prediction (IPR)

Table 2 shows that of the 164 documents processed in the systematic review, 99 contain a highly accurate prediction (HAP). Additionally, more ML documents with an HAP were found than MSC documents. Regarding the sources of variability considered by the documents that contained HAPs, it can be seen that multivariate models have a higher recurrence than univariate models. As explained in [10], despite the introduction of artificial intelligence, each of the individual methods are still not able to produce the desired outcomes because of their disadvantages. For instance, neural networks attain local optimal results instead of global optimal results. Expert systems excessively rely on knowledge and cannot always obtain optimal results, whereas grey prediction systems are suitable for exponential growth models. Thus, by considering every method's merits and taking full advantage of them, the concept of hybrid and combination methods developed rapidly.

Table 2. Systematic review documents. Techniques type used in electric power forecast and qualitative values of the average MAPE.

MAPE	Type	Multivariate Model		Univariate Model		Total
		Not Hybrid	Hybrid	Not Hybrid	Hybrid	
HAP	ML	[33–72]	[6,27,73–91]	[1,14,17,18,92–107]	[29,32,85,108–122]	99
HAP	MSC	[15,123–147]	[148,149]	[1,14,17,18,92–107]	[19,150–152]	52
GPR	ML	[153–157]	[158]	[159–162]	—	10
GPR	MSC	[16,163]	—	—	—	2
RP	ML	[164]	—	—	—	1
Total		74	24	44	22	164

3.1. Forecasting Horizon

Figure 5 shows that the minimum MAPE values (<2) were reached more frequently when the forecast time horizon is 5 min.

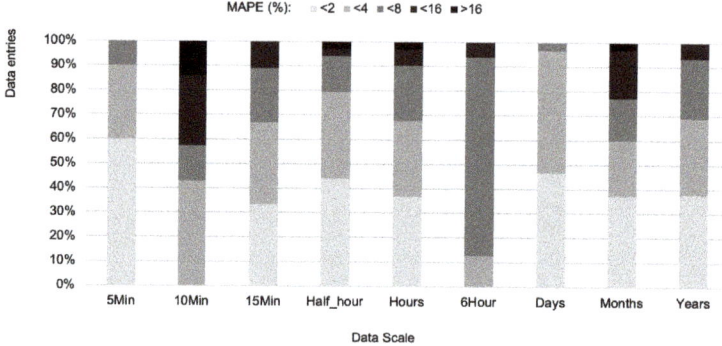

Figure 5. MAPE value interval and the percent of forecasts extracted from the 164 documents considered by time horizon.

3.2. Exogenous Influence

Because forecasting electric power demand is typically based on historical electricity consumption and its relationship with exogenous influences, such as gross domestic product (GDP), population, urbanization, income and exports, research on forecasting electric power demand has evolved using both univariate and multivariate time-series models [15].

Similarly, weather associated variables such as humidity, temperature, and dew point are pertinent for electric power forecasting for extensive time scales. For short-term forecasting such as minutes ahead, the climate changes are already captured in the electric power series [165]. Forecasting models using only previous electricity data (univariate) have been shown to provide HAP and to perform better than models that also use weather variables as exogenous influences (multivariate) [166]. Nevertheless, the use of weather influences was found to be beneficial for electric power forecasting horizons beyond several hours [166,167]. In Figure 6, it can be seen that the precision of the electric energy forecast on average tends to improve when various sources of variability are considered. In our sample of filtered documents, the average MAPE is lower for the forecasts whose models consider sources of variability from the calendar information, weather information, and economic or sociodemographic information, as in [60,125,130,146].

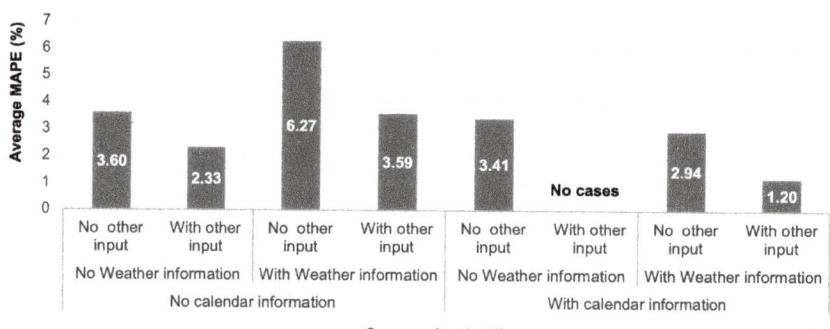

Figure 6. Average MAPE according to the considered source of variability.

4. Classes of Forecasting Models

From the multitude of methods that have been tested and evaluated, the ML and MSC classes seem to be the main competitors.

4.1. Classical Statistical Models

A popular technique such as time series forecasting is applied in several areas [25]. The most widely used statistical method is the ARIMA of Box and Jenkins, which was applied with more force during the eighties, when intelligent systems began to appear [168]. Several time series models make use of the high autocorrelation for small lags in the time series of electric power, and supply electric power forecasts using only previously measured values of electric power as input.

From the multitude of methods that have been tested and evaluated in this review, in this class, regression analysis and ARIMA modeling seem to be the main competitors (Figure 7).

4.2. Classical Regression in the Time Series Context

To explain linear regression in the the context of time series, we assume some output or dependent time series. Assume x_t for $t = 1, \cdots, n$, is being influenced by a collection of possible inputs or independent series, such as $z_{t_1}, z_{t_2}, \cdots, z_{t_q}$, where we first regard the inputs as fixed and known [169]. We express this relation through the linear regression model:

$$x_t = \beta_0 + \beta_1 z_{t1} + \beta_2 z_{t2} + \cdots + \beta_q z_{tq} + w_t \qquad (2)$$

where $\beta_0, \beta_1, \cdots, \beta_q$ are unknown fixed regression coefficients, and w_t is a random error or noise process consisting of independent and identically distributed (*iid*) normal variables with a zero mean and variance σ_w^2. For time series regression, it is rarely the case that the noise is white, and we will need to eventually relax that assumption.

Classical regression models have been used in several academic papers for electric power forecasting [97,98,102,124,130,134,138,140], reaching an accuracy in the forecast with an average MAPE value of 1.569%. Classical regression is often insufficient for explaining all of the interesting dynamics of a time series; instead, the introduction of correlations that may be generated through lagged linear relations led to the autoregressive (AR) and autoregressive moving average (ARMA) models that were presented in [169,170]. Adding nonstationary models to the mix led to the autoregressive integrated moving average (ARIMA) model popularized in the landmark work by Box and Jenkins [169,171].

4.3. Autoregressive Integrated Moving Average

Autoregressive models are based on the idea that the current value of the series, x_t, can be explained as a function of p past values, $x_{t-1}, x_{t-2}, \cdots, x_{t-p}$, where p determines the number of previous steps required to forecast the current value [169].

The acronym ARIMA refers to an autoregressive integrated moving average model. ARIMA models can be applied to non-stationary data, and when the data are seasonal, the SARIMA model can be implemented. The ARIMA and SARIMA models have been used in many studies for forecasting [14–16,99,100,103,127,136], reaching forecast accuracies with an average MAPE value of 3.214%. A typical ARIMA (p, d, q) model can be expressed by Equation (3), where the variable u_t is replaced by a new variable w_t obtained by differencing u_t d times [25]:

$$w_t = (1 - B)^d u_t. \qquad (3)$$

4.4. Machine Learning (ML) Models

ML methods have been suggested in the academic literature as an alternative to MSC methods for time series prediction, with the same objective. They attempt to improve the forecasting accuracy precision by minimizing some loss functions, as for example the sum of squared errors. The distinction between ML and MSC is in how the minimization is performed: the ML methods use nonlinear algorithms while the MSC method use linear processes. The ML methods require a greater dependence on computer science to be implemented and are more demanding than MSC methods, as they are positioned at the intersection of MSC and computer science [172]. There are several approaches developed under ML theory. In this review, artificial neural networks (ANNs), support vector machines (SVMs), decision trees (DTs), adaptive neuro fuzzy inference systems (ANFISs), and recurrent neural networks (RNNs) were found to support the bases of the models that were implemented more frequently in electric power forecasting (Figure 7).

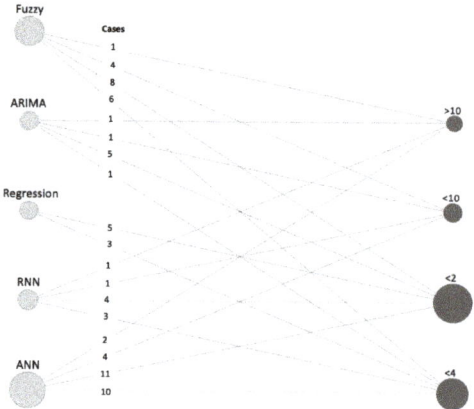

Figure 7. Graph with nodes weighted by sample size for the case of the techniques (light-gray node) and probability of occurrence for the MAPE intervals (dark-gray node).

4.4.1. Artificial Neural Networks (ANN)

Neural networks have been the subject of great interest for many decades due to the desire to understand the brain and to build learning machines [173]. *A neural network is an interconnected assembly of simple processing elements, units or nodes whose functionalities are loosely based on animal neurons. The processing ability of a network is stored in the inter-unit connection strengths, or weights, obtained by a process of adaptation to, or learning from, a set of training patterns* [174].

The ANN models have been used in many studies for electric power forecasting [6,39,44,46,48,53–55,57–62,65,70,75,77,81,82,86,108,153,155,165,175,176] and have reached a forecasting accuracy with an average MAPE value of 3.781%.

4.4.2. Recurrent Neural Networks (RNN)

Models known as a recurrent neural networks allow feedback connections; these models define nonlinear dynamical systems but do not have simple probabilistic interpretations [173]. RNN models have been used in many studies for electric power forecasting [64,69,71,73,88,90,157,177,178] and have reached a forecasting accuracy with an average MAPE value of 3.610%.

4.4.3. Fuzzy Neural Network-Based Forecasting Methods

Fuzzy logic systems (or, simply, fuzzy systems (FSs)) and neural networks are universal approximators; that is, they can approximate any nonlinear function (mapping) with any desired accuracy and have found wide application in the identification, planning, and model-free control of complex nonlinear systems, such as robotic systems and industrial processes. Fuzzy logic offers a linguistic (approximate) approach to drawing conclusions from uncertain data, and neural networks offer the capability of learning and training with or without a teacher (supervisor) [179].

Fuzzy logic algorithms have been used in many studies for electric power forecasting [10,27,33,45,50,52,63,79,80,83,85,91,113,118,122,154,160,180,181] and have reached a forecasting accuracy with an average MAPE value of 4.013%.

4.4.4. Support Vector Machines (SVMs)

Support vector machines are supervised learning algorithms used for solving binary classification and regression problems. The main idea of support vector machines is to construct a hyperplane such that the margin of separation between the two classes is maximized. In this algorithm, each of the data points is plotted as a data point in n-dimensional hyperspace. Then, a hyperplane that maximizes the separation between the two classes is constructed [182]. *This technique was originally designed for binary classification but can be extended to regression and multiclass classification* [173]. Support vector regression algorithms have been used in many studies for electric power forecasting [41,47,49,76,109,110,112,114, 119,164,183–185] and have reached a forecasting accuracy with an average MAPE value of 4.326%.

5. Evaluation of Model Accuracy

As can be seen in Table 3, as mentioned in [77], there are many factors, such as economic development, regional industrial production, holiday periods, weather conditions, social change, electricity price, and population, that are unavoidable, affect electric power randomly, and allow the data to demonstrate different features.

Short-term load forecast models that rely on weather information require the prediction of weather parameters for the next few hours or at most the next few days [75]. Similarly, economic indicators and electrical infrastructure measures are usually useful in forecasting electric power with a long forecast horizon, e.g., a prediction of the annual peak load at least one year in advanced [39]. However, in the daily peak load forecasting for the following month, these indicators are not effective, since the forecast step and horizon are too short to observe their effect [75]; this behavior is shown in Table 3.

Similarly, from Figure 8, it can be seen that among the records of documents that reach HAP, the average MAPE value is lower in the frameworks that implement hybrid models of ML and multivariate dependency, such as those developed in [6,27,73–91]. To verify the hypotheses of the differences in the means and variances in the MAPE, three hypothesis tests are carried out. Table 4 shows that for small and medium effects, the alternative hypothesis on the minor indicates that the MAPE is accepted for the ML model, based on a hybrid method and with multivariate dependencies.

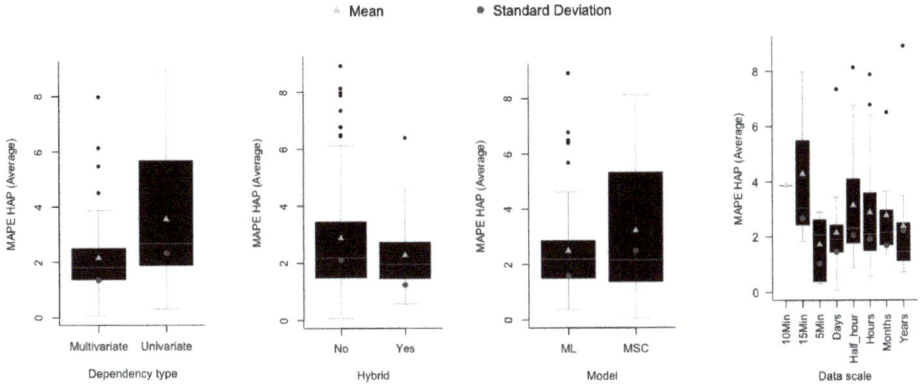

Figure 8. Boxplot of the papers included in the systematic review, with HAP-MAPE.

Table 3. Summary of the papers included in the systematic review, with a HAP-MAPE, hybrid ML forecasting approach and multivariate model. The abbreviations are displayed in Abbreviations Section. [1]/ Average MAPE; [2]/ Approximate sample size. The Classification of the forecasting models can be seen in Figure A1.

Ref	Year	Country	Energy Type	Technique	Forecast	Other Input	MAPE [1]/	N [2]/	Scale	Date Sample
[73]	2006	Australia	No Specific	ERNN; WT	Electricity Load	TM, HM, WS	0.794	26,297	Hours	1999 2002
[74]	2013	Iran	Wind	PSO; ACO	Wind Power	TM, WS	3.513	8736	Hours	2010 2011
[75]	2008	Iran	No Specific	ANN; EA	Peak Load	CI	1.760	26,280	Hours	1997 1999
[76]	2008	EEUU	No Specific	SVR; BT	Electricity Load	CI, TM, HM	1.960	30,144	Hours	2001 2004
[77]	2015	Australia	No specific	ANN	Electrical power	CI	3.710	70,080	Half-Hour	2006 2009
[78]	2017	UK	No Specific	PSO; ANN	Load Demand	CI, TM	1.723	8760	Hours	2008 2008
[79]	2016	Algeria	No Specific	HW-ES; KNN; WD; Fuzzy-CM; ANFIS	Peak Electricity	TM	2.796	1064	Days	2012 2014
[80]	2010	Iran	No Specific	ANFIS	Electricity	GDP, POP, EXP, CPI	2.789	37	Years	1971 2007
[81]	2017	Poland	No Specific	ANN; PCA	Power Load	CI, TM	1.235	26,280	Hours	2009 2012
[82]	2018	India	No Specific	ANN; PSO; GA	Electricity Demand	CPI, GDP	0.220	25	Years	1991 2015
[83]	2017	UK	No Specific	ELM; Fuzzy	Electricity Load	CI, TM, DP	1.435	43,852	Hours	2004 2008
[84]	2019	EEUU	Wind	NWP; WD; CNN	Wind Power	CI, TM, WS, DP	2.550	26,280	Hours	2015 2017
[85]	2008	Iran	No Specific	BNN; MCM; Fuzzy	Load	CI, TM	2.421	1460	Days	2004 2007
[27]	2018	Vietnam	No Specific	WT; ANFIS; COA	Electricity	CI, TM, HM, PRS, RFL, RT, WS	4.330	132	Months	2003 2013
[86]	2017	UK	No Specific	ANN; JOA	Electricity Load	CI, TM, DP	5.710	52,560	Hours	2004 2009
[87]	2015	India	No Specific	ANN; BBO	Electrical Energy	GDP, POP	2.510	33	Years	1980 2012
[88]	2019	China	Wind	GM; ERNN; BP	Power Generation	TM, HM, WS, WDD, PRS	3.730	1441	15 min	2016 2016
[6]	2019	Australia	No Specific	ANN; BOOT	Electricity	57 Index	5.290	4300	6 h	2014 2017
[89]	2019	Uganda	No Specific	PSO; ABC	Electricity	POP, GDP, EP, NS	1.306	17	Years	1990 2016
[90]	2020	Australia	Photovoltaic	WD; LSTM	Power	TM, HM, WS, HR	1.868	213,984	5 min	2014 2016
[91]	2018	Turkey	No Specific	ANFIS	Electrical Load	CI, TM	8.869	8760	Hours	2017 2017

Table 4. Hypothesis Test for Difference in Means. [1/] Levene Test (p-value); [2/] test (p-value).

Variable	Hypothesis (H_0)	Homogeneity of Variance [1/]	Difference in Means [2/]	Effect Size (Cohen's)
Model	$\mu_{ML} \geq \mu_{MSC}$	0.00386	0.07252	Small
Hybrid	$\mu_{Yes} \geq \mu_{No}$	0.09063	0.04321	Small
Dependency	$\mu_{Multi} \geq \mu_{Uni}$	0.00125	0.00059	Medium

In this sense, a summary of the documents found in the review with MAPEs and HAPs that base their models on ML with a hybrid approach and multivariate dependence is presented in the Table 3. When we analyzed these documents, we observed that there are common elements; for example, when building a word cloud from the abstracts, keywords and titles of these documents, we can identify that in 24% of the cases multiple scale decomposition and wavelet theory (WT) were mentioned.

The wavelet transform, including filtering and forecasting, has been suggest for detailed examination of the elements or structure of time series in several academic papers in recent years [73]. WT has been extensively implemented in electric power forecasting for decomposing electricity series into series with particular characteristics that can be predicted more accurately than the original time series [186–188].

6. Case Study

In this section, we propose a hybrid model to forecast the electric power by using a type of recurrent artificial neural network known as long short-term memory (LSTM), developed by [189]; we also implemented wavelet decomposition for the data preprocessing (WD-LSTM), as was used in [90]. We use the acronym WD-LSTM for the proposed hybrid model. The results were compared with those of traditional neural network models (LSTM), as was applied in [71,177] and with results of the lagged regression analysis as in [96].

The performance of this methods is demonstrated with a case study using an actual dataset collected from Chile (Table 5). The objective is to illustrated the approach that allows the electric power demand forecasting, in terms of its lagged values, identifying the type of model that tends to show better forecast accuracy.

Maximum daily and hourly electric power demand data over a diverse period were used (Table 5). Figure 9 shows that there is a regularity in electric power demand data. We observe a clear pattern based on the year and day of the week. The electric power demand also follows a group of patterns within any day and depending on the time of the day.

Table 5. Electric power demand object of forecasting.

Type		Variable	Date	Set Size		
				Training	Validation	Test
Local	$Dmax$	Maximum daily electricity demand (MW).	2006–2019	2475	1516	1062
Energy	Hed	Hourly electricity demand (MW).	2016–2020	865	371	530

The values of four performance evaluation indicators—RMSE: root mean square error, MAE: mean absolute error, R^2: coefficient of determination, and MAPE: mean absolute percentage error—showed that the hybrid deep learning model (WD-LSTM) exhibits superior performance in both forecasting accuracy and stability.

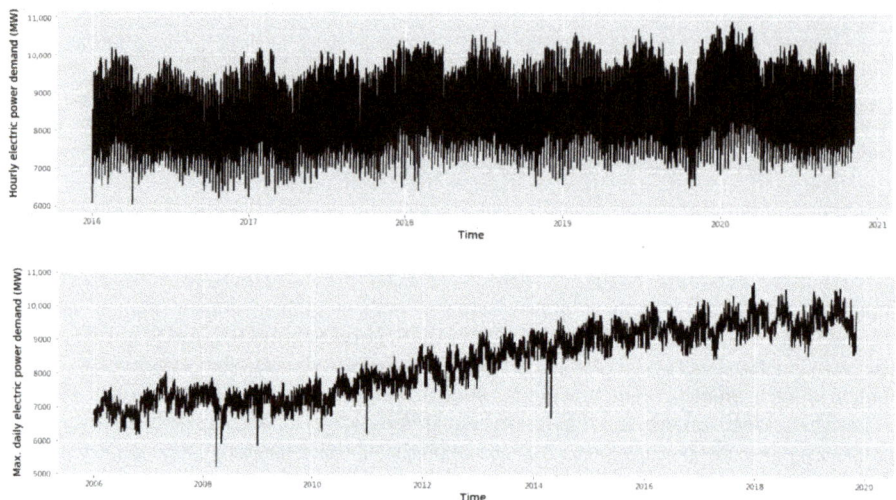

Figure 9. Electric power demand series (MW) in Chile.

Figures 10 and 11 provide the comparative hourly and daily-ahead performance results for three types of days (weekday, weekend, and all days). They likewise provide the performance evaluation results of regression, LSTM, and WD-LSTM applied to each dataset (training, validation, and test). The hybrid deep learning model (WD-LSTM) had the best performance of all forecasting models. The WD-LSTM method generated forecasting results with the lowest MAE, MAPE, and RMSE and with the higher R^2 in most cases. The results further reveal the robustness of the hybrid deep learning model. The superior accuracy of the hybrid model is primarily due to the deep learning framework comprising between two and four independent LSTM networks, which provide an effective means to approximate inherent invariant features and structures. In addition, the low- and high-frequency components exhibited in the electric power datasets can be better extracted by wavelet decomposition. Likewise, each LSTM network managed to focus more on capturing the linear and nonlinear relationships in the energy series, which could not be done with the lagged regression, at least in non-linear cases.

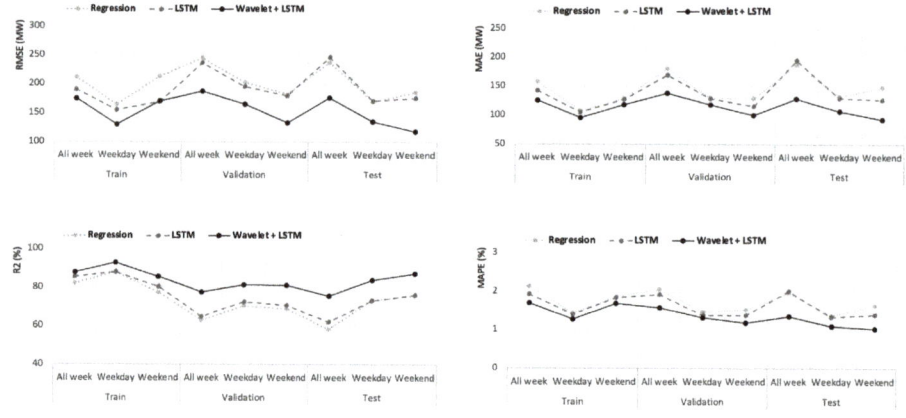

Figure 10. Performance evaluations of different methods for each type of day.

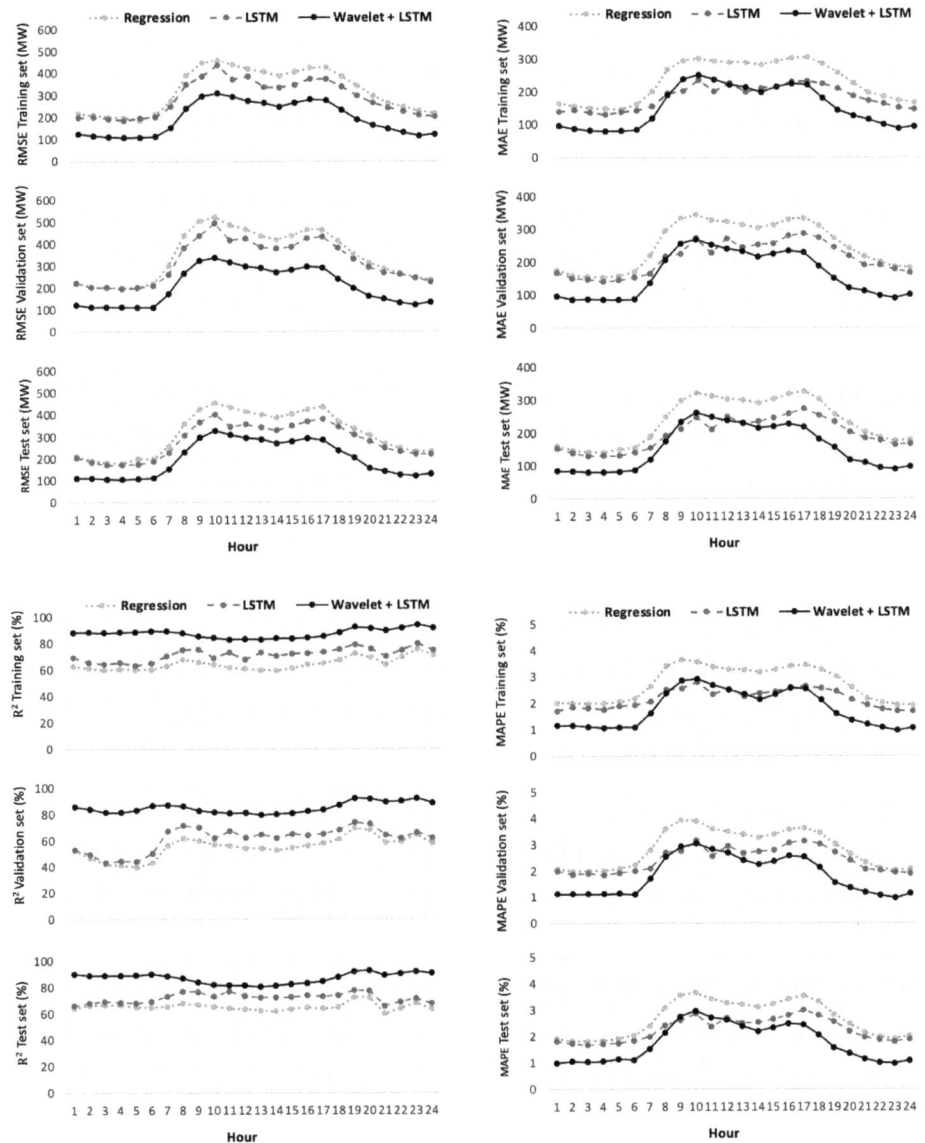

Figure 11. Performance evaluations of different methods for hour.

7. Discussion and Conclusions

This paper presented a systematic review of the forecasting models for electric power from the last 15 years based on ML and MSC techniques. We presented an in-depth analysis of the performance of electric power forecasting models and compared different forecasting models based on their MAPE values. A rigorous framework for comparing different classes of models was introduced, thus generating a reliable picture of the state-of-the-art models' accuracy of electric power forecasting. We were able to identify that a large number of techniques are being used and are aimed at forecasting electrical energy; the techniques with the greatest use are in the fields of ML and ANNs, followed by those that implement algorithms with fuzzy logic and RNNs, while in the MSC area, the use of ARIMA models and regression analysis predominates.

The results can be stratified from three perspectives. The forecasting models (I) from the hybrid class, (II) of multivariate dependency, and (III) based on the machine learning approach demonstrate the best performance for electric power forecasting. Regarding the hybrid models, it is highlighted that 24% of the adjustments with the greatest forecasting precision merged wavelet theory into their models. With regard to multivariate models, we were able to identify that those models that incorporate various sources of variability in their adjustment tend to have, on average, greater precision in their forecasts.

A case of study was presented, in which the implementation of MSC and ML models was compared; we found that the linear models, such as lagged regression, are relatively simple and cannot capture with precision the inherent nonlinear structure of the electric power time series, whereas the deep learning models implemented have a better performance.

Likewise, it was observed that when decomposing the series according to the type of day of electricity consumption (workday or weekend), the models tend to have better forecast accuracy and, in the same way, forecasting errors are minimized by reducing the time horizon (hourly).

Due to electric power systems' participation in the growing trend of environmental optimization around the world, a substantial increase in the contribution of diverse sources to the energy generation is observed. This trend brings about challenges in terms of electric power generation and distribution system operation, because the dimension and complexity of such advances, among other aspects, require the use of a computational intelligence systems that act as sources of data and deal with the control, management, and trading needs at the distribution level in an efficient and robust manner. In this sense, further research could deepen the understanding of the relationship between the type of energy, climate, preprocessing techniques, and performance of machine learning models under various normalized metrics of residuals.

Author Contributions: Conceptualization, E.V., H.A.-C. and R.S.; methodology, E.V., H.A.-C. and R.S.; software, E.V. and H.A.-C.; validation, all authors; writing—original draft preparation, E.V.; writing—review and editing, all authors; supervision, H.A.-C. All authors have read and agreed to the published version of the manuscript.

Funding: This research received internal funding form the Pontificia Universidad Católica de Valparaíso (PUCV).

Acknowledgments: We are grateful to the School of Informatic Engineering at Pontificia Universidad Católica de Valparaíso (PUCV) of Chile for the Scholarship Programme under which the first author is funded for her Doctoral study. Héctor Allende-Cid's work was funded by project 039.457/2020 from VRIEA PUCV.

Conflicts of Interest: The authors declare no conflict of interest. The funders had no role in the design of the study; in the collection, analyses, or interpretation of data; in the writing of the manuscript; or in the decision to publish the results.

Abbreviations

The following abbreviations are used in this manuscript:

Nomenclature

Artificial bee colony	ABC	Gray model	GM
Ant Colony Optimization	ACO	Gross domestic product	GDP
Adaptive Neuro Fuzzy Inference System	ANFIS	Hybrid Monte Carlo	HCM
Artificial Neural Network	ANN	Humidity	HM
Autoregressive Integrated Moving Aevrage	ARIMA	Horizontal Radiation	HR
Bayesian Clustering by Dynamics	BCD	Holt Winters	HW
Bayesian neural network	BNN	Jaya optimization algorithm	JOA
Biogeography based optimization	BOA	K-nearest neighbors	KNN
Back propagation	BP	Long short term memory	LSTM
Calendar information	CI	Number of subscribers	NS
Convolution Neural Network	CNN	Numerical Weather Prediction	NWP

Cuckoo Search Algorithm	COA	Principal component analysis	PCA
Consumer Price Index	CPI	Population	POP
Deep Belief Network	DBN	Air pressure	PRS
Dew point	DP	Particle Swarm Optimization	PSO
Evolutionary Algorithm	EA	Radial basis function network	RBF
Extreme learning machine	ELM	Rainfall	RFL
Electricity price	EP	Rainy time	RT
Elman Recurrent Neural Network	ERNN	Recurrent Neural Network	RNN
Exponential smoothing	ES	Regression Analysis	RA
Exports	EXP	Support Vector Regression	SVR
Fuzzy Neural Network	FNN	Temperature	TM
Gaussian Process	GP	Wind direction	WDD
Genetic algorithm	GA	Wind speed	WS
Generalized Additive Model	GAM	Wavelet theory	WT

Appendix A

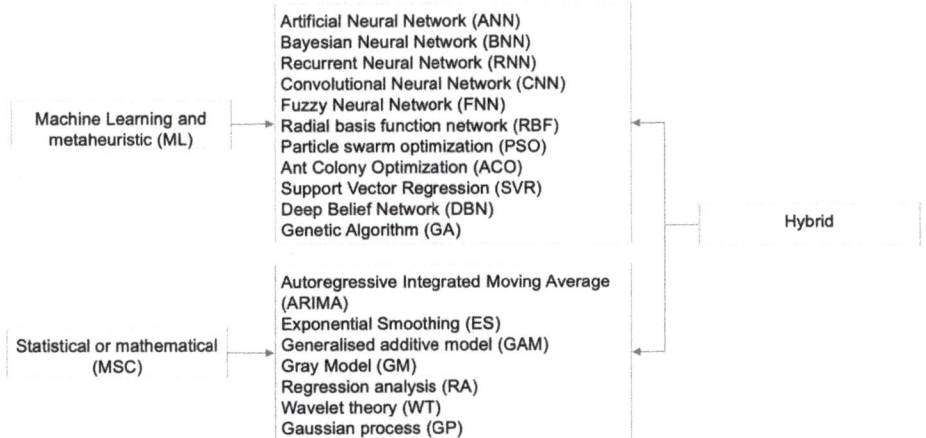

Figure A1. Classification of the forecasting models.

References

1. Taylor, J.W.; De Menezes, L.M.; McSharry, P.E. A comparison of univariate methods for forecasting electricity demand up to a day ahead. *Int. J. Forecast.* **2006**, *22*, 1–16. [CrossRef]
2. Banos, R.; Manzano-Agugliaro, F.; Montoya, F.; Gil, C.; Alcayde, A.; Gómez, J. Optimization methods applied to renewable and sustainable energy: A review. *Renew. Sustain. Energy Rev.* **2011**, *15*, 1753–1766. [CrossRef]
3. Sharma, A.; Kakkar, A. Forecasting daily global solar irradiance generation using machine learning. *Renew. Sustain. Energy Rev.* **2018**, *82*, 2254–2269. [CrossRef]
4. Li, J.; Ward, J.K.; Tong, J.; Collins, L.; Platt, G. Machine learning for solar irradiance forecasting of photovoltaic system. *Renew. Energy* **2016**, *90*, 542–553. [CrossRef]
5. Rahman, M.N.; Esmailpour, A.; Zhao, J. Machine learning with big data an efficient electricity generation forecasting system. *Big Data Res.* **2016**, *5*, 9–15. [CrossRef]
6. Al-Musaylh, M.S.; Deo, R.C.; Adamowski, J.F.; Li, Y. Short-term electricity demand forecasting using machine learning methods enriched with ground-based climate and ECMWF Reanalysis atmospheric predictors in southeast Queensland, Australia. *Renew. Sustain. Energy Rev.* **2019**, *113*, 109293. [CrossRef]
7. Kim, S.G.; Jung, J.Y.; Sim, M.K. A Two-Step Approach to Solar Power Generation Prediction Based on Weather Data Using Machine Learning. *Sustainability* **2019**, *11*, 1501. [CrossRef]

8. Abedinia, O.; Raisz, D.; Amjady, N. Effective prediction model for Hungarian small-scale solar power output. *IET Renew. Power Gener.* **2017**, *11*, 1648–1658. [CrossRef]
9. Abuella, M.; Chowdhury, B. Improving combined solar power forecasts using estimated ramp rates: Data-driven post-processing approach. *IET Renew. Power Gener.* **2018**, *12*, 1127–1135. [CrossRef]
10. Yang, Y.; Chen, Y.; Wang, Y.; Li, C.; Li, L. Modelling a combined method based on ANFIS and neural network improved by DE algorithm: A case study for short-term electricity demand forecasting. *Appl. Soft Comput.* **2016**, *49*, 663–675. [CrossRef]
11. Fiot, J.B.; Dinuzzo, F. Electricity demand forecasting by multi-task learning. *IEEE Trans. Smart Grid* **2016**, *9*, 544–551. [CrossRef]
12. Wang, F.; Yu, Y.; Zhang, Z.; Li, J.; Zhen, Z.; Li, K. Wavelet decomposition and convolutional LSTM networks based improved deep learning model for solar irradiance forecasting. *Appl. Sci.* **2018**, *8*, 1286. [CrossRef]
13. Liu, Y.; Guan, L.; Hou, C.; Han, H.; Liu, Z.; Sun, Y.; Zheng, M. Wind Power Short-Term Prediction Based on LSTM and Discrete Wavelet Transform. *Appl. Sci.* **2019**, *9*, 1108. [CrossRef]
14. Jin, X.; Dong, Y.; Wu, J.; Wang, J. An improved combined forecasting method for electric power load based on autoregressive integrated moving average model. In Proceedings of the 2010 International Conference of Information Science and Management Engineering, Xi'an, China, 7–8 August 2010; Volume 2, pp. 476–480.
15. Sarkodie, S.A. Estimating Ghana's electricity consumption by 2030: An ARIMA forecast. *Energy Sources Part B Econ. Plan. Policy* **2017**, *12*, 936–944. [CrossRef]
16. Barzola-Monteses, J.; Mite-León, M.; Espinoza-Andaluz, M.; Gómez-Romero, J.; Fajardo, W. Time Series Analysis for Predicting Hydroelectric Power Production: The Ecuador Case. *Sustainability* **2019**, *11*, 6539. [CrossRef]
17. Filik, U.B.; Gerek, O.N.; Kurban, M. Hourly forecasting of long term electric energy demand using a novel modeling approach. In Proceedings of the 2009 Fourth International Conference on Innovative Computing, Information and Control (ICICIC), Kaohsiung, Taiwan, 7–9 December 2009; pp. 115–118.
18. Filik, Ü.B.; Gerek, Ö.N.; Kurban, M. A novel modeling approach for hourly forecasting of long-term electric energy demand. *Energy Convers. Manag.* **2011**, *52*, 199–211. [CrossRef]
19. Liang, J.; Liang, Y. Analysis and modeling for China's electricity demand forecasting based on a new mathematical hybrid method. *Information* **2017**, *8*, 33. [CrossRef]
20. Khan, A.R.; Mahmood, A.; Safdar, A.; Khan, Z.A.; Khan, N.A. Load forecasting, dynamic pricing and DSM in smart grid: A review. *Renew. Sustain. Energy Rev.* **2016**, *54*, 1311–1322. [CrossRef]
21. Mosavi, A.; Salimi, M.; Faizollahzadeh Ardabili, S.; Rabczuk, T.; Shamshirband, S.; Varkonyi-Koczy, A.R. State of the art of machine learning models in energy systems, a systematic review. *Energies* **2019**, *12*, 1301. [CrossRef]
22. Qasem, S.N.; Samadianfard, S.; Sadri Nahand, H.; Mosavi, A.; Shamshirband, S.; Chau, K.w. Estimating daily dew point temperature using machine learning algorithms. *Water* **2019**, *11*, 582. [CrossRef]
23. Kalogirou, S.A. Applications of artificial neural-networks for energy systems. *Appl. Energy* **2000**, *67*, 17–35. [CrossRef]
24. Voyant, C.; Notton, G.; Kalogirou, S.; Nivet, M.L.; Paoli, C.; Motte, F.; Fouilloy, A. Machine learning methods for solar radiation forecasting: A review. *Renew. Energy* **2017**, *105*, 569–582. [CrossRef]
25. Blaga, R.; Sabadus, A.; Stefu, N.; Dughir, C.; Paulescu, M.; Badescu, V. A current perspective on the accuracy of incoming solar energy forecasting. *Prog. Energy Combust. Sci.* **2019**, *70*, 119–144. [CrossRef]
26. Verdejo, H.; Awerkin, A.; Becker, C.; Olguin, G. Statistic linear parametric techniques for residential electric energy demand forecasting. A review and an implementation to Chile. *Renew. Sustain. Energy Rev.* **2017**, *74*, 512–521. [CrossRef]
27. Chen, J.F.; Do, Q.H.; Nguyen, T.V.A.; Doan, T.T.H. Forecasting monthly electricity demands by wavelet neuro-fuzzy system optimized by heuristic algorithms. *Information* **2018**, *9*, 51. [CrossRef]
28. Lewis, C.D. *Industrial and Business Forecasting Methods: A Practical Guide to Exponential Smoothing and Curve Fitting*; Butterworth-Heinemann: Oxford, UK, 1982.
29. Zhao, H.; Guo, S. An optimized grey model for annual power load forecasting. *Energy* **2016**, *107*, 272–286. [CrossRef]
30. Ma, X.; Liu, Z. Application of a novel time-delayed polynomial grey model to predict the natural gas consumption in China. *J. Comput. Appl. Math.* **2017**, *324*, 17–24. [CrossRef]

31. Lee, Y.S.; Tong, L.I. Forecasting energy consumption using a grey model improved by incorporating genetic programming. *Energy Convers. Manag.* **2011**, *52*, 147–152. [CrossRef]
32. Wang, J.; Du, P.; Lu, H.; Yang, W.; Niu, T. An improved grey model optimized by multi-objective ant lion optimization algorithm for annual electricity consumption forecasting. *Appl. Soft Comput.* **2018**, *72*, 321–337. [CrossRef]
33. Chang, P.C.; Fan, C.Y.; Lin, J.J. Monthly electricity demand forecasting based on a weighted evolving fuzzy neural network approach. *Int. J. Electr. Power Energy Syst.* **2011**, *33*, 17–27. [CrossRef]
34. Dedinec, A.; Filiposka, S.; Dedinec, A.; Kocarev, L. Deep belief network based electricity load forecasting: An analysis of Macedonian case. *Energy* **2016**, *115*, 1688–1700. [CrossRef]
35. Chang, G.; Lu, H.; Chang, Y.; Lee, Y. An improved neural network-based approach for short-term wind speed and power forecast. *Renew. Energy* **2017**, *105*, 301–311. [CrossRef]
36. Kaboli, S.H.A.; Selvaraj, J.; Rahim, N. Long-term electric energy consumption forecasting via artificial cooperative search algorithm. *Energy* **2016**, *115*, 857–871. [CrossRef]
37. El-Telbany, M.; El-Karmi, F. Short-term forecasting of Jordanian electricity demand using particle swarm optimization. *Electr. Power Syst. Res.* **2008**, *78*, 425–433. [CrossRef]
38. Ardakani, F.; Ardehali, M. Long-term electrical energy consumption forecasting for developing and developed economies based on different optimized models and historical data types. *Energy* **2014**, *65*, 452–461. [CrossRef]
39. Tsekouras, G.J.; Hatziargyriou, N.D.; Dialynas, E.N. An optimized adaptive neural network for annual midterm energy forecasting. *IEEE Trans. Power Syst.* **2006**, *21*, 385–391. [CrossRef]
40. López, M.; Valero, S.; Senabre, C.; Aparicio, J.; Gabaldon, A. Application of SOM neural networks to short-term load forecasting: The Spanish electricity market case study. *Electr. Power Syst. Res.* **2012**, *91*, 18–27. [CrossRef]
41. Setiawan, A.; Koprinska, I.; Agelidis, V.G. Very short-term electricity load demand forecasting using support vector regression. In Proceedings of the 2009 International Joint Conference on Neural Networks, Atlanta, GA, USA, 14–19 June 2009; pp. 2888–2894.
42. Al-Qahtani, F.H.; Crone, S.F. Multivariate k-nearest neighbour regression for time series data—A novel algorithm for forecasting UK electricity demand. In Proceedings of the 2013 International Joint Conference on Neural Networks (IJCNN), Dallas, TX, USA, 4–9 August 2013; pp. 1–8.
43. Guo, Z.; Zhou, K.; Zhang, X.; Yang, S. A deep learning model for short-term power load and probability density forecasting. *Energy* **2018**, *160*, 1186–1200. [CrossRef]
44. Ghanbari, A.; Naghavi, A.; Ghaderi, S.; Sabaghian, M. Artificial Neural Networks and regression approaches comparison for forecasting Iran's annual electricity load. In Proceedings of the 2009 International Conference on Power Engineering, Energy and Electrical Drives, Lisbon, Portugal, 18–20 March 2009; pp. 675–679.
45. Elias, C.N.; Hatziargyriou, N.D. An annual midterm energy forecasting model using fuzzy logic. *IEEE Trans. Power Syst.* **2009**, *24*, 469–478. [CrossRef]
46. Sahay, K.B.; Tripathi, M. Day ahead hourly load forecast of PJM electricity market and ISO New England market by using artificial neural network. In Proceedings of the ISGT 2014, Istanbul, Turkey, 15–18 October 2014; pp. 1–5.
47. Wang, J.; Zhou, Y.; Chen, X. Electricity load forecasting based on support vector machines and simulated annealing particle swarm optimization algorithm. In Proceedings of the 2007 IEEE International Conference on Automation and Logistics, Jinan, China, 18–21 August 2007; pp. 2836–2841.
48. Ehsan, R.M.; Simon, S.P.; Venkateswaran, P. Day-ahead forecasting of solar photovoltaic output power using multilayer perceptron. *Neural Comput. Appl.* **2017**, *28*, 3981–3992. [CrossRef]
49. De Leone, R.; Pietrini, M.; Giovannelli, A. Photovoltaic energy production forecast using support vector regression. *Neural Comput. Appl.* **2015**, *26*, 1955–1962. [CrossRef]
50. Vahabie, A.; Yousefi, M.R.; Araabi, B.; Lucas, C.; Barghinia, S.; Ansarimehr, P. Mutual information based input selection in neuro-fuzzy modeling for short term load forecasting of iran national power system. In Proceedings of the 2007 IEEE International Conference on Control and Automation, Guangzhou, China, 30 May–1 June 2007; pp. 2710–2715.
51. Ouyang, T.; He, Y.; Li, H.; Sun, Z.; Baek, S. Modeling and forecasting short-term power load with copula model and deep belief network. *IEEE Trans. Emerg. Top. Comput. Intell.* **2019**, *3*, 127–136. [CrossRef]
52. Barzamini, R.; Menhaj, M.; Khosravi, A.; Kamalvand, S. Short term load forecasting for iran national power system and its regions using multi layer perceptron and fuzzy inference systems. In Proceedings of the 2005

IEEE International Joint Conference on Neural Networks, Montreal, QC, Canada, 31 July–4 August 2005; Volume 4, pp. 2619–2624.
53. Dilhani, M.S.; Jeenanunta, C. Effect of Neural Network structure for daily electricity load forecasting. In Proceedings of the 2017 Moratuwa Engineering Research Conference (MERCon), Moratuwa, Sri Lanka, 29–31 May 2017; pp. 419–424.
54. Chen, J.F.; Lo, S.K.; Do, Q.H. Forecasting monthly electricity demands: An application of neural networks trained by heuristic algorithms. *Information* **2017**, *8*, 31. [CrossRef]
55. Azadeh, A.; Ghaderi, S.F.; Sheikhalishahi, M.; Nokhandan, B.P. Optimization of short load forecasting in electricity market of Iran using artificial neural networks. *Optim. Eng.* **2014**, *15*, 485–508. [CrossRef]
56. Raza, M.Q.; Baharudin, Z.; Nallagownden, P. A comparative analysis of PSO and LM based NN short term load forecast with exogenous variables for smart power generation. In Proceedings of the 2014 5th International Conference on Intelligent and Advanced Systems (ICIAS), Kuala Lumpur, Malaysia, 3–5 June 2014; pp. 1–6.
57. Ishik, M.Y.; Göze, T.; Özcan, İ.; Güngör, V.Ç.; Aydın, Z. Short term electricity load forecasting: A case study of electric utility market in Turkey. In Proceedings of the 2015 3rd International Istanbul Smart Grid Congress and Fair (ICSG), Istanbul, Turkey, 29–30 April 2015; pp. 1–5.
58. Sahay, K.B.; Kumar, N.; Tripathi, M. Short-term load forecasting of Ontario Electricity Market by considering the effect of temperature. In Proceedings of the 2014 6th IEEE Power India International Conference (PIICON), Delhi, India, 5–7 December 2014; pp. 1–6.
59. Azadeh, A.; Ghadrei, S.; Nokhandan, B.P. One day ahead load forecasting for electricity market of Iran by ANN. In Proceedings of the 2009 International Conference on Power Engineering, Energy and Electrical Drives, Lisbon, Portugal, 18–20 March 2009; pp. 670–674.
60. Akole, M.; Tyagi, B. Artificial neural network based short term load forecasting for restructured power system. In Proceedings of the 2009 International Conference on Power Systems, Kharagpur, India, 27–29 December 2009; pp. 1–7.
61. Chuang, F.K.; Hung, C.Y.; Chang, C.Y.; Kuo, K.C. Deploying arima and artificial neural networks models to predict energy consumption in Taiwan. *Sens. Lett.* **2013**, *11*, 2333–2340. [CrossRef]
62. Kavanagh, K.; Barrett, M.; Conlon, M. Short-term electricity load forecasting for the integrated single electricity market (I-SEM). In Proceedings of the 2017 52nd International Universities Power Engineering Conference (UPEC), Heraklion, Greece, 28–31 August 2017; pp. 1–7.
63. Sarkar, M.R.; Rabbani, M.G.; Khan, A.R.; Hossain, M.M. Electricity demand forecasting of Rajshahi City in Bangladesh using fuzzy linear regression model. In Proceedings of the 2015 International Conference on Electrical Engineering and Information Communication Technology (ICEEICT), Savar, Dhaka, Bangladesh, 21–23 May 2015; pp. 1–3.
64. Kelo, S.M.; Dudul, S.V. Development of an intelligent system for short-term electrical power load forecasting in Maharashtra state. In Proceedings of the 2008 Joint International Conference on Power System Technology and IEEE Power India Conference, New Delhi, India, 12–15 October 2008; pp. 1–8.
65. Gireeshma, K.; Atla, C.R.; Rao, K.L. New Correlation Technique for RE Power Forecasting using Neural Networks. In Proceedings of the 2019 Fifth International Conference on Electrical Energy Systems (ICEES), Chennai, India, 21–22 February 2019; pp. 1–6.
66. Barzamini, R.; Menhaj, M.; Kamalvand, S.; Tajbakhsh, A. Short term load forecasting of Iran national power system using artificial neural network generation two. In Proceedings of the 2005 IEEE Russia Power Tech, St. Petersburg, Russia, 27–30 June 2005; pp. 1–5.
67. Viet, D.T.; Van Phuong, V.; Duong, M.Q.; Kies, A.; Schyska, B.U.; Wu, Y.K. A Short-Term Wind Power Forecasting Tool for Vietnamese Wind Farms and Electricity Market. In Proceedings of the 2018 4th International Conference on Green Technology and Sustainable Development (GTSD), Ho Chi Minh City, Vietnam, 23–24 November 2018; pp. 130–135.
68. Sakunthala, K.; Iniyan, S.; Mahalingam, S. Forecasting energy consumption in Tamil Nadu using hybrid heuristic based regression model. *Therm. Sci.* **2019**, *23*, 2885–2894. [CrossRef]
69. Siridhipakul, C.; Vateekul, P. Multi-step Power Consumption Forecasting in Thailand Using Dual-Stage Attentional LSTM. In Proceedings of the 2019 11th International Conference on Information Technology and Electrical Engineering (ICITEE), Pattaya, Thailand, 10–11 October 2019; pp. 1–6.

70. Tanidir, Ö.; Tör, O. Accuracy of ANN based day-ahead load forecasting in Turkish power system: Degrading and improving factors. *Neural Netw. World* **2015**, *25*, 443. [CrossRef]
71. Liu, J.; Wang, X.; Zhao, Y.; Dong, B.; Lu, K.; Wang, R. Heating Load Forecasting for Combined Heat and Power Plants Via Strand-Based LSTM. *IEEE Access* **2020**, *8*, 33360–33369. [CrossRef]
72. Chen, S.; Huang, J. Forecasting China's primary energy demand based on an improved AI model. *Chin. J. Popul. Resour. Environ.* **2018**, *16*, 36–48. [CrossRef]
73. Benaouda, D.; Murtagh, F.; Starck, J.L.; Renaud, O. Wavelet-based nonlinear multiscale decomposition model for electricity load forecasting. *Neurocomputing* **2006**, *70*, 139–154. [CrossRef]
74. Rahmani, R.; Yusof, R.; Seyedmahmoudian, M.; Mekhilef, S. Hybrid technique of ant colony and particle swarm optimization for short term wind energy forecasting. *J. Wind. Eng. Ind. Aerodyn.* **2013**, *123*, 163–170. [CrossRef]
75. Amjady, N.; Keynia, F. Mid-term load forecasting of power systems by a new prediction method. *Energy Convers. Manag.* **2008**, *49*, 2678–2687. [CrossRef]
76. Fan, S.; Chen, L.; Lee, W.J. Machine learning based switching model for electricity load forecasting. *Energy Convers. Manag.* **2008**, *49*, 1331–1344. [CrossRef]
77. Xiao, L.; Wang, J.; Yang, X.; Xiao, L. A hybrid model based on data preprocessing for electrical power forecasting. *Int. J. Electr. Power Energy Syst.* **2015**, *64*, 311–327. [CrossRef]
78. Raza, M.Q.; Nadarajah, M.; Hung, D.Q.; Baharudin, Z. An intelligent hybrid short-term load forecasting model for smart power grids. *Sustain. Cities Soc.* **2017**, *31*, 264–275. [CrossRef]
79. Laouafi, A.; Mordjaoui, M.; Laouafi, F.; Boukelia, T.E. Daily peak electricity demand forecasting based on an adaptive hybrid two-stage methodology. *Int. J. Electr. Power Energy Syst.* **2016**, *77*, 136–144. [CrossRef]
80. Ghanbari, A.; Ghaderi, S.F.; Azadeh, M.A. Adaptive Neuro-Fuzzy Inference System vs. Regression based approaches for annual electricity load forecasting. In Proceedings of the 2010 2nd International Conference on Computer and Automation Engineering (ICCAE), Singapore, 26–28 February 2010; Volume 5, pp. 26–30.
81. Brodowski, S.; Bielecki, A.; Filocha, M. A hybrid system for forecasting 24-h power load profile for Polish electric grid. *Appl. Soft Comput.* **2017**, *58*, 527–539. [CrossRef]
82. Anand, A.; Suganthi, L. Hybrid GA-PSO optimization of artificial neural network for forecasting electricity demand. *Energies* **2018**, *11*, 728. [CrossRef]
83. Yeom, C.U.; Kwak, K.C. Short-term electricity-load forecasting using a TSK-based extreme learning machine with knowledge representation. *Energies* **2017**, *10*, 1613. [CrossRef]
84. Mujeeb, S.; Alghamdi, T.A.; Ullah, S.; Fatima, A.; Javaid, N.; Saba, T. Exploiting Deep Learning for Wind Power Forecasting Based on Big Data Analytics. *Appl. Sci.* **2019**, *9*, 4417. [CrossRef]
85. Barghinia, S.; Kamankesh, S.; Mahdavi, N.; Vahabie, A.; Gorji, A. A combination method for short term load forecasting used in Iran electricity market by NeuroFuzzy, Bayesian and finding similar days methods. In Proceedings of the 2008 5th International Conference on the European Electricity Market, Lisbon, Portugal, 28–30 May 2008; pp. 1–6.
86. Singh, P.; Mishra, K.; Dwivedi, P. Enhanced hybrid model for electricity load forecast through artificial neural network and Jaya algorithm. In Proceedings of the 2017 International Conference on Intelligent Computing and Control Systems (ICICCS), Madurai, India, 15–16 June 2017; pp. 115–120.
87. Kumaran, J.; Ravi, G. Long-term sector-wise electrical energy forecasting using artificial neural network and biogeography-based optimization. *Electr. Power Compon. Syst.* **2015**, *43*, 1225–1235. [CrossRef]
88. Zhou, J.; Xu, X.; Huo, X.; Li, Y. Forecasting models for wind power using extreme-point symmetric mode decomposition and artificial neural networks. *Sustainability* **2019**, *11*, 650. [CrossRef]
89. Kasule, A.; Ayan, K. Forecasting Uganda's net electricity consumption using a hybrid PSO-ABC Algorithm. *Arab. J. Sci. Eng.* **2019**, *44*, 3021–3031. [CrossRef]
90. Li, P.; Zhou, K.; Lu, X.; Yang, S. A hybrid deep learning model for short-term PV power forecasting. *Appl. Energy* **2020**, *259*, 114216. [CrossRef]
91. Kaysal, A.; Köroglu, S.; Oguz, Y.; Kaysal, K. Artificial Neural Networks and Adaptive Neuro-Fuzzy Inference Systems Approaches to Forecast the Electricity Data for Load Demand, an Analysis of Dinar District Case. In Proceedings of the 2018 2nd International Symposium on Multidisciplinary Studies and Innovative Technologies (ISMSIT), Ankara, Turkey, 19–21 October 2018; pp. 1–6.
92. Soares, L.J.; Medeiros, M.C. Modeling and forecasting short-term electricity load: A comparison of methods with an application to Brazilian data. *Int. J. Forecast.* **2008**, *24*, 630–644. [CrossRef]

93. Almeshaiei, E.; Soltan, H. A methodology for electric power load forecasting. *Alex. Eng. J.* **2011**, *50*, 137–144. [CrossRef]
94. Hamzacebi, C.; Es, H.A. Forecasting the annual electricity consumption of Turkey using an optimized grey model. *Energy* **2014**, *70*, 165–171. [CrossRef]
95. Zhao, W.; Wang, J.; Lu, H. Combining forecasts of electricity consumption in China with time-varying weights updated by a high-order Markov chain model. *Omega* **2014**, *45*, 80–91. [CrossRef]
96. Wang, C.h.; Grozev, G.; Seo, S. Decomposition and statistical analysis for regional electricity demand forecasting. *Energy* **2012**, *41*, 313–325. [CrossRef]
97. Yukseltan, E.; Yucekaya, A.; Bilge, A.H. Forecasting electricity demand for Turkey: Modeling periodic variations and demand segregation. *Appl. Energy* **2017**, *193*, 287–296. [CrossRef]
98. Koprinska, I.; Rana, M.; Agelidis, V.G. Yearly and seasonal models for electricity load forecasting. In Proceedings of the 2011 International Joint Conference on Neural Networks, San Jose, CA, USA, 31 July–5 August 2011; pp. 1474–1481.
99. Rallapalli, S.R.; Ghosh, S. Forecasting monthly peak demand of electricity in India—A critique. *Energy Policy* **2012**, *45*, 516–520. [CrossRef]
100. Damrongkulkamjorn, P.; Churueang, P. Monthly energy forecasting using decomposition method with application of seasonal ARIMA. In Proceedings of the 2005 International Power Engineering Conference, Singapore, 29 November–2 December 2005; pp. 1–229.
101. Alamaniotis, M.; Bargiotas, D.; Tsoukalas, L.H. Towards smart energy systems: Application of kernel machine regression for medium term electricity load forecasting. *SpringerPlus* **2016**, *5*, 58. [CrossRef] [PubMed]
102. Koprinska, I.; Sood, R.; Agelidis, V. Variable selection for five-minute ahead electricity load forecasting. In Proceedings of the 2010 20th International Conference on Pattern Recognition, Istanbul, Turkey, 23–26 August 2010; pp. 2901–2904.
103. He, H.; Liu, T.; Chen, R.; Xiao, Y.; Yang, J. High frequency short-term demand forecasting model for distribution power grid based on ARIMA. In Proceedings of the 2012 IEEE International Conference on Computer Science and Automation Engineering (CSAE), Zhangjiajie, China, 25–27 May 2012; Volume 3, pp. 293–297.
104. An, Y.; Zhou, Y.; Li, R. Forecasting India's Electricity Demand Using a Range of Probabilistic Methods. *Energies* **2019**, *12*, 2574. [CrossRef]
105. Kantanantha, N.; Runsewa, S. Forecasting of electricity demand to reduce the inventory cost of imported coal. In Proceedings of the 2017 4th International Conference on Industrial Engineering and Applications (ICIEA), Nagoya, Japan, 21–23 April 2017; pp. 336–340.
106. Chen, H.; Li, F.; Wang, Y. Component GARCH-M type models for wind power forecasting. In Proceedings of the 2015 IEEE Power & Energy Society General Meeting, Denver, CO, USA, 26–30 July 2015; pp. 1–5.
107. Kusakci, A.O.; Ayvaz, B. Electrical energy consumption forecasting for Turkey using grey forecasting technics with rolling mechanism. In Proceedings of the 2015 2nd International Conference on Knowledge-Based Engineering and Innovation (KBEI), Tehran, Iran, 5–6 November 2015; pp. 8–13.
108. González-Romera, E.; Jaramillo-Morán, M.; Carmona-Fernández, D. Monthly electric energy demand forecasting with neural networks and Fourier series. *Energy Convers. Manag.* **2008**, *49*, 3135–3142. [CrossRef]
109. Cao, G.; Wu, L. Support vector regression with fruit fly optimization algorithm for seasonal electricity consumption forecasting. *Energy* **2016**, *115*, 734–745. [CrossRef]
110. Li, C.; Li, S.; Liu, Y. A least squares support vector machine model optimized by moth-flame optimization algorithm for annual power load forecasting. *Appl. Intell.* **2016**, *45*, 1166–1178. [CrossRef]
111. Koprinska, I.; Rana, M.; Troncoso, A.; Martínez-Álvarez, F. Combining pattern sequence similarity with neural networks for forecasting electricity demand time series. In Proceedings of the 2013 International Joint Conference on Neural Networks (IJCNN), Dallas, TX, USA, 4–9 August 2013; pp. 1–8.
112. Al-Musaylh, M.S.; Deo, R.C.; Li, Y.; Adamowski, J.F. Two-phase particle swarm optimized-support vector regression hybrid model integrated with improved empirical mode decomposition with adaptive noise for multiple-horizon electricity demand forecasting. *Appl. Energy* **2018**, *217*, 422–439. [CrossRef]
113. Puspitasari, I.; Akbar, M.S.; Lee, M.H. Two-level seasonal model based on hybrid ARIMA-ANFIS for forecasting short-term electricity load in Indonesia. In Proceedings of the 2012 International Conference on Statistics in Science, Business and Engineering (ICSSBE), Langkawi, Kedah, Malaysia, 10–12 September 2012; pp. 1–5.

114. Li, W.; Yang, X.; Li, H.; Su, L. Hybrid forecasting approach based on GRNN neural network and SVR machine for electricity demand forecasting. *Energies* **2017**, *10*, 44. [CrossRef]
115. Benaouda, D.; Murtagh, F. Electricity load forecast using neural network trained from wavelet-transformed data. In Proceedings of the 2006 IEEE International Conference on Engineering of Intelligent Systems, Islamabad, Pakistan, 22–23 April 2006; pp. 1–6.
116. Saroha, S.; Aggarwal, S. Wind power forecasting using wavelet transforms and neural networks with tapped delay. *CSEE J. Power Energy Syst.* **2018**, *4*, 197–209. [CrossRef]
117. Ni, K.; Wang, J.; Tang, G.; Wei, D. Research and Application of a Novel Hybrid Model Based on a Deep Neural Network for Electricity Load Forecasting: A Case Study in Australia. *Energies* **2019**, *12*, 2467. [CrossRef]
118. Jadidi, A.; Menezes, R.; de Souza, N.; de Castro Lima, A.C. Short-Term Electric Power Demand Forecasting Using NSGA II-ANFIS Model. *Energies* **2019**, *12*, 1891. [CrossRef]
119. Panapakidis, I.P.; Christoforidis, G.C.; Asimopoulos, N.; Dagoumas, A.S. Combining wavelet transform and support vector regression model for day-ahead peak load forecasting in the Greek power system. In Proceedings of the 2017 IEEE International Conference on Environment and Electrical Engineering and 2017 IEEE Industrial and Commercial Power Systems Europe (EEEIC/I&CPS Europe), Milan, Italy, 6–9 June 2017; pp. 1–6.
120. Jiang, W.; Wu, X.; Gong, Y.; Yu, W.; Zhong, X. Holt–Winters smoothing enhanced by fruit fly optimization algorithm to forecast monthly electricity consumption. *Energy* **2020**, *193*, 116779. [CrossRef]
121. Sulandari, W.; Lee, M.H.; Rodrigues, P.C. Indonesian electricity load forecasting using singular spectrum analysis, fuzzy systems and neural networks. *Energy* **2020**, *190*, 116408. [CrossRef]
122. Laouafi, A.; Mordjaoui, M.; Medoued, A.; Boukelia, T.E.; Ganouche, A. Wind power forecasting approach using neuro-fuzzy system combined with wavelet packet decomposition, data preprocessing, and forecast combination framework. *Wind Eng.* **2017**, *41*, 235–244. [CrossRef]
123. Pao, H.T. Forecast of electricity consumption and economic growth in Taiwan by state space modeling. *Energy* **2009**, *34*, 1779–1791. [CrossRef]
124. Vu, D.H.; Muttaqi, K.M.; Agalgaonkar, A. A variance inflation factor and backward elimination based robust regression model for forecasting monthly electricity demand using climatic variables. *Appl. Energy* **2015**, *140*, 385–394. [CrossRef]
125. Tsekouras, G.; Dialynas, E.; Hatziargyriou, N.; Kavatza, S. A non-linear multivariable regression model for midterm energy forecasting of power systems. *Electr. Power Syst. Res.* **2007**, *77*, 1560–1568. [CrossRef]
126. Clements, A.E.; Hurn, A.; Li, Z. Forecasting day-ahead electricity load using a multiple equation time series approach. *Eur. J. Oper. Res.* **2016**, *251*, 522–530. [CrossRef]
127. Kim, M.S. Modeling special-day effects for forecasting intraday electricity demand. *Eur. J. Oper. Res.* **2013**, *230*, 170–180. [CrossRef]
128. Shang, H.L. Functional time series approach for forecasting very short-term electricity demand. *J. Appl. Stat.* **2013**, *40*, 152–168. [CrossRef]
129. Bessec, M.; Fouquau, J. Short-run electricity load forecasting with combinations of stationary wavelet transforms. *Eur. J. Oper. Res.* **2018**, *264*, 149–164. [CrossRef]
130. Chui, F.; Elkamel, A.; Surit, R.; Croiset, E.; Douglas, P. Long-term electricity demand forecasting for power system planning using economic, demographic and climatic variables. *Eur. J. Ind. Eng.* **2009**, *3*, 277–304. [CrossRef]
131. Truong, N.V.; Wang, L.; Wong, P.K. Modelling and short-term forecasting of daily peak power demand in Victoria using two-dimensional wavelet based SDP models. *Int. J. Electr. Power Energy Syst.* **2008**, *30*, 511–518. [CrossRef]
132. Mestekemper, T.; Kauermann, G.; Smith, M.S. A comparison of periodic autoregressive and dynamic factor models in intraday energy demand forecasting. *Int. J. Forecast.* **2013**, *29*, 1–12. [CrossRef]
133. Massidda, L.; Marrocu, M. Use of Multilinear Adaptive Regression Splines and numerical weather prediction to forecast the power output of a PV plant in Borkum, Germany. *Sol. Energy* **2017**, *146*, 141–149. [CrossRef]
134. Duan, L.; Niu, D.; Gu, Z. Long and medium term power load forecasting with multi-level recursive regression analysis. In Proceedings of the 2008 Second International Symposium on Intelligent Information Technology Application, Washington, DC, USA, 20–22 December 2008; Volume 1, pp. 514–518.

135. Guo, H.; Chen, Q.; Xia, Q.; Kang, C.; Zhang, X. A monthly electricity consumption forecasting method based on vector error correction model and self-adaptive screening method. *Int. J. Electr. Power Energy Syst.* **2018**, *95*, 427–439. [CrossRef]
136. Elamin, N.; Fukushige, M. Modeling and forecasting hourly electricity demand by SARIMAX with interactions. *Energy* **2018**, *165*, 257–268. [CrossRef]
137. Sigauke, C. Forecasting medium-term electricity demand in a South African electric power supply system. *J. Energy S. Afr.* **2017**, *28*, 54–67. [CrossRef]
138. Tanrisever, F.; Derinkuyu, K.; Heeren, M. Forecasting electricity infeed for distribution system networks: An analysis of the Dutch case. *Energy* **2013**, *58*, 247–257. [CrossRef]
139. Bernardi, M.; Petrella, L. Multiple seasonal cycles forecasting model: The Italian electricity demand. *Stat. Methods Appl.* **2015**, *24*, 671–695. [CrossRef]
140. Angelopoulos, D.; Psarras, J.; Siskos, Y. Long-term electricity demand forecasting via ordinal regression analysis: The case of Greece. In Proceedings of the 2017 IEEE Manchester PowerTech, Manchester, UK, 18–22 June 2017; pp. 1–6.
141. Bermúdez, J.D. Exponential smoothing with covariates applied to electricity demand forecast. *Eur. J. Ind. Eng.* **2013**, *7*, 333–349. [CrossRef]
142. Sigauke, C.; Chikobvu, D. Peak electricity demand forecasting using time series regression models: An application to South African data. *J. Stat. Manag. Syst.* **2016**, *19*, 567–586. [CrossRef]
143. Angelopoulos, D.; Siskos, Y.; Psarras, J. Disaggregating time series on multiple criteria for robust forecasting: The case of long-term electricity demand in Greece. *Eur. J. Oper. Res.* **2019**, *275*, 252–265. [CrossRef]
144. Kumru, M.; Kumru, P.Y. Calendar-based short-term forecasting of daily average electricity demand. In Proceedings of the 2015 International Conference on Industrial Engineering and Operations Management (IEOM), Hyatt Regency, Dubai, 3–5 March 2015; pp. 1–5.
145. He, Y.; Zheng, Y.; Xu, Q. Forecasting energy consumption in Anhui province of China through two Box-Cox transformation quantile regression probability density methods. *Measurement* **2019**, *136*, 579–593. [CrossRef]
146. İlseven, E.; Göl, M. Medium-term electricity demand forecasting based on MARS. In Proceedings of the 2017 IEEE PES Innovative Smart Grid Technologies Conference Europe (ISGT-Europe), Torino, Italy, 26–29 September 2017; pp. 1–6.
147. Lebotsa, M.E.; Sigauke, C.; Bere, A.; Fildes, R.; Boylan, J.E. Short term electricity demand forecasting using partially linear additive quantile regression with an application to the unit commitment problem. *Appl. Energy* **2018**, *222*, 104–118. [CrossRef]
148. Cho, H.; Goude, Y.; Brossat, X.; Yao, Q. Modeling and forecasting daily electricity load curves: A hybrid approach. *J. Am. Stat. Assoc.* **2013**, *108*, 7–21. [CrossRef]
149. Sigauke, C.; Chikobvu, D. Prediction of daily peak electricity demand in South Africa using volatility forecasting models. *Energy Econ.* **2011**, *33*, 882–888. [CrossRef]
150. Li, S.; Yang, X.; Li, R. Forecasting China's coal power installed capacity: A comparison of MGM, ARIMA, GM-ARIMA, and NMGM models. *Sustainability* **2018**, *10*, 506. [CrossRef]
151. He, Y.; Zheng, Y. Short-term power load probability density forecasting based on Yeo-Johnson transformation quantile regression and Gaussian kernel function. *Energy* **2018**, *154*, 143–156. [CrossRef]
152. Cheng, C.T.; Miao, S.M.; Luo, B.; Sun, Y.J. Forecasting monthly energy production of small hydropower plants in ungauged basins using grey model and improved seasonal index. *J. Hydroinform.* **2017**, *19*, 993–1008. [CrossRef]
153. Monteiro, C.; Ramirez-Rosado, I.J.; Fernandez-Jimenez, L.A. Short-term forecasting model for electric power production of small-hydro power plants. *Renew. Energy* **2013**, *50*, 387–394. [CrossRef]
154. Moreno, J. Hydraulic plant generation forecasting in Colombian power market using ANFIS. *Energy Econ.* **2009**, *31*, 450–455. [CrossRef]
155. Perez-Mora, N.; Canals, V.; Martinez-Moll, V. Short-term Spanish aggregated solar energy forecast. In *International Work-Conference on Artificial Neural Networks*; Springer: Berlin/Heidelberg, Germany, 2015; pp. 307–319.
156. Lahouar, A.; Mejri, A.; Slama, J.B.H. Importance based selection method for day-ahead photovoltaic power forecast using random forests. In Proceedings of the 2017 International Conference on Green Energy Conversion Systems (GECS), Hammamet, Tunisia, 23–25 March 2017; pp. 1–7.

157. Jung, Y.; Jung, J.; Kim, B.; Han, S. Long short-term memory recurrent neural network for modeling temporal patterns in long-term power forecasting for solar PV facilities: Case study of South Korea. *J. Clean. Prod.* **2020**, *250*, 119476. [CrossRef]
158. Lahouar, A.; Slama, J.B.H. Hour-ahead wind power forecast based on random forests. *Renew. Energy* **2017**, *109*, 529–541. [CrossRef]
159. Rana, M.; Koprinska, I.; Troncoso, A. Forecasting hourly electricity load profile using neural networks. In Proceedings of the 2014 International Joint Conference on Neural Networks (IJCNN), Beijing, China, 6–11 July 2014; pp. 824–831.
160. Lu, S.L. Integrating heuristic time series with modified grey forecasting for renewable energy in Taiwan. *Renew. Energy* **2019**, *133*, 1436–1444. [CrossRef]
161. Colombo, T.; Koprinska, I.; Panella, M. Maximum Length Weighted Nearest Neighbor approach for electricity load forecasting. In Proceedings of the 2015 International Joint Conference on Neural Networks (IJCNN), Killarney, Ireland, 12–17 July 2015; pp. 1–8.
162. Wang, C.H.; Lin, K.P.; Lu, Y.M.; Wu, C.F. Deep Belief Network with Seasonal Decomposition for Solar Power Output Forecasting. *Int. J. Reliab. Qual. Saf. Eng.* **2019**, *26*, 1950029. [CrossRef]
163. Lahouar, A. Gaussian Process Based Method for Point and Probabilistic Short-Term Wind Power Forecast. In *International Conference on the Sciences of Electronics, Technologies of Information and Telecommunications*; Springer: Berlin/Heidelberg, Germany, 2018; pp. 134–147.
164. da Silva Fonseca, J.G., Jr.; Oozeki, T.; Takashima, T.; Koshimizu, G.; Uchida, Y.; Ogimoto, K. Use of support vector regression and numerically predicted cloudiness to forecast power output of a photovoltaic power plant in Kitakyushu, Japan. *Prog. Photovoltaics Res. Appl.* **2012**, *20*, 874–882. [CrossRef]
165. Koprinska, I.; Rana, M.; Agelidis, V.G. Correlation and instance based feature selection for electricity load forecasting. *Knowl. Based Syst.* **2015**, *82*, 29–40. [CrossRef]
166. Taylor, J.W. An evaluation of methods for very short-term load forecasting using minute-by-minute British data. *Int. J. Forecast.* **2008**, *24*, 645–658. [CrossRef]
167. Taylor, J.W. Short-term load forecasting with exponentially weighted methods. *IEEE Trans. Power Syst.* **2012**, *27*, 458–464. [CrossRef]
168. Lotufo, A.; Minussi, C. Electric power systems load forecasting: A survey. In Proceedings of the PowerTech Budapest 99. Abstract Records., (Cat. No. 99EX376), Budapest, Hungary, 29 August–2 September 1999; p. 36.
169. Shumway, R.H.; Stoffer, D.S. *Time Series Analysis and Its Applications: With R Examples*; Springer: Berlin/Heidelberg, Germany, 2017.
170. Whittle, P. *Hypothesis Testing in Time Series Analysis*; Almqvist & Wiksells boktr.: Stockholm, Sweden, 1951; Volume 4.
171. Box, G.E.; Jenkins, G.M.; Reinsel, G.C. *Time Series Analysis: Forecasting and Control*; John Wiley & Sons: Hoboken, NJ, USA, 2011; Volume 734.
172. Makridakis, S.; Spiliotis, E.; Assimakopoulos, V. Statistical and Machine Learning forecasting methods: Concerns and ways forward. *PLoS ONE* **2018**, *13*, e0194889. [CrossRef] [PubMed]
173. Murphy, K.P. *Machine Learning: A Probabilistic Perspective*; MIT press: Cambridge, MA, USA, 2012.
174. Tosh, C.R.; Ruxton, G.D. *Modelling Perception with Artificial Neural Networks*; Cambridge University Press: Cambridge, UK, 2010.
175. Azadeh, A.; Ghaderi, S.; Sohrabkhani, S. A simulated-based neural network algorithm for forecasting electrical energy consumption in Iran. *Energy Policy* **2008**, *36*, 2637–2644. [CrossRef]
176. Hamzaçebi, C.; Es, H.A.; Çakmak, R. Forecasting of Turkey's monthly electricity demand by seasonal artificial neural network. *Neural Comput. Appl.* **2017**, *31*, 2217–2231. [CrossRef]
177. Ma, J.; Oppong, A.; Acheampong, K.N.; Abruquah, L.A. Forecasting renewable energy consumption under zero assumptions. *Sustainability* **2018**, *10*, 576. [CrossRef]
178. Li, G.; Wang, H.; Zhang, S.; Xin, J.; Liu, H. Recurrent neural networks based photovoltaic power forecasting approach. *Energies* **2019**, *12*, 2538. [CrossRef]
179. Tzafestas, S.G. *Introduction to Mobile Robot Control*; Elsevier: Amsterdam, The Netherlands, 2013.
180. Efendi, R.; Ismail, Z.; Deris, M.M. A new linguistic out-sample approach of fuzzy time series for daily forecasting of Malaysian electricity load demand. *Appl. Soft Comput.* **2015**, *28*, 422–430. [CrossRef]

181. Marwala, L.; Twala, B. Forecasting electricity consumption in South Africa: ARMA, neural networks and neuro-fuzzy systems. In Proceedings of the 2014 International Joint Conference on Neural Networks (IJCNN), Beijing, China, 6–11 July 2014; pp. 3049–3055.
182. Rebala, G.; Ravi, A.; Churiwala, S. *An Introduction to Machine Learning*; Springer: Berlin/Heidelberg, Germany, 2019.
183. Al-Musaylh, M.S.; Deo, R.C.; Adamowski, J.F.; Li, Y. Short-term electricity demand forecasting with MARS, SVR and ARIMA models using aggregated demand data in Queensland, Australia. *Adv. Eng. Inform.* **2018**, *35*, 1–16. [CrossRef]
184. Lin, K.P.; Pai, P.F. Solar power output forecasting using evolutionary seasonal decomposition least-square support vector regression. *J. Clean. Prod.* **2016**, *134*, 456–462. [CrossRef]
185. Sood, R.; Koprinska, I.; Agelidis, V.G. Electricity load forecasting based on autocorrelation analysis. In Proceedings of the 2010 International Joint Conference on Neural Networks (IJCNN), Barcelona, Spain, 18–23 July 2010; pp. 1–8.
186. Mourad, M.; Bouzid, B.; Mohamed, B. A hybrid wavelet transform and ANFIS model for short term electric load prediction. In Proceedings of the 2012 2nd International Conference on Advances in Computational Tools for Engineering Applications (ACTEA), Beirut, Lebanon, 12–15 December 2012; pp. 292–295.
187. Sudheer, G.; Suseelatha, A. Pronóstico de carga a corto plazo usando la transformación wavelet combinada con Holt—Winters y modelos vecinos más próximos ponderados. *Int. J. Electr. Power Energy Syst.* **2015**.
188. Guan, C.; Luh, P.B.; Michel, L.D.; Wang, Y.; Friedland, P.B. Very short-term load forecasting: Wavelet neural networks with data pre-filtering. *IEEE Trans. Power Syst.* **2012**, *28*, 30–41. [CrossRef]
189. Hochreiter, S.; Schmidhuber, J. Long short-term memory. *Neural Comput.* **1997**, *9*, 1735–1780. [CrossRef] [PubMed]

Publisher's Note: MDPI stays neutral with regard to jurisdictional claims in published maps and institutional affiliations.

© 2020 by the authors. Licensee MDPI, Basel, Switzerland. This article is an open access article distributed under the terms and conditions of the Creative Commons Attribution (CC BY) license (http://creativecommons.org/licenses/by/4.0/).

Article

Modeling Spectral Properties in Stationary Processes of Varying Dimensions with Applications to Brain Local Field Potential Signals

Raanju R. Sundararajan [1,*]**, Ron Frostig** [2] **and Hernando Ombao** [3]

1. Department of Statistical Science, Southern Methodist University, Dallas, TX 75275, USA
2. School of Biological Sciences, University of California Irvine, Irvine, CA 92697, USA; rfrostig@uci.edu
3. Statistics Program, King Abdullah University of Science and Technology, Thuwal 23955, Saudi Arabia; hernando.ombao@kaust.edu.sa
* Correspondence: rsundararajan@smu.edu

Received: 22 October 2020; Accepted: 3 December 2020; Published: 5 December 2020

Abstract: In some applications, it is important to compare the stochastic properties of two multivariate time series that have unequal dimensions. A new method is proposed to compare the spread of spectral information in two multivariate stationary processes with different dimensions. To measure discrepancies, a frequency specific spectral ratio (FS-ratio) statistic is proposed and its asymptotic properties are derived. The FS-ratio is blind to the dimension of the stationary process and captures the proportion of spectral power in various frequency bands. Here we develop a technique to automatically identify frequency bands that carry significant spectral power. We apply our method to track changes in the complexity of a 32-channel local field potential (LFP) signal from a rat following an experimentally induced stroke. At every epoch (a distinct time segment from the duration of the experiment), the nonstationary LFP signal is decomposed into stationary and nonstationary latent sources and the complexity is analyzed through these latent stationary sources and their dimensions that can change across epochs. The analysis indicates that spectral information in the Beta frequency band (12–30 Hertz) demonstrated the greatest change in structure and complexity due to the stroke.

Keywords: multivariate time series; nonstationary; spectral matrix; local field potential

MSC: 62M10; 62M15

1. Introduction

Numerous applications require comparing two multivariate time series of unequal dimensions. Neuroscience experiments result in a stationary or nonstationary multivariate signal from different epochs (distinct non-overlapping successive time segments of the duration of the experiment). A popular approach to modeling such data decomposes the observed signal at every epoch into useful latent sources that can be stationary or nonstationary. These latent sources are lower dimensional time series obtained by linear transforms of the components of the observed multivariate series and they aim to capture important statistical properties of the observed series. At these epochs, dimension reduction techniques such as principal component analysis (PCA), factor modeling, independent component analysis (ICA), stationary subspace analysis (SSA) are often applied to extract useful lower-dimensional latent sources. Artificially setting the dimension of these latent sources to be the same across the epochs results in loss of important information since these changes could be indicative of useful brain processes such as learning (Fiecas and Ombao [1]).

Indeed brain processes evolve across the entire recording period (Fiecas and Ombao [1], Ombao et al. [2]) leading to changes in the dimension of the latent sources across epochs. Moreover, the evolution of the dimension can itself serve as a feature in understanding how the brain function evolves during an experiment. As another example in neuroscience, the aim in functional connectivity is to model dependence between different brain regions at various epochs in an experiment; Cribben et al. [3], Cribben et al. [4], Cribben and Yu [5], Zhu and Cribben [6]. To mitigate the problem of high-dimensionality arising due to signal from densely voxelated cortical surface, parcellation leads to disjoint regions of interest (ROI) of the brain and signal summaries are obtained in each of these regions. Dependence measures between these ROIs are then computed using their respective signal summaries. In the above pursuit of region-wise comparison of the brain, it is natural to encounter the problem of comparing multivariate processes, say from two different regions that have unequal dimensions. In Wang et al. [7] the problem of modeling effective connectivity in high-dimensional cortical surface signal is pursued wherein a factor analysis is carried out on each ROI and vector autoregressive (VAR) models are used to jointly model the latent factors. Here again, one can potentially end up with unequal number of optimal latent factors from different ROIs thereby making the comparisons challenging.

The application that motivates our methodology is the analysis of local field potentials (LFP) in an experiment that simulates ischemic stroke in humans (Data source: Stroke experiment conducted in the lab of co-author (Ron Frostig) at his Neurobiology lab; http://frostiglab.bio.uci.edu/Home.html). The dataset comprises of 600 epochs worth of LFP recordings (each epoch is 1 s long) from 32 microelectrodes implanted in a rat's cortex. Figure 1 below depicts the rat's cortex and the locations of the 32 sensors implanted on the cortical surface from which the LFP signal is recorded. This 32-dimensional signal is our observed time series. A stroke is induced midway through the experiment (epoch 300) by clamping the medial cerebral artery. The goal is to develop a method that tracks changes in the complexity of signals following the stroke. From the observed LFP signal, useful lower-dimensional sources are extracted at each epoch and we shall characterize complexity in LFP through these useful latent sources and their varying dimensions across epochs.

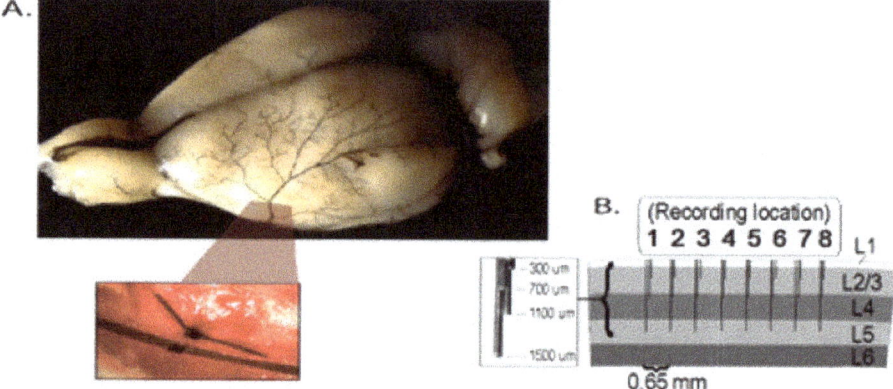

Figure 1. (**A**) Visual representation of the 32 microelectrodes on the rat's cortex from which the local field potential (LFP) signal is recorded. (**B**) The distance between microelectrodes is 0.65 mm and the total distance between microelectrode 1 and microelectrode 8 is 3.9 mm.

Motivated by such applications, we propose a new method to compare spectral information in different multivariate stationary processes of varying dimensions. More specifically, the aim is to capture the amount of spectral information in various frequency bands in different stationary processes of unequal dimensions. There are already many methods and models that discuss evolution of spectral information but the key contribution of this paper is in modeling evolution of the spectrum while allowing dimension to also evolve over time. We introduce a frequency-specific spectral ratio, which we call the FS-ratio, statistic that measures the proportion of spectral power in various frequency bands. FS-ratio can be used to (i). identify frequency bands where there is significant discrepancies between pre and post stroke epochs, (ii). identify frequency bands that account for most variation within pre (and post) stroke epochs and (iii). identify the frequency bands that are consistent (vs inconsistent) across all the 600 epochs. One of the key features of this statistic is that it is blind to the dimension of the multivariate stationary process and can be used to compare successive epochs with possibly different dimensions in the stationary sources. Thus, the proposed FS-ratio is very useful in (a). discriminating between the pre and post stroke onset and (b). tracking changes over the entire course of the experiment while allowing for varying dimensions. In Section 2 we develop our FS-ratio statistic and derive its asymptotic properties. We return to the LFP dataset in Section 3 and discuss the usefulness of the proposed ratio statistic in discriminating between pre and post stroke onset. Section 4 concludes. Finally, we evaluate the performance of the proposed FS-ratio statistic through several simulation examples and the results are provided in Appendix A.

2. Methodology

In this section, we first describe our FS-ratio statistic and the method to analyze the evolution of spectral information in stationary processes with varying dimensions. Using the FS-Ratio statistic, a technique to locate the frequency bands carrying significant spectral power is discussed in Section 2.1.1. The theoretical properties of the proposed statistic along with the required assumptions are discussed in Section 2.1.2.

2.1. The FS-Ratio Statistic

Let X_t be a d_1-variate time series and Y_t be a d_2-variate time series where $d_1 \neq d_2$ and $t = 1, 2, \ldots, T$. The spectral matrices of the two zero-mean multivariate stationary series are given by $f_X(\omega) \in \mathbb{C}^{d_1 \times d_1}$ and $f_Y(\omega) \in \mathbb{C}^{d_2 \times d_2}$ for $\omega \in (-\pi, \pi)$. Here $(-\pi, \pi)$ represents the normalized frequency range used to take care of aliases in the frequency components outside this range. This range $(-\pi, \pi)$ is sometimes referred to as angular frequency scale with frequency 2π being called the Nyquist or folding frequency. With the discrete Fourier transforms of X_t and Y_t expressed as $J_{X,T}(\omega) = \frac{1}{\sqrt{2\pi T}} X_t e^{-it\omega}$ and $J_{Y,T}(\omega) = \frac{1}{\sqrt{2\pi T}} Y_t e^{-it\omega}$, respectively, the periodogram matrices $I_{X,T}(\omega) \in \mathbb{C}^{d_1 \times d_1}$ and $I_{Y,T}(\omega) \in \mathbb{C}^{d_2 \times d_2}$ of the two series are obtained by

$$I_{X,T}(\omega) = J_{X,T}(\omega) J_{X,T}(\omega)^* \quad \text{and} \quad I_{Y,T}(\omega) = J_{Y,T}(\omega) J_{Y,T}(\omega)^*, \tag{1}$$

where $J_{X,T}(\omega)^*$ denotes the conjugate transpose. The estimated spectral matrices, for $\omega \in (-\pi, \pi)$, are given by

$$\hat{f}_X(\omega) = \frac{1}{T} \sum_{j=-\lfloor \frac{T}{2} \rfloor+1}^{\lfloor \frac{T}{2} \rfloor} K_h(\omega - \omega_j) I_{X,T}(\omega_j) \quad \text{and} \quad \hat{f}_Y(\omega) = \frac{1}{T} \sum_{j=-\lfloor \frac{T}{2} \rfloor+1}^{\lfloor \frac{T}{2} \rfloor} K_h(\omega - \omega_j) I_{Y,T}(\omega_j), \tag{2}$$

where $\omega_j = \frac{2\pi}{T} j$ and $K_h(\cdot) = \frac{1}{h} K(\frac{\cdot}{h})$ where $K(\cdot)$ is a nonnegative symmetric kernel function and h denotes the bandwidth. Assumptions on the kernel and bandwidth to ensure uniform consistency in $\omega \in (-\pi, \pi)$ of the estimated spectral matrices are listed in Section 2.1.2.

The aim of this work is to compare the two spectral matrices $f_X(\omega)$ and $f_Y(\omega)$ over a specific frequency range (a,b) for some $0 < a < b < \pi$. The challenge here, however, is that the dimensions of the processes X_t and Y_t are unequal and hence their spectral matrices have varying dimensions. We thus focus on the spread or distribution of spectral power in each of these stationary processes across different frequency ranges. More precisely, for the d_1-variate series X_t we define the frequency-specific spectral (FS-ratio) parameter as

$$R_{X,a,b} = \frac{r_{X,a,b}}{r_{X,0,\pi}} = \frac{\int_a^b ||vec(f_X(\omega))||_2^2 d\omega}{\int_0^\pi ||vec(f_X(\omega))||_2^2 d\omega} \tag{3}$$

for some frequency band $(a,b) \subset (0,\pi)$ where $vec(\cdot)$ denotes vectorization of a matrix into a single column vector and $||\cdot||_2^2$ is the squared Euclidean norm. Observe that $R_{X,a,b} \in (0,1)$ can be viewed as a measure that captures the proportion of spectral power found in the frequency range (a,b). Similarly using the spectral matrix $f_Y(\omega)$, $R_{Y,a,b} \in (0,1)$ can be defined for the d_2-variate series Y_t. Comparisons can now be made between the parameters $R_{X,a,b}$ and $R_{Y,a,b}$ to understand the amount of spectral power in the frequency range (a,b) for the two multivariate series with unequal dimensions.

The data analogue of the FS-ratio parameter in (3) is then given by the FS-ratio statistic:

$$\widehat{R}_{X,a,b} = \frac{\hat{r}_{X,a,b}}{\hat{r}_{X,0,\pi}} = \frac{\int_a^b ||vec(\widehat{f}_X(\omega))||_2^2 d\omega}{\int_0^\pi ||vec(\widehat{f}_X(\omega))||_2^2 d\omega} \tag{4}$$

for some $0 < a < b < \pi$. Similarly, the data analogue $\widehat{R}_{Y,a,b}$ can be obtained for the d_2-variate series Y_t. The asymptotic properties of the quantities $\hat{r}_{X,a,b}$ and $\widehat{R}_{X,a,b}$ are discussed in Section 2.1.2. In neuroscience applications such as the one in Section 3, pre-defined frequency bands such as Theta, Alpha, Beta and Gamma are often used to understand the distribution of spectral power across these frequency bands. As opposed to using pre specified frequency bands, in Section 2.1.1 below we provide a data-driven technique to locate the various frequency ranges (a,b) that carry significant spectral power.

2.1.1. Finding Frequency Bands of Interest

In this section we describe our technique that uses the FS-Ratio statistic to find the frequency bands of interest. More precisely, we aim to locate the intervals (a,b) used in (3) and (4) wherein the multivariate time series has significant proportions of spectral power.

Let X_t be a d_1-variate zero-mean second order stationary time series with its $d_1 \times d_1$ spectral matrix given by $f_X(\omega)$. With the FS-Ratio parameter defined in (3), we consider the scan parameter

$$\lambda_{X,a} = 1 - \frac{R_{X,0,a-\Delta}}{R_{X,0,a}} = 1 - \frac{\int_0^{a-\Delta} ||vec(f_X(\omega))||_2^2 d\omega}{\int_0^a ||vec(f_X(\omega))||_2^2 d\omega} \tag{5}$$

for a small $\Delta > 0$ and $0 < a < \pi$. For the data analogue of the parameter above we consider a discretized sequence of frequency points $0 < a_1 < a_2 < \ldots < a_Q < \pi$ and evaluate the scan statistic as

$$\widehat{\lambda}_{X,a_j} = 1 - \frac{\widehat{R}_{X,0,a_j}}{\widehat{R}_{X,0,a_{j+1}}} \tag{6}$$

for $j = 1, 2, \ldots, Q - 1$. A plot of the scan statistic $\widehat{\lambda}_{X,a_j}$ across the various frequency points a_j will indicate the frequency ranges over which the spectral matrix of X_t has significant proportions of spectral power. Typically, one notices upward bumps in these plots over frequency ranges that carry significant spectral power; see Example 1 below and the top panel of Figure 2. Similarly for the d_2-variate series Y_t one can define $\lambda_{Y,a}$, find the estimated version $\widehat{\lambda}_{Y,a_j}$ and obtain the plot of it across the various frequency points a_j. Comparisons can then be made between the series X_t and Y_t using these plots. The choice for Δ in (5) and number of points Q in (6) depends on the application under consideration. Certain applications demand attention to spectral power in very small frequency ranges and certain others might not. A multiscale approach can also be used where one obtains plots of the scan statistic $\widehat{\lambda}_{X,a_j}$ across frequency points a_j for a sequence of Δ values. Visual inspection of these plots will help detect frequency ranges wherein the upward bumps are consistent across most values of Δ. If we let the interval $(0, 0.5)$ correspond to the interval $(0, \pi)$, our simulation study and real data analysis indicate a choice of $\Delta = 0.01$ and $Q = 49$ as reasonable.

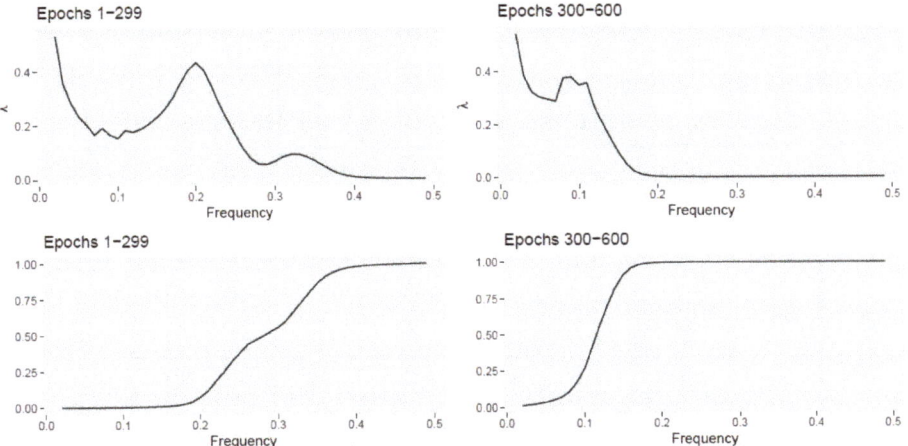

Figure 2. Example 1 (Top) Plot of average scan statistic $\widehat{\lambda}_{X,a_j}$ for epochs $i < 300$ (left) and $i \geq 300$ (right) at a discretized sequence of frequency points $0 < a_1 = 0.01 < a_2 = 0.02 < \ldots < a_{49} = 0.49 < 0.5$ ($\Delta = 0.01$). Here $(0, 0.5)$ corresponds to the interval $(0, \pi)$. (Bottom) Plot of the average of the statistic $\widehat{R}_{X,0,a_j}$ at the same frequency points.

We next provide a simple illustration of the scan statistic $\widehat{\lambda}_{X,a}$ through the following simulation example and show how it is useful in detecting important frequency ranges. The simulation scheme in this illustration is designed to mimic the real data situation in Section 3. There the entire duration of the neuroscience experiment is divided into non-overlapping successive time segments (a total of N epochs). The multivariate stationary processes of interest in these N epochs tend to have different dimensions and we attempt to mimic that scenario.

Example 1. *We consider N stationary processes, $X_{1,t}, X_{2,t}, \ldots, X_{N,t}$, with the series $X_{i,t}$ given by*

$$X_{i,t} = \begin{cases} V^{(1)}_{i,t} & \text{if } i < \frac{N}{2} \\ V^{(2)}_{i,t} & \text{if } i \geq \frac{N}{2} \end{cases} \quad (7)$$

where $i = 1, 2, \ldots, N = 600$ epochs, $t = 1, 2, \ldots, T = 1000$. Here $V^{(1)}_{i,t} \in \mathbb{R}^3$ and its components are given by $v_{0,t+k-1} + v_{1,t+k-1}$ for $k = 1, 2, 3$ and $v_{0,t}$ follows a AR(2) with $(-0.8, -0.7)$ and $v_{1,t}$ follows a AR(2) with

$(0.25, -0.75)$. The components of $V_{i,t}^{(2)} \in \mathbb{R}^2$ are given by $v_{2,t+k-1}$ for $k = 1, 2$ and $v_{2,t}$ follows a AR(2) with $(1.25, -0.75)$.

We consider a discretized set of frequency points $\{a_1, a_2, \ldots, a_Q\}$ of the interval $(0, \pi)$. At each point a_j we evaluate the average of the scan statistic $\widehat{\lambda}_{X,a_j}$ over epochs 1-299 and the average of the scan statistic over epochs 300–600. More precisely at each frequency point a_j and at each epoch, we obtain $\widehat{\lambda}_{X,a_j}$ and compute averages of these quantities over the respective epochs. In the top panel of Figure 2 we plot this average scan statistic. For epochs 1–299, $V_{i,t}^{(1)}$ from (7) is a combination of two AR(2) processes with spectral density peaks at roughly 0.22 and 0.33. The top left plot in Figure 2 witnesses the scan statistic exhibiting bumps around those frequencies. Similarly for epochs 300–600, $V_{i,t}^{(2)}$ from (7) is generated from an AR(2) process with peak at roughly 0.12. The top right plot in Figure 2 witnesses the scan statistic exhibiting a bump around that frequency.

In the bottom panel of Figure 2 we plot averages of the statistic $\widehat{R}_{X,0,a_j}$ for $j = 1, 2, \ldots, Q - 1$. We observe that this statistic is not as capable as the scan statistic $\widehat{\lambda}_{X,a_j}$ in bringing out the frequency ranges of significant spectral proportions.

2.1.2. Theoretical Properties of the FS-Ratio Statistic

In this section we list the required assumptions and discuss the asymptotic properties of the statistics $\widehat{r}_{X,a,b}$ and FS-ratio $\widehat{R}_{X,a,b}$.

Assumption 1. *Let $Z_t = (X_t, Y_t)'$, $t \in \mathbb{Z}$ be a $(d_1 + d_2)$-variate zero-mean second-order stationary time series. For any $k > 0$, the kth order cumulants of Z_t satisfy*

$$\sum_{u_1, u_2, \ldots, u_{k-1} \in \mathbb{Z}} [1 + |u_j|^2] \, c_{b_1, b_2, \ldots, b_k}(u_1, u_2, \ldots, u_{k-1}) < \infty$$

for $j = 1, 2, \ldots, k - 1$ and $b_1, b_2, \ldots, b_k = 1, 2, \ldots, d = d_1 + d_2$ where $c_{b_1, b_2, \ldots, b_k}(u_1, u_2, \ldots, u_{k-1})$ is the kth order joint cumulant of $Z_{b_1, u_1}, \ldots, Z_{b_{k-1}, u_{k-1}}, Z_{b_k, 0}$ as defined in Brillinger [8].

Please note that the kth order cumulant is given by $c_{b_1, b_2, \ldots, b_k}(u_1, u_2, \ldots, u_{k-1}) = cum\{Z_{b_1, u_1}, \ldots, Z_{b_{k-1}, u_{k-1}}, Z_{b_k, 0}\}$ where Z_{b_r, u_s} refers to component b_r of the vector Z_{u_s} with u_s being the time point; see Theorem 2.3.2 of Brillinger [8]. For example when $k = 2$, the 2nd order cumulant $cum\{Z_{b_1, u_1}, Z_{b_2, u_2}\} = cov(Z_{b_1, u_1}, Z_{b_2, u_2})$ is the covariance between those two random variables.

Assumption 2. *(a). The kernel function $K(\cdot)$ is bounded, symmetric, nonnegative and Lipschitz-continuous with compact support $[-\pi, \pi]$ and*

$$\int_{-\pi}^{\pi} K(\omega) d\omega = 1.$$

where $K(\omega)$ has a continuous Fourier transform $k(u)$ such that

$$\int k^2(u) du < \infty \quad \text{and} \quad \int k^4(u) du < \infty.$$

(b). The bandwidth h is such that $h^{9/2} T \to 0$ and $h^2 T \to \infty$ as $T \to \infty$.

Remark 1. *Assumptions 1 and 2 above are the same as in Eichler [9] where the first requires existence of all order moments of Y_t and the second ensures consistency of the estimated spectral matrix. It must be noted that* the assumptions on the kernel and bandwidth are primarily for establishing asymptotic result in (13) and can be weakened for Theorem 1.

Theorem 1. *Suppose that Assumptions 1,2 are satisfied. Then as $T \to \infty$,*

$$(a). \qquad \widehat{r}_{X,a,b} \xrightarrow{P} \int_a^b \sum_{r,s=1}^{d_1} f_{X,rs}(\omega) \overline{f_{X,rs}(\omega)} \, d\omega, \qquad (8)$$

where $f_X(\omega) = \left(f_{X,rs}\right)_{r,s=1,2,\ldots,d_1}$ is the $d_1 \times d_1$ spectral matrix of X_t and $\xrightarrow{P}$ denotes convergence in probability. Furthermore, let $\overline{\Pi}_{(a,b)} = (0, \pi) \setminus (a,b)$ for some $0 < a < b < \pi$. If $r_{X,a,b} > 0$ and $r_{X,\overline{\Pi}_{(a,b)}} > 0$,

$$(b). \qquad \widehat{R}_{X,a,b} \xrightarrow{P} \left(1 + \frac{r_{X,\overline{\Pi}_{(a,b)}}}{r_{X,a,b}}\right)^{-1} \qquad (9)$$

where $r_{X,\overline{\Pi}_{(a,b)}} = \int_{\overline{\Pi}_{(a,b)}} f_X(\omega)^2 d\omega$.

Proof. See Appendix B for details of the proof. □

Please note that in finite sample situations explored using simulation examples in Appendix A and the real data application in Section 3, we use the block bootstrap technique of Politis and Romano [10] for resampling from a stationary process. This is done to obtain sample quantiles of the FS-ratio statistic $\widehat{R}_{X,a,b}$.

Remark 2. *In a special case wherein the dimensions of the two processes are the same ($d_1 = d_2$), we wish to test for the equality of spectral matrices of same dimensions over an interval $0 < a < b < \pi$. Let us assume $d_1 = d_2$ and $d = d_1 + d_2$ and define the $d \times d$ spectral matrix of $Z_t = (X_t, Y_t)'$ as*

$$f_Z(\omega) = \begin{bmatrix} f_{Z,11}(\omega) & f_{Z,12}(\omega) \\ f_{Z,21}(\omega) & f_{Z,22}(\omega) \end{bmatrix} \qquad (10)$$

where the $d_1 \times d_1$ matrix $f_{Z,12}(\omega)$ is the cross-spectral matrix of the processes X_t and Y_t and $f_{Z,11}(\omega)$ and $f_{Z,22}(\omega)$ are the spectral matrices of X_t and Y_t respectively. We consider testing

$$H_0 : f_X(\omega) = f_Y(\omega) \; \forall \; \omega \in (a,b) \qquad (11)$$

where $0 < a < b < \pi$. The test statistic is

$$\widehat{D}_{X,Y} = \int_a^b ||vec(\widehat{f}_X(\omega) - \widehat{f}_Y(\omega))||_2^2 d\omega. \qquad (12)$$

The L_2 norm above in (12) on the spectral matrices is similar to the statistics considered in Eichler [9] and Dette and Paparoditis [11] wherein the problem of testing equality of spectral matrices is discussed. Suppose that Assumptions 1,2 are satisfied, an application of Theorem 3.5 of Eichler [9] yields, under H_0,

$$2\pi T \sqrt{h} \, \widehat{D}_{X,Y} - \frac{\mu_{XY}}{\sqrt{h}} \xrightarrow{D} N(0, \sigma_{XY}^2) \qquad (13)$$

where

$$\mu_{XY} = A_K \int_{-\pi}^{\pi} 1_{\omega \in (a,b)} \Big(\sum_{p_1,p_2=1}^{2} (-1 + 2\delta_{p_1 p_2}) tr(f_{Z,p_1 p_2}(\omega))^2 \Big) d\omega \qquad (14)$$

and

$$\sigma_{XY}^2 = B_K \int_{-\pi}^{\pi} \mathbf{1}_{\omega \in (a,b)} \Big(\sum_{p_1,p_2,p_3,p_4=1}^{2} (-1+2\delta_{p_1p_2})(-1+2\delta_{p_3p_4}) tr(f_{Z,p_1p_3}^{ij}(\omega) \overline{(f_{Z,p_2p_4}^{ij}(\omega))}^T)^2 \Big) d\omega. \tag{15}$$

where $\xrightarrow{D}$ denotes convergence in distribution,

$$A_K = \int_{-\pi}^{\pi} K^2(v)dv, \quad B_K = 4 \int_{a-\pi}^{b+\pi} \left(\int_{-\pi}^{\pi} K(u)K(u+v)du \right)^2 dv$$

and $\delta_{rs} = I(r=s)$ is the Kronecker delta and $tr(\cdot)$ denotes the trace of a matrix.

Remark 3. *In the neuroscience application in Section 3, the entire duration of the experiment is divided into non-overlapping successive time segments (a total of N epochs). Each epoch results in a multivariate stationary process of interest with the dimensions of these processes varying across epochs. Letting $X_{1,t}, X_{2,t}, \ldots, X_{N,t}$ be the N stationary processes at these epochs, one can obtain the FS-ratio statistics $\widehat{R}_{X_i,a,b}$, for $i = 1, 2, \ldots, N$, and view this is a series with time index being the epoch index i. Applying change point detection to this series to formally test for the significance of change points would require use of a divergence measure that measures distance between $\widehat{R}_{X_i,a,b}$ and $\widehat{R}_{X_j,a,b}$ when $i \neq j$. Different norms can be used to construct this divergence measure and this would serve as the test statistic. Large sample distributions of this statistic would provide critical values necessary for the test. One of the issues here would be in dealing with differing errors in estimating the FS-ratio statistics when the dimensions of the two series are very different and this needs further investigation.*

3. Analysis of Complexity of Rat Local Field Potentials in a Stroke Experiment

In this section, we investigate the ability of the FS-ratio statistic to identify changes in the spectral properties of the local field potential (LFP) of a rat (Local field potential data on the experimental rat comes from the stroke experiment conducted at Frostig laboratory at University of California Irvine: http://frostiglab.bio.uci.edu/Home.html). The aim is to identify changes in complexity and structure of the multivariate cortex signal over the course of the experiment. It is also of interest to understand the differential roles of frequency bands and determine the specific bands that demonstrate the most significant changes that occurred due to the stroke.

At 32 locations on the rat's cortex, microelectrodes are inserted: 4 layers in the cortex, at 300 μm, 700 μm, 1100 μm and 1500 μm and 8 microelectrodes lined up in each of the 4 layers. We look at the field potential specific to the 32 locations recorded for a total duration of 10 min. This 10 min duration is divided into $N = 600$ epochs (distinct successive non-overlapping time segments of the duration of the experiment) with each epoch comprising of 1 s worth of data. The sampling rate here is 1000 Hz resulting in $T = 1000$ observations per epoch. Midway through the recording period (after epoch 300) a stroke is artificially induced by clamping the medial cerebral artery that supplied blood to the recorded area.

As a first step in our analysis, we applied a component-wise univariate test of second-order stationarity (Dwivedi and Subba Rao [12]) of the LFP signal at each epoch. In Figure 3, we present the p-values from a test of second-order stationarity carried out on each of the $p = 32$ microelectrodes at each epoch. We notice that these individual microelectrodes are more stationary after the stroke than before.

Next, we model the observed 32-dimensional signal as a multivariate nonstationary time series using the stationary subspace analysis (SSA) setup. We assume the observed $p = 32$ dimensional LFP signal $S_{i,t}$ is linearly generated by stationary and nonstationary sources in the cortex. More precisely we have,

$$S_{i,t} = A_i X_{i,t} + \varepsilon_{i,t}, \quad i = 1, 2, \ldots, N = 600, \tag{16}$$

where $X_{i,t} \in \mathbb{R}^{d_i}$ is latent stationary source, A_i is a $p \times d_i$ unknown demixing matrix, $\varepsilon_{i,t}$ are the nonstationary sources. This setup of starting with an observed nonstationary time series and, after some transformation, getting to a lower dimensional stationary time series has interesting applications in neuroscience. For instance, EEG signals measuring brain activity appear often as a multivariate nonstationary time series; see Ombao et al. [13], Srinivasan [14], Nunez and Srinivasan [15], von Bünau et al. [16], Wu et al. [17], Gao et al. [18], Euán et al. [19] for examples. Kaplan et al. [20] regard the nonstationarity as background activity in the brain signal and removing this nonstationarity was seen to improve prediction accuracy in neuroscience experiments; von Bünau et al. [21] and von Bünau et al. [16]. Thus, the aim of SSA is to separate the stationary from the nonstationary sources within each epoch and focus attention on the stationary sources. From a stroke neuroscientist's perspective, the stationary sources within a short epoch of 1 s are considered to be the "stable" components of the signal since they are consistent within that short interval. The word consistent here refers to the statistical properties of the signal remaining the same within an epoch. Of course the transient components (nonstationary components) may also be of interest in other applications.

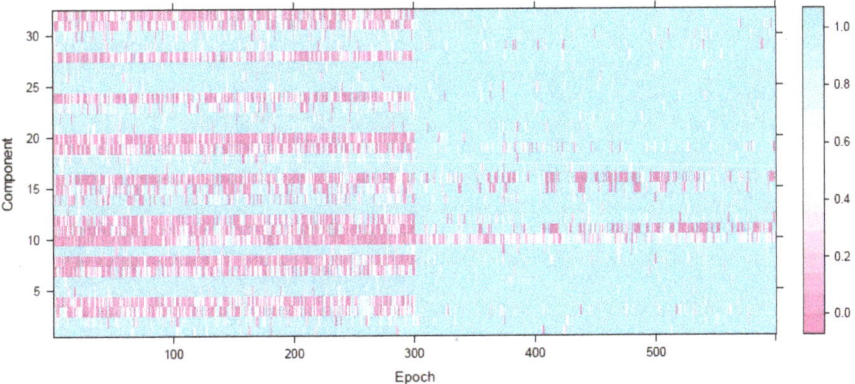

Figure 3. p-values from the test of second-order stationarity on each of the $p = 32$ LFP microelectrodes (y-axis) for all 600 epochs (x-axis).

The next goal in the data analysis is to estimate the epoch-evolving dimension d_i and the latent stationary time series $X_{i,t} \in \mathbb{R}^{d_i}$ where $d_i < p$. In Figure 4, we apply SSA and plot the estimates of the stationary subspace dimension d_i across $N = 600$ epochs using the method in Sundararajan et al. [22].

The evolutionary dimension d_i of the latent stationary sources were presented in Figure 4. The plot indicates increase in the number of stationary sources in post-stroke epochs (after epoch 300) and this agrees with the results in Figure 3 wherein more epochs after the stroke witness stationary behavior in the individual LFP components. It is indeed interesting that immediately post-occlusion (or immediately after stroke onset), the LFPs are highly synchronized: the plots of the observed LFP $S_{i,t}$ and the estimated squared coherence between the 32 components (Figure 5) suggest that different electrodes look very similar and there is high coherence in between the entire network of electrodes at various frequency bands. Please note that for the observed 32 dimensional signal $S_{i,t}$ in epoch i, the squared coherence between two components $S_{p,i,t}$ and $S_{q,i,t}$, for $p \neq q$, at frequency ω is given by

$$C_{p,q}(\omega) = \frac{|f_{S,pq}(\omega)|^2}{f_{S,pp}(\omega) f_{S,qq}(\omega)} \qquad (17)$$

where $f_{S,pq}(\omega)$ denotes the cross-spectrum between those two components and $f_{S,pp}(\omega)$ and $f_{S,qq}(\omega)$ are the univariate spectra of the components series $S_{p,i,t}$ and $S_{q,i,t}$ respectively. This observation of high coherence across electrodes immediately post-occlusion was confirmed by the neuroscientists and also reported in Ellen Wann's PhD dissertation (Wann [23]). Next, we investigate further into the lead-lag cross-dependence between microelectrodes. We pre-whitened the observed time series to make the lag-0 covariance matrix identity. More precisely, one considers $\Sigma_i^{-1/2} S_{i,t}$ where $\Sigma_i^{-1/2}$ is the inverse square root of the lag-0 covariance matrix $V(S_{i,t})$. We observe, in Figure 5, the significant drop in the magnitude of squared coherence after pre-whitening indicating that the dependence among the 32 components is predominantly due to a contemporaneous (i.e., lag-0) dependence. One can also notice, from the right plot in Figure 5, a drop in the coherence in the gamma frequency band after the stroke.

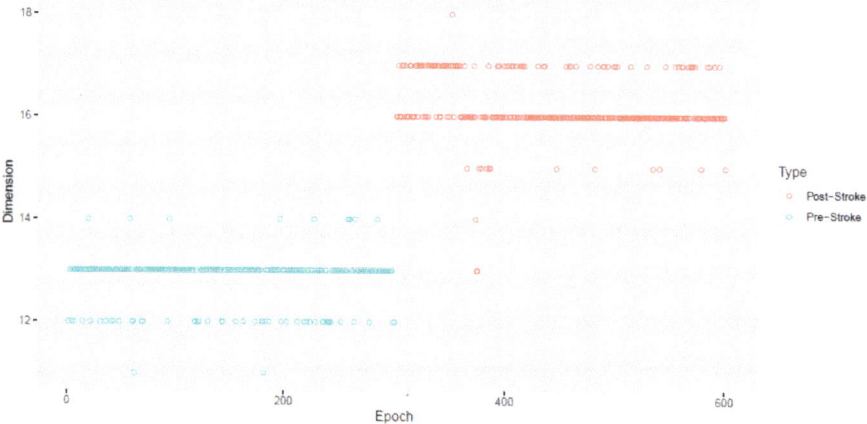

Figure 4. Plot of estimated stationary subspace dimensions $\widehat{d}_i$ for the $i = 1, 2, \ldots, N = 600$ epochs in the stroke experiment. Please note that for each epoch i there is a single estimated dimension $\widehat{d}_i$ that is plotted.

We then estimated the latent stationary sources $X_{i,t}$ for the $i = 1, 2, \ldots, N = 600$ epochs using the DSSA method in Sundararajan and Pourahmadi [24]. In order to overcome identifiability issues in the model in (16), SSA and PCA methods for time series assume an identity lag-0 covariance matrix for $X_{i,t}$ and resort to a pre-whitening technique to achieve this. Figure 6 plots the average squared coherence in the non pre-whitened and pre-whitened stationary sources across different frequency bands. Similar to the coherence pattern in the observed LFP in Figure 5, the left plot in Figure 6 witnesses an increase in the coherence after the occurrence of the stroke. This indicates the importance of the stationary components in explaining the high degree of synchronicity. Also, the right plot in Figure 6 indicates a substantial drop in the magnitude of coherence in the stationary sources. The pre-whitened stationary sources have lower coherence than the coherence of the stationary sources based on the non pre-whitened. As noted, previous findings have already indicated an increased coherence post stroke onset. Our analysis provided an additional insight that the increase in the coherence post-stroke is due only to contemporaneous (or lag-0) dependence. This indicates perfect temporal synchrony in a sense that there is no lead-lag cross-dependence between the electrodes. This was suggested by visual inspection of the LFP traces and hypothesized by neuroscientists though never formally confirmed until now with our analysis.

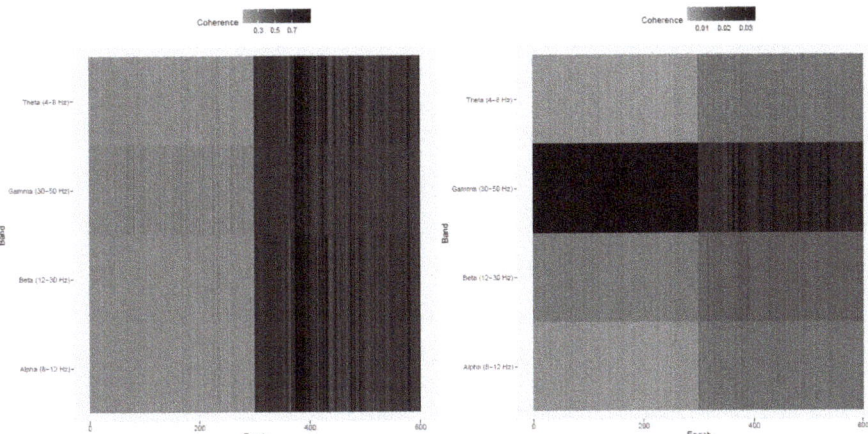

Figure 5. (**Left**) average squared coherence among the 32 components of the observed LFP signal across 600 epochs. The averages are computed across the specified frequency bands. (**Right**) average squared coherence among the 32 components of the pre-whitened LFP signal across 600 epochs.

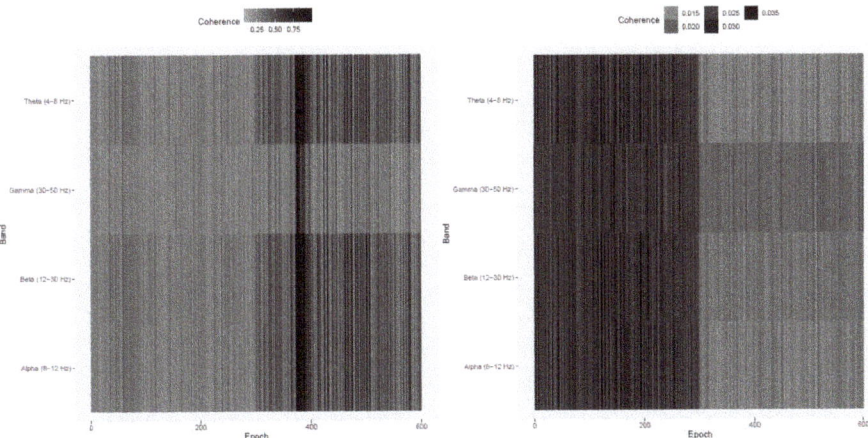

Figure 6. (**Left**) average squared coherence in the estimated stationary sources across 600 epochs. The averages are computed across the specified frequency bands. (**Right**) average squared coherence in the pre-whitened stationary sources across 600 epochs.

Next, the FS-ratio statistic was evaluated on these estimated stationary sources at each of the 600 epochs at various frequency bands. Figure 7 plots the estimated FS-ratio statistic $\widehat{R}_{X_i,a,b}$, $i = 1, 2, \ldots, N = 600$, for the known frequency bands: theta (4–8 Hertz), alpha (8–12 Hertz), beta (12–30 Hertz) and gamma (30–50 Hertz). At each epoch i, we obtained a 95% confidence interval for the FS-ratio statistic using the block bootstrap technique of Politis and Romano [10]. To select the block length, we follow the procedure in Politis and White [25], Patton et al. [26]. Please note that this procedure is for the univariate case and hence we apply it to each component of the multivariate process $X_{i,t}$ and obtain the block length as the average over all components. The confidence intervals are the blue shaded regions in Figures 7 and 8.

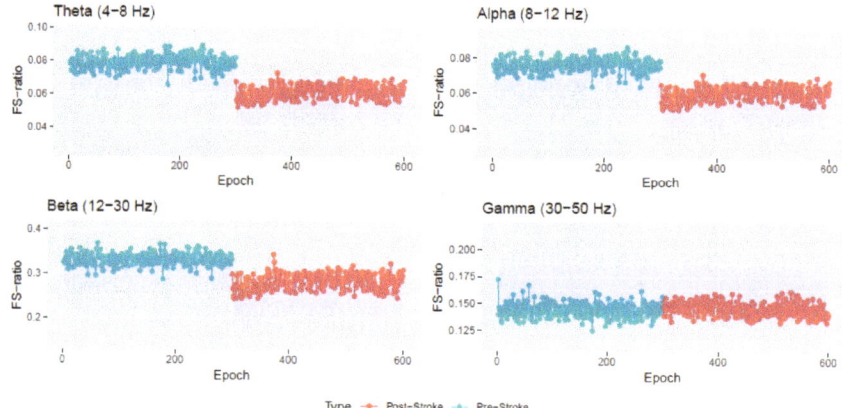

Figure 7. Plot of the FS-ratio statistic $\widehat{R}_{X_i,a,b}$ for $i = 1, 2, \ldots, N = 600$ for various frequency bands. The blue shaded region corresponds to a 95% confidence interval.

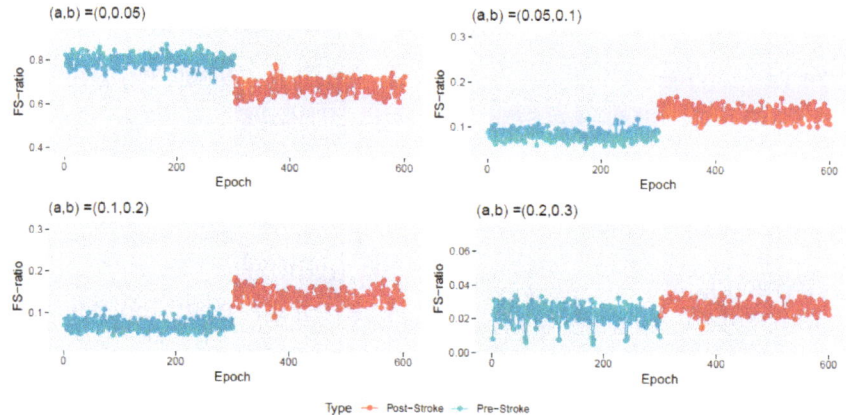

Figure 8. Plot of the FS-ratio statistic $\widehat{R}_{X_i,a,b}$ for $i = 1, 2, \ldots, N = 600$ for specified frequency ranges (a, b). Here $(a, b) \subset (0, 0.5)$ and $(0, 0.5)$ corresponds to the interval $(0, \pi)$. The blue shaded region corresponds to a 95% confidence interval.

The FS-ratio statistic is seen to have differences in the pre and post stroke epochs in the Theta, Alpha, and Beta bands but not in the Gamma band. It can also be seen that the biggest difference in FS-ratio between pre and post stroke is in the Beta band wherein there is a decrease in the amount of spectral information after the stroke. Figure 8 also presents the FS-ratio statistic on other specified frequency bands wherein one notices differences between the pre and post stroke epochs.

Tables 1 and 2 contain numerical summaries of the FS-ratio statistic for the pre and post stroke epochs at various frequency bands. We notice that the Beta band is where there is maximum difference observed between the pre and post stroke epochs. The Gamma band is consistent throughout the experiment's 600 epochs. Within the pre stroke epochs (and also within the post-stroke epochs), the most variation in FS-ratio is observed in the Beta band.

Table 1. Numerical summaries of FS-ratio statistic $\widehat{R}_{X_i,a,b}$ for **pre stroke** epochs $i = 1, 2, \ldots, 300$.

Frequency Band	Mean	Median	SD	Lower CI	Upper CI
Theta (4–8 Hz)	0.079	0.079	0.004	0.061	0.081
Alpha (8–12 Hz)	0.076	0.077	0.0035	0.059	0.078
Beta (12–30 Hz)	0.332	0.332	0.0129	0.267	0.341
Gamma (30–50 Hz)	0.144	0.144	0.006	0.141	0.191

Table 2. Numerical summaries of FS-ratio statistic $\widehat{R}_{X_i,a,b}$ for **post stroke** epochs $i = 301, 302, \ldots, 600$.

Frequency Band	Mean	Median	SD	Lower CI	Upper CI
Theta (4–8 Hz)	0.062	0.062	0.004	0.0422	0.0669
Alpha (8–12 Hz)	0.060	0.061	0.004	0.041	0.064
Beta (12–30 Hz)	0.283	0.285	0.018	0.202	0.292
Gamma (30–50 Hz)	0.146	0.146	0.006	0.135	0.187

Discussion

The *p*-values presented in Figure 3 represent a test of second-order stationarity carried out on each of the $p = 32$ microelectrodes at each epoch. We noticed that immediately after stroke the individual microelectrodes behaved in a more stationary manner and this was visibly different from what was observed before the stroke. Based on this analysis, it might be plausible that the LFP signal, under normal circumstances, exhibits nonstationary behavior and immediately post stroke the signal behaves in a more stationary manner thereby showing that the brain's typical functions are affected. The plots of the observed LFP $S_{i,t}$ and the estimated squared coherence between the 32 components (Figure 5) indicate high cross-electrode coherence at various frequency bands immediately post stroke. This observation was also confirmed by the neuroscientists and also reported in Ellen Wann's PhD dissertation (Wann [23]).

In Fontaine et al. [27], a univariate LFP microelectrode-wise change point analysis was performed on the same dataset. In their work, for various frequency bands, changes in the non-linear spectral dependence of the LFP signal is modeled using parametric copulas. They detected change-points for a fixed microelectrode and fixed frequency band. One can notice the detection of numerous change points in the Delta, Theta, Alpha, Beta and Gamma bands for individual microelectrodes 1, 9 and 17. The detected change points include several epochs with very few of them being close to the time of the occlusion (or induced stroke) which was epoch $i = 300$.

In contrast, the advantages of our method are as follows: (i). The method treats the observed LFP signal as a multivariate nonstationary time series. Using (16), we model this observed multivariate signal as a mixture of stationary and nonstationary components. Figure 4 presents the dimension of stationary subspace (dimension of $X_{i,t}$) across the 600 epochs and this is seen to be a useful feature in understanding changes in the cortical signal after the occurrence of the induced stroke (epoch 300). In other words, an increase in the dimension d_i after the stroke points to a more stationary behavior of the LFP signal after the stroke. (ii). The FS-ratio statistic, with the ability to compare two multivariate processes with unequal dimensions, is applied on the estimated processes $X_{i,t}$ for each of the 600 epochs and frequency band specific numerical summaries are presented. The Beta frequency band is seen to be display the greatest changes within the pre stroke and post stroke epochs and also between the pre stroke and post stroke epochs. Also, from Figures 7 and 8, it is very easy to spot a change point at epoch 300 when the stroke was induced.

4. Concluding Remarks

In this work, we proposed a new frequency-specific spectral ratio statistic FS-ratio that is demonstrated to be useful in comparing spectral information in two multivariate stationary processes of different dimensions. The method is motivated by applications in neuroscience wherein brain signal is recorded across several epochs and the widely used tactic is to assume the observed signal be linearly generated by latent sources of interest in lower dimensions. Applying PCA/ICA/SSA and other dimension reduction methods to the observed signal in different epochs in the experiment results in different estimates of the dimensions of latent sources. In these situations, the FS-ratio is seen to be useful because (i). It captures the proportion of spectral power in various frequency bands by means of a L_2-norm on the spectral matrices and (ii). It is blind to the dimension of the stationary process as it only looks at the proportion of spectral power at frequency bands. Under mild assumptions, the asymptotic properties of FS-ratio statistic are derived. We also provide a data-driven technique to locate the frequency bands that carry significant proportions of spectral power. In the application of our method to the LFP dataset, we witness the ability of our method in (i). identifying frequency bands where the pre and post stroke epochs are different, (ii). identifying frequency bands that accounts for most discrepancies within pre (and post) stroke epochs, (iii). identifying the frequency bands that are consistent across all the 600 epochs of the experiment and (iv). understanding the importance of contemporaneous dependence, both in the observed LFP and the stationary sources, across the 600 epochs and this indicated perfect synchrony (no lead-lag cross-dependence) among microelectrodes immediately after the stroke.

Topological data analysis (TDA) methods for characterizing complexity and detecting phase transitions exist in the literature; M. Piangerelli [28], Rucco et al. [29], Wang et al. [30]. Topological features from the observed series are extracted using techniques such as persistent entropy, persistence diagrams and Betti numbers and this can be viewed as another approach to identify changes in the multivariate time series due to events such as epilepsy and seizure.

Author Contributions: The stroke experiment was conducted at the R.F.'s laboratory (Frostig laboratory at University of California Irvine: http://frostiglab.bio.uci.edu/Home.html) and the LFP data came from this experiment. The statistical methodology, theory and computations were performed by R.R.S. and H.O. All authors have read and agreed to the published version of the manuscript.

Funding: This work is support in part by KAUST, NIH NS066001, Leducq Foundation 15CVD02 and NIH MH115697.

Conflicts of Interest: The authors declare no conflict of interest.

Appendix A. Simulation Study

In this section, we illustrate the performance of the FS-ratio statistic in capturing spread of spectral information using simulated examples. We consider four simulation schemes and report the key summaries of the FS-ratio statistic across repetitions of each of the four schemes. In addition, 95% bootstrap confidence limits for the FS-ratio statistic are computed from $B = 500$ bootstrap replications. Here we use the block bootstrap procedure of Politis and White [25], Patton et al. [26]. For an estimate of the spectral matrix defined in (2), the Bartlett-Priestley kernel with bandwidth $h = T^{-0.3}$ and the Daniell kernel, see Example 10.4.1 in Brockwell and Davis [31] with $m = \sqrt{T}$ were implemented. Similar results were obtained for the two kernel choices and only the results from the latter are presented.

The simulation schemes presented below are designed to mimic the real data situation in Section 3. There, the entire duration of the neuroscience experiment is divided into non-overlapping successive time segments (N epochs). Then from these N epochs, lower dimensional stationary sources of varying dimensions were extracted. Similarly, in our simulations below we shall simulate N stationary processes with varying dimensions and investigate the evolution of the FS-ratio statistic. We now present 4 simulation schemes to assess the behavior of our FS-ratio statistic. Scheme 1 simulates N multivariate stationary VAR(2) processes, $X_{i,t}$ where $i = 1, 2, \ldots, N = 500$, with dimensions randomly chosen from $\{2, 3, \ldots, 30\}$ and the error vectors are component-wise i.i.d $N(0,1)$. The phase parameter θ of the AR components are allowed to vary across epochs i.e two different choices, θ_1 and θ_2, are included with $\theta_1 = \frac{4\pi}{25}$ for epochs $i < N/2$ and $\theta_2 = \frac{4\pi}{5}$ for epochs $i \geq N/2$. This causes a shift in the frequency bands of interest as we move across the epochs. Scheme 2 is similar to Scheme 1 but an interaction between the dimension and frequency is included. More precisely, lower dimensional signals simulated in epochs $i < N/2$ have a peak at frequency $\theta_1 = \frac{4\pi}{25}$, and the higher dimensional signals simulated in epochs $i \geq N/2$ have a peak at frequency $\theta_2 = \frac{4\pi}{5}$. Scheme 3 is also similar to Scheme 1 but the error vectors are allowed to have contemporaneous dependence. The N multivariate processes in Scheme 4 are first simulated as in Scheme 1 but are then pre-multiplied with a half-orthogonal matrix. This is in line with the model assumption (16) made in Section 3.

Scheme 1: We simulate N stationary process $X_{1,t}, X_{2,t}, \ldots, X_{N,t}$, for $t = 1, 2, \ldots, T$, where the ith process includes a d_i-variate process $X_{i,t} = (X_{1,i,t}, X_{2,i,t}, \ldots, X_{d_i,i,t})'$ where each $X_{k,i,t}$, $k = 1, 2, \ldots, d_i$, are independently generated univariate stationary AR(2) process given by

$$X_{k,i,t} = \phi_{i,1} X_{k,i,t-1} + \phi_{i,2} X_{k,i,t-2} + \epsilon_{k,i,t}$$

$\phi_{i,1} = 2\xi_i \cos(\theta_i)$, $\phi_{i,2} = -\xi_i^2$, $\epsilon_{k,i,t}$ are i.i.d $N(0,1)$ and $k = 1, 2, \ldots, d_i$, $i = 1, 2, \ldots, N = 500$, $t = 1, 2, \ldots, T = 1000$. The dimension d_i for $X_{i,t}$ is randomly chosen from $\{2, 3, \ldots, 30\}$. Here $\xi_i \sim U(0.8, 0.98)$ and θ_i is given by

$$\theta_i = \begin{cases} \cos(\frac{4\pi}{25}) & \text{if } i < \frac{N}{2} \\ \cos(\frac{4\pi}{5}) & \text{if } i \geq \frac{N}{2} \end{cases}$$

Scheme 2: We follow Scheme 1 in generating N process $X_{1,t}, X_{2,t}, \ldots, X_{N,t}$, for $t = 1, 2, \ldots, T = 1000$ and $i = 1, 2, \ldots, N = 500$. Unlike Scheme 1, the dimension d_i for $X_{i,t}$ is chosen such that

$$d_i = \begin{cases} d_{1,i} & \text{if } i < \frac{N}{2} \\ d_{2,i} & \text{if } i \geq \frac{N}{2} \end{cases}$$

where $d_{1,i}$ is simulated from discrete uniform distribution over $\{1, 2, \ldots, 14\}$ and $d_{2,i}$ is simulated from discrete uniform distribution over $\{15, 16, \ldots, 30\}$. Observe that in Scheme 2 there is an interaction between the dimension and frequency. The lower dimensional signals simulated in epochs $i < N/2$ has a peak at frequency $\frac{4\pi}{25}$, and the higher dimensional signals simulated in epochs $i \geq N/2$ has a peak at frequency $\frac{4\pi}{5}$.

Scheme 3: Similar to Scheme 1, the d_i-variate process in the ith epoch $X_{i,t} = (X_{1,i,t}, X_{2,i,t}, \ldots, X_{d_i,i,t})'$ are where each $X_{k,i,t}$ are independently generated univariate stationary AR(2) process given by

$$X_{k,i,t} = \phi_{i,1} X_{k,i,t-1} + \phi_{i,2} X_{k,i,t-2} + \epsilon_{k,i,t}$$

$\phi_{i,1} = 2\xi_i \cos(\theta_i)$, $\phi_{i,2} = -\xi_i^2$. The $d_i \times d_i$ variance matrix of the Gaussian noise $\epsilon_{i,t}$ is given by

$$V(\epsilon_{i,t}) = \begin{bmatrix} 1 & \rho & \rho^2 & \cdots & \rho^{p_i-1} \\ \rho & 1 & \rho & \cdots & \rho^{p_i-2} \\ \vdots & & & & \\ \rho^{p_i-1} & \rho^{p_i-2} & \rho^{p_i-3} & \cdots & 1 \end{bmatrix}$$

$\rho = 0.4$ and $k = 1, 2, \ldots, d_i$, $i = 1, 2, \ldots, N = 500$, $t = 1, 2, \ldots, T = 1000$. The dimension d_i for $X_{i,t}$ is randomly chosen from $\{2, 3, \ldots, 30\}$. Here again, $\xi_i \sim U(0.8, 0.98)$ and θ_i is given by

$$\theta_i = \begin{cases} \cos(\frac{4\pi}{25}) & \text{if } i < \frac{N}{2} \\ \cos(\frac{4\pi}{5}) & \text{if } i \geq \frac{N}{2} \end{cases}$$

Scheme 4: Here we let $\sum_{i=1}^{N} d_i = 30$ and follow Scheme 1 in generating the N process $X_{1,t}, X_{2,t}, \ldots, X_{N,T}$, for $t = 1, 2, \ldots, T = 1000$ and $i = 1, 2, \ldots, N = 500$. Then we obtain $M_{i,t} = A_i X_t$ where $A_i = 1_{(i < \frac{N}{2})} I^{d_i} A_1 + 1_{(i \geq \frac{N}{2})} I^{d_i} A_2$ and A_1 and A_2 are two 30×30 randomly generated orthogonal matrices and I^{d_i} is the $d_i \times d$ matrix containing the first d_i rows of the identity matrix I_{30}. We consider $M_{i,t} \in \mathbb{R}^{d_i}$ and study the spread of spectral properties across the $N = 500$ epochs.

Tables A1 and A2 contain the numerical summaries of the FS-ratio statistic over 100 replications of Scheme 1. Please note that the phase parameter θ_i for $i < N/2$ in Scheme 1 is at $4\pi/25$ on a $(0, \pi)$ scale or equivalently at 0.0796 on a $(0, 0.5)$ scale. We see from Table A1 that almost all of the spectral information is contained in the first two chosen frequency ranges around this peak. Similarly for $i \geq N/2$, the phase parameter is at $4\pi/5$ on a $(0, \pi)$ scale or equivalently at 0.3981 on a $(0, 0.5)$ scale. Figure A1 plots a histogram density of the FS-ratio statistic from the 100 replications and similar histogram densities for Schemes 2, 3 and 4 can be found in Figures A2–A4. From Table A2 we notice that the last two chosen frequency ranges have all of the spectral information.

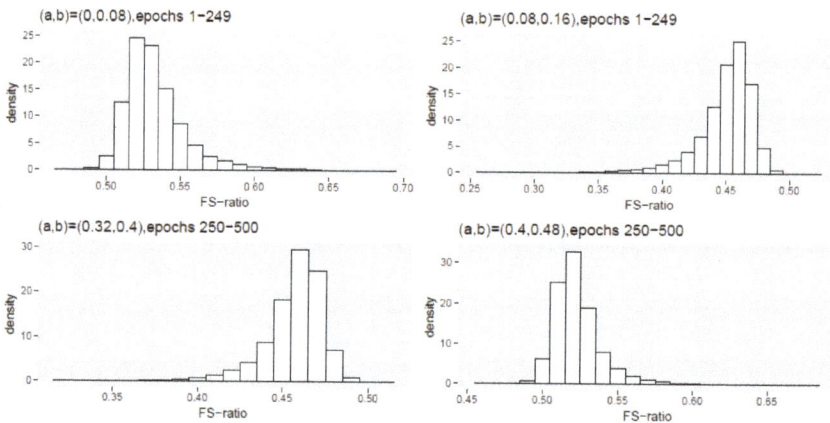

Figure A1. Scheme 1: Histogram density of the FS-ratio statistic for different frequency ranges $(a, b) \subset (0, 0.5)$.

Table A1. Scheme 1, epochs 1–249: Numerical summaries of FS-ratio statistic $\hat{R}_{X_i,a,b}$ for epochs $i = 1, 2, \ldots, 249$ for specified frequency ranges (a, b). Here $(a, b) \subset (0, 0.5)$ and $(0, 0.5)$ corresponds to the interval $(0, \pi)$.

Frequency Range (a,b)	Mean	Median	SD	Lower CI	Upper CI
(0,0.08)	0.5342	0.5391	0.0221	0.4984	0.6253
(0.08,0.16)	0.4486	0.4544	0.0227	0.3566	0.4831
(0.16,0.24)	0.0002	0.0002	0.0001	0.0005	0.0019
(0.24,0.32)	0	0	0	0	0.0002
(0.32,0.40)	0	0	0	0	0
(0.40,0.48)	0	0	0	0	0

Table A2. Scheme 1, epochs 250–500: Numerical summaries of FS-ratio statistic $\hat{R}_{X_i,a,b}$ for epochs $i = 250, 2, \ldots, 500$ for specified frequency ranges (a, b). Here $(a, b) \subset (0, 0.5)$ and $(0, 0.5)$ corresponds to the interval $(0, \pi)$.

Frequency Range (a,b)	Mean	Median	SD	Lower CI	Upper CI
(0,0.08)	0	0	0	0	0
(0.08,0.16)	0	0	0	0	0
(0.16,0.24)	0	0	0	0	0
(0.24,0.32)	0.0003	0.0002	0.0002	0.0005	0.0017
(0.32,0.40)	0.4561	0.4595	0.0181	0.3786	0.4903
(0.40,0.48)	0.5205	0.5210	0.0169	0.4759	0.5826

Tables A3 and A4 include numerical summaries of the FS-ratio statistic over 100 replications of the model in Scheme 2. Similar to Scheme 1, the phase parameter θ_i for $i < N/2$ is at 0.0796 on a $(0, 0.5)$ scale for $i < N/2$ and at 0.3981 on a $(0, 0.5)$ scale for $i \geq N/2$. The results from Table A3 indicate most of the spectral information are present in the first two chosen frequency ranges. Similarly for $i \geq N/2$, Table A4 shows that the last two chosen frequency ranges have all of the spectral information.

Table A3. Scheme 2, epochs 1–249: Numerical summaries of FS-ratio statistic $\hat{R}_{X_i,a,b}$ for epochs $i = 1, 2, \ldots, 249$ for specified frequency ranges (a, b). Here $(a, b) \subset (0, 0.5)$ and $(0, 0.5)$ corresponds to the interval $(0, \pi)$.

Frequency Range (a,b)	Mean	Median	SD	Lower CI	Upper CI
(0,0.08)	0.5407	0.5349	0.0286	0.5041	0.6360
(0.08,0.16)	0.4423	0.4482	0.0284	0.3475	0.4778
(0.16,0.24)	0.0002	0.0003	0.0002	0.0006	0.0022
(0.24,0.32)	0	0	0	0	0.0002
(0.32,0.40)	0	0	0	0	0
(0.40,0.48)	0	0	0	0	0

Table A4. Scheme 2, epochs 250–500: Numerical summaries of FS-ratio statistic $\widehat{R}_{X_i,a,b}$ for epochs $i = 250, 2, \ldots, 500$ for specified frequency ranges (a,b). Here $(a,b) \subset (0, 0.5)$ and $(0, 0.5)$ corresponds to the interval $(0, \pi)$.

Frequency Range (a,b)	Mean	Median	SD	Lower CI	Upper CI
(0,0.08)	0	0	0	0	0
(0.08,0.16)	0	0	0	0	0
(0.16,0.24)	0	0	0	0	0.0001
(0.24,0.32)	0.0003	0.0002	0.0001	0.0005	0.0016
(0.32,0.40)	0.4605	0.4616	0.0108	0.3862	0.4938
(0.40,0.48)	0.5194	0.5186	0.0096	0.4800	0.5863

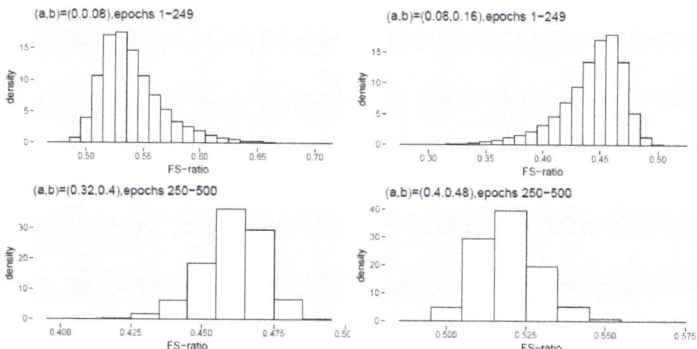

Figure A2. Scheme 2: Histogram density of the FS-ratio statistic for different frequency ranges $(a,b) \subset (0, 0.5)$.

Tables A5 and A6 contain the numerical summaries of the FS-ratio statistic over 100 replications of the model in Scheme 3. As in Scheme 1, the phase parameter θ_i for $i < N/2$ is at 0.0796 on a $(0, 0.5)$ scale for $i < N/2$ and at 0.3981 on a $(0, 0.5)$ scale for $i \geq N/2$. As in Scheme 1, results from Table A5 indicate most of the spectral information are present in the first two chosen frequency ranges. Similarly for $i \geq N/2$, Table A6 shows that the last two chosen frequency ranges have all of the spectral information.

Table A5. Scheme 3, epochs 1–249: Numerical summaries of FS-ratio statistic $\widehat{R}_{X_i,a,b}$ for epochs $i = 1, 2, \ldots, 249$ for specified frequency ranges (a,b). Here $(a,b) \subset (0, 0.5)$ and $(0, 0.5)$ corresponds to the interval $(0, \pi)$.

Frequency Range (a,b)	Mean	Median	SD	Lower CI	Upper CI
(0,0.08)	0.5371	0.5327	0.0238	0.4977	0.6284
(0.08,0.16)	0.4459	0.4504	0.0239	0.3549	0.4843
(0.16,0.24)	0.0002	0.0002	0.0001	0.0005	0.0020
(0.24,0.32)	0	0	0	0	0.0002
(0.32,0.40)	0	0	0	0	0
(0.40,0.48)	0	0	0	0	0

Table A6. Scheme 3, epochs 250–500: Numerical summaries of FS-ratio statistic $\widehat{R}_{X_i,a,b}$ for epochs $i = 250, 2, \ldots, 500$ for specified frequency ranges (a, b). Here $(a, b) \subset (0, 0.5)$ and $(0, 0.5)$ corresponds to the interval $(0, \pi)$.

Frequency Range (a,b)	Mean	Median	SD	Lower CI	Upper CI
(0,0.08)	0	0	0	0	0
(0.08,0.16)	0	0	0	0	0
(0.16,0.24)	0	0	0	0	0.0001
(0.24,0.32)	0.0003	0.0003	0.0002	0.0005	0.0018
(0.32,0.40)	0.4531	0.4566	0.0196	0.3758	0.4907
(0.40,0.48)	0.5252	0.5225	0.0172	0.4810	0.5948

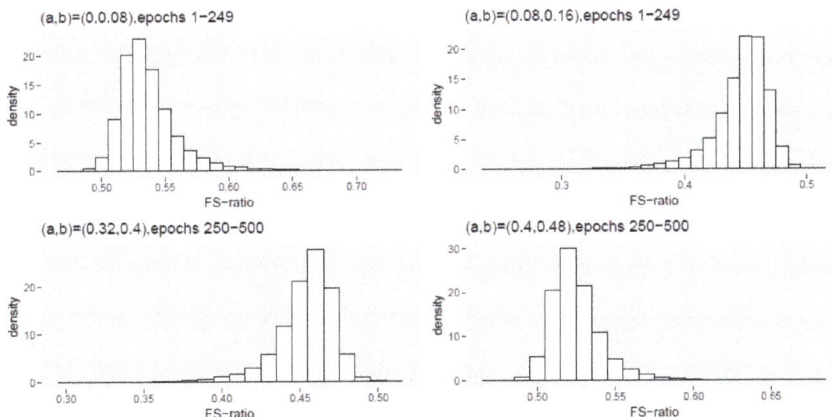

Figure A3. Scheme 3: Histogram density of the FS-ratio statistic for different frequency ranges $(a, b) \subset (0, 0.5)$.

Tables A7 and A8 contain the numerical summaries of the FS-ratio statistic over 100 replications of the model in Scheme 4. Here we look at $M_{i,t} = A_i X_{i,t}$ which is a mixture of the components of $X_{i,t}$ generated as in Scheme 1. Please note that the peak of the spectral densities of the components of $M_{i,t}$ is still at the phase parameter θ_i defined in Scheme 1. Hence, the results from Tables A7 and A8 are similar to the results from Scheme 1.

Table A7. Scheme 4, epochs 1–249: Numerical summaries of FS-ratio statistic $\widehat{R}_{M_{i,t},a,b}$ for epochs $i = 1, 2, \ldots, 249$ for specified frequency ranges (a, b). Here $(a, b) \subset (0, 0.5)$ and $(0, 0.5)$ corresponds to the interval $(0, \pi)$.

Frequency Range (a,b)	Mean	Median	SD	Lower CI	Upper CI
(0,0.08)	0.5342	0.5321	0.0155	0.4871	0.5955
(0.08,0.16)	0.4489	0.4510	0.0159	0.3872	0.4949
(0.16,0.24)	0.0002	0.0002	0.0001	0.0003	0.0011
(0.24,0.32)	0.0001	0.0001	0	0	0.0001
(0.32,0.40)	0	0	0	0	0
(0.40,0.48)	0	0	0	0	0

Table A8. Scheme 4, epochs 250–500: Numerical summaries of FS-ratio statistic $\widehat{R}_{M_i,t,a,b}$ for epochs $i = 250, 2, \ldots, 500$ for specified frequency ranges (a,b). Here $(a,b) \subset (0, 0.5)$ and $(0, 0.5)$ corresponds to the interval $(0, \pi)$.

Frequency Range (a,b)	Mean	Median	SD	Lower CI	Upper CI
(0,0.08)	0	0	0	0	0
(0.08,0.16)	0	0	0	0	0
(0.16,0.24)	0	0	0	0	0.0001
(0.24,0.32)	0.0003	0.0003	0.0001	0.0003	0.0011
(0.32,0.40)	0.4553	0.4570	0.0130	0.4021	0.4987
(0.40,0.48)	0.5234	0.5219	0.0119	0.4782	0.5738

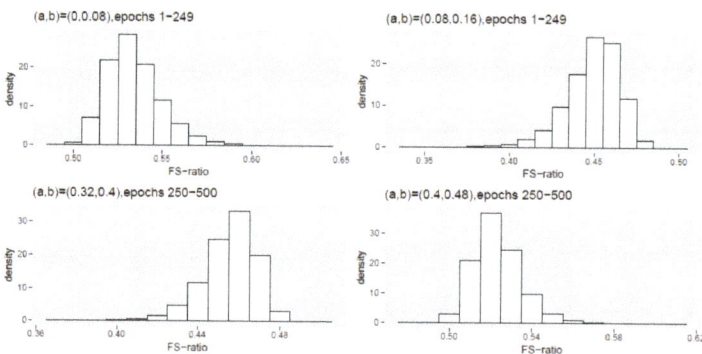

Figure A4. Scheme 4: Histogram density of the FS-ratio statistic for different frequency ranges $(a,b) \subset (0, 0.5)$.

Appendix B. Proofs

Here we present the proofs of the theoretical results in Section 2.1.2.

Proof of Theorem 1 (a). Recall that for some $0 < a < b < \pi$,

$$\widehat{r}_{X,a,b} = \int_a^b ||vec(\widehat{f}_X(\omega))||_2^2 d\omega = \int_a^b ||\frac{1}{T} \sum_{j=-\lfloor T/2 \rfloor}^{\lfloor T/2 \rfloor} K_h(\omega - \omega_j)vec(I_{X,T}(\omega_j))||^2 d\omega$$

$$= \int_a^b \frac{1}{T^2} \sum_{j_1,j_2=-\lfloor T/2 \rfloor}^{\lfloor T/2 \rfloor} K_h(\omega - \omega_{j_1}) K_h(\omega - \omega_{j_2}) \sum_{r,s=1}^{d_i} I_{X,T,rs}(\omega_{j_1}) \overline{I_{X,T,rs}(\omega_{j_2})} d\omega.$$

We first consider the expected value of this quantity.

$$E(\widehat{r}_{X,a,b}) = \int_a^b \frac{1}{T^2} \sum_{j_1,j_2=-\lfloor T/2 \rfloor}^{\lfloor T/2 \rfloor} K_h(\omega - \omega_{j_1}) K_h(\omega - \omega_{j_2}) \sum_{r,s=1}^{d_i} E\left(I_{X,T,rs}(\omega_{j_1}) \overline{I_{X,T,rs}(\omega_{j_2})}\right) d\omega$$

$$= \int_a^b \frac{1}{T^2} \sum_{j_1,j_2=-\lfloor T/2 \rfloor}^{\lfloor T/2 \rfloor} K_h(\omega - \omega_{j_1}) K_h(\omega - \omega_{j_2}) \sum_{r,s=1}^{d_i} f_{X,rs}(\omega_{j_1}) \overline{f_{X,rs}(\omega_{j_2})} d\omega + o(1).$$

It can be seen that as $T \to \infty$, $h \to 0$ and $Th \to \infty$ the above quantity converges to

$$\int_a^b \sum_{r,s=1}^{d_i} \Big(\int_{-\pi}^{\pi} K(v)dv\Big)^2 f_{X,rs}(\omega)\overline{f_{X,rs}(\omega)}\, d\omega = \int_a^b \sum_{r,s=1}^{d_i} f_{X,rs}(\omega)\overline{f_{X,rs}(\omega)}\, d\omega.$$

Next, for the variance we have $V(\widehat{r}_{X,a,b}) = A_1 - A_2$, where

$$A_1 = \int_a^b \int_a^b \frac{1}{T^4} \sum_{j_1,j_2,j_3,j_4} K_h(\omega - \omega_{j_1}) K_h(\omega - \omega_{j_2}) K_h(\lambda - \omega_{j_3}) K_h(\lambda - \omega_{j_4}) \times$$

$$\sum_{r,s,t,u=1}^{d_i} E\Big(I_{X,T,rs}(\omega_{j_1})\overline{I_{X,T,rs}(\omega_{j_2})} I_{X,T,tu}(\omega_{j_3})\overline{I_{X,T,tu}(\omega_{j_4})}\Big) d\omega\, d\lambda \text{ and}$$

$$A_2 = \int_a^b \int_a^b \frac{1}{T^4} \sum_{j_1,j_2,j_3,j_4} K_h(\omega - \omega_{j_1}) K_h(\omega - \omega_{j_2}) K_h(\lambda - \omega_{j_3}) K_h(\lambda - \omega_{j_4}) \times$$

$$\sum_{r,s,t,u=1}^{d_i} E\Big(I_{X,T,rs}(\omega_{j_1})\overline{I_{X,T,rs}(\omega_{j_2})}\Big) E\Big(I_{X,T,tu}(\omega_{j_3})\overline{I_{X,T,tu}(\omega_{j_4})}\Big) d\omega\, d\lambda.$$

For the difference in the expectations between A_1 and A_2 we discuss the relevant cases and their convergence to 0. Firstly, it can be seen that for the following three cases the difference in the expectations is asymptotically 0: (a). $\omega_{j_1} \neq \omega_{j_2} \neq \omega_{j_3} \neq \omega_{j_4}$, (b). $\omega_{j_1} = \omega_{j_2} \neq \omega_{j_3} \neq \omega_{j_4}$, (c). $\omega_{j_1} = \omega_{j_2} = \omega_{j_3} = \omega_{j_4}$. Next, when $\omega_{j_1} = \omega_{j_3} \neq \omega_{j_2} = \omega_{j_4}$ we have,

$$\int_a^b \int_a^b \frac{1}{T^4} \sum_{j_1,j_2} K_h(\omega - \omega_{j_1}) K_h(\omega - \omega_{j_2}) K_h(\lambda - \omega_{j_1}) K_h(\lambda - \omega_{j_2}) \sum_{r,s,t,u=1}^{d_i} \Big[E\Big(I_{X,T,rs}(\omega_{j_1})\overline{I_{X,T,rs}(\omega_{j_2})} \times$$

$$I_{X,T,tu}(\omega_{j_1})\overline{I_{X,T,tu}(\omega_{j_2})}\Big) - E\Big(I_{X,T,rs}(\omega_{j_1})\overline{I_{X,T,rs}(\omega_{j_2})}\Big) E\Big(I_{X,T,tu}(\omega_{j_1})\overline{I_{X,T,tu}(\omega_{j_2})}\Big) \Big] d\omega\, d\lambda$$

$$= \int_a^b \int_a^b \frac{1}{T^4} \sum_{j_1,j_2} K_h(\omega - \omega_{j_1}) K_h(\omega - \omega_{j_2}) K_h(\lambda - \omega_{j_1}) K_h(\lambda - \omega_{j_2}) \sum_{r,s,t,u=1}^{d_i} \Big[E\Big(I_{X,T,rs}(\omega_{j_1}) I_{X,T,tu}(\omega_{j_1})\Big) \times$$

$$E\Big(\overline{I_{X,T,rs}(\omega_{j_2})} \overline{I_{X,T,tu}(\omega_{j_2})}\Big) - E\Big(I_{X,T,rs}(\omega_{j_1})\Big) E\Big(\overline{I_{X,T,rs}(\omega_{j_2})}\Big) E\Big(I_{X,T,tu}(\omega_{j_1})\Big) E\Big(\overline{I_{X,T,tu}(\omega_{j_2})}\Big) \Big] d\omega\, d\lambda + o(1)$$

$$= \frac{1}{T^4} \sum_{j_1,j_2} \Big(\int_a^b K_h(\omega - \omega_{j_1}) K_h(\omega - \omega_{j_2}) d\omega \Big)^2 \sum_{r,s,t,u=1}^{d_i} \Big[\big(f_{X,rt}(\omega_{j_1}) f_{X,su}(\omega_{j_1}) + f_{X,rs}(\omega_{j_1}) f_{X,tu}(\omega_{j_1})\big) \times$$

$$\big(\overline{f_{X,rt}(\omega_{j_2})}\,\overline{f_{X,su}(\omega_{j_2})} + \overline{f_{X,rs}(\omega_{j_2})}\,\overline{f_{X,tu}(\omega_{j_2})}\big) - \big(f_{X,rs}(\omega_{j_1})\overline{f_{X,rs}(\omega_{j_2})} f_{X,tu}(\omega_{j_1})\overline{f_{X,tu}(\omega_{j_2})}\big) \Big]$$

$$+ o(1) = \frac{1}{T^4 h^2} \sum_{j_1,j_2} \Big(\int_{\frac{a-\omega_{j_1}}{h}}^{\frac{b-\omega_{j_1}}{h}} K(u) K(u + \frac{\omega_{j_1} - \omega_{j_2}}{h}) du \Big)^2 \sum_{r,s,t,u=1}^{d_i} [\cdots] = O\Big(\frac{1}{T^2 h}\Big).$$

The case when $\omega_{j_1} = \omega_{j_2} = \omega_{j_3} \neq \omega_{j_4}$ would have the same rate of decay as above. Next, when $\omega_{j_1} = \omega_{j_3} \neq \omega_{j_2} \neq \omega_{j_4}$ we have,

$$\int_a^b \int_a^b \frac{1}{T^4} \sum_{j_1,j_2,j_4} K_h(\omega - \omega_{j_1}) K_h(\omega - \omega_{j_2}) K_h(\lambda - \omega_{j_1}) K_h(\lambda - \omega_{j_4}) \sum_{r,s,t,u=1}^{d_i} \Big[E\Big(I_{X,T,rs}(\omega_{j_1}) \overline{I_{X,T,rs}(\omega_{j_2})} \times$$

$$I_{X,T,tu}(\omega_{j_1}) \overline{I_{X,T,tu}(\omega_{j_4})}\Big) - E\Big(I_{X,T,rs}(\omega_{j_1}) \overline{I_{X,T,rs}(\omega_{j_2})}\Big) E\Big(I_{X,T,tu}(\omega_{j_1}) \overline{I_{X,T,tu}(\omega_{j_4})}\Big) \Big] d\omega \, d\lambda$$

$$= \int_a^b \int_a^b \frac{1}{T^4} \sum_{j_1,j_2,j_4} K_h(\omega - \omega_{j_1}) K_h(\omega - \omega_{j_2}) K_h(\lambda - \omega_{j_1}) K_h(\lambda - \omega_{j_4}) \sum_{r,s,t,u=1}^{d_i} \Big[E\Big(I_{X,T,rs}(\omega_{j_1}) I_{X,T,tu}(\omega_{j_1})\Big) \times$$

$$E\Big(\overline{I_{X,T,rs}(\omega_{j_2})}\Big) E\Big(\overline{I_{X,T,tu}(\omega_{j_4})}\Big) - E\Big(I_{X,T,rs}(\omega_{j_1})\Big) E\Big(\overline{I_{X,T,rs}(\omega_{j_2})}\Big) E\Big(I_{X,T,tu}(\omega_{j_1})\Big) E\Big(\overline{I_{X,T,tu}(\omega_{j_4})}\Big) \Big] d\omega \, d\lambda + o(1)$$

$$= \frac{1}{T^4} \sum_{j_1,j_2,j_4} \Big(\int_a^b K_h(\omega - \omega_{j_1}) K_h(\omega - \omega_{j_2}) \, d\omega \Big) \Big(\int_a^b K_h(\lambda - \omega_{j_1}) K_h(\lambda - \omega_{j_4}) \, d\lambda \Big) \sum_{r,s,t,u=1}^{d_i} \Big[\Big(f_{X,rt}(\omega_{j_1}) \times$$

$$f_{X,su}(\omega_{j_1}) + f_{X,rs}(\omega_{j_1}) f_{X,tu}(\omega_{j_1})\Big) \times \Big(\overline{f_{X,rs}(\omega_{j_2})} \, \overline{f_{X,tu}(\omega_{j_4})}\Big) - \Big(f_{X,rs}(\omega_{j_1}) \overline{f_{X,rs}(\omega_{j_2})} f_{X,tu}(\omega_{j_1}) \overline{f_{X,tu}(\omega_{j_4})}\Big) \Big]$$

$$+ o(1) = \frac{1}{T^4 h^2} \sum_{j_1,j_2,j_4} \Big(\int_{\frac{a-\omega_{j_1}}{h}}^{\frac{b-\omega_{j_1}}{h}} K(u) K(u + \frac{\omega_{j_1} - \omega_{j_2}}{h}) \, du \Big) \Big(\int_{\frac{a-\omega_{j_1}}{h}}^{\frac{b-\omega_{j_1}}{h}} K(v) K(v + \frac{\omega_{j_1} - \omega_{j_4}}{h}) \, dv \Big) \times$$

$$\sum_{r,s,t,u=1}^{d_i} \Big[\cdots \Big] + o(1) = O(\frac{1}{T}).$$

Finally, we look at the case $\omega_{j_1} = \omega_{j_2} = \omega_{j_3} = \omega_{j_4}$. We have

$$\int_a^b \int_a^b \frac{1}{T^4} \sum_{j_1} K_h^2(\omega - \omega_{j_1}) K_h^2(\lambda - \omega_{j_1}) \sum_{r,s,t,u=1}^{d_i} \Big[E\Big(I_{X,T,rs}(\omega_{j_1}) \overline{I_{X,T,rs}(\omega_{j_1})} \times$$

$$I_{X,T,tu}(\omega_{j_1}) \overline{I_{X,T,tu}(\omega_{j_1})}\Big) - E\Big(I_{X,T,rs}(\omega_{j_1}) \overline{I_{X,T,rs}(\omega_{j_1})}\Big) E\Big(I_{X,T,tu}(\omega_{j_1}) \overline{I_{X,T,tu}(\omega_{j_1})}\Big) \Big] d\omega \, d\lambda$$

$$= \frac{1}{T^4} \sum_{j_1} \Big(\int_a^b K_h^2(\omega - \omega_{j_1}) \, d\omega \Big)^2 \sum_{r,s,t,u=1}^{d_i} \Big[\cdots \Big] = \frac{1}{T^4 h^4} \sum_{j_1} \Big(\int_a^b K^2(\frac{\omega - \omega_{j_1}}{h}) d\omega \Big)^2 \sum_{r,s,t,u=1}^{d_i} \Big[\cdots \Big]$$

$$= \frac{1}{T^4 h^2} \sum_{j_1} \Big(\int_{\frac{a-\omega_{j_1}}{h}}^{\frac{b-\omega_{j_1}}{h}} K^2(u) du \Big)^2 \sum_{r,s,t,u=1}^{d_i} \Big[\cdots \Big] = O(\frac{1}{T^3 h^2})$$

□

Proof of Theorem 1 (b). First, we observe that

$$\widehat{R}_{X,a,b} = \frac{\int_a^b ||vec(\widehat{f}_X(\omega))||_2^2 d\omega}{\int_0^\pi ||vec(\widehat{f}_X(\omega))||_2^2 d\omega} = \frac{\int_a^b ||vec(\widehat{f}_X(\omega))||_2^2 d\omega}{\int_a^b ||vec(\widehat{f}_X(\omega))||_2^2 d\omega + \int_{\overline{\Pi}_{(a,b)}} ||vec(\widehat{f}_X(\omega))||_2^2 d\omega}$$

$$= \Big(1 + \frac{\int_{\overline{\Pi}_{(a,b)}} ||vec(\widehat{f}_X(\omega))||_2^2 d\omega}{\int_a^b ||vec(\widehat{f}_X(\omega))||_2^2 d\omega} \Big)^{-1}. \tag{A1}$$

A sufficient condition for joint consistency of $(\hat{r}_{X,a,b}, \hat{r}_{X,\overline{\Pi}_{(a,b)}})^\top$. Following the proof of Theorem 1 (a), we have $cov(\hat{r}_{X,a,b}, \hat{r}_{X,\overline{\Pi}_{(a,b)}}) = C_1 - C_2$, where

$$C_1 = \int_{\overline{\Pi}_{(a,b)}} \int_a^b \frac{1}{T^4} \sum_{j_1,j_2,j_3,j_4} K_h(\omega - \omega_{j_1}) K_h(\omega - \omega_{j_2}) K_h(\lambda - \omega_{j_3}) K_h(\lambda - \omega_{j_4}) \times$$

$$\sum_{r,s,t,u=1}^{d_i} E\left(I_{X,T,rs}(\omega_{j_1}) \overline{I_{X,T,rs}(\omega_{j_2})} I_{X,T,tu}(\omega_{j_3}) \overline{I_{X,T,tu}(\omega_{j_4})}\right) d\omega \, d\lambda \text{ and}$$

$$C_2 = \int_{\overline{\Pi}_{(a,b)}} \int_a^b \frac{1}{T^4} \sum_{j_1,j_2,j_3,j_4} K_h(\omega - \omega_{j_1}) K_h(\omega - \omega_{j_2}) K_h(\lambda - \omega_{j_3}) K_h(\lambda - \omega_{j_4}) \times$$

$$\sum_{r,s,t,u=1}^{d_i} E\left(I_{X,T,rs}(\omega_{j_1}) \overline{I_{X,T,rs}(\omega_{j_2})}\right) E\left(I_{X,T,tu}(\omega_{j_3}) \overline{I_{X,T,tu}(\omega_{j_4})}\right) d\omega \, d\lambda.$$

As in the proof of Theorem 1 (a), it can be seen that, for the various cases, the covariance terms are of $O(\frac{1}{T^{\delta_1} h^{\delta_2}})$ where $\delta_1, \delta_2 \in \{0,1,2,3\}$ and $\delta_1 > \delta_2$. The result above along with Theorem 1 implies

$$\left(\hat{r}_{X,a,b}, \hat{r}_{X,\overline{\Pi}_{(a,b)}}\right)^\top \xrightarrow{P} \left(r_{X,a,b}, r_{X,\overline{\Pi}_{(a,b)}}\right)^\top.$$

Finally, an application of the continuous mapping theorem yields the result. □

References

1. Fiecas, M.; Ombao, H. Modeling the Evolution of Dynamic Brain Processes During an Associative Learning Experiment. *J. Am. Stat. Assoc.* **2016**, *111*, 1440–1453. [CrossRef]
2. Ombao, H.; Fiecas, M.; Ting, C.M.; Low, Y.F. Statistical models for brain signals with properties that evolve across trials. *NeuroImage* **2018**, *180*, 609–618. [CrossRef] [PubMed]
3. Cribben, I.; Haraldsdottir, R.; Atlas, L.Y.; Wager, T.D.; Lindquist, M.A. Dynamic connectivity regression: Determining state-related changes in brain connectivity. *NeuroImage* **2012**, *61*, 907–920. [CrossRef]
4. Cribben, I.; Wager, T.; Lindquist, M. Detecting functional connectivity change points for single-subject fMRI data. *Front. Comput. Neurosci.* **2013**, *7*, 143. [CrossRef] [PubMed]
5. Cribben, I.; Yu, Y. Estimating whole-brain dynamics by using spectral clustering. *J. R. Stat. Soc. Ser. (Appl. Stat.)* **2016**, *66*, 607–627. [CrossRef]
6. Zhu, Y.; Cribben, I. Sparse Graphical Models for Functional Connectivity Networks: Best Methods and the Autocorrelation Issue. *Brain Connect.* **2018**, *8*, 139–165. [CrossRef] [PubMed]
7. Wang, Y.; Ting, C.; Ombao, H. Modeling Effective Connectivity in High-Dimensional Cortical Source Signals. *IEEE J. Sel. Top. Signal Process.* **2016**, *10*, 1315–1325. [CrossRef]
8. Brillinger, D. *Time Series*; Society for Industrial and Applied Mathematics: Philadelphia, PA, USA, 2001. [CrossRef]
9. Eichler, M. Testing nonparametric and semiparametric hypotheses in vector stationary processes. *J. Multivar. Anal.* **2008**, *99*, 968–1009. [CrossRef]
10. Politis, D.N.; Romano, J.P. The Stationary Bootstrap. *J. Am. Stat. Assoc.* **1994**, *89*, 1303–1313. [CrossRef]
11. Dette, H.; Paparoditis, E. Bootstrapping frequency domain tests in multivariate time series with an application to comparing spectral densities. *J. R. Stat. Soc. Ser. B* **2009**, *71*, 831–857. [CrossRef]
12. Dwivedi, Y.; Subba Rao, S. A test for second-order stationarity of a time series based on the discrete Fourier transform. *J. Time Ser. Anal.* **2011**, *32*, 68–91. [CrossRef]
13. Ombao, H.; von Sachs, R.; Guo, W. SLEX Analysis of Multivariate Nonstationary Time Series. *J. Am. Stat. Assoc.* **2005**, *100*, 519–531. [CrossRef]

14. Srinivasan, R. *High-Resolution EEG: Theory and Practice in Event-Related Potentials: A Methods Handbook*; Handy, T.C., Ed.; MIT Press: Cambridge, MA, USA, 2003.
15. Nunez, P.; Srinivasan, R. *Electric Fields of the Brain: The Neurophysics of EEG*, 2nd ed.; Ocford University Press: New York, NY, USA, 2003.
16. von Bünau, P.; Meinecke, F.C.; Scholler, S.; Müller, K.R. Finding stationary brain sources in EEG data. In Proceedings of the 2010 Annual International Conference of the IEEE Engineering in Medicine and Biology, Buenos Aires, Argentina, 31 August–4 September 2010; pp. 2810–2813. [CrossRef]
17. Wu, J.; Srinivasan, R.; Quinlan, E.B.; Solodkin, A.; Small, S.L.; Cramer, S.C. Utility of EEG measures of brain function in patients with acute stroke. *J. Neurophysiol.* **2016**, *115*, 2399–2405. [CrossRef] [PubMed]
18. Gao, X.; Shababa, B.; Fortin, N.; Ombao, H. Evolutionary State-Space Models With Applications to Time-Frequency Analysis of Local Field Potentials. *arXiv* **2016**, arXiv:1610.07271.
19. Euán, C.; Sun, Y.; Ombao, H. Coherence-based time series clustering for statistical inference and visualization of brain connectivity. *Ann. Appl. Stat.* **2019**, *13*, 990–1015. [CrossRef]
20. Kaplan, A.Y.; Fingelkurts, A.A.; Fingelkurts, A.A.; Borisov, S.V.; Darkhovsky, B.S. Nonstationary nature of the brain activity as revealed by EEG/MEG: Methodological, practical and conceptual challenges. *Signal Process.* **2005**, *85*, 2190–2212. [CrossRef]
21. von Bünau, P.; Meinecke, F.C.; Király, F.C.; Müller, K.R. Finding Stationary Subspaces in Multivariate Time Series. *Phys. Rev. Lett.* **2009**, *103*, 214101. [CrossRef]
22. Sundararajan, R.; Pipiras, V.; Pourahmadi, M. Stationary subspace analysis of nonstationary covariance processes: eigenstructure description and testing. *arXiv* **2019**, arXiv:1904.09420.
23. Wann, E.G. Large-Scale Spatiotemporal Neuronal Activity Dynamics Predict Cortical Viability in a Rodent Model of Ischemic Stroke. Ph.D. Thesis, UC Irvine, Irvine, CA, USA, 2017.
24. Sundararajan, R.R.; Pourahmadi, M. Stationary subspace analysis of nonstationary processes. *J. Time Ser. Anal.* **2018**, *39*, 338–355. [CrossRef]
25. Politis, D.N.; White, H. Automatic Block-Length Selection for the Dependent Bootstrap. *Econom. Rev.* **2004**, *23*, 53–70. [CrossRef]
26. Patton, A.; Politis, D.N.; White, H. Correction to "Automatic Block-Length Selection for the Dependent Bootstrap" by D. Politis and H. White. *Econom. Rev.* **2009**, *28*, 372–375. [CrossRef]
27. Fontaine, C.; Frostig, R.D.; Ombao, H. Modeling non-linear spectral domain dependence using copulas with applications to rat local field potentials. *Econom. Stat.* **2020**, *15*, 85–103. [CrossRef]
28. Piangerelli, M.; Rucco, M.; Tesei, L.; Merelli, E. Topological classifier for detecting the emergence of epileptic seizures. *BMC Res. Notes* **2018**, *11*, 392. [CrossRef] [PubMed]
29. Rucco, M.; Concettoni, E.; Cristalli, C.; Ferrante, A.; Merelli, E. Topological classification of small DC motors. In Proceedings of the 2015 IEEE 1st International Forum on Research and Technologies for Society and Industry Leveraging a Better Tomorrow (RTSI), Turin, Italy, 16–18 September 2015; pp. 192–197. [CrossRef]
30. Wang, Y.; Ombao, H.; Chung, M.K. Topological data analysis of single-trial electroencephalographic signals. *Ann. Appl. Stat.* **2018**, *12*, 1506–1534. [CrossRef] [PubMed]
31. Brockwell, P.J.; Davis, R.A. *Time Series: Theory and Methods*; Springer: New York, NY, USA, 1991. [CrossRef]

Publisher's Note: MDPI stays neutral with regard to jurisdictional claims in published maps and institutional affiliations.

© 2020 by the authors. Licensee MDPI, Basel, Switzerland. This article is an open access article distributed under the terms and conditions of the Creative Commons Attribution (CC BY) license (http://creativecommons.org/licenses/by/4.0/).

entropy

Article

Asymptotic Properties of Estimators for Seasonally Cointegrated State Space Models Obtained Using the CVA Subspace Method

Dietmar Bauer * and Rainer Buschmeier

Department of Business Administration and Economics, Bielefeld University, Universitaetsstrasse 25, 33615 Bielefeld, Germany; RBuschmeier@uni-bielefeld.de
* Correspondence: Dietmar.Bauer@uni-bielefeld.de

Citation: Bauer, D.; Buschmeier, R. Asymptotic Properties of Estimators for Seasonally Cointegrated State Space Models Obtained Using the CVA Subspace Method. *Entropy* **2021**, 23, 436. https://doi.org/10.3390/e23040436

Academic Editor: Christian H. Weiss

Received: 19 February 2021
Accepted: 31 March 2021
Published: 8 April 2021

Publisher's Note: MDPI stays neutral with regard to jurisdictional claims in published maps and institutional affiliations.

Copyright: © 2021 by the authors. Licensee MDPI, Basel, Switzerland. This article is an open access article distributed under the terms and conditions of the Creative Commons Attribution (CC BY) license (https://creativecommons.org/licenses/by/4.0/).

Abstract: This paper investigates the asymptotic properties of estimators obtained from the so called CVA (canonical variate analysis) subspace algorithm proposed by Larimore (1983) in the case when the data is generated using a minimal state space system containing unit roots at the seasonal frequencies such that the yearly difference is a stationary vector autoregressive moving average (VARMA) process. The empirically most important special cases of such data generating processes are the I(1) case as well as the case of seasonally integrated quarterly or monthly data. However, increasingly also datasets with a higher sampling rate such as hourly, daily or weekly observations are available, for example for electricity consumption. In these cases the vector error correction representation (VECM) of the vector autoregressive (VAR) model is not very helpful as it demands the parameterization of one matrix per seasonal unit root. Even for weekly series this amounts to 52 matrices using yearly periodicity, for hourly data this is prohibitive. For such processes estimation using quasi-maximum likelihood maximization is extremely hard since the Gaussian likelihood typically has many local maxima while the parameter space often is high-dimensional. Additionally estimating a large number of models to test hypotheses on the cointegrating rank at the various unit roots becomes practically impossible for weekly data, for example. This paper shows that in this setting CVA provides consistent estimators of the transfer function generating the data, making it a valuable initial estimator for subsequent quasi-likelihood maximization. Furthermore, the paper proposes new tests for the cointegrating rank at the seasonal frequencies, which are easy to compute and numerically robust, making the method suitable for automatic modeling. A simulation study demonstrates by example that for processes of moderate to large dimension the new tests may outperform traditional tests based on long VAR approximations in sample sizes typically found in quarterly macroeconomic data. Further simulations show that the unit root tests are robust with respect to different distributions for the innovations as well as with respect to GARCH-type conditional heteroskedasticity. Moreover, an application to Kaggle data on hourly electricity consumption by different American providers demonstrates the usefulness of the method for applications. Therefore the CVA algorithm provides a very useful initial guess for subsequent quasi maximum likelihood estimation and also delivers relevant information on the cointegrating ranks at the different unit root frequencies. It is thus a useful tool for example in (but not limited to) automatic modeling applications where a large number of time series involving a substantial number of variables need to be modelled in parallel.

Keywords: cointegration; subspace algorithms; VARMA models; seasonality

JEL Classification: C13; C32

1. Introduction

Many time series show seasonal patterns that, according to [1] for example, cannot be modeled appropriately using seasonal dummies because they exhibit a slowly trending behavior typical for unit root processes.

To model such processes in the vector autoregressive (VAR) framework, Ref. [2] (abbreviated as JS in the following) extend the error correction representation for seasonally integrated autoregressive processes pioneered by [3] to the multivariate case. This vector error correction formulation (VECM) models the yearly differences of a process observed S times per year. The model includes systems having unit roots at some or all of the possible locations $z_j = \exp(\frac{2\pi j}{S}i), j = 0, ..., S-1$ of seasonal unit roots. In JS all unit roots are assumed to be simple such that the process of yearly differences is stationary.

In this setting JS propose an estimator for the autoregressive polynomial subject to restrictions on its rank (the so-called cointegrating rank) at the unit roots z_j based on an iterative scheme focusing on a pair of complex-conjugated unit roots (or the unit roots $z_j = 1$ or $z_j = -1$ respectively) at a time. The main idea here is the reformulation of the model using the so called vector error correction representation. Beside estimators JS also derived likelihood ratio tests for the cointegrating rank at the various unit roots.

Refs. [4,5] propose simpler estimation schemes based on complex reduced rank regression (cRRR in the following). They also show that their numerically simpler algorithm leads to test statistics for the cointegrating rank that are asymptotically equivalent to the quasi maximum likelihood tests of JS. These schemes still typically alternate between cRRR problems corresponding to different unit roots until convergence, although a one step version estimating only once at each unit root exists. Ref. [6] provides updating equations for quasi maximum likelihood estimation in situations where constraints on the parameters prohibit focusing on one unit root at a time.

The leading case here is that of quarterly data ($S = 4$) where potential unit roots are located at ± 1 and $\pm i$, implying that the VECM representation contains four potentially rank restricted matrices. However, increasingly time series of much higher sampling frequency such as hourly, daily or weekly observations are available. In such cases it is unrealistic that all unit roots are present. If a unit root is not present, the corresponding matrix in the VECM is of full rank. Therefore in situations with only a few unit roots being present, the VECM requires a large number of parameters to be estimated. Also in cases with a long period length (such as for example hourly data with yearly cycles) usage of the VECM involves the estimation of all coefficient matrices for lags for at least one year.

In general, for processes of moderate to large dimension the VAR framework involves estimation of a large number of parameters which potentially can be avoided by using the more flexible vector autoregressive moving average (VARMA) or the—in a sense—equivalent state space framework. This setting has been used in empirical research for the modeling of electricity markets, see the survey [7] for a long list of contributions. In particular, ref. [8] use the model described below without formal verification of the asymptotic theory for the quasi maximum likelihood estimation.

Recently, ref. [9] show that in the setting of dynamic factor models, typically used for observation processes of high dimension, the common assumption that the factors are generated using a vector autoregression jointly with the assumption that the idiosyncratic component is white noise (or more generally generated using a VAR or VARMA model independent of the factors) leads to a VARMA process. Also a number of papers (see for example [10–12]) show that in their empirical application the usage of VARMA models instead of approximations using the VAR model leads to superior prediction performance. This, jointly with the fact that the linearization of dynamic stochastic general equilibrium models (DSGE) leads to state space models, see e.g., [13], has fuelled recent interest in VARMA—and thus state space—modeling in particular in macroeconomics, see for example [14].

In this respect, quasi maximum likelihood estimation is the most often used approach for inference. Due to the typically highly non-convex nature of the quasi likelihood function (using the Gaussian density) in the VARMA setting, the criterion function shows many local maxima where the optimization can easily get stuck. Randomization alone does not solve the problem efficiently, as typically the parameter space is high-dimensional causing problems of the curse of dimensionality type.

Moreover, VARMA modeling requires a full specification of the state space unit root structure of the process, see [15]. The state space unit root structure specifies the number of common trends at each seasonal frequency (see below for definitions). For data of weekly or higher sampling frequency it is unlikely that the state space unit root structure is known prior to estimation. Testing all possible combinations is numerically infeasible in many cases.

As an attractive alternative in this respect the class of subspace algorithms is investigated in this paper. One particular member of this class, the so called canonical variate analysis (CVA) introduced by [16] (in the literature the algorithm is often called canonical correlation analysis; CCA), has been shown to provide system estimators which (under the assumption of known system order) are asymptotically equivalent to quasi maximum likelihood estimation (using the Gaussian likelihood) in the stationary case [17]. CVA shares a number of robustness properties in the stationary case with VAR estimators: [18] shows that CVA produces consistent estimators of the underlying transfer function in situations where the innovations are conditionally heteroskedastic processes of considerable generality. Ref. [19] shows that CVA provides consistent estimators of the transfer function even for stationary fractionally integrated processes, if the order of the system tends to infinity as a function of the sample size at a sufficient rate.

In the I(1) case [20] introduce a heuristic adaptation of the algorithm using the assumption of known cointegrating rank in order to show consistency for the corresponding transfer function estimators. However, the specification of the cointegrating rank is no easy task in itself. In case of misspecification of the cointegrating rank the properties of this approach are unclear. Ref. [21] states without proof that also the original CVA algorithm delivers consistent estimates in the I(1) case without the need to impose the true cointegrating rank.

Furthermore for I(1) processes [20] proposed various tests for the cointegrating rank and compared them to tests in the Johansen framework showing superior finite sample performance in particular for multivariate data sets with a large dimension of the modeled process.

This paper builds on these results and shows that CVA can also be used in the seasonally integrated case. The main contributions of the paper are:

(i) It is shown that the original CVA algorithm in the seasonally integrated case provides strongly consistent system estimators under the assumption of known system order (thus delivering the currently unpublished proof of the claim in the I(1) case in [21]).
(ii) Upper bounds for the order of convergence for the estimated system matrices are given, establishing the familiar superconsistency for the estimation of the cointegrating spaces at all unit roots.
(iii) Several tests for separate (that is for each unit root irrespective of the specification at the other potential unit roots) determination of the seasonal cointegrating ranks are proposed which are based on the estimated systems and are simple to implement.

The derivation of the asymptotic properties of the estimators is complemented by a simulation study and an application, both demonstrating the potential of CVA and one of the suggested tests. Jointly our results imply that CVA constitutes a very reasonable initial estimate for subsequent quasi likelihood maximization in the VARMA case. Moreover the method provides valuable information on the number of unit roots present in the process, which can be used for subsequent investigation at the very least by providing upper bounds on the number of common trends present at each unit root frequency. Contrary to the JS approach in the VAR framework these tests can be performed in parallel for all unit roots, eliminating the interdependence of the results inherent in the VECM representation. Moreover, they do not use the VECM representation involving a large number of parameters in the case of high sampling rates.

These properties make CVA a useful tool in automatic modeling of multivariate (with a substantial number of variables) seasonally (co-)integrated processes.

The paper is organized as follows: in the next section the model set and the main assumptions of the paper are presented. The estimation methods are described in Section 3. Section 4 states the consistency results. Inference on the cointegrating ranks is proposed in Section 5. Data preprocessing is discussed in Section 6. The simulations are contained in Section 7, while Section 8 discusses an application to real world data. Section 9 concludes the paper. Appendix A contains supporting material, Appendix C provides the proofs of the main results of this paper, which are based on preliminary results presented in Appendix B.

Throughout the paper we will use the symbols $o(g_T)$ and $O(g_T)$ to denote orders of almost sure convergence where T denotes the sample size, i.e., $x_T = o(g_T)$ if $x_T/g_T \to 0$ almost surely and $x_T = O(g_T)$ if x_T/g_T is bounded almost surely for large enough T (that is there exists a constant $M < \infty$ such that $\limsup_{T\to\infty} x_T/g_T \leq M$ a.s.). Furthermore, $o_P(g_T), O_P(g_T)$ denote the corresponding in probability versions.

2. Model Set and Assumptions

In this paper state space processes $(y_t)_{t\in\mathbb{Z}}, y_t \in \mathbb{R}^s$, are considered which are defined as the solutions to the following equations for given white noise $(\varepsilon_t)_{t\in\mathbb{Z}}, \varepsilon_t \in \mathbb{R}^s, \mathbb{E}\varepsilon_t = 0, \mathbb{E}\varepsilon_t\varepsilon_t' = \Omega > 0$:

$$\begin{aligned} x_{t+1} &= Ax_t + K\varepsilon_t, \\ y_t &= Cx_t + \varepsilon_t. \end{aligned} \quad (1)$$

Here $x_t \in \mathbb{R}^n$ denotes the unobserved state and $A \in \mathbb{R}^{n\times n}, C \in \mathbb{R}^{s\times n}$ and $K \in \mathbb{R}^{n\times s}$ define the state space system typically written as the tuple (A, C, K).

In this paper we consider without restriction of generality only minimal state space systems in innovations representation. For a minimal system the integer n is called the order of the system. As is well known (cf. e.g., [22]) minimal systems are only identified up to the choice of the basis of the state space. Two minimal systems (A, C, K) and $(\tilde{A}, \tilde{C}, \tilde{K})$ are observationally equivalent if and only if there exists a nonsingular matrix $\mathcal{T} \in \mathbb{R}^{n\times n}$ such that $A = \mathcal{T}\tilde{A}\mathcal{T}^{-1}, C = \tilde{C}\mathcal{T}^{-1}, K = \mathcal{T}\tilde{K}$. For two observationally equivalent systems the impulse response sequences $k_0 = I_s, k_{j+1} = CA^jK = \tilde{C}\tilde{A}^j\tilde{K}, j = 0, 1, \ldots$ coincide.

Ref. [15] shows that the structure of the Jordan normal form of the matrix A determines the properties (such as stationarity) of the solutions to (1) for $t \in \mathbb{Z}$. Eigenvalues of A on the unit circle lead to unit root processes in the sense of [15] who also define a *state space unit root structure* indicating the location and multiplicity of unit roots. A process $(y_t)_{t\in\mathbb{Z}}$ with state space unit root structure $\Omega_S = \{(0, (c_0)), (2\pi/S, (c_1)), \ldots, (\pi, (c_{S/2}))\}$ for some even integer S is called multi frequency I(1) (in short MFI(1)). Even S is chosen because it simplifies the notation by implying that $S/2$ also is an integer and $z = -1$ is a seasonal unit root. By adjusting the notation appropriately all results hold true for odd S as well).

If, moreover, such a process is observed for S periods per year, it is called *seasonal MFI(1)*. In this case the canonical form in [15] takes the following form:

$$\begin{aligned} A &= \text{diag}(A_0, A_1, \ldots, A_{S/2}, A_\bullet), \\ A_0 &= I_{c_0}, \\ A_j &= \begin{bmatrix} \cos(\omega_j)I_{c_j} & \sin(\omega_j)I_{c_j} \\ -\sin(\omega_j)I_{c_j} & \cos(\omega_j)I_{c_j} \end{bmatrix}, \quad 0 < j < S/2, \\ A_{S/2} &= -I_{c_{S/2}}, \\ C &= \begin{bmatrix} C_{0,R} \mid C_{1,R} & C_{1,I} \mid \ldots & \ldots \mid C_{S/2-1,R} & C_{S/2-1,I} \mid C_{S/2} \mid C_\bullet \end{bmatrix} \\ &= \begin{bmatrix} C_0 \mid C_1 \mid \ldots \mid C_{S/2-1} \mid C_{S/2} \mid C_\bullet \end{bmatrix}, \\ K &= \begin{bmatrix} K'_{0,R} \mid K'_{1,R} & K'_{1,I} \mid \ldots & \ldots \mid K'_{S/2-1,R} & K'_{S/2-1,I} \mid K'_{S/2} \mid K'_\bullet \end{bmatrix}' \end{aligned} \quad (2)$$

where $\omega_j = 2\pi j/S, j = 0, \ldots, S/2$ denote the unit root frequencies and $C_{j,R} \in \mathbb{R}^{s\times c_j}, C_{j,I} \in \mathbb{R}^{s\times c_j}, K_{j,R} \in \mathbb{R}^{c_j\times s}, K_{j,I} \in \mathbb{R}^{c_j\times s}$ where $0 \leq c_j \leq s, 0 \leq j \leq S/2$. Furthermore for $C_{j,\mathbb{C}} := C_{j,R} - iC_{j,I}$ it holds that $C'_{j,\mathbb{C}}C_{j,\mathbb{C}} = I_{c_j}$ and $K_{j,\mathbb{C}} = K_{j,R} + iK_{j,I}$ is of full row rank and positive upper triangular ($C_{0,I} = C_{S/2,I} = 0, K_{0,I} = K_{S/2,I} = 0$), see [15] for details. Finally

$|\lambda_{max}(\mathcal{A}_\bullet)| < 1$, where $\lambda_{max}(\mathcal{A})$ denotes an eigenvalue of the matrix $\mathcal{A}$ with maximal modulus. The stable subsystem $(\mathcal{A}_\bullet, \mathcal{C}_\bullet, \mathcal{K}_\bullet)$ is assumed to be in echelon canonical form (see [22]).

Using this notation the assumptions on the data generating process (dgp) in this paper can be stated as follows:

Assumption 1. $(y_t)_{t \in \mathbb{Z}}$ *has a minimal state space representation* $(\mathcal{A}_\circ, \mathcal{C}_\circ, \mathcal{K}_\circ)$, $\mathcal{A}_\circ \in \mathbb{R}^{n \times n}$ *of the form* (2) *with minimal* $(\mathcal{A}_{\circ,\bullet}, \mathcal{C}_{\circ,\bullet}, \mathcal{K}_{\circ,\bullet})$, $\mathcal{A}_{\circ,\bullet} \in \mathbb{R}^{n_\bullet \times n_\bullet}$ *in echelon canonical form where* $c = n - n_\bullet > 0$.

Furthermore the stability assumption $|\lambda_{max}(\mathcal{A}_{\circ,\bullet})| < 1$ *and the strict minimum-phase condition* $\rho_0 := |\lambda_{max}(\mathcal{A}_\circ - \mathcal{K}_\circ \mathcal{C}_\circ)| < 1$ *hold.*

The state at time $t = 1$ *is given by* $x_1 = [x'_{1,0}, \ldots, x'_{1,S/2}, x'_{1,\bullet}]'$ *where* $x_{1,j} \in \mathbb{R}^{\delta_j c_j}$ *(for* $\delta_j = 2, 0 < j < S/2$ *and* $\delta_j = 1$ *else) is deterministic and* $x_{1,\bullet} = \sum_{j=1}^{\infty} \mathcal{A}_{\circ,\bullet}^{j-1} \mathcal{K}_{\circ,\bullet} \varepsilon_{1-j}$ *is such that* $(x_{t,\bullet})_{t \in \mathbb{Z}}$ *is stationary.*

The noise process $(\varepsilon_t)_{t \in \mathbb{Z}}$ *is assumed to be a strictly stationary ergodic martingale difference sequence with respect to the filtration* $\mathcal{F}_t$ *with zero conditional mean* $\mathbb{E}(\varepsilon_t | \mathcal{F}_{t-1}) = 0$, *deterministic conditional variance* $\mathbb{E}(\varepsilon_t \varepsilon'_t | \mathcal{F}_{t-1}) = \Omega > 0$ *and finite fourth moments.*

Due to the block diagonal form of A the state equations are in a convenient form such that partitioning the state vector accordingly as

$$x_t = \begin{pmatrix} x_{t,0} \\ x_{t,1} \\ \vdots \\ x_{t,S/2} \\ x_{t,\bullet} \end{pmatrix}, \quad (3)$$

the blocks $(x_{t,j})_{t \in \mathbb{Z}}, x_{t,j} \in \mathbb{R}^{\delta_j c_j}$ for $c_j > 0$ are unit root processes with state space unit root structure $\{(\omega_j, (c_j))\}$. Finally $(x_{t,\bullet})_{t \in \mathbb{Z}}$ is assumed to be stationary due to the assumptions on $x_{1,\bullet}$. If $(\tilde{y}_t)_{t \in \mathbb{N}}$ denotes a different solution to the state space equations corresponding to $\tilde{x}_1$ then (for $t > 1$)

$$\tilde{y}_t - y_t = CA^{t-1}(\tilde{x}_1 - x_1) = \sum_{j=0}^{S/2} C_j A_j^{t-1}(\tilde{x}_{1,j} - x_{1,j}) + C_\bullet A_\bullet^{t-1}(\tilde{x}_{1,\bullet} - x_{1,\bullet}).$$

Note that $C_j A_j^{t-1} z_{12} = \cos(\omega_j t) z_1 + \sin(\omega_j t) z_2$, $0 < j < S/2$ (for appropriate vectors z_{12}, z_1, z_2),

$$C_0 A_0^{t-1} = C_0, \quad C_{S/2} A_{S/2}^{t-1} = (-1)^{t-1} C_{S/2}.$$

Therefore the sum $\sum_{j=0}^{S/2} C_j A_j^{t-1}(\tilde{x}_{1,j} - x_{1,j})$ can be modeled using a constant and seasonal dummies. The term $C_\bullet A_\bullet^{t-1}(\tilde{x}_{1,\bullet} - x_{1,\bullet})$ tends to zero with an exponential rate as $t \to \infty$ and hence does not influence the asymptotics.

Assumption 1 implies that the yearly difference

$$\begin{aligned} y_t - y_{t-S} &= CA^S x_{t-S} + \varepsilon_t + \sum_{i=1}^{S} CA^{i-1} K \varepsilon_{t-i} - C x_{t-S} - \varepsilon_{t-S} \\ &= (CA^S - C) x_{t-S} + v_t = (C_\bullet A_\bullet^S - C_\bullet) x_{t-S,\bullet} + v_t \end{aligned}$$

is a stationary VARMA process where $v_t = \varepsilon_t + \sum_{i=1}^{S} CA^{i-1} K \varepsilon_{t-i} - \varepsilon_{t-S}$ since $A_j^S = I_{\delta_j c_j}$. Thus the process according to Assumption 1 is a unit root process in the sense of [15]. Note that we do not assume that all unit roots are contained such that the spectral density of the stationary process $(y_t - y_{t-S})_{t \in \mathbb{Z}}$ may contain zeros due to overdifferentiation and hence the process potentially is not stably invertible. The special form of A_0 implies that $I(1)$ processes are a special case of our dgp while $I(d), d > 1, d \in \mathbb{N}$, processes are not allowed for.

3. Canonical Variate Analysis

The main idea of CVA is that, given the state, the system equations (1) are linear in the system matrices. Therefore, based on an estimate of the state sequence, the system can be estimated using least squares regression. The estimate of the state is based on the following equation (for details see for example [17]):

Let $Y_{t,f}^+ := [y_t', y_{t+1}', \ldots, y_{t+f-1}']'$ denote the vector of stacked observations for some integer $f \geq n$ and let $E_{t,f}^+ := [\varepsilon_t', \varepsilon_{t+1}', \ldots, \varepsilon_{t+f-1}']'$. Further define $Y_{t,p}^- := [y_{t-1}', \ldots, y_{t-p}']'$. Then (for $t > p$)

$$\begin{aligned} Y_{t,f}^+ &= \mathcal{O}_f x_t + \mathcal{E}_f E_{t,f}^+ = \mathcal{O}_f \mathcal{K}_p Y_{t,p}^- + \mathcal{O}_f (\mathcal{A}_\circ - \mathcal{K}_\circ \mathcal{C}_\circ)^p x_{t-p} + \mathcal{E}_f E_{t,f}^+ \\ &= \beta_1 Y_{t,p}^- + N_{t,f}^+ \end{aligned} \quad (4)$$

where $\mathcal{K}_p := [\mathcal{K}_\circ, \bar{\mathcal{A}}_\circ \mathcal{K}_\circ, \bar{\mathcal{A}}_\circ^2 \mathcal{K}_\circ, \ldots, \bar{\mathcal{A}}_\circ^{p-1} \mathcal{K}_\circ]$ for $\bar{\mathcal{A}}_\circ := \mathcal{A}_\circ - \mathcal{K}_\circ \mathcal{C}_\circ$ and $\mathcal{O}_f := [\mathcal{C}_\circ', \mathcal{A}_\circ' \mathcal{C}_\circ', \ldots, (\mathcal{A}_\circ^{f-1})' \mathcal{C}_\circ']'$. The strict minimum-phase assumption implies $\bar{\mathcal{A}}_\circ^p \to 0$ for $p \to \infty$.

Let $\langle a_t, b_t \rangle := T^{-1} \sum_{t=p+1}^{T-f+1} a_t b_t'$ for sequences $(a_t)_{t \in \mathbb{N}}$ and $(b_t)_{t \in \mathbb{N}}$. Then an estimate of β_1 is obtained from the reduced rank regression (RRR) $Y_{t,f}^+ = \beta_1 Y_{t,p}^- + N_{t,f}^+$ under the rank constraint rank$(\beta_1) = n$. This results in the estimate $\hat{\mathcal{O}}_f \hat{\mathcal{K}}_p := [(\hat{\Xi}_f^+)^{-1} \hat{U}_n \hat{S}_n][\hat{V}_n'(\hat{\Xi}_p^-)^{-1}]$ of β_1 using the singular value decomposition (SVD)

$$\hat{\Xi}_f^+ \hat{\beta}_1 \hat{\Xi}_p^- = \hat{U} \hat{S} \hat{V}' = \hat{U}_n \hat{S}_n \hat{V}_n' + \hat{R}_n.$$

Here $\hat{\beta}_1 = \langle Y_{t,f}^+, Y_{t,p}^- \rangle \langle Y_{t,p}^-, Y_{t,p}^- \rangle^{-1}$ denotes the unrestricted least squares estimate of β_1 and

$$\hat{\Xi}_f^+ := \langle Y_{t,f}^+, Y_{t,f}^+ \rangle^{-1/2}, \quad \hat{\Xi}_p^- := \langle Y_{t,p}^-, Y_{t,p}^- \rangle^{1/2}. \quad (5)$$

Here the symmetric matrix square root is used. The definition is, however, not of importance and other square roots such as Cholesky factors could be used. $\hat{U}_n \in \mathbb{R}^{fs \times n}$ denotes the matrix whose columns are the left singular vectors to the n largest singular values which are the diagonal entries in $\hat{S}_n := \text{diag}(\hat{\sigma}_1, \hat{\sigma}_2, \ldots, \hat{\sigma}_n), \hat{\sigma}_1 \geq \cdots \geq \hat{\sigma}_n > 0$ and $\hat{V}_n \in \mathbb{R}^{ps \times n}$ contains the corresponding right singular vectors as its columns. $\hat{R}_n$ denotes the approximation error.

The system estimate $(\hat{A}, \hat{C}, \hat{K})$ is then obtained using the estimated state $\hat{x}_t := \hat{\mathcal{K}}_p Y_{t,p}^-, t = p+1, \ldots, T+1$ via regression in the system equations.

In the algorithm a specific decomposition of the rank n matrix $\hat{\mathcal{O}}_f \hat{\mathcal{K}}_p$ into the two factors $\hat{\mathcal{O}}_f$ and $\hat{\mathcal{K}}_p$ is given such that $\hat{\mathcal{K}}_p \hat{\Xi}_p^- (\hat{\Xi}_p^-)' \hat{\mathcal{K}}_p' = I_n$. It is obvious that every other decomposition of $\hat{\mathcal{O}}_f \hat{\mathcal{K}}_p$ produces an estimated state sequence in a different coordinate system, leading to a different observationally equivalent representation of the same transfer function estimator. Therefore, with respect to consistency of the transfer function estimator it is sufficient to show that there exists a factorization of $\hat{\mathcal{O}}_f \hat{\mathcal{K}}_p$ leading to convergent system matrix estimators $(\tilde{A}, \tilde{C}, \tilde{K})$, even if this factorization cannot be used in actual computations, as it requires information not known at the time of estimation.

In order to generate a consistent initial guess for subsequent quasi likelihood optimization in the set of all state space systems corresponding to processes with state space unit root structure $\Omega_S := \{(\omega_0, (c_0)), \ldots, (\omega_{S/2}, (c_{S/2}))\}$, however, we will derive a realizable (for known integers c_j and matrices E_j such that $E_j' \mathcal{C}_{o,j,\mathbb{C}} = I_{c_j}$) consistent system estimate. To this end note that consistency of the transfer function implies (see for example [23]) that the eigenvalues $\tilde{\lambda}_l$ of $\hat{A}$ converge (in a specific sense) to the eigenvalues λ_j of $\mathcal{A}_\circ$. Therefore transforming $\hat{A}$ into complex Jordan normal form (where $\hat{A}$ is almost surely diagonalizable), ordering the eigenvalues such that groups of eigenvalues $\tilde{\lambda}_l(j), l = 1, \ldots, c_j$ converging to λ_j are grouped together, we obtain a realizable system $(\check{A}, \check{C}, \check{K})$ where the diagonal blocks of the block diagonal matrix $\check{A}$ corresponding to the unit roots converge to a diagonal matrix with the eigenvalues z_j on the diagonal:

$$\check{A}_{j,\mathbb{C}} = \begin{bmatrix} \tilde{\lambda}_1(j) & 0 & \cdots & 0 \\ 0 & \tilde{\lambda}_2(j) & \ddots & \vdots \\ \vdots & \ddots & \ddots & 0 \\ 0 & \cdots & 0 & \tilde{\lambda}_{c_j}(j) \end{bmatrix} \to A_{j,\mathbb{C}} = \begin{bmatrix} z_j & 0 & \cdots & 0 \\ 0 & z_j & \ddots & \vdots \\ \vdots & \ddots & \ddots & 0 \\ 0 & \cdots & 0 & z_j \end{bmatrix}.$$

Replacing $\check{A}_{j,\mathbb{C}}$ by the limit $A_{j,\mathbb{C}}$ and transforming the estimates to the real Jordan normal form, we obtain estimates $(\check{A}, \check{C}, \check{K})$ corresponding to unit root processes with state space unit root structure Ω_S.

Note, however, that this representation not necessarily converges as perturbation analysis only implies convergence of the eigenspaces. Therefore in the final step the estimate $(\check{A}, \check{C}, \check{K})$ is converted such that we obtain convergence of the system matrix estimates. In the class of observationally equivalent systems with the matrix

$$\check{A}_{\mathbb{C}} = \text{diag}(A_{0,\mathbb{C}}, A_{1,\mathbb{C}}, \overline{A_{1,\mathbb{C}}} \ldots, \overline{A_{S/2-1,\mathbb{C}}}, A_{S/2,\mathbb{C}}, \check{A}_\bullet), \quad A_{j,\mathbb{C}} = I_{c_j} z_j,$$

block diagonal transformations of the form $\mathcal{T} = \text{diag}(\mathcal{T}_0, \mathcal{T}_1, \overline{\mathcal{T}_1}, \ldots, \mathcal{T}_{S/2}, I)$ do not change the matrix $\check{A}_{\mathbb{C}}$. Here the basis of the stable subsystem can be chosen such that the corresponding transformed $(\check{A}_\bullet, \check{C}_\bullet, \check{K}_\bullet)$ is uniquely defined using an overlapping echelon form (see [22], Section 2.6). The impact of such transformations on the blocks of C is given by $\check{C}_{j,\mathbb{C}} \mathcal{T}_j^{-1}$. Therefore, if for each $j = 0, \ldots, S/2$ a matrix $E_j \in \mathbb{C}^{s \times c_j}$ is known such that $E_j' \mathcal{C}_{\circ,j,\mathbb{C}} \in \mathbb{C}^{c_j \times c_j}$ is nonsingular, the restriction $E_j' \check{C}_{j,\mathbb{C}} = I_{c_j}$ picks a unique representative $(\check{A}, \check{C}, \check{K})$ of the class of systems observationally equivalent to $(\check{A}, \check{C}, \check{K})$.

Note that this estimate $(\check{A}, \check{C}, \check{K})$ is realizable if the integers c_j (needed to identify the c_j eigenvalues of $\hat{A}$ closest to z_j), the matrices E_j (needed to fix a basis for $x_{t,j}$) and the index corresponding to the overlapping echelon form for the stable part are known. Furthermore, this estimate corresponds to a process with state space unit root structure Ω_S and hence can be used as a starting value for quasi likelihood maximization.

Finally in this section it should be noted that the estimate of the state $\hat{x}_t$ here mainly serves the purpose of obtaining an estimator for the state space system. Based on this estimate, Kalman filtering techniques can be used to obtain different estimates of the state sequence. The relation between these different estimates is unclear and so is their usage for inference. For this paper the state estimates $\hat{x}_t$ are only an intermediate step in the CVA algorithm.

4. Asymptotic Properties of the System Estimators

As follows from the last section, the central step in the CVA procedure is a RRR problem involving stationary and nonstationary components. The asymptotic properties of the solution to such RRR problems are derived in Theorem 3.2. of [24]. Using these results the following theorem can be proved (see Appendix C.1):

Theorem 1. *Let the process $(y_t)_{t \in \mathbb{Z}}$ be generated according to Assumption 1. Let $(\hat{A}, \hat{C}, \hat{K})$ denote the CVA estimators of the system matrices using the assumption of correctly specified order n with $f \geq n$ not depending on the sample size and finite and $p = o((\log T)^{\bar{a}})$ for some real $0 < \bar{a} < \infty, p \geq -d \log T / \log \rho_0$ for some $d > 1$ where $0 < \rho_0 = |\lambda_{max}(\mathcal{A}_\circ - \mathcal{K}_\circ \mathcal{C}_\circ)| < 1$. Let $(\mathcal{A}_\circ, \mathcal{C}_\circ, \mathcal{K}_\circ)$ be in the form given in (2) where $(\mathcal{A}_{\circ,\bullet}, \mathcal{C}_{\circ,\bullet}, \mathcal{K}_{\circ,\bullet})$ is in echelon canonical form and for each $j = 0, \ldots, S/2$ there exists a row selector matrix $E_j \in \mathbb{R}^{s \times c_j}$ such that $E_j' \mathcal{C}_{\circ,j,\mathbb{C}}$ is non-singular. Then for some integer a:*

(I) $\hat{C} \hat{A}^j \hat{K} - \mathcal{C}_\circ \mathcal{A}_\circ^j \mathcal{K}_\circ = O_P((\log T)^a / \sqrt{T})$ *for each $j \geq 0$.*

(II) *Using $D_x = \text{diag}(T^{-1} I_c, T^{-1/2} I_{n-c})$ where $c = \sum_{j=0}^{S/2} c_j \delta_j$ we have*

$$(\check{A} - \mathcal{A}_\circ) D_x^{-1} = O_P((\log T)^a), \sqrt{T}(\check{K} - \mathcal{K}_\circ) = O_P((\log T)^a), (\check{C} - \mathcal{C}_\circ) D_x^{-1} = O_P((\log T)^a)$$

for some integer $a < \infty$.

(III) If the noise is assumed to be an iid sequence, then results (I) and (II) hold almost surely.

Beside stating consistency in the seasonal integration case, the theorem also improves on the results of [20] in the I(1) case by showing that no adaptation of CVA is needed in order to obtain consistent estimators of the impulse response sequence or the system matrices. Note that this consistency result for the impulse response sequence concerns both the short and the long-run dynamics. In particular it implies that short-run prediction coefficients are consistent. Moreover the theorem establishes strong consistency rather than weak consistency as opposed to [20]. (II) establishes orders of convergence which, however, apply only to a transformed system that requires knowledge of the integers c_j and matrices E_j to be realized. No tight bounds for the integer a are derived, since they do not seem to be of much value.

Note that the assumptions on the innovations rule out conditionally heteroskedastic processes. However, since the proof mostly relies on convergence properties for covariance estimators for stationary processes and continuous mapping theorems for integrated processes, it appears likely that the results can be extended to conditionally heteroskedastic processes as well. For the stationary cases this follows directly from the arguments in [18], while for integrated processes results (using different assumptions on the innovations) given for example in [25] can be used. The conditions of [25] hold for example in a large number of GARCH type processes, see [26]. The combination of the different sets of assumptions on the innovations is not straightforward, however, and hence would further complicate the proofs. We refrain from including them.

It is worth pointing out that due to the block diagonal structure of $\mathcal{A}_\circ$ the result $(\check{\mathcal{C}} - \mathcal{C}_\circ)D_x^{-1} = O_P((\log T)^a)$ implies consistency of the blocks $\check{\mathcal{C}}_j$ corresponding to unit root z_j (or the corresponding complex pair) of order almost T^{-1}. Using the complex valued canonical form this implies consistent estimation of $\mathcal{C}_{\circ,j,\mathbb{C}}$ by the corresponding $\check{\mathcal{C}}_{j,\mathbb{C}}$. In the canonical form this matrix determines the cointegrating relations (both the static as well as the dynamic ones, for details see [15]) as the unitary complement to this matrix. It thus follows that CVA delivers estimators for the cointegrating relations at the various unit roots that are (super-)consistent. In fact, the proof can be extended to show convergence in distribution of $(\check{\mathcal{C}} - \mathcal{C}_\circ)D_x^{-1}$. This distribution could be used in order to derive tests for cointegrating relations. However, preliminary simulations indicate that these estimates and hence the corresponding tests are not optimal and can be improved upon by quasi maximum likelihood estimation in the VARMA setting initialized by the CVA estimates. Therefore we refrain from presenting these results.

Note that the assumptions impose the restriction $\rho_0 > 0$ excluding VAR systems. This is done solely for stating a uniform lower bound on the increase of p as a function of T. This bound is related to the lag length selection achieved using BIC, see [27]. In the VAR case the lag length estimator using BIC will converge to the true order and thus remain finite. All results hold true if in the VAR case a fixed (that is independent of the sample size) $p \geq n$ is used.

5. Inference Based on the Subspace Estimators

Beside consistency of the impulse response sequence also the specification of the integers $c_0, ..., c_{S/2}$ is of interest. First, following [20] this information can be obtained by detecting the unity singular values in the RRR step of the procedure. Second, from the system representation (2) it is clear that the location of the unit roots is determined by the eigenvalues of $\mathcal{A}_\circ$ on the unit circle: The integers c_j denote the number of eigenvalues at the corresponding locations on the unit circle (provided the eigenvalues are simple). Due to perturbation theory (see for example Lemma A2) we know that the eigenvalues of $\hat{A}$ will converge (for $T \to \infty$) to the eigenvalues of $\mathcal{A}_\circ$ and the distribution of the mean of all eigenvalues of $\hat{A}$ converging to an eigenvalue of $\mathcal{A}_\circ$ can be derived based on the distribution of the estimation error $\hat{A} - \mathcal{A}_\circ$. This can be used to derive tests on the number of eigenvalues at a particular location on the unit circle. Third, if $n \leq s$ the state process is a VAR(1) process and hence in some cases allows for inference on the number of cointegrating relations and thus also on the integers c_j as outlined in [4]. Tests based on these three arguments will be discussed below.

Theorem 2. *Under the assumptions of Theorem 1 the test statistic $T \sum_{i=1}^{c}(1 - \hat{\sigma}_i^2)$ converges in distribution to the random variable*

$$Z = tr\left[\mathbb{E}(\tilde{\varepsilon}_{t,\perp}\tilde{\varepsilon}'_{t,\perp})\left(\int_0^1 W(r)W(r)'\right)^{-1}\right]$$

where $\tilde{\varepsilon}_{t,\perp} = \tilde{\varepsilon}_{t,1} - \mathbb{E}\tilde{\varepsilon}_{t,1}\tilde{\varepsilon}'_{t,\bullet}(\mathbb{E}\tilde{\varepsilon}_{t,\bullet}\tilde{\varepsilon}'_{t,\bullet})^{-1}\tilde{\varepsilon}_{t,\bullet}$ (for definition of $\tilde{\varepsilon}_{t,1}$ and $\tilde{\varepsilon}_{t,\bullet}$ see the proof in Appendix C.2) and where $W(r)$ denotes a c-dimensional Brownian motion with variance

$$\sum_{i=0}^{S-1} \mathcal{A}_u^i \mathcal{K}_u \Omega \mathcal{K}_u' (\mathcal{A}_u^i)'$$

with $\mathcal{A}_u$ denoting the $c \times c$ heading submatrix of $\mathcal{A}$ and $\mathcal{K}_u$ denoting the submatrix of $\mathcal{K}$ composed of the first c rows such that $(\mathcal{A}_u, \mathcal{C}_u, \mathcal{K}_u)$ denotes the unstable subsystem.

The theorem is proved in Appendix C.2, where also the many nuisance parameters of the limiting random variable are explained and defined. The proof also corrects an error in Theorem 4 of [20], where the wrong distribution is given since the second order terms were neglected.

As the distribution is not pivotal and in particular contains information that is unknown when performing the RRR step, it is not of much interest for direct application. Nevertheless the order of convergence allows for the derivation of simple consistent estimators of the number of common trends: Let $\hat{c}_T$ denote the number of singular values calculated in the RRR that exceed $\sqrt{1 - h(T)/T}$ for arbitrary $h(T) \to \infty, h(T) < T, h(T)/T \to 0$, for $T \to \infty$. Then it is a direct consequence of Theorem 2 in combination with $\hat{\sigma}_j \to \sigma_j < 1, j > c$, that $\hat{c}_T \to c$ in probability, implying consistent estimation of c. Based on these results also estimators for c could be derived, for example along the lines of [28]. However, as [29] shows, such estimators have not performed well in simulations and thus are not considered subsequently.

The singular values do not provide information on the location of the unit roots. This additional information is contained in the eigenvalues of the matrix $\mathcal{A}_\circ$:

Theorem 3. *Under the assumptions of Theorem 1 let $\hat{\lambda}_i(m), i = 1, ..., c_m$ denote the c_m eigenvalues of $\hat{\mathcal{A}}$ closest to the unit root $z_m, |z_m| = 1$. Then defining $\hat{\mu}_m = \sum_{i=1}^{c_m}(\hat{\lambda}_i(m) - z_m)$ we obtain*

$$T\hat{\mu}_m \xrightarrow{d} tr\left[\left(\int B(r)B(r)dr\right)^{-1}\int B(r)dB(r)'\right]$$

where $B(r)$ denotes a c_m-dimensional Brownian motion with zero expectation and variance I_{c_m} for $z_m = \pm 1$ and a complex Brownian motion with expectation zero and variance equal to the identity matrix else.

Further if $\tilde{\mathcal{A}} := \langle x_{t+1}, x_t \rangle \langle x_t, x_t \rangle^{-1}$ using the true state x_t and $\tilde{\mu}_m = \sum_{i=1}^{c_m}(\tilde{\lambda}_i(m) - z_m)$ where $\tilde{\lambda}_i(m), i = 1, ..., c_m$ denote the c_m eigenvalues of $\tilde{\mathcal{A}}$ closest to z_m, then $T(\hat{\mu}_m - \tilde{\mu}_m) = o_P(1)$.

Therefore the estimated eigenvalues can be used in order to obtain a test on the number of common trends at a particular frequency for each frequency separately. The test distribution is obtained as the limit to

$$T tr[\langle \mathcal{K}_{\circ,m,\mathbb{C}}\varepsilon_t, x_{t,m,\mathbb{C}}\rangle\langle x_{t,m,\mathbb{C}}, x_{t,m,\mathbb{C}}\rangle^{-1}]$$

where $x_{t,m,\mathbb{C}} = \overline{z_m} x_{t-1,m,\mathbb{C}} + \mathcal{K}_{\circ,m,\mathbb{C}} \varepsilon_{t-1}, x_{1,m,\mathbb{C}} = 0$. The distribution thus does not depend on the presence of other unit roots or stationary components of the state. Furthermore it can be seen that it is independent of the noise variance or the matrix $\mathcal{K}_{\circ,m,\mathbb{C}}$. Hence critical values are easily obtained from simulations. Also note that the limiting distribution is identical for all complex unit roots.

Therefore, for each seasonal unit root location z_m we can order the eigenvalues of the estimated matrix $\hat{\mathcal{A}}$ with increasing distance to z_m. Then starting from the assumption of $H_0: c_m = \bar{c}$ (for a reasonable $\bar{c}$ obtained, e.g., from a plot of the eigenvalues) one can perform the test with statistic $T\hat{\mu}_m$. If the test rejects, then the hypothesis $H_0: c_m = \bar{c} - 1$ is tested, until the hypothesis is not rejected anymore, or $H_0: c_m = 1$ is reached. This is then the last test. If H_0 is rejected again, no unit root is found at this location. Otherwise we do not have evidence against $c_m = 1$. In any case, the system needs to be estimated only once and the calculation of the test statistics is easy even for all seasonal unit roots jointly.

The third option for obtaining tests is to use the tests derived in [4] based on the JS framework for VARs. In the case $n \leq s$ the state process $x_{t+1} = \mathcal{A} x_t + \mathcal{K} \varepsilon_t$ is a seasonally integrated VAR(1) process (for $n > s$ the noise variance is singular). The corresponding VECM representation equals

$$p(L)x_t = \sum_{m=1}^{S}(I_n - \mathcal{A}z_m)X_{t-1}^{(m)} + \mathcal{K}\varepsilon_{t-1} = \sum_{m=1}^{S}\alpha_m \beta'_m X_{t-1}^{(m)} + \mathcal{K}\varepsilon_{t-1}$$

where $z_m = \exp(\frac{2\pi m}{S}i), m = 1, ..., S$ and

$$p(L) = 1 - L^S \quad , \quad p_t = p(L)x_t = x_t - x_{t-S},$$
$$p_m(L) = \frac{p(L)}{1 - \overline{z_m}L} \quad , \quad X_t^{(m)} = -\frac{p_m(L)}{p_m(z_m)z_m}x_t.$$

Note that in this VAR(1) setting no additional stationary regressors of the form $p(L)x_{t-j}$ occur. Also no seasonal dummies are needed but could be added to the equation. In this setting [4] suggests to use the eigenvalues $\hat{\lambda}_i$ (ordered with increasing modulus) of the matrix (the superscript $(.)^\pi$ denotes the residuals with respect to the remaining regressors $X_{t-1}^{(j)}, j \neq m$)

$$\langle X_{t-1}^{(m),\pi}, p_t^\pi \rangle \langle p_t^\pi, p_t^\pi \rangle^{-1} \langle p_t^\pi, X_{t-1}^{(m),\pi} \rangle \langle X_{t-1}^{(m),\pi}, X_{t-1}^{(m),\pi} \rangle^{-1}$$

as the basis for a test statistic

$$\tilde{C}_m := -\delta_m \sum_{i=1}^{c_m} \log(1 - \hat{\lambda}_i).$$

where $\delta_m = 2$ for complex unit roots and $\delta_m = 1$ for real unit roots. In the $I(1)$ case this leads to the familiar Johansen trace test, for seasonal unit roots a different asymptotic distribution is obtained.

Theorem 4. *Under the assumptions of Theorem 1 let $\hat{C}_m$ be calculated based on the estimated state and let $\tilde{C}_m$ denote the same statistic based on the true state. Then for $n \leq s$ it holds that $\hat{C}_m - \tilde{C}_m = o_P(T^{-1})$ and*

$$T\hat{C}_m \xrightarrow{d} tr\left[\int dB(r)B(r)' \left(\int B(r)B(r)dr\right)^{-1} \int B(r)dB(r)'\right]$$

where $B(r)$ is a real Brownian motion for $z_m = \pm 1$ or a complex Brownian motion else.

Thus again under the null hypothesis the test statistic based on the estimated state and the one based on the true state reject jointly asymptotically with probability one. Therefore for $n \leq s$ the tests of JS can be used to obtain information on the number of common cycles, ignoring the fact that the estimated state is used in place of the true state process.

After presenting three disjoint ideas for providing information on the number and location of unit roots, the question arises, which one to use in practice. In the following a number of ideas are given in this respect.

The criterion based on the singular values given in Theorem 2 is of limited information as it only provides the overall number of unit roots. Since the limiting distribution is not pivotal it cannot be used for tests and the choice of the cutoff value $h(T)$ is somewhat arbitrary. Nevertheless, using a relatively large value one obtains a useful upper bound on c which can be included in the typical sequential procedures for tests for c_j.

Using the results of Theorem 4 has the advantage of using a framework that is well known to many researchers. It is remarkable that in terms of the asymptotic distributions there is no difference involved in using the estimated state in place of the true state. The assumption $n \leq s$, however, is somewhat restrictive except in situations with a large s.

Finally the results of Theorem 3 provide simple to use tests for all unit roots, independently of the specification of the model for the remaining unit roots. Again it is remarkable that, under the null, inference is identical for known and for estimated state.

Since our estimators are not quasi maximum likelihood estimators the question of a comparison with the usual likelihood ratio tests arises. For VAR models simulation exercises documented in Section 7 below demonstrate that there are situations where the proposed tests outperform tests in the VAR framework. Comparisons with tests in the state space framework (or equivalently in the VARMA framework) are complicated by the fact that no results are currently available in the literature of this framework. One difference, however, is given by the fact that quasi likelihood ratio tests in the VARMA setting require a full specification of the c_j values for all unit roots. This introduces interdependencies such that the tests for one unit root depend on the specification of the cointegrating rank at the other roots. The interdependencies can be broken by performing tests based on alternative specifications for each unit root. The test based on Theorem 3 does not require this but can be performed based on the same estimate $\hat{A}$. This is seen as an advantage.

The question of the comparison of the empirical size in finite samples as well as power to local alternatives between the CVA based tests and tests based on quasi-likelihood ratios is left as a research question.

6. Deterministic Terms

Up to now it has been assumed that no deterministic terms appear in the model contrary to common practice. In the VAR framework dealing with trends is complicated by the usage of the VECM representation, see e.g., [30]. In the state space framework used in this paper, however, deterministic terms are easily incorporated.

Theorem 5. *Let the process $(y_t)_{t \in \mathbb{Z}}$ be generated according to Assumption 1 and assume that the process $(\tilde{y}_t)_{t \in \mathbb{Z}}$ is observed where $\tilde{y}_t = y_t + \Phi d_t$ with*

$$d_t = [\ 1,\ \cos(\tfrac{2\pi}{5}t),\ \sin(\tfrac{2\pi}{5}t),\ \cdots\ (-1)^t\ ]' \in \mathbb{R}^S$$

and $\Phi \in \mathbb{R}^{s \times S}$.

Then if the CVA estimation is applied to

$$\tilde{y}_t^\pi := y_t - \left(\sum_{t=1}^{T} y_t d_t' \right) \left(\sum_{t=1}^{T} d_t d_t' \right)^{-1} d_t, \quad t = 1, \ldots, T,$$

the results of Theorem 1 hold, i.e., the system is estimated consistently and the orders of convergence for the transformed system $(\breve{A}, \breve{C}, \breve{K})$ hold true.

Furthermore the convergence in distribution results in Theorems 2–4 hold true where in the limits the Brownian motions $B(r)$ occurring in the distributions must be replaced by their demeaned versions $B(r) - \int_0^1 B(s)ds$.

In this sense the results are robust to some operations typically termed preprocessing of data such as demeaning and deseasonalizing using seasonal dummies. More general preprocessing steps such as detrending or the extraction of more general deterministic terms analogous to [30] can be investigated along the same lines.

7. Simulations

The estimation of the seasonal cointegration ranks and spaces is usually carried out via quasi maximum likelihood methods that originated from the VAR model class. Typical estimators in this setting are those of [2,4,5,31]. In the first two experiments we focus on the estimation of the cointegrating spaces and the specification of the cointegration ranks in the classical situation of quarterly data and show that there are certain situations in which CVA estimators and the test in Theorem 3 possess finite sample properties superior to those of the methods above. In a third experiment the test performance is evaluated for a daily sampling rate. Moreover, the prediction accuracy of CVA is investigated as well as its robustness to innovations exhibiting behaviors often encountered at such higher sampling rates. All simulations are carried out using 1000 replications.

To investigate the practical usefulness of the proposed procedures we generate quarterly data using two VAR dgps of dimension $s = 2$ first and then two more general VARMA dgps with $s = 8$. Each pair contains dgps with different state space unit root structures

$$\{(0, (1)), (\pi/2, (c_{\pi/2})), (\pi, (1))\}, \quad c_{\pi/2} = 1, 2.$$

From all four dgps samples of size $T \in \{50, 100, 200, 500\}$ are generated with initial values set to zero. Although none of the dgps contains deterministics, the data is adjusted for a constant and quarterly seasonal dummies as in [5]. For reasons of comparability, the adjustment for deterministic terms is done before estimation.

In the third experiment we generate daily data with dimension $s = 4$ from a state space system including unit roots corresponding to weekly frequencies (that is a period length of seven days). In the simulations we use several years of data (excluding new year's day to account for 52 weeks of seven days each). The first 200 observations are discarded to include the effects of different starting values. In this example the focus lies on a comparison of the prediction accuracy. Furthermore we investigate the robustness of the test procedures to conditional heteroskedasticity of the GARCH type as well as to non-normality of the innovations.

To assess the performance of specifying the cointegrating rank at unit root z using CVA, the following test statistic is constructed from the results in Theorem 3

$$\Lambda(c) = T|(\frac{1}{c}\sum_{i=1}^{c}\hat{\lambda}_i) - z|. \tag{6}$$

Here $\hat{\lambda}_1, \ldots, \hat{\lambda}_n$ are the eigenvalues of $\hat{A}$ ordered increasingly according to the distance from z. Note that a similar test in [20] only uses the c-th largest eigenvalue, whereas here the average over the nearest c eigenvalues is taken. Critical values have been obtained by simulation using large sample sizes (sample size 2000 (JS) and 5000 (CVA), 10,000 replications).

In our first two experiments usage of $\Lambda(c)$ is compared with variants of the likelihood ratio test from [2] (JS), [4] (Q_1), and [5] (Q_2, Q_3). Q_1 is Cubadda's trace test for complex-valued data, Q_2 takes the information at frequency $\pi/2$ into account when the analysis is carried out at frequency $3\pi/2$, and Q_3 iterates between $\pi/2$ and $3\pi/2$ in the alternating reduced rank regression (ARR) of [5]. For the procedure of [2] the likelihood maximization at frequency $\pi/2$ is carried out using numerical optimization (BFGS) with initial values obtained from an unrestricted regression.

All tests are evaluated by comparing the percentages of correctly detected common trends, or *hit rates*, with 0.95, the hit rate to be expected from a nominal significance level of 0.05. The testing procedure employed for all tests is the same: at each of the frequencies it is started from a null hypothesis of s unit roots against less than s unit roots. In case of rejection, $s-1$ unit roots are tested versus less than $s-1$ and so on, until there are zero unit roots under the alternative.

For the first two experiments the estimation performance of CVA for the simultaneous estimation of the seasonal cointegrating spaces is compared with the maximum likelihood estimates of [2,4,31] (cRRR), and also with an iterative procedure (Generalized ARR or GARR) of [5]. The comparison is carried out by means of the gap metric, measuring the distance between the true and the estimated cointegrating space as in [32]. The smaller the mean gap over all replications, the better is the estimation performance. Throughout a difference between two mean gaps or two hit rates is considered statistically significant if it is larger than twice the Monte Carlo standard error.

For all procedures used in this section, an AR lag length has to be chosen first. For CVA this can be done using the AIC as in ([33], Section 5), as is done in the third experiment.

In the first two experiments where sample sizes are rather small, we estimate the lag length via minimization of the corrected AIC (AICc) ([34], p. 432), $\hat{k}_{AICc}$, benefitting the simulation results. For larger sample sizes the two criteria lead to the same choices. Due to the quarterly data we work with, the lag length is then chosen to be $\hat{k} = \max\{\hat{k}_{AICc}, 4\}$.

Other information criteria could be chosen here. An anonymous referee also suggested the application of the Modified Akaike Information Criterion (MAIC) of [35], proposed there for the I(1)-case. In an attempt to apply it to the seasonally integrated case considered here, it performed considerably worse than the AICc. Thus we refrain from using the MAIC in the following and also omit the results of that attempt. They can be obtained from the authors upon request.

For CVA the truncation indices f and p are chosen as $\hat{f} = \hat{p} = 2\hat{k}$ ([33], Section 5). The system order n is estimated by minimizing ([33], Section 5)

$$SVC(n) = \hat{\sigma}_{n+1}^2 + 2ns\frac{\log T}{T}. \tag{7}$$

Here $\hat{\sigma}_i$ denotes the i-th largest singular value from the singular value decomposition of $\hat{\Xi}_f^+ \hat{\beta}_1 \hat{\Xi}_p^-$ (Step 2 of CVA). Note that selecting the number of states by SVC is made less influential insofar as $\hat{n} = \max\{c_0 + 2c_{\pi/2} + c_\pi, \hat{n}_{SVC}\}$, where $\hat{n}_{SVC}$ denotes the SVC estimated system order.

In Section 7.1 we start with the two VAR dgps and find that the likelihood-based procedures are mostly superior. Continuing with the VARMA dgps in Section 7.2, CVA performs better and is superior for the smaller sample sizes in terms of size and gap and better for all sample sizes in terms of power. Section 7.3 evaluates the performance of the tests for unit roots for larger sample sizes together with the prediction performance in this setting. We find that the tests are robust to the distribution of the innovations as well as to conditional heteroskedasticity of the GARCH type. Furthermore the empirical size of the tests lies close to the size already for moderate sample sizes, where the tests also show almost perfect power properties.

7.1. VAR Processes

The VAR dgps considered in this paper are given by,

$$X_t = \Pi_1 X_{t-1} + \Pi_2 X_{t-2} + \Pi_3 X_{t-3} + \Pi_4 X_{t-4} + \varepsilon_t, \qquad \varepsilon_t \sim N\left(\begin{bmatrix} 0 \\ 0 \end{bmatrix}, \begin{bmatrix} 1 & 0.5 \\ 0.5 & 1 \end{bmatrix}\right) \tag{8}$$

where $(\varepsilon_t)_{t \in \mathbb{Z}}$ is white noise and the coefficient matrices are

$$\Pi_1 = \begin{bmatrix} \gamma & 0 \\ 0 & 0 \end{bmatrix}, \Pi_2 = \begin{bmatrix} -0.4 & 0.4 - \gamma \\ 0 & 0 \end{bmatrix},$$

$$\Pi_3 = \begin{bmatrix} -\gamma & 0 \\ 0 & 0 \end{bmatrix}, \Pi_4 = \begin{bmatrix} 0.6 - (\gamma/10) & 0.4 + \gamma \\ 0 & 1 \end{bmatrix}.$$

This dgp is adopted from [5] with a slight adjustment to Π_4. The corresponding VECM representation in the notation of [5] equals

$$X_{0,t} = \begin{bmatrix} -0.2 \\ 0 \end{bmatrix} \begin{bmatrix} 1 + \gamma/8 & -1 \end{bmatrix} X_{1,t-1} + \begin{bmatrix} 0.2 \\ 0 \end{bmatrix} \begin{bmatrix} 1 + \gamma/8 & -1 \end{bmatrix} X_{2,t-1} + \begin{bmatrix} \gamma \\ 0 \end{bmatrix} \begin{bmatrix} 1 + 0.05L & -L \end{bmatrix} X_{3,t-1} + \varepsilon_t.$$

As can be seen from Table 1, the dgps possess unit roots at frequencies 0, π, and $\pi/2$, where $c_{\pi/2} = 2[1]$ for $\gamma = 0[0.2]$, respectively. Note that in all cases the order of integration equals 1, while the number of common cycles at $\pi/2$ is varied.

Table 1. Eigenvalues of the coefficient matrix of the companion form.

		j							
		1	2	3	4	5	6	7	8
$\gamma = 0.2$	z_j	−1	1	i	−i	0.126 + i0.99	0.126 − i0.99	−0.790	0.737
	$\|z_j\|$	1	1	1	1	0.998	0.998	0.790	0.737
$\gamma = 0$	μ_j	−1	i	−i	1	i	−i	0.775	−0.775
	$\|\mu_j\|$	1	1	1	1	1	1	0.775	0.775

Table 2 exhibits the hit rates from the application of the different test statistics. At frequencies 0 and π, Λ is compared with the trace test of Johansen (J; based on [31] for unit roots $z = -1$), whereas at $\pi/2$ it is competing with JS, Q_1, Q_2, and Q_3. All competitors are likelihood-based tests which is the term we are referring to when we compare Λ to them as a whole.

Table 2. Hit rates for the different tests (VAR dgp). Twice the maximum (over all entries) Monte Carlo standard error is 0.005.

		0		$\pi/2$					π	
	T	Λ	J	Λ	JS	Q1	Q2	Q3	Λ	J
$\gamma = 0$	50	0.685	0.348	0.351	0.903	0.844	0.851	0.844	0.681	0.343
	100	0.841	0.732	0.490	0.925	0.900	0.902	0.900	0.831	0.724
	200	0.897	0.951	0.841	0.934	0.925	0.924	0.925	0.876	0.936
	500	0.931	0.938	0.916	0.949	0.941	0.942	0.941	0.927	0.948
$\gamma = 0.2$	50	0.550	0.367	0.811	0.796	0.777	0.778	0.788	0.604	0.297
	100	0.711	0.801	0.087	0.920	0.913	0.908	0.908	0.799	0.806
	200	0.907	0.922	0.855	0.954	0.949	0.948	0.947	0.854	0.939
	500	0.944	0.953	0.927	0.939	0.938	0.938	0.936	0.924	0.942

The results for 0 and π are very similar for both dgps in that Λ scores more hits than the likelihood-based tests when the sample size is small, $T \in \{50, 100\}$. Convergence of its finite sample distribution is slower than for the other test statistics, however, as J is closer to 0.95 from $T = 200$ on. For $T = 500$ the distribution of Λ only seems to have converged to its asymptotic distribution when $c_{\pi/2} = 2$ at frequency 0, whereas convergence of the likelihood-based tests has occurred in all cases.

At $\pi/2$ the likelihood ratio test of JS strictly dominates all implementations of [5] for all sample sizes and both dgps. It strictly dominates the CVA-based test procedure as well, except for one case, it seems: when $c_{\pi/2} = 1$ and $T = 50$ Λ scores slightly, but significantly, more hits than the likelihood ratio test of JS. Surprisingly, Λ is drastically worse when $T = 100$ with only 8.7%, only to be up at 85% for $T = 200$.

The behavior of Λ is explained by z_5 and z_6 being close to $\pm i$ when $c_{\pi/2} = 1$, cf. Table 1. For future reference we will call the corresponding roots *false unit roots*.

For $T = 50$ the estimates of eigenvalues corresponding to actual unit roots are rather not very close to $\pm i$ in contrast to the false unit roots. Thus the latter are mistaken for actual unit roots (cf. the first panel in Figure 1), leading to a hit rate of 81.1%, one that is even larger than the rates at 0 and π. As the sample size increases, the eigenvalue estimates of the true unit roots become more and more accurate, visible from the second and third panel in Figure 1. Accordingly they can be detected correctly more often. Unfortunately however, for $T = 100$ the false unit roots remain to be detected such that often two instead of just one unit root are found by Λ, resulting in a hit rate of only 8.7%. For $T \in \{200, 500\}$ Λ is able to distinguish the false unit roots from the true ones and the detection rate is getting closer to the asymptotic rate, 85.5% and 92.7%, respectively.

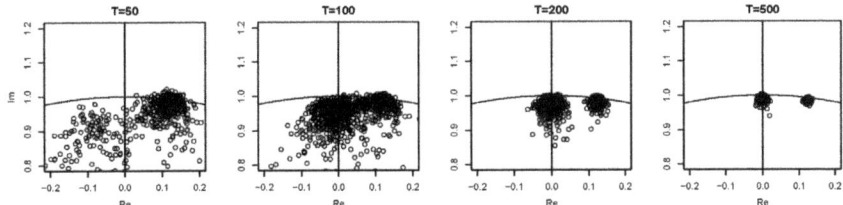

Figure 1. Eigenvalues around $z = i$ of 1000 replications when $\gamma = 0.2$ ($c_{\pi/2} = 1$).

When the VAR dgp without false unit roots and $c_{\pi/2} = 2$ is considered, it is visible that the hit rates of Λ at $\pi/2$ are monotonously increasing in the sample size again. The rates are smaller than those of the likelihood-based tests, however, and also clearly worse than those of Λ at 0 and π, cf. Table 2 again.

Taken together, at frequencies 0 and π which correspond to real-valued unit roots, the use of Λ was advantageous for $T = 50$. It also scored more hits for $T = 100$ and $c_{\pi/2} = 1$. For higher sample sizes the likelihood-based tests clearly dominate Λ at these two frequencies. At $\pi/2$ this superiority of the likelihood-based tests for all sample sizes and both dgps continues. The example also points to a general weakness: if the sample size is low and *false unit roots* are present, it can be difficult for Λ to distinguish them from actual unit roots.

7.2. VARMA Processes

The second setup consists of VARMA data generated by a state space system (A_r, C_r, K_r), $r = 1, 2$, as in (1), where the matrices A_1 and A_2 are constructed as in (2) and are taken to be

$$A_1 = \begin{bmatrix} 1 & 0 & 0 & 0 \\ 0 & -1 & 0 & 0 \\ 0 & 0 & 0 & 1 \\ 0 & 0 & -1 & 0 \end{bmatrix}, \quad A_2 = \begin{bmatrix} 1 & 0 & 0 & 0 & 0 & 0 \\ 0 & -1 & 0 & 0 & 0 & 0 \\ 0 & 0 & 0 & 1 & 0 & 0 \\ 0 & 0 & -1 & 0 & 0 & 0 \\ 0 & 0 & 0 & 0 & 0 & 1 \\ 0 & 0 & 0 & 0 & -1 & 0 \end{bmatrix}. \tag{9}$$

These two choices yield the same state space unit root structures as those of the two VAR dgps with $c_{\pi/2} = 1$ and $c_{\pi/2} = 2$ for A_1 and A_2, respectively. The other two system matrices $K_r \in \mathbb{R}^{(2+2r) \times s}$ and $C_r \in \mathbb{R}^{s \times (2+2r)}$ with $s = 8$ are drawn randomly from a standard normal distribution in each replication and $(\varepsilon_t)_{t \in \mathbb{Z}}$ is multivariate normal white noise with an identity covariance matrix.

Note that these systems are within the VARMA model class such that the dgp is contained in the VAR setting only by increasing the lag length as a function of the sample size. While superiority of the CVA approach in such a setting might be expected, this is far from obvious. Moreover, using a long VAR approximation is the industry norm in such situations.

From the hit rates in Table 3 it can be seen that the combination of large s, small T, and a minimal lag length of four render the likelihood-based tests useless at all frequencies with hit rates below ten percent for $T = 50$. Λ in contrast does not suffer from this problem and is already close to 95% for this sample size. Only when $T = 200$ do the likelihood-based tests appear to work, exhibiting hit rates close to 95%.

Table 3. Hit rates for the different tests (VARMA dgp). Twice the maximum (over all entries) Monte Carlo standard error is 0.005.

		0		$\pi/2$					π	
	T	Λ	J	Λ	JS	Q1	Q2	Q3	Λ	J
A_1	50	0.890	0.003	0.906	0.024	0.027	0.032	0.025	0.897	0.008
	100	0.928	0.434	0.944	0.755	0.783	0.783	0.761	0.930	0.440
	200	0.936	0.937	0.923	0.925	0.915	0.916	0.915	0.925	0.924
	500	0.852	0.901	0.853	0.919	0.906	0.904	0.904	0.853	0.894
A_2	50	0.863	0.008	0.785	0.062	0.047	0.063	0.039	0.867	0.006
	100	0.917	0.500	0.880	0.582	0.596	0.596	0.571	0.916	0.518
	200	0.931	0.927	0.882	0.908	0.915	0.913	0.911	0.919	0.922
	500	0.824	0.882	0.786	0.878	0.860	0.859	0.861	0.812	0.865

For all tests alike, however, it is striking that hit rates move away from 95% when $T = 500$. This behavior is most pronounced for Λ, e.g., from $T = 200$ to $T = 500$ its hit rate drops from 93.1% to 82.4% at 0 when A_2 is used. This phenomenon is a consequence of the fact that f and k in the algorithm are chosen data dependent. An inspection of how the hit rates depend on f and k and a comparison with the actually selected $\hat{f}, \hat{k}$ reveals that for $T = 500$ too large values of f and k are chosen too often and leave room for improvement in the hit rates, cf. Figure 2. The figure stresses an important point: The performance of the unit root tests is heavily influenced by the selected lag lengths for all procedures. We tested a number of different information criteria in this respect. AICc turned out to be the best criterion overall, but not uniformly. Figure 2 indicates advantages for this example of BIC over AIC as it on average selects smaller lag lengths, associated here with higher hit rates.

To study the power of the different procedures, the transition dynamics A_r in (9) are multiplied by $\rho \in \{0.8, 0.85, 0.9, 0.95\}$ so that the systems do not contain unit roots at any of the frequencies. Here empirical power is defined as the frequency of choosing zero common trends. This is why for $\rho = 1$, when there are in fact common trends present in our specifications, the empirical power values plotted in Figure 3 are not equal to the actual size we could define as one minus the hit rate: our measure of empirical power in this situation only counts the false test conclusion of zero common trends, but there are of course multiple ways the testing procedure could conclude falsely.

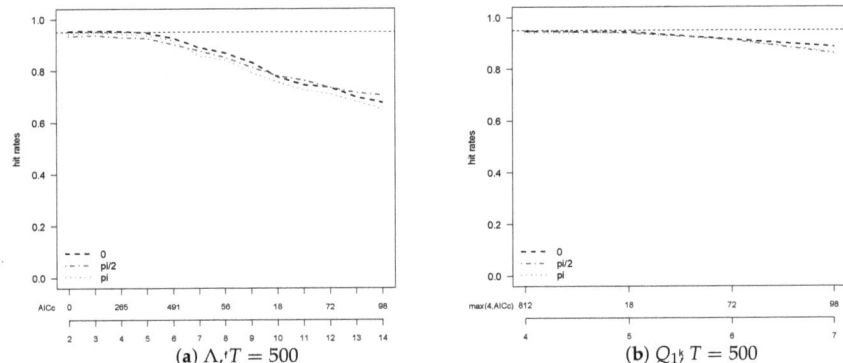

Figure 2. Relationship between hit rates and chosen values of f and k, illustration for the VARMA dgp using A_2. The lower x-axes show f or k, above are the choice frequencies of the selection criteria.

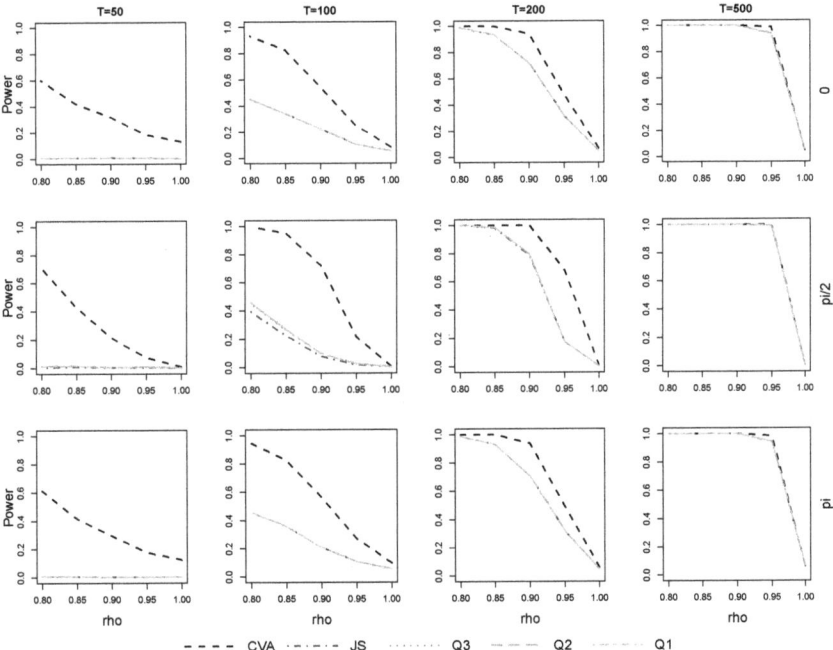

Figure 3. Empirical power of the different test procedures (VARMA dgp with A_2). Twice the Monte Carlo standard error is 0.005.

As expected, rejection of the null hypothesis is easiest when ρ is small and is very difficult when it is close to 1, cf. Figure 3 for the case of A_2.

Further, there are almost no differences among the likelihood-based tests over all combinations of sample size and frequency, only for $T = 100$ is JS significantly worse than the $Q_i, i = 1, 2, 3$ at $\pi/2$. It is also clearly visible at all frequencies that the likelihood-based tests possess no or only very limited power when $T = 50$ and $T = 100$, respectively. Λ, in contrast, is clearly more powerful in these cases. As the sample size increases to $T = 200$, the power of each test improves, still Λ remains the most powerful option. Only for $T = 500$ have the differences almost vanished with small, but significant, advantages for Λ at 0 and π.

The results are the same when A_1 is used and $c_{\pi/2} = 1$ and all of the differences described here are statistically significant.

Next the estimation performance of CVA is evaluated by calculation of the gaps between the true and the estimated cointegrating spaces. At all frequencies these gaps are compared with the GARR procedure of [5] which cycles through frequencies. At $\pi/2$ CVA and GARR are also compared with our implementation of JS and cRRR of [4], whereas it is also compared with the usual Johansen procedure at 0 and π. All estimates are conditional on the true state space unit root structure in the sense that the minimal number of states used is larger or equal to the number of unit roots over all frequencies. Other than imposing a minimum state dimension, the estimation of the order using SVC is not influenced. The likelihood-based procedures, on the other hand, take the unit root structure as given, i.e., do not perform CI rank testing for this estimation exercise.

From the results in Table 4 it can be noted first that the likelihood-based procedures show mostly equal mean gaps. Only for $\pi/2$ and $T = 50$ and both dgps does JS possess significantly larger gaps than cRRR and GARR and other differences are not statistically significant. Thus it does not matter in our example whether the iterative procedure is used or not.

Second, CVA is again superior for $T = 50$ where it exhibits mean gaps that are significantly smaller than those of the other estimators at all frequencies. This advantage is turned around for higher sample sizes, though: mean gaps are smaller for the likelihood-based procedures when $T \in \{100, 200, 500\}$ and A_2 is used, if only slightly. When A_1 is used instead, mean gaps do not differ significantly from each other at $\pi/2$ when $T > 50$ and at $0, \pi$ when $T = 100$ and those of CVA are only very modestly worse when $T \in \{200, 500\}$ at $0, \pi$.

Table 4. Mean gaps between estimated and true cointegrating spaces (VARMA dgp). 2MCse denotes twice the maximal Monte Carlo standard error for the corresponding row.

	T	2MCse	0			$\pi/2$				π		
			CVA	J	GARR	CVA	JS	cRRR	GARR	CVA	J	GARR
A_1	50	0.016	0.116	0.189	0.192	0.091	0.147	0.130	0.130	0.111	0.192	0.197
	100	0.004	0.047	0.048	0.048	0.039	0.035	0.035	0.035	0.047	0.046	0.046
	200	0.003	0.023	0.019	0.019	0.019	0.016	0.016	0.016	0.024	0.019	0.019
	500	0.003	0.012	0.007	0.007	0.008	0.008	0.006	0.006	0.011	0.007	0.007
A_2	50	0.016	0.174	0.245	0.242	0.250	0.349	0.331	0.331	0.165	0.231	0.234
	100	0.004	0.072	0.061	0.061	0.098	0.080	0.078	0.078	0.069	0.060	0.060
	200	0.003	0.031	0.026	0.026	0.047	0.036	0.034	0.034	0.032	0.027	0.027
	500	0.003	0.016	0.011	0.010	0.021	0.015	0.013	0.013	0.017	0.011	0.011

Thus, when it comes to estimating the cointegrating spaces, CVA is superior for $T = 50$ and equally good or only slightly worse than the likelihood-based procedures for higher sample sizes. For the systems analyzed, decreasing $c_{\pi/2}$ leads to gaps that are smaller for all methods and these improvements are slightly larger for CVA than for the other estimators.

7.3. Robustness of Unit Root Tests for Daily Data

In this last simulation example we examine the robustness of the proposed procedures with regard to test performance and prediction accuracy with respect to the innovation distribution and the existence of conditional heteroskedasticity of the GARCH-type, as these features are often observed in data of higher frequency, for example in financial applications. While our asymptotic results do not depend on the distribution of the innovations (subject to the assumptions), the assumptions do not include GARCH effects. Nevertheless, the theory in [25,26] suggests that the tests might be robust also in this respect.

We generate a state space system of order $n = 8$ using the matrix $A = [A_{i,j}]_{i,j=1,...,8}$ where $A_{i,i+1} = 1, i = 1,...,6, A_{7,1} = 1, A_{8,8} = 0.8$ and $A_{i,j} = 0$ else. This implies that the eigenvalues of this matrix are $\lambda_j = \exp(2\pi ij/7), j = 1,...,7, \lambda_8 = 0.8$. Therefore the corresponding process has state space unit root structure

$$((0,(1)),(2\pi/7,(1)),(4\pi/7,(1)),(6\pi/7,(1))).$$

The entries of the matrices C and K are chosen as independent standard normally distributed random variables as before.

A process $(y_t)_{t=1,...,T}$ is generated from filtering an independent identically distributed innovation process $(\varepsilon_t)_{t=-199,...,T+1}$ through the system (A, C, K). The first 200 observations are discarded, the last are used for validation purposes. A total of 1000 replications are generated where in each replication a different system is chosen.

With the generated data three different estimates are obtained: An autoregressive model (called AR in the following) is estimated with lag length chosen using AIC of maximal lag length equal to $\lfloor \sqrt{T} \rfloor$. Second, an autoregressive model with large lag length (called ARlong) is estimated. This estimate is used to hint at the behavior of an autoregression using the lag length equal to a full year. This would correspond to estimating a VECM without rank restrictions, when accounting for yearly differences. The third method consists of the CVA estimates, where $f = p = 2\hat{k}_{AIC}$ is chosen. The order is estimated by minimizing SVC. However, we correct for orders smaller than $n = 7$ which would limit the possibilities of finding all unit roots.

First, we compare the prediction accuracy for the three methods for two different distributions of the innovations: Beside the standard normal distribution also the student t-distribution with $v = 5$ degrees of freedom (scaled to unit variance) is used. This distribution shows considerably heavier tails than the normal distribution but nevertheless is covered by our assumptions.

Figure 4 provides the results for out-of-sample one day ahead mean absolute prediction error (over all coordinates) for the sample sizes $T = 364$ days (one year), $T = 1092$ (3 years) and $T = 3276$ (nine years). The long AR model is estimated with lag lengths of 8 weeks for the smallest sample size, 10 weeks for the medium sample size and 12 weeks for the largest sample size.

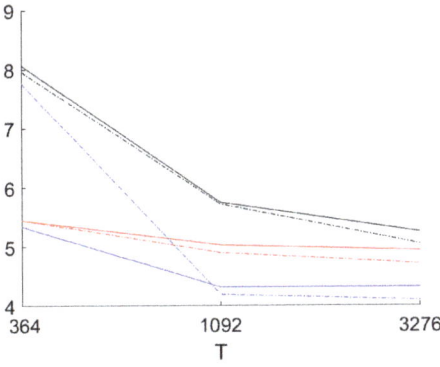

Figure 4. Mean of absolute value of one day ahead prediction error over all four components. CVA (blue), AR (red) and long AR (black). Dash-dot lines refer to the t-distribution.

In the figure the results for the normally distributed innovations are presented as well as the ones for the t-distributed residuals (scaled to unit variance). It can be seen that for the two larger sample sizes the mean absolute error for the residuals for CVA is smaller in all cases. For the smallest sample size, by contrast, results are mixed. For CVA the results for the heavy tailed distribution in this case are much worse than for the normal distribution. For the larger sample sizes the differences are small. The maximal standard error of the estimated means over 1000 replications for $T = 1092$ and $T = 3276$ amounts to 0.05. This allows the conclusion that CVA performs better for the two larger sample sizes. For $T = 364$ there are no statistically significant differences between the performance of the three methods: CVA seems to suffer more from few very large errors (using the root mean square errors the CVA results are worse for $T = 364$ in comparison; if one uses the 95% percentiles CVA performs best also for the smallest sample size). This results in a standard error over the replications of the mean absolute error for $T = 364$ of 0.18 for normally distributed innovations and 3.4 for t-distributed innovations. The long AR models are clearly worse than the two other approaches. This happens even if we are still far from using a full year as the lag length.

With regard to the unit root tests we investigate results for the tests of the hypotheses $H_0 : c_m = 1$ versus $H_1 : c_m = 0$ at all frequencies $2\pi m/364, m = 0, ..., 363$. The data generating process features unit roots with $c_m = 1$ at the seven frequencies $2\pi k/7, k = 0, ..., 6$. Therefore the tests should not reject at these frequencies, but should reject at all others.

Consequently we compare the minimum of the non-rejection rates for the seven unit roots (called empirical size below) as well as the maximum of the non-rejection rates for the non-unit root frequencies $\omega_j = 2\pi j/364, j \neq 52k, k = 0, 1, 2, ..., 6$ (called empirical power below).

For the larger sample sizes the empirical size is practically 95% while the empirical power is 100%. For $T = 364$ we obtain an empirical size of 90% for the normal distribution and 91.5% for the t-distribution. The worst empirical power equals 89.3% (normal) and 87.6% (t-distribution). Hence even for one year of data the discrimination properties of the unit root tests are good and we do not observe differences between the normal distribution for the innovations and the heavy tailed t-distribution.

Finally we compare the empirical size and power of the tests for the various unit roots for smaller sample sizes $T \in \{104, 208, 312, 416, 520\}$. For the experiments we consider univariate GARCH models of the form

$$\varepsilon_{t,i} = h_{t,i}\eta_{t,i}, \quad h_{t,i}^2 = 1 + \alpha \varepsilon_{t-1,i}^2 + \beta h_{t-1,i}^2, \quad i = 1, .., 4,$$

where $(\eta_{t,i})_{t \in \mathbb{Z}}$ is independent and identically standard normally distributed. $\alpha, \beta \geq 0$ are reals. It follows that the component processes $(\varepsilon_{t,i})_{t \in \mathbb{Z}}$ show conditional heteroskedasticity, the persistence of which is governed by $\alpha + \beta$. Here $0 < \alpha + \beta < 1$ implies stationarity while $\alpha + \beta = 1$ implies persistent conditional heteroskedasticity usually termed I-GARCH. We include five different processes for the innovations:

1. norm: $\alpha = \beta = 0$, no GARCH effects
2. G1: $\alpha = 0.8, \beta = 0.1$
3. IG1: $\alpha = 0.8, \beta = 0.2$
4. IG2: $\alpha = 0.5, \beta = 0.5$
5. IG3: $\alpha = 0.2, \beta = 0.8$

For the five different sample sizes 1000 replications of the estimates using the CVA algorithm are obtained. For each estimate we calculate the test statistic for testing $H_0 : c_m = 1$ versus $H_0 : c_m = 0$ for $m = 0, ..., 363$ corresponding to the unit roots $z_m = \exp(2\pi i m/364)$. This set of unit roots contains all seven unit roots $\exp(2\pi i k/7), k = 0, ..., 6$.

Figure 5 provides the mean over the 1000 replications of the test statistics $\Lambda(1)$ for $z_j, j = 0, ..., 363$ and the five sample sizes. It can be seen that the test $\Lambda(1)$ is able to pinpoint the seven unit roots present in the data generating process fairly accurately even for sample size $T = 104$. The zoom on the region around the unit root frequency $2\pi/7$ shows that the mean value is larger than the cutoff value of the test (the dashed horizontal line) for the adjacent frequency $2\pi\frac{53}{364}$ already for $T=312$.

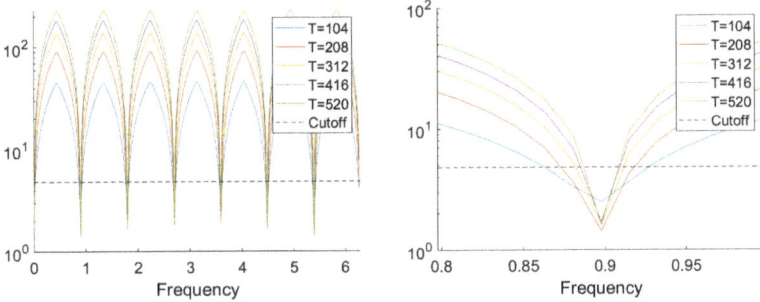

(a) Mean of unit root test statistics. (b) Zoom of mean unit root tests.

Figure 5. Results of the unit root tests for all seasonal unit roots jointly.

Table 5 lists the minimum of the achieved percentages of non-rejections of the test statistic for the seven unit root frequencies as well as the maximum over all non-unit root frequencies. It can be seen that for all GARCH models for $T=312$ the test rejects unit roots at all non unit root frequencies every time, while the empirical size is close to the nominal 5%. For small sample sizes the tests are slightly undersized while for $T=208$ a slight oversizing is observed. The two larger sample sizes are omitted as the tests perform perfectly there.

Table 5. Percentage of accept (minimum for all unit root frequencies) and reject (maximum for non unit root frequencies) of $\Lambda(1)$ test statistic.

	Unit Root Frequencies					Non Unit Root Frequencies				
T	norm	G1	IG1	IG2	IG3	norm	G1	IG1	IG2	IG3
104	0.94	0.89	0.87	0.88	0.87	0.87	0.82	0.79	0.82	0.79
208	0.98	0.96	0.95	0.94	0.96	0.78	0.75	0.72	0.72	0.69
312	0.97	0.96	0.96	0.95	0.95	0.00	0.00	0.00	0.00	0.00

It follows from the examples presented in this subsection that the test is robust also in small samples with respect to heavy tailed distributions of the innovations (subject to the assumptions). Furthermore also a remarkable robustness with respect to GARCH-type conditional heteroskedasticity is observed.

8. Application

In this section we apply CVA to the modeling of electricity consumption using a data set from [36]. The dataset contains hourly consumption data (in megawatts) from a number of US regions, scraped from the webpage of PJM Interconnection LLC, a regional transmission organization. The number of regions have changed over time, thus the data set contains many missing values. It also contains data aggregated into regions called east and west, which are not used subsequently.

In order to avoid problems with missing values, we restrict the analysis to four regions, for which data over the same time period is available: American Electric Power (AEP; in the following printed in blue), the Dayton Power and Light Company (DAYTON; black), Dominion Virginia Power (DOM; red) and Duquesne Light Co. (DUQ; green). We use data from 1 May 2005 until 31 July 2018. In this period only 3 data points are missing for the four regions and their imputation is handled by interpolation of the corresponding previous values. One observation in this sample is an obvious outlier which is corrected for analogously.

The data is split into an estimation sample covering observations up to the end of 2016 (102,291 observations on 4263 days) and a validation sample containing data in 2017 and 2018 (13,845 observations on 577 days). Data is equally sampled, but contains two hour segments when switching from winter to summer time or back. Table 6 contains some summary statistics.

Table 6. Summary of data sets.

Region	Daily Obs. (4263 est., 577 val.)					Hourly Obs. (102,291 est., 13,845 val.)			
	Mean	Mean(log)	Std.(log)	AIC	BIC	Mean(log)	Std.(log)	AIC	BIC
AEP	371,844	12.82	0.127	43	12	9.63	0.168	782	532
DAYTON	48,897	10.79	0.144	43	3	7.60	0.193	772	531
DOM	262,727	12.47	0.158	17	3	9.28	0.215	795	554
DUQ	39,837	10.58	0.130	23	7	7.40	0.177	800	529

Figure 6 provides an overview of the data: Panel (a) shows the full data on an hourly basis, while (b) presents aggregation to daily frequency. Panel (c) zooms in on a two year stretch of daily consumption. Panel (d) finally provides hourly data for the first month in the validation data. The figures clearly document strong daily, weekly and yearly patterns. From these figures it appears that these seasonal fluctuations are somewhat regular with changes throughout time. It is hence not clear whether a fixed seasonal pattern is appropriate. Also note that the sampling frequency is on an hourly basis such that a year roughly covers 8760 observations.

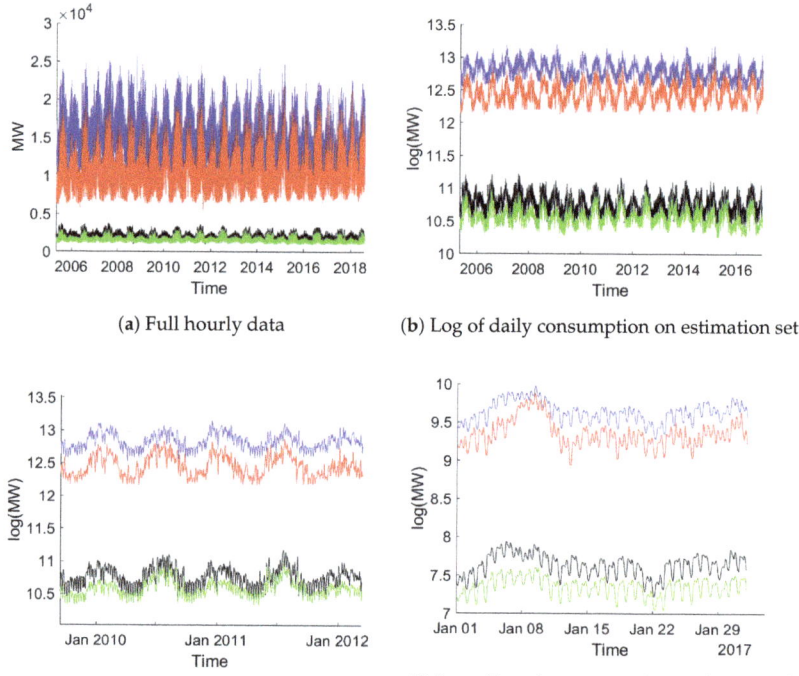

(a) Full hourly data
(b) Log of daily consumption on estimation set
(c) Log of daily consumption from 2010 to 2012
(d) Log of hourly consumption on first month of validation set

Figure 6. Electricity consumption data.

In the following we estimate (on the estimation part) and compare (on the validation part) a number of different models, first for the full hourly data set and afterwards for the aggregated daily data. As a benchmark we will use univariate AR models including deterministic seasonal patterns for daily, weekly and yearly variations. Subsequently we estimate models using CVA including different sets of such seasonal patterns.

First in the analysis using dummy variables fixed periodic patterns have been estimated. We model the natural logarithm of consumption (to reduce problems due to heteroskedasticity) and include dummies for weekdays, hours and sine and cosine terms corresponding to the first 20 Fourier frequencies with respect to annual periodicity. The corresponding results can be viewed in Figure 7. It is obvious that there is quite some periodic variation. Also the four data sets show very similar patterns as expected.

After the extraction of these deterministic terms the next step is univariate autoregressive (AR) modeling. Figure 8 shows the BIC values of AR models of lag lengths zero to 800 for the four series as well as the BIC of a multivariate AR model for the same number of lags. The chosen values are given in Table 6.

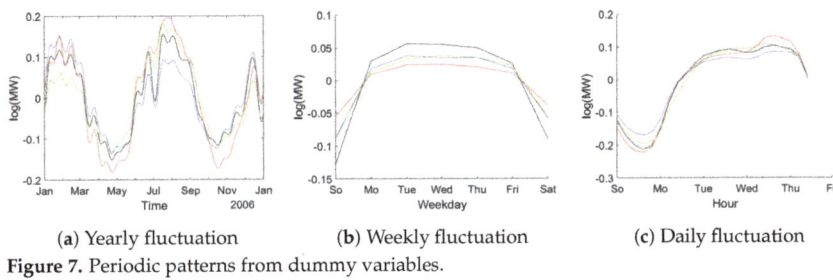

(a) Yearly fluctuation   (b) Weekly fluctuation   (c) Daily fluctuation

Figure 7. Periodic patterns from dummy variables.

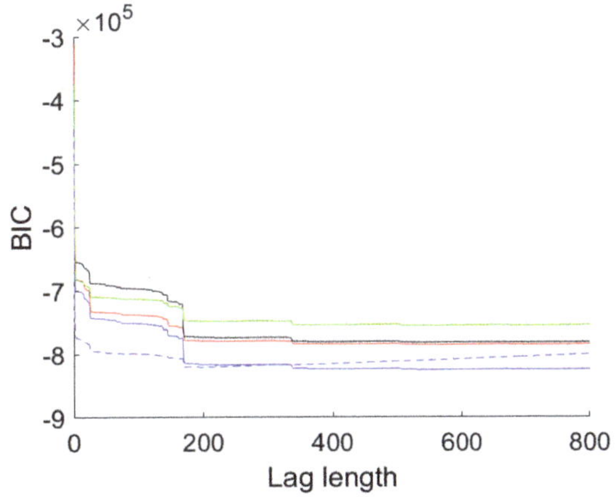

Figure 8. BIC values for univariate models and multivariate model (dashed line; divided by four to fit).

The BIC curve is extremely flat for the univariate models. Noticeable drops in BIC occur around lag 24 (one day), 144 (six days), 168 (one week), 336 (two weeks), 504 (three weeks). BIC selects large lag lengths from 529 (DUQ) up to 554 (DOM). AIC selects lag lengths close to the maximum allowed with a minimum at 772 lags. The BIC pattern of the multivariate model differs in that the two drops at two and three weeks are missing. Instead, the optimal BIC value is obtained at lag 194, well below the optimal lag lengths in the univariate cases. AIC here opts for lag length 531, just over 22 days.

Subsequently CVA is applied with $f = \hat{k}_{BIC}, p = \hat{k}_{AIC}$ as estimated for the multivariate model. This differs from the usual recommendation of $f = p = 2\hat{k}_{AIC}$ in order to avoid numerical problems with huge matrices. The order is chosen according to SVC, resulting in $\hat{n} = 240$. The corresponding model is termed Mod 1 in the following. Note that this configuration of $f, \hat{n}$ does not fulfill the requirements of our asymptotic theory. The bound $f \geq n$ ensures that the matrix $\mathcal{O}_f$ has full column rank. Generically this will be the case for $fs \geq n$ leading to a less restrictive assumption. In practice too low values of f will be detected by $\hat{n}$ estimated close to the maximum, which is not the case here.

As a second model we only use weekday dummies but neglect the other deterministics. Again AIC ($\hat{k}_{AIC} = 531$) and BIC ($\hat{k}_{BIC} = 195$) are used to determine the optimal lag length in the multivariate AR model. The corresponding CVA estimated model uses $\hat{n} = 245$ according to SVC, resulting in Mod 2.

The third model uses only a constant as deterministic term. Again similar AIC (555) and BIC (195) values are selected. A state space model, Mod 3, using CVA is estimated with $\hat{n} = 209$.

Figure 9 provides information on the results. Panel (a) shows the coefficients of the univariate AR models. It can be seen that lags around one day and one to three weeks play the biggest role for all four datasets. Panel (b) shows that the multivariate models lead to better one step ahead predictions in terms of the root mean square error (RMSE). Mod 1 and Mod 2 show practically equivalent out of sample prediction error for all four data sets, while Mod 3 delivers the best out of sample fit for all four regions.

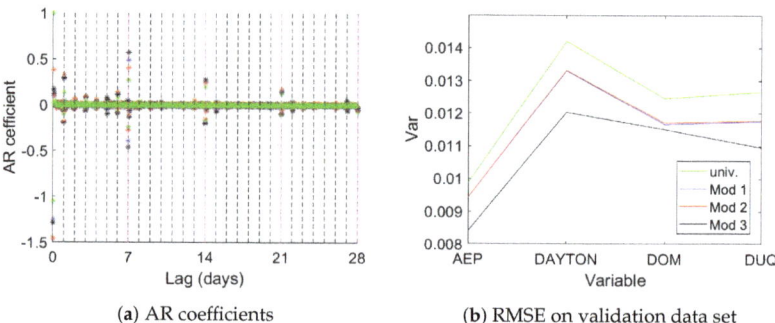

(a) AR coefficients (b) RMSE on validation data set

Figure 9. Results for the hourly datasets.

In particular in financial applications data of high sampling frequency shows persistent behaviour, also in terms of conditional heteroskedasticity, as well as heavy tailed distributions of the innovations. For our data sets Figure 10 below provides some information in this respect for the residuals according to Mod 3. Panel (a) provides a plot of the residuals in the year 2018 (contained in the validation period). It can be seen that large deviations occur occasionally, while else residuals vary in a tight band around 0. The kernel density estimates for the normalized (to unit variance) residuals on the full validation data set in panel (b) show the typical heavy tailed distributions. Panel (c) contains an ACF plot for the four regions again calculated using the full validation sample. It demonstrates that the model successfully eliminates all autocorrelations with only a few ACF values occurring outside the confidence interval. Panel (d) provides the ACF plot for the squared innovations to examine GARCH-type effects. While GARCH-effects are clearly visible, the ACF drops to zero fast with occasional positive values (except maybe for the Duquesne data).

Applying the eigenvalue based test $\Lambda(1)$ for $c = 1$ and all Fourier frequencies $\omega_j = 2\pi j/(365 * 24)$ we find that for Mod 2 and Mod 3 the largest p-value is obtained for ω_{365} corresponding to a period length of one day with 0.0187 for Mod 2 (test statistic 6.6) and 0.02 for Mod 3 (with a statistic of 6.5). This implies that the unit root at frequency ω_{365} is not rejected for a significance level of 1%, but is rejected for 5%. All other unit roots are rejected at every usual significance level. For Mod 1 the test statistic for ω_{365} equals 41.2 corresponding to a p-value of practically 0. This implies that on top of a deterministic daily pattern the series show strong persistence at the daily period. Excluding the hourly dummies pulls the roots closest to ω_{365} closer to the unit circle resulting in insignificant unit root tests and improves the one step ahead forecasts. Including the dummies weakens the evidence of a unit root while leading to worse predictions.

The analysis is repeated with data aggregated to daily sampling frequency. The aggregation reduces the required lag lengths, as is visible from Table 6 in the univariate case, and hence we use CVA with the recommended $f = p = 2\hat{k}_{AIC}$. Beside the univariate models, in this case also a naive model of predicting the consumption for today as yesterday's consumption is used. Three multivariate models are estimated: Mod 1 contains weekday dummies and sine and cosine terms for the first twenty Fourier frequencies corresponding to a period of one year. Mod 2 only contains the weekday dummies, while Mod 3 only uses the constant. Figure 11 provides the out-of-sample RMSE for one day ahead predictions (panel (a)) and seven days ahead predictions (panel (b)).

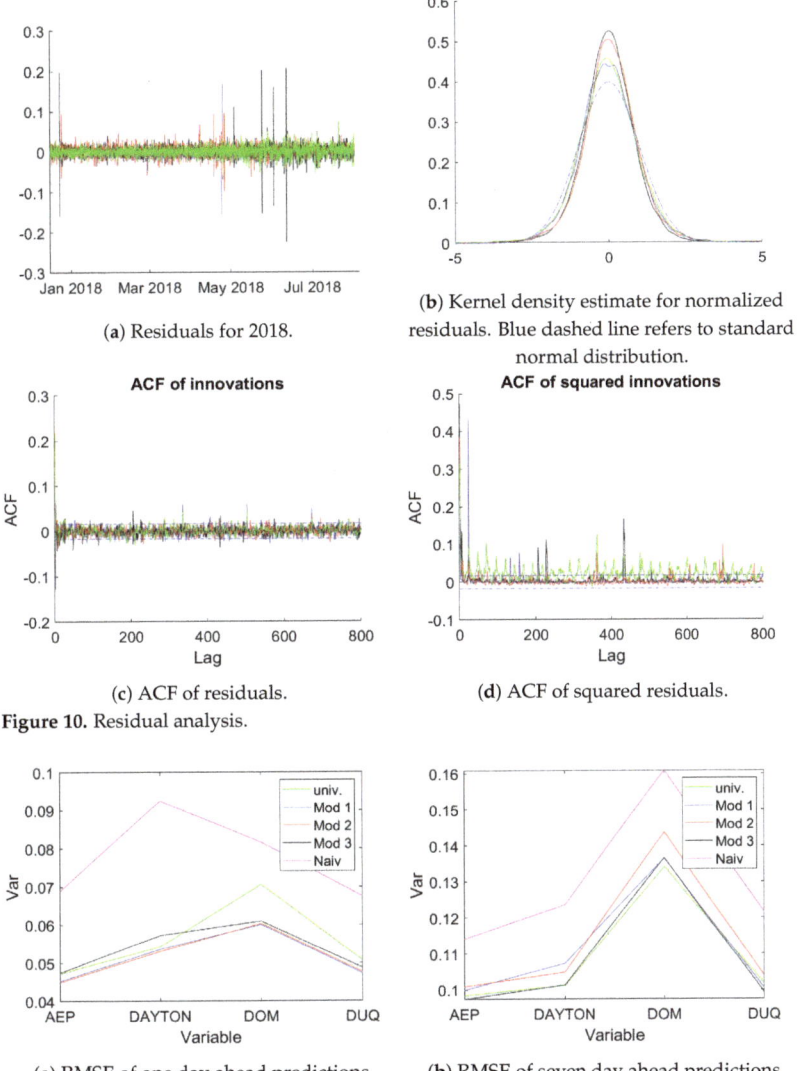

(a) Residuals for 2018.

(b) Kernel density estimate for normalized residuals. Blue dashed line refers to standard normal distribution.

(c) ACF of residuals.

(d) ACF of squared residuals.

Figure 10. Residual analysis.

(a) RMSE of one day ahead predictions

(b) RMSE of seven day ahead predictions

Figure 11. Results for the hourly datasets.

It can be seen that both Mod 1 and Mod 2 beat the univariate AR models in terms of one step ahead prediction error, while Mod 3 performs better for seven days ahead prediction. Mod 1 performs on par with Mod 2 for one step ahead prediction but performs better in predicting seven steps ahead. In Figure 12 poles and zeros for the three estimated state space models are plotted. Here the poles (marked with 'x') are the eigenvalues of the matrix A. These are the inverses of the determinantal roots of the autoregressive matrix polynomial in the equivalent VARMA representation. The zeros are the inverses of zeros of the determinant of the MA polynomial. We can see that for Mod 3 with only a constant, poles close to $2\pi j/7, j = 1, ..., 6$ arise to capture the weekly pattern. The other two models only show one pole close to the unit circle, a real pole of almost $z = 1$. The pole corresponding to Mod 1 is closer to the unit circle than the one for Mod 2 (see (b)).

For Mod 3 we obtain p-values for the tests of three complex unit roots of 0.05 ($\omega = 2\pi/7$), 0.165 ($4\pi/7$) and 0.01 ($6\pi/7$), which are hence all not statistically significant for significance level $\alpha = 0.01$. The corresponding test for $z = 1$ shows a p-value of 0.004. This provides evidence against the null hypothesis of the root being present. For Mod 1 the p-value for the test of $z = 1$ is 0.28 and hence we cannot reject the null. Mod 2 provides a p-value of 0.023 and hence weak evidence for the presence of the unit root. This can be seen from the distance of the nearest pole from the point $z = 1$ in Figure 12.

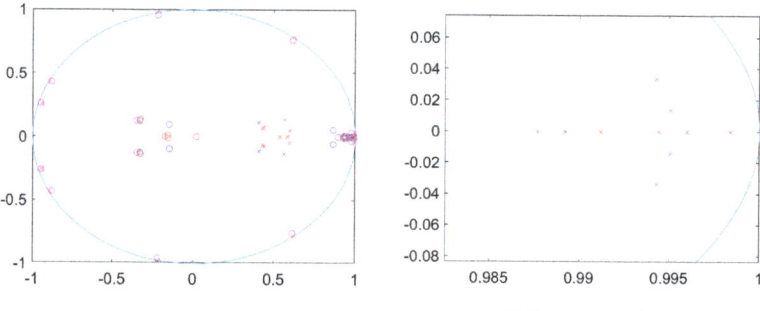

(a) Poles and zeros of the three models.   (b) Zoom around $z = 1$.

Figure 12. Poles (x) and zeros (o) of the transfer functions corresponding to the three models: Mod 1 (red), Mod 2 (blue), Mod 3 (magenta).

Jointly this indicates that the location and strength of persistence due to the estimated roots is influenced by the presence of deterministic terms: if the deterministic terms are not included in the model, the cyclical patterns are generated by poles situated close to the unit circle.

The decision whether on top of the deterministic seasonality unit roots exist, is not easy in all cases: for the daily data the locations of the poles indicate that deterministic seasonality is enough to capture weekly fluctuations while a unit root at $z = 1$ appears to be needed to capture yearly variations. For hourly data there is evidence that the daily cycle is best captured with a unit root at frequency ω_{365}. This leads to the best predictive fit. Finally note that temporal aggregation from hourly data to daily data implies that the frequency ω_{365} for hourly data aliases to the frequency $\omega = 0$ in the daily data. Therefore the higher evidence of a unit root at $z = 1$ found in daily data might be a consequence of the unit root at frequency ω_{365} found for hourly data, compare [37].

The system matrix estimates as well as the evidence in support of unit roots at ω_{365} for hourly data and $z = 1$ for daily data that we obtain from the CVA modeling can be taken as starting points in subsequent quasi maximum likelihood estimation.

9. Conclusions

In this paper the asymptotic properties of CVA estimators for seasonally integrated unit root processes are investigated. The main results can be summarized as follows:

- CVA provides consistent estimators for long-run and short-run dynamics without knowledge of the location and number of unit roots. Hence the algorithm is robust with respect to the presence of trending components at frequency zero as well as at the other seasonal unit root frequencies.
- The singular values calculated in the RRR step reveal information on the total number of unit roots. The distance of the singular values to one can be used to construct a consistent estimator of this quantity.
- The eigenvalues of $\hat{A}$ can be used in order to test for the number of common trends. Under the null hypothesis these tests are asymptotically equivalent to the corresponding tests using the true state, making the derivation of asymptotic results and the simulation of the test distribution simple.
- An analogous statement holds for the Johansen trace test in the $I(1)$ case and analogous tests in the $MFI(1)$ case calculated on the basis of the estimated state in the restrictive setting of $n \leq s$. Under the null hypothesis these tests reject and accept asymptotically jointly with the corresponding tests calculated using the true state.
- From the simulation exercises we conclude that CVA performs best when the dgp is of the more general VARMA type, the process dimension is moderate to large and the sample size is small. Then it is superior to the likelihood-based procedures based on VAR approximations in terms of the estimation performance and the size and power of Λ, the test developed from CVA. For higher sample sizes the likelihood-based procedures are clearly superior when it comes to the size of the corresponding tests, whereas Λ remains the best test choice in terms of empirical power. The estimation performance is about equal for all procedures when the sample size is high with slight advantages for the likelihood-based procedures.
- The simulations also demonstrate that the unit root test results are robust with respect to the distribution of the innovation sequence as well as some forms of conditional heteroskedasticity of the GARCH-type.

Because of the promising performance of CVA and in particular its robustness it can be recommended as a simple way to extract information on the number of common trends from the estimated matrix of transition dynamics. This information can be used in order to reduce the uncertainty in a subsequent likelihood ratio analysis where quasi maximum likelihood estimates can be obtained starting from the CVA estimates. Since the CVA estimates can be obtained for a range of orders numerically fast they are seen as a valuable starting point for the empirical modeling of time series potentially including seasonal cointegration. Moreover they can also be used in situations where the number of seasons is large or even unclear as in hourly data sets as demonstrated in the case study.

Author Contributions: Conceptualization, D.B. and R.B.; methodology, D.B.; software, R.B.; formal analysis, D.B. and R.B.; writing—original draft preparation, D.B. and R.B.; writing—review and editing, D.B. and R.B.; visualization, D.B. and R.B.; supervision, D.B. Both authors have read and agreed to the published version of the manuscript.

Funding: This research was funded by the Deutsche Forschungsgemeinschaft (DFG, German Research Foundation—Projektnummer 276051388) which is gratefully acknowledged. We acknowledge support for the publication costs by the Deutsche Forschungsgemeinschaft and the Open Access Publication Fund of Bielefeld University.

Institutional Review Board Statement: Not applicable.

Informed Consent Statement: Not applicable.

Conflicts of Interest: The authors declare no conflict of interest.

Appendix A. Supporting Material

Appendix A.1. Complex Valued Canonical Form

Additionally to the real valued canonical form (2) we will also use the corresponding complex valued representation obtained by transforming each block corresponding to unit root $z_j = \cos(\omega_j) + i\sin(\omega_j)$ with the transformation matrix

$$\mathcal{T}_j = \begin{bmatrix} I_{c_j} & iI_{c_j} \\ I_{c_j} & -iI_{c_j} \end{bmatrix}$$

leading to the triple of system matrices in the j-th block as:

$$\mathcal{A}_{j,\mathbb{C}} = \begin{bmatrix} \overline{z_j} I_{c_j} & 0 \\ 0 & z_j I_{c_j} \end{bmatrix}, \quad \mathcal{K}_{j,\mathbb{C}} = \begin{bmatrix} K_{j,\mathbb{C}} \\ \overline{K_{j,\mathbb{C}}} \end{bmatrix}, \quad \mathcal{C}_{j,\mathbb{C}} = \begin{bmatrix} C_{j,\mathbb{C}}/2 & \overline{C_{j,\mathbb{C}}}/2 \end{bmatrix},$$

such that

$$x_{t+1,j,\mathbb{C}} = \overline{z_j} x_{t,j,\mathbb{C}} + K_{j,\mathbb{C}} \varepsilon_t, \quad x_{t,j} = \mathcal{T}_j^{-1} \begin{bmatrix} x_{t,j,\mathbb{C}} \\ \overline{x_{t,j,\mathbb{C}}} \end{bmatrix}.$$

Lemma A1. *Let $x_t = [x'_{t,0}, x'_{t,1}, \ldots, x'_{t,S/2}, x'_{t,\bullet}]'$ where $x_{t,j}$ is generated according to $x_{t+1,j} = A_j x_{t,j} + K_j \varepsilon_t, t \in \mathbb{N}$ with A_j as in (2) and $K_j = [K'_{j,R}, K'_{j,I}]' \in \mathbb{R}^{\delta_j c_j \times s}$ using iid white noise process $(\varepsilon_t)_{t \in \mathbb{N}}$ where $x_{0,j}$ is deterministic. Further let $(x_{t,\bullet})_{t \in \mathbb{N}}$ denote the stationary solution to the equation $x_{t+1,\bullet} = A_\bullet x_{t,\bullet} + K_\bullet \varepsilon_t$ such that $M_\bullet = \mathbb{E} x_{t,\bullet} x'_{t,\bullet} > 0$.*
(I) Then using $Q_T = \sqrt{(\log \log T)/T}$ for $u_t = \sum_{i=0}^{q} \varphi_i \varepsilon_{t+i}$ for arbitrary $q \in \mathbb{N}, q < \infty$, and coefficients $\varphi_i, i = 0, \ldots, q$ we have

$$\begin{array}{ll} \langle x_{t,\bullet}, u_t \rangle = O(Q_T) & , \quad \langle u_{t-j}, u_t \rangle - \mathbb{E} u_{t-j} u'_t = O(Q_T), \\ \langle x_{t,j}, x_{t,\bullet} \rangle = O(\log T) & , \quad \langle x_{t,j}, u_t \rangle = O(\log T) \\ \langle x_{t,j}, x_{t,k} \rangle / T = O(\log \log T) & , \quad j, k = 0, \ldots, S/2. \end{array}$$

If $(\varepsilon_t)_{t \in \mathbb{Z}}$ only fulfills Assumptions 1 then the order bounds hold in probability rather than almost surely.
(II) Furthermore for $0 < j, k < S/2$

$$\begin{array}{ll}
\langle x_{t,j,\mathbb{C}}, \varepsilon_t \rangle & \xrightarrow{d} \quad \frac{1}{2} \int_0^1 W_j dB'_{j,\mathbb{C}} =: M_j, \\
\langle x_{t,j,\mathbb{C}}, x_{t,k,\mathbb{C}} \rangle / T & \xrightarrow{d} \quad \begin{cases} \frac{1}{2} \int_0^1 W_j W'_j := N_j &, \quad j = k, \\ 0 &, \quad j \neq k \end{cases} \\
\langle x_{t,j}, \varepsilon_t \rangle & \xrightarrow{d} \quad \begin{bmatrix} \frac{1}{2} \int_0^1 (W_{j,R} dB'_{j,R} + W_{j,I} dB'_{j,I}) \\ \frac{1}{2} \int_0^1 (W_{j,I} dB'_{j,R} - W_{j,R} dB'_{j,I}) \end{bmatrix}, \\
\langle x_{t,k}, x_{t,j} \rangle / T & \xrightarrow{d} \quad \begin{cases} \frac{1}{2} \begin{bmatrix} \int_0^1 (W_{k,R} W'_{k,R} + W_{k,I} W'_{k,I}) & \int_0^1 (W_{k,R} W'_{k,I} - W_{k,I} W'_{k,R}) \\ -\int_0^1 (W_{k,R} W'_{k,I} - W_{k,I} W'_{k,R}) & \int_0^1 (W_{k,R} W'_{k,R} + W_{k,I} W'_{k,I}) \end{bmatrix} &, \quad j = k \\ 0 &, \quad j \neq k \end{cases}
\end{array}$$

where $W_j = W_{j,R} + iW_{j,I} = K_{j,\mathbb{C}} B_{j,\mathbb{C}}, K_{j,\mathbb{C}} = K_{j,R} + iK_{j,I}, B_{j,\mathbb{C}} = B_{j,R} + iB_{j,I}$ and $B_{j,R}, B_{j,I}$ are two independent Brownian motions with covariance matrix Ω. For $j = 0$ and $j = S/2$ the results hold analogously:

$$\langle x_{t,0}, \varepsilon_t \rangle \xrightarrow{d} \int_0^1 W_{0,R} dW'_{0,R} \quad , \quad \langle x_{t,0}, x_{t,0} \rangle / T \xrightarrow{d} \int_0^1 W_{0,R} W'_{0,R},$$

$$\langle x_{t,S/2}, \varepsilon_t \rangle \xrightarrow{d} \int_0^1 W_{S/2,R} dW'_{S/2,R} \quad , \quad \langle x_{t,S/2}, x_{t,S/2} \rangle / T \xrightarrow{d} \int_0^1 W_{S/2,R} W'_{S/2,R}.$$

Proof. Most evaluations in (I) are standard, see for example Lemma 4 in [38].
(II) follows from the results in Section 4 of [2] for the complex valued representations or [39] for the corresponding real case. □

Appendix A.2. Perturbation of Eigendecompositions

Lemma A2 (Rayleigh-Schrödinger expansion). *Let $\hat{A}_t = A - \delta A_t$ where $\|\delta A_t\| \to 0$ and where $A = U\Lambda U^{-1} \in \mathbb{R}^{n\times n}$, $\Lambda = \text{diag}(\lambda_1 I_{c_1}, ..., \lambda_J I_{c_J})$, $\sum_{j=1}^J c_j = n$ is diagonalizable. $U = [U_1, ..., U_J] \in \mathbb{C}^{n\times n}$ is a nonsingular matrix such that for $U^{-1} = [V_1, ..., V_J]'$ we have $V_j' U_j = I_{c_j}$.*

Then for each circle $B(\lambda_j, \delta)$ around λ_j not containing any other eigenvalue of A there exist from some t onwards

- *c_j eigenvalues of $\hat{A}_t$ in the circle $B(\lambda_j, \delta)$ around λ_j*
- *a basis $\hat{U}_{t,j}$ for the space spanned by the eigenspaces to these c_j eigenvalues such that $V_j' \hat{U}_{t,j} = I_{c_j}$,*
- *a sequence of matrices $\hat{B}_{t,j} = V_j' \hat{A}_t \hat{U}_{t,j} \in \mathbb{C}^{c_j \times c_j}$.*

Then $\hat{U}_{t,j} = \sum_{k=0}^\infty Z_k$, $\hat{B}_{t,j} = \sum_{k=0}^\infty C_k$ where

$$Z_0 = U_j, \quad C_0 = \lambda_j I_{c_j},$$

$$Z_k = \Sigma(\delta A_t Z_{k-1} + \sum_{i=1}^{k-1} Z_{k-i} C_i), \quad C_k = -V_j' \delta A_t Z_{k-1}.$$

Here $\Sigma = U(\Lambda - I_n \lambda_j)^+ U^{-1}$ where $\text{diag}(s_1, ..., s_n)^+ = \text{diag}(s_1^+, ..., s_n^+)$ and $x^+ = 1/x, x \neq 0$ and zero else, that is $(\Lambda - I_n \lambda_j)^+$ denotes a quasi-inverse.
Furthermore for $\rho = \|\delta A_t\| < 1$ we obtain: $\|C_k\| \leq \mu_C \rho^k$, $\|Z_k\| \leq \mu_Z \rho^k$, $k \geq 0$.

The results follow directly from Section 2.9 of [23], see in particular Proposition 2.9.1 and the discussion below this proposition. Further note that the results hold for each root separately and hence the restriction $\ell_j = 1$ needs to hold only for the investigated root for the results to apply. Finally note that a second order approximation $\hat{U}_{t,j} = Z_0 + Z_1 + Z_2$ and $\hat{B}_{t,j} = C_0 + C_1 + C_2$ is accurate to the order $o(\|\delta A_t\|^2)$.

Appendix A.3. Random Transformation of Systems

Lemma A3. *Let the assumptions of Theorem 1 hold and use the same notation as given there. Let $(\tilde{\mathcal{A}}, \tilde{\mathcal{C}}, \tilde{\mathcal{K}})$ denote a sequence of systems converging a.s. to $(\mathcal{A}, \mathcal{C}, \mathcal{K})$ such that $(\tilde{\mathcal{A}} - \mathcal{A})D_x^{-1} = O((\log T)^a)$, $\sqrt{T}(\tilde{\mathcal{K}} - \mathcal{K}) = O((\log T)^a)$, $(\tilde{\mathcal{C}} - \mathcal{C})D_x^{-1} = O((\log T)^a)$ and let $\mathcal{A}_0 = S_0 \mathcal{A} S_0^{-1} = \text{diag}(\mathcal{A}_{0,11}, \mathcal{A}_{0,22})$, $\mathcal{K}_0 = S_0 \mathcal{K}$, $\mathcal{C}_0 = \mathcal{C} S_0^{-1}$. Further let*

$$S_T = \begin{bmatrix} S_{T,11} & S_{T,12} \\ 0 & S_{T,22} \end{bmatrix} \to S_0$$

such that $(S_T - S_0)D_x^{-1} = O((\log T)^a)$. Let $\Delta S = (S_T - S_0)D_x^{-1}$, $\Delta \mathcal{A} = (\tilde{\mathcal{A}} - \mathcal{A})D_x^{-1}$ and denote the sequence of transformed systems as $(\hat{\mathcal{A}}, \hat{\mathcal{C}}, \hat{\mathcal{K}}) = (S_T \tilde{\mathcal{A}} S_T^{-1}, \tilde{\mathcal{C}} S_T^{-1}, S_T \tilde{\mathcal{K}})$. Let the block entries of S_0 be denoted as S_{ij} and the blocks of ΔS be denoted as ΔS_{ij}. Then:

$$T(\hat{\mathcal{A}}_{11} - \mathcal{A}_{0,11}) = (\Delta S_{11} \mathcal{A}_{11} - \mathcal{A}_{0,11} \Delta S_{11} + S_{11} \Delta \mathcal{A}_{11} + S_{12} \Delta \mathcal{A}_{21}) S_{11}^{-1} + o(1),$$

$$\sqrt{T}(\hat{\mathcal{A}}_{12} - \mathcal{A}_{0,12}) = (S_{11} \Delta \mathcal{A}_{12} + S_{12} \Delta \mathcal{A}_{22}) S_{22}^{-1} + \Delta S_{12} S_{22}^{-1} \mathcal{A}_{0,22} - \mathcal{A}_{0,11} \Delta S_{12} S_{22}^{-1} + o(1),$$

$$T(\hat{\mathcal{A}}_{21} - \mathcal{A}_{0,21}) = S_{22} \Delta \mathcal{A}_{21} S_{11}^{-1} + o(1),$$

$$\sqrt{T}(\hat{\mathcal{A}}_{22} - \mathcal{A}_{0,22}) = \Delta S_{22} S_{22}^{-1} \mathcal{A}_{0,22} + S_{22} \Delta \mathcal{A}_{22} S_{22}^{-1} - \mathcal{A}_{0,22} \Delta S_{22} S_{22}^{-1} + o(1),$$

$$\sqrt{T}(\hat{\mathcal{K}} - \mathcal{K}_0) = \begin{bmatrix} \Delta S_{12} \mathcal{K}_2 + S_{11} \sqrt{T}(\tilde{\mathcal{K}}_1 - \mathcal{K}_1) + S_{12} \sqrt{T}(\tilde{\mathcal{K}}_2 - \mathcal{K}_2) \\ \Delta S_{22} \mathcal{K}_2 + S_{22} \sqrt{T}(\tilde{\mathcal{K}}_2 - \mathcal{K}_2) \end{bmatrix} + o(1),$$

$$(\hat{\mathcal{C}} - \mathcal{C}_0) D_x^{-1} = (\tilde{\mathcal{C}} - \mathcal{C}) D_x^{-1} \begin{bmatrix} S_{11}^{-1} & 0 \\ 0 & S_{22}^{-1} \end{bmatrix} - \mathcal{C}_0 \begin{bmatrix} \Delta S_{11} S_{11}^{-1} & \Delta S_{12} S_{22}^{-1} \\ 0 & \Delta S_{22} S_{22}^{-1} \end{bmatrix} + o(1).$$

Proof. The proof follows from straightforward computations using the various orders of convergence by neglecting higher order terms. □

Appendix B. Reduced Rank Regression with Integrated Variables

The main results of this paper are based on a more general result documented in [24] (henceforth called BRRR). BRRR uses a slightly different setting and in particular a different dgp. The following lemma provides the essence of the results of BRRR that will be used below.

Lemma A4. *Let* $(y_t)_{t \in \mathbb{N}}, (z_t^r)_{t \in \mathbb{N}}, (z_t^u)_{t \in \mathbb{N}}, y_t \in \mathbb{R}^s, z_t^r \in \mathbb{R}^m, z_t^u \in \mathbb{R}^l$ *be three processes related via*

$$y_t = b_r z_t^r + b_u z_t^u + u_t$$

where the zero mean stationary process $(u_t)_{t \in \mathbb{N}}$ *is such that* $\mathbb{E} u_t (z_t^r)' = 0, \mathbb{E} u_t (z_t^u)' = 0, \mathbb{E} u_t u_t' > 0$ *and where* $n = \text{rank}(b_r) < \min(s, m)$, *that is* b_r *is of reduced rank.*

Further assume that there exist square nonsingular matrices $\mathcal{T}_y \in \mathbb{R}^{s \times s}, \mathcal{T}_r \in \mathbb{R}^{m \times m}, \mathcal{T}_u \in \mathbb{R}^{n \times n}$ *such that*

$$\tilde{y}_t = \mathcal{T}_y y_t = (\mathcal{T}_y b_r \mathcal{T}_r^{-1})(\mathcal{T}_r z_t^r) + (\mathcal{T}_y b_r \mathcal{T}_r^{-1})(\mathcal{T}_r z_t^r) + \mathcal{T}_y u_t = \tilde{b}_r \tilde{z}_t + \tilde{b}_u \tilde{z}_t^u + \tilde{u}_t$$

such that with $c_\bullet = n - c$ *we have*

$$\tilde{b}_r = \begin{bmatrix} I_c & 0 & 0 \\ 0 & 0 & \tilde{b}_{r,\bullet} \end{bmatrix}, \quad \tilde{b}_{r,\bullet} = \tilde{O}_\bullet \Gamma'_\bullet, \quad \tilde{O}_\bullet \in \mathbb{R}^{(s-c) \times c_\bullet}, \quad \Gamma_\bullet \in \mathbb{R}^{m_\bullet \times c_\bullet}.$$

Here the partitioning corresponds to $\tilde{z}_t' = [\tilde{z}_{t,1}', \tilde{z}_{t,2}', \tilde{z}_{t,\bullet}']$ *where* $\tilde{z}_{t,1} \in \mathbb{R}^c, \tilde{z}_{t,2} \in \mathbb{R}^{m-c-m_\bullet}$ *are MFI(1) processes and* $(\tilde{z}_{t,\bullet})_{t \in \mathbb{N}}, \tilde{z}_{t,\bullet} \in \mathbb{R}^{m_\bullet}$ *is stationary,* $\tilde{z}_t^u = [(\tilde{z}_{t,1}^u)', (\tilde{z}_{t,\bullet}^u)']'$ *where* $(\tilde{z}_{t,1}^u)_{t \in \mathbb{N}}$ *is a MFI(1) process and* $(\tilde{z}_{t,\bullet}^u)_{t \in \mathbb{N}}$ *is stationary and where the following bounds hold* $(\tilde{z}_{t,:} := [\tilde{z}_{t,1}', \tilde{z}_{t,2}']')$:

$$\langle \tilde{u}_t, \tilde{u}_t \rangle = O(1) \quad , \quad \langle \tilde{u}_t, \tilde{z}_{t,\bullet} \rangle = O(Q_T) \quad , \quad \langle \tilde{u}_t, \tilde{z}_{t,\bullet}^u \rangle = O(Q_T),$$

$$\langle \tilde{u}_t, \tilde{u}_t \rangle - \mathbb{E} \tilde{u}_t \tilde{u}_t' = O(Q_T) \quad , \quad \langle \tilde{u}_t, \tilde{z}_{t,:} \rangle = O(\log T) \quad , \quad \langle \tilde{u}_t, \tilde{z}_{t,1}^u \rangle = O(\log T),$$

$$\hat{M}_\bullet = \langle \begin{pmatrix} \tilde{z}_{t,\bullet} \\ \tilde{z}_{t,\bullet}^u \end{pmatrix}, \begin{pmatrix} \tilde{z}_{t,\bullet} \\ \tilde{z}_{t,\bullet}^u \end{pmatrix} \rangle \quad , \quad \hat{M}_\bullet^{-1} = O(1) \quad , \quad \hat{M}_\bullet = O(1), M_\bullet > 0$$

$$\hat{M}_1 = \langle \begin{pmatrix} \tilde{z}_{t,:} \\ \tilde{z}_{t,1}^u \end{pmatrix}, \begin{pmatrix} \tilde{z}_{t,:} \\ \tilde{z}_{t,1}^u \end{pmatrix} \rangle \quad , \quad \hat{M}_1/T = O(\log \log T) \quad , \quad (\hat{M}_1)^{-1} = O(Q_T^2),$$

$$\langle \begin{pmatrix} \tilde{z}_{t,\bullet} \\ \tilde{z}_{t,\bullet}^u \end{pmatrix}, \begin{pmatrix} \tilde{z}_{t,:} \\ \tilde{z}_{t,1}^u \end{pmatrix} \rangle = O(\log T) \quad , \quad \hat{M}_\bullet - M_\bullet = O(Q_T).$$

Then the reduced rank regression estimator $\hat{b}_{RRR} = [\hat{b}_{r,RRR}, \hat{b}_{u,RRR}]$ *maximizing the Gaussian likelihood subject to* $\text{rank}(\beta_r) = n = c + c_\bullet$ *is consistent:* $\hat{b}_{RRR} - b = O((\log T)^a / \sqrt{T})$ *for some* $a < \infty$. *Furthermore* $\tilde{b}_{RRR,r} - \tilde{b}_r = [O((\log T)^a / T), O((\log T)^a / \sqrt{T})]$ *with* $\tilde{b}_{RRR,r} = \mathcal{T}_y \hat{b}_{RRR,r} \mathcal{T}_r^{-1}$, *where the second block has* $m_\bullet$ *columns and corresponds to the stationary components of the regressor vector.*

Proof. The theorem slightly extends the results of BRRR by adding high level assumptions instead of low level assumptions on the data generating process. The proof hence consists in adjusting the proof in BRRR. In the following we only indicate where arguments in BRRR need to be replaced. A detailed proof would replicate much of the arguments in BRRR and hence is omitted.

The representation of Theorem 3.1 in BRRR is contained in the assumptions. Then consistency follows from examining the proof of the first part of Theorem 3.2 in BRRR: essential for the norm bounds are Lemma A.1 (I) and (III). The norm bounds stated under point (I) are directly assumed in this lemma except for the filtered version using n_t in place of x_t. Instead, here the results for n_t which are needed in the proof of Theorem 3.2 of BRRR are directly assumed. (III) then follows. Lemmas A.3–A.5 in BRRR do not depend on the assumptions on the various processes and hence continue to hold. Then the proof for consistency in Appendix A.3.1 of BRRR only uses these norm bounds referring also to [38] (which is also only based on the norm bounds contained in the assumptions of this lemma) and hence continues to hold. □

Appendix C. Proofs of the Theorems

Appendix C.1. Proof of Theorem 1

For proving consistency of the transfer function estimators it is sufficient to find a (possibly) random matrix $\breve{S}_T$ such that the least squares estimates $(\breve{\mathcal{A}}, \breve{\mathcal{C}}, \breve{\mathcal{K}})$ of one representation $(\mathcal{A}, \mathcal{C}, \mathcal{K})$ of the true system obtained using $\tilde{x}_t := \breve{S}_T \hat{x}_t$ converges (a.s.) to $(\mathcal{A}, \mathcal{C}, \mathcal{K})$. This will be done in two steps: First a particular basis (which is not realizable in practice) will be chosen such that $\breve{\mathcal{K}}_p - \mathcal{K}_p = o(1)$ sufficiently fast such that in the second step the regressions in the system equations based on the resulting state estimator $\tilde{x}_t$ are consistent. The derivation of the first step will also provide an approximation of the error term which can be used in order to derive the asymptotic distribution.

Appendix C.1.1. Proof of Theorem 1 (I)

The central step in CVA is the solution to the RRR problem. The following proof heavily draws on the results contained in [24] (henceforth called BRRR) collected in Lemma A4 for easier reference. As in BRRR, in order to derive the asymptotic properties we first transform the vectors in order to separate stationary and nonstationary terms. In order to achieve the separation let $Z_t = [y'_{t-1}, y'_{t-2}, ..., y'_{t-S}]' \in \mathbb{R}^{sS}$. Then for $p = kS$ we obtain

$$Y^-_{t,p} = \begin{pmatrix} y_{t-1} \\ y_{t-2} \\ \vdots \\ y_{t-S} \\ y_{t-S-1} \\ \vdots \\ y_{t-kS} \end{pmatrix} = \begin{pmatrix} Z_t \\ Z_{t-S} \\ \vdots \\ Z_{t-(k-1)S} \end{pmatrix}.$$

It is easy to see that for each j the process $(Z_{rS-j})_{r \in \mathbb{N}}$ is an $I(1)$ process. Moreover the strict minimum-phase condition for $(\mathcal{A}_\circ, \mathcal{C}_\circ, \mathcal{K}_\circ)$ implies that also for the system corresponding to $(Z_{rS-j})_{r \in \mathbb{N}}$ the strict minimum-phase condition holds.

Define the transformation $\mathcal{T}_S := [\mathcal{O}_{S,1}, \mathcal{O}_{S,\perp}]'$ where $\mathcal{O}_{S,1} \in \mathbb{R}^{sS \times c}$ denotes the matrix containing the first c columns of $\mathcal{O}_S$ for the system $(\mathcal{A}_\circ, \mathcal{C}_\circ, \mathcal{K}_\circ)$ in the canonical form. Further $\mathcal{O}_{S,\perp}$ is a block column of an orthonormal matrix such that $\mathcal{O}'_{S,\perp} \mathcal{O}_{S,1} = 0$. Then the argument of [20] shows that in $\mathcal{T}_S Z_t$ the first c components are integrated while the remaining $sS - c$ components are stationary. Then consider for $p = kS < t \leq T - f + 1$ (using $\mathcal{O}^\dagger_{S,1} = (\mathcal{O}'_{S,1} \mathcal{O}_{S,1})^{-1} \mathcal{O}'_{S,1}$)

$$\tilde{z}_{t,p} := \begin{bmatrix} \mathcal{O}^\dagger_{S,1} \mathcal{O}_S (x_t - \mathcal{A}^p_\circ x_{t-p}) \\ \mathcal{O}'_{S,\perp} Z_t \\ \mathcal{O}^\dagger_{S,1} (Z_t - Z_{t-S}) \\ \mathcal{O}'_{S,\perp} Z_{t-S} \\ \vdots \\ \mathcal{O}^\dagger_{S,1} (Z_{t-(k-2)S} - Z_{t-(k-1)S}) \\ \mathcal{O}'_{S,\perp} Z_{t-(k-1)S} \end{bmatrix}, \quad \tilde{y}_t := \begin{bmatrix} \mathcal{O}^\dagger_{f,1} \\ \mathcal{O}'_{f,\perp} \end{bmatrix} Y^+_{t,f}. \quad (A1)$$

Here $\mathcal{O}_{f,\perp}$ is a matrix such that $\mathcal{O}'_{f,\perp} \mathcal{O}_{f,1} = 0, \mathcal{O}'_{f,\perp} \mathcal{O}_{f,\perp} = I$. Obviously $\tilde{z}_{t,p}$ is a linear transformation of $Y^-_{t,p}$ and $\tilde{y}_t$ of $Y^+_{t,f}$. It can be shown that the linear transformation is nonsingular such that there is a one-one relation between $Y^-_{t,p}$ and $\tilde{z}_{t,p}$. In $\tilde{z}_{t,p}$ and $\tilde{y}_t$ only the first c components are unit root processes, the remaining components being stationary.

For $p \neq kS$ the final $p - kS$ block rows of $\tilde{z}_{t,p}$ are defined as $y_{t-(k-1)S-j} - y_{t-kS-j}, j = 1, ..., p - kS$. Clearly also these components are stationary.

Partition $\tilde{z}_{t,p} = [\tilde{z}'_{t,1}, \tilde{z}'_{t,\bullet}]', \tilde{z}_{t,1} \in \mathbb{R}^c$, into its first c and the remaining coordinates (omitting the subscript p on the right hand side for notational convenience). Similarly partition $\tilde{y}_t = [\tilde{y}'_{t,1}, \tilde{y}'_{t,\bullet}]'$, $\tilde{y}_{t,1} \in \mathbb{R}^c$. Using these transformed matrices, $Y^+_{t,f} = \beta_1 Y^-_{t,p} + N^+_{t,f}$ can be written as

$$\tilde{y}_t = \tilde{b}_1 \tilde{z}_{t,p} + \tilde{N}^+_{t,f,p} = \begin{bmatrix} \tilde{y}_{t,1} \\ \tilde{y}_{t,\bullet} \end{bmatrix} = \begin{bmatrix} I_c & 0 \\ 0 & \tilde{b}_{\bullet,p} \end{bmatrix} \begin{bmatrix} \tilde{z}_{t,1} \\ \tilde{z}_{t,\bullet} \end{bmatrix} + \tilde{\mathcal{O}}_f \mathcal{A}^p_\circ x_{t-p} + \begin{bmatrix} \tilde{\varepsilon}_{t,1} \\ \tilde{\varepsilon}_{t,\bullet} \end{bmatrix} \quad (A2)$$

where

$$\tilde{b}_1 = \begin{bmatrix} I_c & 0 \\ 0 & \tilde{b}_{\bullet,p} \end{bmatrix}, \quad \tilde{b}_{\bullet,p} = \mathbb{E}\tilde{y}_{t,\bullet} \tilde{z}'_{t,\bullet} (\mathbb{E}\tilde{z}_{t,\bullet} \tilde{z}'_{t,\bullet})^{-1} = O_{\bullet,p} \Gamma'_{\bullet,p}, \quad \tilde{b}_1 = O_p \Gamma'_p$$

and where $\tilde{b}_{\bullet,p}$ is of rank $n - c$ providing a representation of the form given in Theorem 3.1 of BRRR except that the error term $\tilde{N}^+_{t,f,p}$ (defined by the equation) is not white. Finally (A2) also defines the sub blocks $\tilde{\varepsilon}_{t,i}$ of $\tilde{N}^+_{t,f,p}$ which are hence linear combinations of $E^+_{t,f}$ and therefore typically MA(f) processes. Note that $\tilde{z}_{t,1}, \tilde{z}_{t,\bullet}, \tilde{y}_{t,\bullet}$ depend on the choice of f and p which is not reflected in the notation.

Here $(\mathbb{E}\tilde{z}_{t,\bullet} \tilde{z}'_{t,\bullet})^{-1}$ and $\mathbb{E}\tilde{y}_{t,\bullet} \tilde{z}'_{t,\bullet}$ are worth a remark: for $p = kS$ the results of [20] can be directly used to obtain upper and lower bounds for the norms of these matrices uniformly in $k \in \mathbb{N}$. For $p \neq kS$ the additional rows in $\tilde{z}_{t,\bullet}$ add entries to $\mathbb{E}\tilde{y}_{t,\bullet} \tilde{z}'_{t,\bullet}$ that are of order $O(\lambda^p)$ for some $0 < \lambda < 1$ as $y_t - y_{t-S}$ is a VARMA process. Similarly the smallest eigenvalue of $\mathbb{E}\tilde{z}_{t,\bullet} \tilde{z}'_{t,\bullet}$ can be bounded from below based on arguments for $p = kS$ following [20] which in turn refer to Theorem 6.6.10 of [22]. The additional terms for $p \neq kS$ correspond to backward innovations with non-singular covariance matrix thus also leading to a lower bound of the smallest eigenvalue uniformly in k. (The backward innovations representation for a stationary VARMA process $(y_t)_{t \in \mathbb{Z}}$ equals $y_t = \sum_{j=1}^\infty k^b_j y_{t+j} + \varepsilon^b_t$ and can be obtained from the complex conjugate of the spectral density. Nonsingularity of the spectral density implies that the backward innovation ε^b_t have nonsingular covariance matrix. This implies a lower bound on the accuracy with which components of $y_{t-(k-1)S-j}$ can be predicted based on $y_{t-i}, i \leq (k-1)S$.)

Furthermore the strict minimum-phase assumption for the state space representation $(\mathcal{A}_\circ, \mathcal{C}_\circ, \mathcal{K}_\circ)$ of the process $(y_t)_{t \in \mathbb{Z}}$ implies the strict minimum-phase assumption for the sub-sampled process $(Z_{kS+j})_{k \in \mathbb{Z}}$. Thus the arguments of [20] show that $[\tilde{b}_{\bullet,p}, 0] \to \tilde{b}_{\bullet,\infty}$ where the norm of the difference is of order $O(\|\mathcal{A}^p_\circ\|)$. The increase of p as a function of the sample size jointly with the strict minimum-phase assumption implies that $O(\|\mathcal{A}^p_\circ\|) = o(T^{-1})$. This also implies that $\tilde{\mathcal{O}}_f \mathcal{A}^p_\circ x_{t-p} = o_p(T^{-1/2})$.

Correspondingly there exists a limiting decomposition $\tilde{b}_{\bullet,\infty} = O_\bullet \Gamma'_\bullet$ such that $\Gamma'_\bullet S_\bullet = I_{n-c}$ where $S_\bullet$ denotes a selector matrix whose columns contain the vectors of the canonical basis of $\mathbb{R}^\infty$. Since $[\mathcal{K}_\circ, (\mathcal{A}_\circ - \mathcal{K}_\circ \mathcal{C}_\circ) \mathcal{K}_\circ, (\mathcal{A}_\circ - \mathcal{K}_\circ \mathcal{C}_\circ)^2 \mathcal{K}_\circ, ..., (\mathcal{A}_\circ - \mathcal{K}_\circ \mathcal{C}_\circ)^{n-1} \mathcal{K}_\circ]$ is of full row rank it can be assumed that $S_\bullet$ only has nonzero entries in its first ns rows. Denoting the submatrix of the first ps rows by $S_{p,2}$ then also $[\Gamma'_\bullet]_{1:p} S_{p,2} = I_{n-c}$ where $[.]_{1:p}$ denotes the first p block columns of a matrix. This fixes a unique decomposition of $\tilde{b}_\bullet$ and hence $O_\bullet$ and $\Gamma_\bullet$ do not depend on p. Convergence of $\tilde{b}_{\bullet,p}$ to $\tilde{b}_\bullet$ jointly with the lower bound on $p(T)$ then implies convergence of order $o(T^{-1})$ of $O_{\bullet,p}$ to $O_\bullet$ and $\Gamma'_{\bullet,p}$ to $[\Gamma'_\bullet]_{1:p}$ using the decomposition of $\tilde{b}_{\bullet,p}$ such that $\Gamma'_{\bullet,p} S_{p,2} = I_{n-c}$. Correspondingly $O_p \to O$ and $\|\Gamma'_p - [\Gamma']_{1:p}\| \to 0$.

Therefore the reduced rank regression in the CVA procedure shows the same structure as investigated in Lemma A4 with the differences that $\tilde{z}_{t,2}$ and $\tilde{z}^u_t$ do not occur, and $\tilde{z}_{t,\bullet}$ has increasing size as a function of the sample size. The next lemma therefore extends the results of BRRR to the RRR used in CVA:

In the following we will use a generic $a \in \mathbb{N}$ in statements like $O((\log T)^a)$, not necessarily the same in each occurrence. In this sense e.g., the product of two terms that are $O((\log T)^a)$ is again taken to be $O((\log T)^a)$.

Lemma A5. *Let the assumptions of Theorem 1 hold where additionally $(\varepsilon_t)_{t\in\mathbb{Z}}$ is iid. Introduce the notation*

$$\tilde{D}_z = \mathrm{diag}(T^{-1/2}I_c, I_{ps-c}), \quad \tilde{D}_y = \mathrm{diag}(T^{-1/2}I_c, I_{fs-c}), \quad \tilde{D}_x = \mathrm{diag}(T^{-1/2}I_c, I_{n-c}).$$

Let $\bar{G}_p$ denote a solution to

$$(\tilde{D}_z\langle \tilde{z}_{t,p}, \tilde{y}_t\rangle \tilde{D}_y)(\tilde{D}_y\langle \tilde{y}_t, \tilde{y}_t\rangle \tilde{D}_y)^{-1}(\tilde{D}_y\langle \tilde{y}_t, \tilde{z}_{t,p}\rangle \tilde{D}_z)\bar{G}_p = (\tilde{D}_z\langle \tilde{z}_{t,p}, \tilde{z}_{t,p}\rangle \tilde{D}_z)\bar{G}_p \bar{R}^2$$

using the notation of (A1) where $\bar{R}^2 \to \Theta^2 = \mathrm{diag}(I_c, \Theta_\bullet) \in \mathbb{R}^{n \times n}$ and where $\bar{G}_p$ is normalized such that $\bar{G}_{1,1,p} = I_c$, $\bar{G}'_{\bullet,2,p} S_{p,2} = I_{n-c}$ for a selector matrix $S_{p,2}$. Further let

$$\bar{\Gamma}_p = \begin{bmatrix} I_c & 0 \\ 0 & \bar{\Gamma}_{\bullet,p} \end{bmatrix}, \bar{\Gamma}'_{\bullet,p} S_{p,2} = I_{n-c}$$

denote the solution to the decoupled problem where the stationary and the nonstationary subproblem are separated:

$$\begin{pmatrix} \langle \tilde{z}_{t,1}, \tilde{y}_{t,1}\rangle\langle \tilde{y}_{t,1}, \tilde{y}_{t,1}\rangle^{-1}\langle \tilde{y}_{t,1}, \tilde{z}_{t,1}\rangle \bar{\Gamma}_{1,1,p} \\ \langle \tilde{z}_{t,\bullet}, \tilde{y}_{t,\bullet}\rangle\langle \tilde{y}_{t,\bullet}, \tilde{y}_{t,\bullet}\rangle^{-1}\langle \tilde{y}_{t,\bullet}, \tilde{z}_{t,\bullet}\rangle \bar{\Gamma}_{\bullet,p} \end{pmatrix} = \begin{pmatrix} \langle \tilde{z}_{t,1}, \tilde{z}_{t,1}\rangle \bar{\Gamma}_{1,1,p} \Theta_1 \\ \langle \tilde{z}_{t,\bullet}, \tilde{z}_{t,\bullet}\rangle \bar{\Gamma}_{\bullet,p} \Theta_\bullet \end{pmatrix}.$$

(I) Then if $f \geq n$ fixed independent of T and $p \geq -d\log T/\log \rho_0, d>1, p = o((\log T)^{\bar{a}})$ for some $\bar{a} < \infty$ the a.s. results of Lemma A.6 (I)-(III) and Lemma A.7 of [24] hold true for $(\log T)^3$ replaced by $(\log T)^a$ for some integer $a < \infty$. In particular $\bar{G}_p - \bar{\Gamma}_p = O((\log T)^a/T^{1/2})$.
(II) Using the notation $\delta G_p := \bar{G}_p - \bar{\Gamma}_p$ define

$$\tilde{S}_T := \begin{bmatrix} I_c & -\sqrt{T}\delta G'_{\bullet,1,p}\langle \tilde{z}_{t,\bullet}, \tilde{z}_{t,\bullet}\rangle \bar{\Gamma}^\dagger_{\bullet,p} \\ 0 & I_{ps-c} \end{bmatrix}, \quad \bar{\Gamma}^\dagger_{\bullet,p} := \bar{\Gamma}_{\bullet,p}(\bar{\Gamma}'_{\bullet,p}\langle \tilde{z}_{t,\bullet}, \tilde{z}_{t,\bullet}\rangle \bar{\Gamma}_{\bullet,p})^{-1}.$$

Then for $\tilde{\Gamma}'_p := \tilde{S}_T \tilde{D}_x^{-1} \bar{G}'_p \tilde{D}_z$ and

$$\Gamma' = \begin{bmatrix} I & 0 \\ 0 & \Gamma'_\bullet \end{bmatrix}$$

we obtain $\tilde{\Gamma}'_p - [\Gamma']_{1:p} = [O((\log T)^a/T), O((\log T)^a/T^{1/2})]$ where the partitioning corresponds to the partitioning of $\tilde{z}_{t,p}$ into $\tilde{z}_{t,1}$ and $\tilde{z}_{t,\bullet}$. Here $\Gamma'_\bullet$ denotes the right factor of $\bar{b}_{\bullet,\infty} = O_\bullet \Gamma'_\bullet$ such that $[\Gamma'_\bullet]_{1:p}S_{p,2} = I_{n-c}$ holds.

(III) Let the assumptions of Theorem 1 hold. Then $\hat{Z}_T := \mathrm{Tvec}\left((\tilde{\Gamma}'_p - [\Gamma']_{1:p})\begin{bmatrix} I_c \\ 0 \end{bmatrix}\right)$ converges in distribution.

Proof. (I) First consider the entries of the vectors $\tilde{y}_{t,\bullet}$ and $\tilde{z}_{t,\bullet}$ (see (A1)) more closely. Since in

$$\mathcal{O}'_{f,\perp} Y^+_{t,f} = \mathcal{O}'_{f,\perp}(\mathcal{O}_{f,\bullet} x_{t,\bullet} + \mathcal{E}_f E^+_{t,f})$$

the nonstationary directions are filtered by definition, $\tilde{y}_{t,\bullet}$ is stationary and does not depend on T.

Further, also $\tilde{z}_{t,\bullet}$ is stationary for fixed p as the nonstationary directions are either filtered by pre-multiplication with $\mathcal{O}'_{S,\perp}$ or by yearly differencing $Z_t - Z_{t-S}$.

Therefore we obtain from stationary theory for fixed $p = kS$ that

$$\|\mathbb{E}\tilde{y}_{t,\bullet}\tilde{z}'_{t,\bullet}(\mathbb{E}\tilde{z}_{t,\bullet}\tilde{z}'_{t,\bullet})^{-1} - \langle \tilde{y}_{t,\bullet}, \tilde{z}_{t,\bullet}\rangle\langle \tilde{z}_{t,\bullet}, \tilde{z}_{t,\bullet}\rangle^{-1}\| = o(1).$$

Here $\sup_p \|(\mathbb{E}\tilde{z}_{t,\bullet}\tilde{z}'_{t,\bullet})^{-1}\| < \infty$ has been discussed before. Now $\mathbb{E}\tilde{y}_{t,\bullet}\tilde{z}'_{t,\bullet}(\mathbb{E}\tilde{z}_{t,\bullet}\tilde{z}'_{t,\bullet})^{-1} = \tilde{\beta}_{\bullet,p} + o(T^{-1/2}) = O_{\bullet,p}[\Gamma'_\bullet]_{1:p} + o(T^{-1/2})$ where the $o(T^{-1/2})$ term appears due to neglecting $\tilde{\mathcal{O}}_f \bar{A}^p x_{t-p}$. It follows that $\det[(\tilde{\beta}_{\bullet,p}S_{p,2})'(\tilde{\beta}_{\bullet,p}S_{p,2})] = \det[O'_{\bullet,p}O_{\bullet,p}] > 0$ and hence $\|\hat{\beta}_{\bullet,p} - \tilde{\beta}_{\bullet,p}\|_{Fr} = o(1)$ implies $\lim_{T \to \infty} \det[(\hat{\beta}_{\bullet,p} S_{p,2})'(\hat{\beta}_{\bullet,p} S_{p,2})] > 0$ a.s. where $\hat{\beta}_{\bullet,p} := \langle \tilde{y}_{t,\bullet}, \tilde{z}_{t,\bullet} \rangle \langle \tilde{z}_{t,\bullet}, \tilde{z}_{t,\bullet} \rangle^{-1}$. Since $\hat{O}_{\bullet,p} \tilde{\Gamma}'_{\bullet,p} - \tilde{\beta}_{\bullet,p} = o(1)$ due to consistency, also

$$\lim_{T \to \infty} \det\left[(\hat{O}_{\bullet,p} \tilde{\Gamma}'_{\bullet,p} S_{p,2})'(\hat{O}_{\bullet,p} \tilde{\Gamma}'_{\bullet,p} S_{p,2})\right] = \lim_{T \to \infty} \det \hat{O}'_{\bullet,p} \hat{O}_{\bullet,p} \det(\tilde{\Gamma}'_{\bullet,p} S_{p,2})^2 > 0 \quad \text{a.s.}$$

Since $\tilde{\Gamma}_{\bullet,p} - \Gamma_{\bullet,p} = o(1)$ due to the definition of $\tilde{\Gamma}_{\bullet,p}$ and the continuity of the solution of the eigenvalue problem it follows that $\hat{O}_{\bullet,p} - O_{\bullet,p} = o(1)$ and therefore $\limsup_T \det \hat{O}'_{\bullet,p} \hat{O}_{\bullet,p} > 0$. As in Lemma 6 of [40] it can be shown that $\Gamma'_{\bullet,p} - [\Gamma'_{\bullet}]_{1:p} = o(T^{-1})$ and $O_{\bullet,p} = O_{\bullet} + o(T^{-1})$ for the range of p given in Theorem 1 since these matrices correspond to a stationary problem. Hence the chosen normalization of $\tilde{\Gamma}_{\bullet,p}$ can be used a.s.

Next in order to obtain the convergence of $\bar{G}$ to $\tilde{\Gamma}_p$, Lemma A.6 of BRRR is slightly extended to the current situation (for details and notation see there). Lemma A.6 of BRRR contains three parts: BRRR(I) gives bounds on the error in the matrices (with $l_T = \log T$)

$$\delta_{yz} = \begin{bmatrix} \frac{1}{T}\langle \tilde{y}_{t,1}, \tilde{z}_{t,1}\rangle & \frac{1}{\sqrt{T}}\langle \tilde{y}_{t,1}, \tilde{z}_{t,\bullet}\rangle \\ \frac{1}{\sqrt{T}}\langle \tilde{y}_{t,\bullet}, \tilde{z}_{t,1}\rangle & \langle \tilde{y}_{t,\bullet}, \tilde{z}_{t,\bullet}\rangle \end{bmatrix} - \begin{bmatrix} \frac{1}{T}\langle \tilde{z}_{t,1}, \tilde{z}_{t,1}\rangle & 0 \\ 0 & \langle \tilde{y}_{t,\bullet}, \tilde{z}_{t,\bullet}\rangle \end{bmatrix} = \begin{bmatrix} O(\frac{1}{T}l_T^a) & O(\frac{1}{\sqrt{T}}l_T^a) \\ O(\frac{1}{\sqrt{T}}l_T^a) & 0 \end{bmatrix},$$

$$\delta_{yy} = \begin{bmatrix} \frac{1}{T}\langle \tilde{y}_{t,1}, \tilde{y}_{t,1}\rangle & \frac{1}{\sqrt{T}}\langle \tilde{y}_{t,1}, \tilde{y}_{t,\bullet}\rangle \\ \frac{1}{\sqrt{T}}\langle \tilde{y}_{t,\bullet}, \tilde{y}_{t,1}\rangle & \langle \tilde{y}_{t,\bullet}, \tilde{y}_{t,\bullet}\rangle \end{bmatrix} - \begin{bmatrix} \frac{1}{T}\langle \tilde{z}_{t,1}, \tilde{z}_{t,1}\rangle & 0 \\ 0 & \langle \tilde{y}_{t,\bullet}, \tilde{y}_{t,\bullet}\rangle \end{bmatrix} = \begin{bmatrix} O(\frac{1}{T}l_T^a) & O(\frac{1}{\sqrt{T}}l_T^a) \\ O(\frac{1}{\sqrt{T}}l_T^a) & 0 \end{bmatrix},$$

$$\delta_{zz} = \begin{bmatrix} \frac{1}{T}\langle \tilde{z}_{t,1}, \tilde{z}_{t,1}\rangle & \frac{1}{\sqrt{T}}\langle \tilde{z}_{t,1}, \tilde{z}_{t,\bullet}\rangle \\ \frac{1}{\sqrt{T}}\langle \tilde{z}_{t,\bullet}, \tilde{z}_{t,1}\rangle & \langle \tilde{z}_{t,\bullet}, \tilde{z}_{t,\bullet}\rangle \end{bmatrix} - \begin{bmatrix} \frac{1}{T}\langle \tilde{z}_{t,1}, \tilde{z}_{t,1}\rangle & 0 \\ 0 & \langle \tilde{z}_{t,\bullet}, \tilde{z}_{t,\bullet}\rangle \end{bmatrix} = \begin{bmatrix} 0 & O(\frac{1}{\sqrt{T}}l_T^a) \\ O(\frac{1}{\sqrt{T}}l_T^a) & 0 \end{bmatrix}.$$

BRRR(II) deals with $J = \bar{Q} - \Phi =$

$$\tilde{D}_z \langle \tilde{z}_t, \tilde{y}_t \rangle \tilde{D}_y (\tilde{D}_y \langle \tilde{y}_t, \tilde{y}_t \rangle \tilde{D}_y)^{-1} \tilde{D}_y \langle \tilde{y}_t, \tilde{z}_t \rangle \tilde{D}_z - \begin{bmatrix} \frac{1}{T}\langle \tilde{z}_{t,1}, \tilde{z}_{t,1}\rangle & 0 \\ 0 & \langle \tilde{z}_{t,\bullet}, \tilde{y}_{t,\bullet}\rangle \langle \tilde{y}_{t,\bullet}, \tilde{y}_{t,\bullet}\rangle^{-1}\langle \tilde{y}_{t,\bullet}, \tilde{z}_{t,\bullet}\rangle \end{bmatrix}$$

and BRRR(III) shows that there exists a solution $\bar{G}_p$ converging to a solution $\tilde{\Gamma}_p$ of the separated problem.

For showing the orders of convergence of δ_{zz} the arguments are unchanged except for noting that in $\langle \tilde{z}_{t,1}, \tilde{z}_{t,\bullet} \rangle$ the number of columns increases as a function of the sample size. Since the a.s. bounds on the entries of this expression hold uniformly (as follows straightforwardly from the arguments of Lemma A.1 of BRRR) this does not change the arguments. With respect to δ_{yz} note that now $\tilde{y}_t = \tilde{\beta}_1 \tilde{z}_{t,p} + \tilde{\epsilon}_t + \tilde{\mathcal{O}}_f \bar{A}^p x_{t-p}$. Due to the increase of p as a function of the sample size, $\bar{A}^p = o(T^{-1-\epsilon})$ for small enough $\epsilon > 0$ and therefore $\tilde{\mathcal{O}}_f \bar{A}^p x_{t-p} = o(T^{-1/2-\epsilon/2})$ since $x_t = o(T^{(1+\epsilon)/2})$ (uniformly in $1 \le t \le T$) whether $(x_t)_{t \in \mathbb{Z}}$ is a unit root process or stationary. Hence $\langle \tilde{\mathcal{O}}_f \bar{A}^p x_{t-p}, \tilde{\mathcal{O}}_f \bar{A}^p x_{t-p}\rangle = o(1)$. Further $\langle \tilde{\mathcal{O}}_f \bar{A}^p x_{t-p}, \tilde{\epsilon}_t \rangle = o(T^{-1/2})$ follows from $\langle x_{t-p}, \tilde{\epsilon}_t \rangle = O(\log T)$ (see Lemma A.1 (I)). This shows that the additional term is always of lower order and can be neglected. The remaining arguments follow exactly as in the proof of Lemma A.6 of BRRR. The proof of Lemma A.7 of BRRR only uses the order bounds derived above and hence follows immediately. This shows (I).

(II) Using the definition of $\tilde{\mathcal{S}}_T$ we obtain:

$$\tilde{\Gamma}'_p = \tilde{\mathcal{S}}_T \tilde{D}_x^{-1} \bar{G}'_p \tilde{D}_z = \tilde{\mathcal{S}}_T \begin{bmatrix} I_c & \sqrt{T}\delta G'_{\bullet,1,p} \\ \delta G'_{1,2,p}/\sqrt{T} & \bar{G}'_{\bullet,2,p} \end{bmatrix}$$
$$= \begin{bmatrix} I_c - \delta G'_{\bullet,1,p}\langle \tilde{z}_{t,\bullet}, \tilde{z}_{t,\bullet}\rangle \tilde{\Gamma}^\dagger_{\bullet,p}\delta G'_{1,2,p} & \sqrt{T}\delta G'_{\bullet,1,p}(I - \langle \tilde{z}_{t,\bullet}, \tilde{z}_{t,\bullet}\rangle \tilde{\Gamma}^\dagger_{\bullet,p}\bar{G}'_{\bullet,2,p}) \\ \delta G'_{1,2,p}/\sqrt{T} & \bar{G}'_{\bullet,2,p} \end{bmatrix}.$$

From (I) and Lemma A.7 of BRRR $\delta G_{\bullet,1,p} = O((\log T)^a/T^{1/2})$, $\delta G_{1,2,p} = O((\log T)^a/T^{1/2})$ and $\bar{G}_{\bullet,2,p} - \tilde{\Gamma}_{\bullet,p} = o((\log T)^a/T^{1/2})$. Finally

$$\delta G'_{\bullet,1,p}(I - \langle \tilde{z}_{t,\bullet}, \tilde{z}_{t,\bullet}\rangle \tilde{\Gamma}^\dagger_{\bullet,p}\bar{G}'_{\bullet,2,p}) = \delta G'_{\bullet,1,p}(I - \langle \tilde{z}_{t,\bullet}, \tilde{z}_{t,\bullet}\rangle \tilde{\Gamma}^\dagger_{\bullet,p}\tilde{\Gamma}'_{\bullet,p}) + O((\log T)^a/T) = O((\log T)^a/T)$$

as in the proof of Lemma A.7 of BRRR. Using Lemma A.5 (III) of BRRR with $\hat{\Xi}_f = \langle \tilde{y}_{t,\bullet}, \tilde{y}_{t,\bullet} \rangle^{-1/2}$ it follows that $\bar{\Gamma}'_{\bullet,p} - \Gamma'_{\bullet,p} = O((\log T)^a T^{-1/2})$. Since $\bar{G}_{\bullet,2,p} - \bar{\Gamma}_{\bullet,p} = o((\log T)^a / T^{1/2})$ the same rate of convergence holds for $\bar{G}'_{\bullet,2,p} - \Gamma'_{\bullet,p} = O((\log T)^a / T^{1/2})$. It follows that $\bar{\Gamma}'_p - [\Gamma']_{1:p} = [O((\log T)^a/T), O((\log T)^a/T^{1/2})]$.

(III) From above we have

$$T(\bar{\Gamma}'_p - [\Gamma']_{1:p})\begin{pmatrix} I_c \\ 0 \end{pmatrix} = \begin{bmatrix} -(\sqrt{T}\delta G'_{\bullet,1,p} \langle \tilde{z}_{t,\bullet}, \tilde{z}_{t,\bullet} \rangle \bar{\Gamma}^\dagger_{\bullet,p} \sqrt{T}\delta G'_{1,2,p}) \\ \sqrt{T}\delta G'_{1,2,p} \end{bmatrix} + o_P(1). \quad (A3)$$

Now from the proof of Lemma A.7 of BRRR we obtain

$$[\sqrt{T}\delta G_{\bullet,1,p}]' = \Xi O_\bullet (I - \Theta_\bullet^2)^{-1} \Gamma'_{\bullet,p} + o_P(1).$$

Furthermore using the expression given in Lemma A.7 of BRRR:

$$\begin{aligned}
\sqrt{T}\delta G_{1,2,p} &= \sqrt{T}Z_{11}^{-1}[\delta_{zz}^{1,\bullet}\Gamma_{\bullet,p}\Theta_\bullet^2 - J_{1,\bullet}\Gamma_{\bullet,p}](I - \Theta_\bullet^2)^{-1} + o_P(1) \\
&= \sqrt{T}Z_{11}^{-1}[\delta_{zz}^{1,\bullet}\Gamma_{\bullet,p}(\Theta_\bullet^2 - I) + [\delta_{zz}^{1,\bullet} - J_{1,\bullet}]\Gamma_{\bullet,p}](I - \Theta_\bullet^2)^{-1} + o_P(1) \\
&= -Z_{11}^{-1}\langle \tilde{z}_{t,1}, x_{t,\bullet}\rangle - Z_{11}^{-1}\sqrt{T}[J_{1,\bullet} - \delta_{zz}^{1,\bullet}]\Gamma_{\bullet,p}(I - \Theta_\bullet^2)^{-1} + o_P(1) \\
&= -Z_{11}^{-1}\langle \tilde{z}_{t,1}, x_{t,\bullet}\rangle - Z_{11}^{-1}\mathbb{E}\tilde{\varepsilon}_{t,1}\tilde{\varepsilon}'_{t,\bullet}(\mathbb{E}\tilde{y}_{t,\bullet}(\tilde{y}_{t,\bullet})')^{-1}\mathbb{E}\tilde{y}_{t,\bullet}x'_{t,\bullet}(I - \Theta_\bullet^2)^{-1} + o_P(1).
\end{aligned}$$

This shows the result. □

The transformations in the representation lead to an estimator $\tilde{G}$ taking the place of $\hat{\mathcal{K}}_p$. Using $\tilde{\mathcal{S}}_T$ as defined in Lemma A5 the corresponding estimator $\tilde{\Gamma}'_p = \tilde{\mathcal{S}}_T \tilde{D}_x^{-1} \tilde{G}'_p \tilde{D}_z$ fulfills $\tilde{\Gamma}'_p - \Gamma'_p = [O((\log T)^a/T), O((\log T)^a/\sqrt{T})]$.

Based on this result let $(\mathcal{A}, \mathcal{C}, \mathcal{K})$ denote the realization of the true transfer function in the state basis corresponding to Γ'_p where $\Gamma'_p S_p = I_n$ and let $(\tilde{\mathcal{A}}, \tilde{\mathcal{C}}, \tilde{\mathcal{K}})$ denote the (unfeasible) CVA estimates using $\tilde{x}_t := \tilde{\Gamma}'_p \tilde{z}_{t,p}$. The next lemma then provides the main ingredients for the rest of the proofs:

Lemma A6. *Let the assumptions of Theorem 1 hold and define $D_x := \text{diag}(I_c T^{-1}, I_{n-c} T^{-1/2})$. Then there exists an integer $a < \infty$ such that*

$$(\tilde{\mathcal{A}} - \mathcal{A})D_x^{-1} = O((\log T)^a), \quad (\tilde{\mathcal{C}} - \mathcal{C})D_x^{-1} = O((\log T)^a), \quad (\tilde{\mathcal{K}} - \mathcal{K}) = O((\log T)^a/T^{1/2}).$$

Proof. First note that the regression of $Y^+_{t,f}$ onto $Y^-_{t,p}$ includes time points $t = p+1, ..., T-f+1$ whereas for estimating the system matrices we can use $\hat{x}_t, t = p+1, ..., T+1$ and $y_t, t = p+1, ..., T$. Thus in this proof we use $\langle a_t, b_t \rangle_{p+1}^T := T^{-1}\sum_{t=p+1}^T a_t b'_t$ instead of $\langle a_t, b_t \rangle = T^{-1}\sum_{t=p+1}^{T-f+1} a_t b'_t$.

The following orders of convergence are straightforward to derive using the results of Lemma A1, $\tilde{\mathcal{A}}^p = o(T^{-1})$, $(\tilde{\Gamma}'_p - [\Gamma']_{1:p})D_z^{-1} = O((\log T)^a)$ and $\tilde{x}_t - x_t = (\tilde{\Gamma}'_p - [\Gamma']_{1:p})\tilde{z}_{t,p} - \tilde{\mathcal{A}}^p x_{t-p}, t > p$ according to Lemma A5 and Lemma A1 for the range of p given in Theorem 1:

$$\langle \varepsilon_t, \tilde{x}_t - x_t \rangle_{p+1}^T = O(p(\log T)^a/T) \quad , \quad \tilde{D}_z \langle \tilde{z}_{t,p}, \tilde{x}_t - x_t \rangle_{p+1}^T = O(p(\log T)^a/T^{1/2})$$

$$\tilde{D}_z \langle \tilde{z}_{t+1,p}, \tilde{x}_t - x_t \rangle_{p+1}^T = O(p(\log T)^a/T^{1/2}) \quad , \quad \tilde{D}_x \langle x_t, \tilde{x}_t - x_t \rangle_{p+1}^T = O(p(\log T)^a/T^{1/2})$$

$$\langle \tilde{x}_t - x_t, \tilde{x}_t - x_t \rangle_{p+1}^T = O(p^2(\log T)^a/T) \quad .$$

Using these orders of convergence we obtain

$$\tilde{D}_x \langle \tilde{x}_t, \tilde{x}_t \rangle_{p+1}^T \tilde{D}_x = \tilde{D}_x \langle x_t, x_t \rangle_{p+1}^T \tilde{D}_x + O(p^2(\log T)^a/T^{1/2}) > 0 \quad \text{a.s.}$$

From Lemma A1 also $(\tilde{D}_x \langle \tilde{x}_t, \tilde{x}_t \rangle_{p+1}^T \tilde{D}_x)^{-1} = (\tilde{D}_x \langle x_t, x_t \rangle_{p+1}^T \tilde{D}_x)^{-1}(1 + o(1)) = O((\log T)^a)$.
Therefore

$$
\begin{aligned}
(\check{\mathcal{C}} - \mathcal{C})D_x^{-1} &= \sqrt{T}\Big(\langle \varepsilon_t, \tilde{x}_t\rangle_{p+1}^T - \mathcal{C}\langle \tilde{x}_t - x_t, \tilde{x}_t\rangle_{p+1}^T\Big)\tilde{D}_x(\tilde{D}_x\langle \tilde{x}_t, \tilde{x}_t\rangle_{p+1}^T \tilde{D}_x)^{-1} \\
&= \sqrt{T}\langle \varepsilon_t, x_t\rangle_{p+1}^T \tilde{D}_x(\tilde{D}_x\langle x_t, x_t\rangle_{p+1}^T \tilde{D}_x + o(1))^{-1} \\
&\quad - \sqrt{T}\mathcal{C}\langle \tilde{x}_t - x_t, x_t\rangle_{p+1}^T \tilde{D}_x(\tilde{D}_x\langle x_t, x_t\rangle_{p+1}^T \tilde{D}_x)^{-1} + o(1) = O(p(\log T)^a).
\end{aligned}
\quad (A4)
$$

This in particular establishes consistency for the estimate. Next analogously (using the notation $\delta x_t = \tilde{x}_t - x_t$) we obtain $(\check{\mathcal{A}} - \mathcal{A})D_x^{-1} =$

$$
\begin{aligned}
&\sqrt{T}\langle \tilde{x}_{t+1} - \mathcal{A}\tilde{x}_t, \tilde{x}_t\rangle_{p+1}^T \tilde{D}_x(\tilde{D}_x\langle \tilde{x}_t, \tilde{x}_t\rangle_{p+1}^T \tilde{D}_x)^{-1} \\
&= \sqrt{T}\Big(\langle(\tilde{x}_{t+1} - x_{t+1}) + (x_{t+1} - \mathcal{A}x_t) + \mathcal{A}(x_t - \tilde{x}_t), \tilde{x}_t\rangle_{p+1}^T \tilde{D}_x\Big)(\tilde{D}_x\langle x_t, x_t\rangle_{p+1}^T \tilde{D}_x + o(1))^{-1} \\
&= \sqrt{T}\Big(\langle \delta x_{t+1}, x_t\rangle_{p+1}^T - \mathcal{A}\langle \delta x_t, x_t\rangle_{p+1}^T + \langle \mathcal{K}\varepsilon_t, x_t\rangle_{p+1}^T\Big)\tilde{D}_x(\tilde{D}_x\langle x_t, x_t\rangle_{p+1}^T \tilde{D}_x)^{-1} + o(1) \\
&= O(p(\log T)^a)
\end{aligned}
\quad (A5)
$$

and therefore consistency for $\check{\mathcal{A}}$ is established. Finally note that for

$$\hat{\varepsilon}_t = y_t - \check{\mathcal{C}}\tilde{x}_t = \varepsilon_t + \mathcal{C}(x_t - \tilde{x}_t) + (\mathcal{C} - \check{\mathcal{C}})\tilde{x}_t$$

it follows that $\langle \hat{\varepsilon}_t, \hat{\varepsilon}_t\rangle_{p+1}^T = \Omega + O(p^2(\log T)^a/T^{1/2})$. Furthermore since $\hat{\varepsilon}_t$ denotes the residuals of the regression of y_t onto $\tilde{x}_t$ it follows that $\langle \hat{\varepsilon}_t, \tilde{x}_t\rangle_{p+1}^T = 0$. From this we obtain

$$
\begin{aligned}
\sqrt{T}(\check{\mathcal{K}} - \mathcal{K}) &= \sqrt{T}(\langle \tilde{x}_{t+1} - \mathcal{K}\hat{\varepsilon}_t, \hat{\varepsilon}_t\rangle_{p+1}^T (\langle \hat{\varepsilon}_t, \hat{\varepsilon}_t\rangle_{p+1}^T)^{-1} \\
&= \sqrt{T}\Big(\langle(\tilde{x}_{t+1} - x_{t+1}) - \mathcal{A}\delta x_t + \mathcal{K}(\varepsilon_t - \hat{\varepsilon}_t), \hat{\varepsilon}_t\rangle_{p+1}^T\Big)(\langle \hat{\varepsilon}_t, \hat{\varepsilon}_t\rangle_{p+1}^T)^{-1} \\
&= \sqrt{T}\Big(\langle \delta x_{t+1} - \mathcal{A}\delta x_t + \mathcal{K}(\varepsilon_t - \hat{\varepsilon}_t), \hat{\varepsilon}_t\rangle_{p+1}^T\Big)\Omega^{-1}(1 + o(1)) \\
&= \sqrt{T}\Big(\langle \delta x_{t+1} - \mathcal{A}\delta x_t + \mathcal{K}(\varepsilon_t - \hat{\varepsilon}_t), \varepsilon_t\rangle_{p+1}^T\Big)\Omega^{-1}(1 + o(1)) + o(1) \\
&= \Big(\sqrt{T}\langle \delta x_{t+1}, \varepsilon_t\rangle_{p+1}^T\Big)\Omega^{-1} + o(1) = \Big(\sqrt{T}\langle(\Gamma'_p - \Gamma'_p)\bar{z}_{t+1,p}, \varepsilon_t\rangle_{p+1}^T\Big)\Omega^{-1} + o(1) \\
&= O(p(\log T)^a).
\end{aligned}
\quad (A6)
$$

□

These expressions do not only show consistency of a specific order, but also give the relevant highest order terms for the asymptotic distribution, which are used below. As $\hat{\mathcal{C}}\hat{\mathcal{A}}^j\hat{\mathcal{K}} = \check{\mathcal{C}}\check{\mathcal{A}}^j\check{\mathcal{K}} \to \mathcal{C}\mathcal{A}^j\mathcal{K} = \mathcal{C}_\circ\mathcal{A}_\circ^j\mathcal{K}_\circ$, Lemma A6 establishes consistency for the impulse response sequence $\hat{\mathcal{C}}\hat{\mathcal{A}}^j\hat{\mathcal{K}}$ (thus proofs Theorem 1 (I)) as well as, jointly with $p = O((\log T)^a)$, the rate of convergence $O((\log T)^a/T^{1/2})$ for the not realizable choice of the basis and the impulse response sequence $\mathcal{C}\mathcal{A}^j\mathcal{K}$.

Appendix C.1.2. Proof of Theorem 1 (II)

In order to arrive at the canonical representation $(\check{\mathcal{A}}, \check{\mathcal{C}}, \check{\mathcal{K}})$ two steps are performed: first the reordered Jordan normal form is calculated, afterwards the matrices $\tilde{C}_{j,\mathbb{C}}$ are transformed such that $E'_j\check{C}_{j,\mathbb{C}} = I_{c_j}$ holds. We will follow these steps below.

In the first step a transformation matrix $\hat{U}$ needs to be found such that $\check{\mathcal{A}} = \hat{U}\tilde{\mathcal{A}}\hat{U}^{-1}$ is in reordered Jordan normal form. In this respect $\tilde{\mathcal{A}}$ and $\mathcal{A}$ are used in Lemma A2. Accordingly $\hat{U}_t = [\hat{U}_{t,1}, ..., \hat{U}_{t,S/2}, \hat{U}_{t,\bullet}]$ can be defined such that $V'_j\hat{U}_{t,j} = I_{c_j}$ where $U \in \mathbb{R}^{n \times n}$ corresponds to the transformation from $\mathcal{A}$ to $\mathcal{A}_\circ$ as given in the theorem. An appropriate choice of $\tilde{z}_{t,1}$ leads to $U = I_n$. Furthermore the basis in the space spanned by the columns of $\hat{U}_{t,\bullet}$ where $\hat{U}'_{t,j}\hat{U}_{t,\bullet} = 0$ can be chosen such that $[0, I]\hat{U}_{t,\bullet} = I$ for large enough T.

A first order approximation according to Lemma A2 then leads to

$$\hat{U}_{t,j} = U_j + Z_1 + O(\|\hat{\mathcal{A}} - \mathcal{A}\|^2) = U_j - \Sigma(\hat{\mathcal{A}} - \mathcal{A})U_j + O(\|\hat{\mathcal{A}} - \mathcal{A}\|^2)$$

for $j = 0, ..., S/2$. Consequently $\|\hat{U}_{t,j} - U_j\| = O((\log T)^a T^{-1})$ and thus also $\hat{U}_t - U = O((\log T)^a T^{-1})$. Consequently the order of convergence for the transformed system $(\hat{A}, \hat{C}, \hat{K})$ is unchanged. In a second step an upper triangular transformation matrix $\tilde{U}$ can be found transforming $(\hat{A}, \hat{C}, \hat{K})$ such that $\tilde{A}$ corresponds to the reordered Jordan normal form. Due to the upper block triangularity of this transform we can apply Lemma A3 to show that the order of convergence remains identical.

For the second step note that Lemma A3 provides the required terms: An application to the block diagonal transformation $\mathcal{S}_T = \text{diag}(E_1' \tilde{C}_{1,C}, ..., E_{S/2}' \tilde{C}_{S/2,C}, \mathcal{S}_{T,\bullet})$, where $\mathcal{S}_{T,\bullet}$ transforms the stationary subsystem to echelon form, concludes the proof.

Appendix C.1.3. Proof of Theorem 1 (III)

The only argument that uses the iid assumption is the almost sure convergence of $(\tilde{D}_x \langle x_t, x_t \rangle \tilde{D}_x)^{-1}$. Weakening the assumptions on the noise implies that this order of convergence still holds in probability while the almost sure version cannot be shown with the tools of this paper. This concludes the proof of Theorem 1.

Appendix C.2. Proof of Theorem 2

Using the notation introduced in (A1),

$$\hat{X} = \tilde{D}_z \langle \tilde{y}_t, \tilde{z}_{t,p} \rangle \tilde{D}_z (\tilde{D}_z \langle \tilde{z}_{t,p}, \tilde{z}_{t,p} \rangle \tilde{D}_z)^{-1} \tilde{D}_z \langle \tilde{z}_{t,p}, \tilde{y}_t \rangle \tilde{D}_z (\tilde{D}_z \langle \tilde{y}_t, \tilde{y}_t \rangle \tilde{D}_z)^{-1} \to X_\circ = \begin{bmatrix} I_c & 0 \\ 0 & X_{\circ,\bullet} \end{bmatrix}$$

for a suitable matrix $X_{\circ,\bullet}$. The eigenvalues of $\hat{X}$ are the squares of the singular values of the RRR problem in the first step of CVA. Therefore

$$T \sum_{i=1}^c (1 - \hat{\sigma}_i^2) = -T \text{tr}\left[U_c'(\hat{X} - X_\circ)[U_c - (X_\circ - I)^\dagger (\hat{X} - X_\circ) U_c] \right] + o_P(1)$$

$$= -T \text{tr}\left[\Delta X_{11} - \Delta X_{12} (X_{\circ,\bullet} - I)^\dagger \Delta X_{21} \right] + o_P(1)$$

according to a second order approximation in the Rayleigh-Schrödinger expansions (Lemma A2). Now, in the current situation we obtain $(I - \hat{X}) \begin{bmatrix} I \\ 0 \end{bmatrix} =$

$$= \left(\tilde{D}_y \langle \tilde{y}_t, \tilde{y}_t \rangle \tilde{D}_y - \tilde{D}_y \langle \tilde{y}_t, \tilde{z}_{t,p} \rangle \tilde{D}_z (\tilde{D}_z \langle \tilde{z}_{t,p}, \tilde{z}_{t,p} \rangle \tilde{D}_z)^{-1} \tilde{D}_z \langle \tilde{z}_{t,p}, \tilde{y}_t \rangle \tilde{D}_y \right) (\tilde{D}_y \langle \tilde{y}_t, \tilde{y}_t \rangle \tilde{D}_y)^{-1} \begin{bmatrix} I \\ 0 \end{bmatrix}$$

$$= \left(\tilde{D}_y \langle \tilde{\varepsilon}_t, \tilde{\varepsilon}_t \rangle \tilde{D}_y - \tilde{D}_y \langle \tilde{\varepsilon}_t, \tilde{z}_{t,p} \rangle \tilde{D}_z (\tilde{D}_z \langle \tilde{z}_{t,p}, \tilde{z}_{t,p} \rangle \tilde{D}_z)^{-1} \tilde{D}_z \langle \tilde{z}_{t,p}, \tilde{\varepsilon}_t \rangle \tilde{D}_y \right) (\tilde{D}_y \langle \tilde{y}_t, \tilde{y}_t \rangle \tilde{D}_y)^{-1} \begin{bmatrix} I \\ 0 \end{bmatrix}.$$

Furthermore $\langle \tilde{\varepsilon}_t, \tilde{z}_{t,p} \rangle \tilde{D}_z (\tilde{D}_z \langle \tilde{z}_{t,p}, \tilde{z}_{t,p} \rangle \tilde{D}_z)^{-1} \tilde{D}_z \langle \tilde{z}_{t,p}, \tilde{\varepsilon}_t \rangle = O_P(T^{-1})$ and

$$(\tilde{D}_y \langle \tilde{y}_t, \tilde{y}_t \rangle \tilde{D}_y)^{-1} \begin{bmatrix} I \\ 0 \end{bmatrix} = \begin{bmatrix} I \\ -\langle \tilde{y}_{t,\bullet}, \tilde{y}_{t,\bullet} \rangle^{-1} \langle \tilde{y}_{t,\bullet}, \tilde{y}_{t,1} \rangle / \sqrt{T} \end{bmatrix} (\langle \tilde{y}_{t,1}^\pi, \tilde{y}_{t,1}^\pi \rangle / T)^{-1}$$

where $\tilde{y}_{t,1}^\pi = \tilde{y}_{t,1} - \langle \tilde{y}_{t,1}, \tilde{y}_{t,\bullet} \rangle \langle \tilde{y}_{t,\bullet}, \tilde{y}_{t,\bullet} \rangle^{-1} \tilde{y}_{t,\bullet}$.

From this we get using $\mathbb{E} \tilde{\varepsilon}_{t,\bullet} \tilde{\varepsilon}_{t,\bullet}' = \mathbb{E} \tilde{y}_{t,\bullet} \tilde{y}_{t,\bullet}' - X_{\circ,\bullet} \mathbb{E} \tilde{y}_{t,\bullet} \tilde{y}_{t,\bullet}'$:

$$T(I_c - \hat{X}_{1,1}) = \left(\langle \tilde{\varepsilon}_{t,1}, \tilde{\varepsilon}_{t,1} \rangle - \langle \tilde{\varepsilon}_{t,1}, \tilde{\varepsilon}_{t,\bullet} \rangle \langle \tilde{y}_{t,\bullet}, \tilde{y}_{t,\bullet} \rangle^{-1} \langle \tilde{y}_{t,\bullet}, \tilde{y}_{t,1} \rangle \right) (\langle \tilde{y}_{t,1}^\pi, \tilde{y}_{t,1}^\pi \rangle / T)^{-1} + o_P(1),$$

$$\sqrt{T} \Delta X_{2,1} = (-\langle \tilde{\varepsilon}_{t,\bullet}, \tilde{\varepsilon}_{t,1} \rangle + (I - X_{\circ,\bullet}) \langle \tilde{y}_{t,\bullet}, \tilde{y}_{t,1} \rangle)(\langle \tilde{y}_{t,1}^\pi, \tilde{y}_{t,1}^\pi \rangle / T)^{-1} + o_P(1)$$

$$\sqrt{T} \Delta X_{1,2} = -\mathbb{E} \tilde{\varepsilon}_{t,1} \tilde{\varepsilon}_{t,\bullet}' (\mathbb{E} \tilde{y}_{t,\bullet} \tilde{y}_{t,\bullet}')^{-1} + o_P(1).$$

Thus $T\sum_{i=1}^c(1-\hat{\sigma}_i^2) =$

$$= \text{tr}\left[\left(\langle\tilde{\varepsilon}_{t,1},\tilde{\varepsilon}_{t,1}\rangle - \mathbb{E}\tilde{\varepsilon}_{t,1}\tilde{\varepsilon}'_{t,\bullet}(\mathbb{E}\tilde{y}_{t,\bullet}\tilde{y}'_{t,\bullet})^{-1}(I-X_{0,\bullet})^{-1}\mathbb{E}\tilde{\varepsilon}_{t,\bullet}\tilde{\varepsilon}'_{t,1}\right)(\langle\tilde{y}_{t,1},\tilde{y}_{t,1}\rangle/T)^{-1}\right] + o_P(1)$$

$$= \text{tr}\left[\left(\langle\tilde{\varepsilon}_{t,1},\tilde{\varepsilon}_{t,1}\rangle - \mathbb{E}\tilde{\varepsilon}_{t,1}\tilde{\varepsilon}'_{t,\bullet}(\mathbb{E}\tilde{\varepsilon}_{t,\bullet}\tilde{\varepsilon}'_{t,\bullet})^{-1}\mathbb{E}\tilde{\varepsilon}_{t,\bullet}\tilde{\varepsilon}'_{t,1}\right)(\langle\tilde{y}_{t,1},\tilde{y}_{t,1}\rangle/T)^{-1}\right] + o_P(1) \xrightarrow{d} Z.$$

Appendix C.3. Proof of Theorem 3

The proof of Theorem 3 follows the same path as the proof of Theorem 1. In (A5) it was shown that the asymptotic distribution of $T(\tilde{\mathcal{A}}_{11} - \mathcal{A}_{\circ,11})$ depends on

$$\langle\tilde{x}_{t+1,j} - x_{t+1,j}, x_{t,k}\rangle, \langle\tilde{x}_{t,j} - x_{t,j}, x_{t,k}\rangle \quad, \quad \langle\varepsilon_t, x_{t,j}\rangle, \langle x_{t,k}, x_{t,j}\rangle/T$$

for $j,k = 0,...,S/2$. Note that

$$\delta x_{t+i} = \tilde{x}_{t+i} - x_{t+i} = (\tilde{\Gamma}'_p - [\Gamma']_{1:p})\tilde{z}_{t+i,p} + o_P(T^{-1})$$

for $i = 0,1$. Then the results of Lemma A5 show that the first c columns of $(\tilde{\Gamma}'_p - [\Gamma']_{1:p})$ converge to a random variable (below denoted as Z_Γ) when multiplied with T while the remaining columns converge in distribution when multiplied with $\sqrt{T}$. Therefore

$$\langle\delta x_{t+i}, x_{t,k}\rangle = T(\tilde{\Gamma}'_p - [\Gamma']_{1:p})\frac{\langle\tilde{z}_{t+i,p}, x_{t,k}\rangle}{T} + o_P(1) = T(\tilde{\Gamma}'_p - [\Gamma']_{1:p})\begin{bmatrix}I_c\\0\end{bmatrix}\frac{\langle\tilde{z}_{t+i,1}, x_{t,k}\rangle}{T} + o_P(1).$$

Due to the definition (A1), $\tilde{z}_{t,1} = [x_{t,j}]_{j=0,...,S/2} + o(T^{-1})$ and hence (using $\mathcal{A}_\circ = \text{diag}(\mathcal{A}_{\circ,u}, \mathcal{A}_{\circ,\bullet})$)

$$\langle\tilde{z}_{t+1,1}, x_{t,k}\rangle/T = \mathcal{A}_{\circ,u}\langle\tilde{z}_{t,1}, x_{t,k}\rangle/T + o(1).$$

Considering now the complex-valued representation and using the notation

$$\Delta\Gamma_1 := T(\tilde{\Gamma}'_p - [\Gamma']_{1:p})\begin{bmatrix}I_c\\0\end{bmatrix}, \quad S_j = [0_{c_j,\sum_{i<j}c_i}, I_{c_j}, 0_{c_j,\sum_{i>j}c_j}]$$

where $S_j\tilde{z}_{t,1} = x_{t,j,\mathbb{C}}$, it follows that the contribution of these two terms to the limiting distribution of the diagonal block corresponding to the unit root z_j amounts to (using $\langle x_{t,j,\mathbb{C}}, x_{t,k,\mathbb{C}}\rangle/T \to 0$ for $k \neq j$ and $\delta x_{t,j,\mathbb{C}} = \tilde{x}_{t,j,\mathbb{C}} - x_{t,j,\mathbb{C}}$)

$$\langle\delta x_{t+1,j,\mathbb{C}}, x_{t,j,\mathbb{C}}\rangle - \mathcal{A}_{\circ,jj}\langle\delta x_{t,j,\mathbb{C}}, x_{t,j,\mathbb{C}}\rangle =$$

$$= S_j\Delta\Gamma_1\mathcal{A}_{\circ,u}\frac{\langle\tilde{z}_{t,1}, x_{t,j,\mathbb{C}}\rangle}{T} - \mathcal{A}_{\circ,jj}S_j\Delta\Gamma_1\frac{\langle\tilde{z}_{t,1}, x_{t,j,\mathbb{C}}\rangle}{T} + o_P(1)$$

$$= S_j\Delta\Gamma_1 S'_j\mathcal{A}_{\circ,jj}\frac{\langle x_{t,j,\mathbb{C}}, x_{t,j,\mathbb{C}}\rangle}{T} - \mathcal{A}_{\circ,jj}S_j\Delta\Gamma_1 S'_j\frac{\langle x_{t,j,\mathbb{C}}, x_{t,j,\mathbb{C}}\rangle}{T} + o_P(1)$$

$$= S_j\Delta\Gamma_1 S'_j\bar{z}_j\frac{\langle x_{t,j,\mathbb{C}}, x_{t,j,\mathbb{C}}\rangle}{T} - \bar{z}_j S_j\Delta\Gamma_1 S'_j\frac{\langle x_{t,j,\mathbb{C}}, x_{t,j,\mathbb{C}}\rangle}{T} + o_P(1) = o_P(1).$$

Therefore, for the diagonal blocks in (A5) these two terms do not contribute and the asymptotic distribution is determined by

$$T\langle\mathcal{K}_{\circ,j}\varepsilon_t, x_{t,j}\rangle\langle x_{t,j}, x_{t,j}\rangle^{-1}$$

for which the asymptotic results are provided in Lemma A1. This also shows that estimating the state does not change the asymptotic distribution in the diagonal blocks as the impact of $\tilde{\Gamma}_p - \Gamma_p$ is of lower order.

In order to derive the distribution of the sum of the eigenvalues note that as in the proof of Theorem 2, according to Lemma A2 the sum of the eigenvalues of $\tilde{\mathcal{A}}$ converging to z_j obeys the following second order approximation:

$$T \sum_{i=1}^{c_j} (\hat{\lambda}_i - z_j) = T\text{tr}\left[U_j'(\hat{\mathcal{A}} - \mathcal{A}_\circ)[U_j - \mathcal{A}_\circ(z_j)^\dagger(\hat{\mathcal{A}} - \mathcal{A}_\circ)U_j]\right] + o_P(T^{-1})$$

$$= T\text{tr}\left[\hat{\mathcal{A}}_{\circ,jj} - z_j I_{c_j}\right] + o_P(1)$$

since $(\hat{\mathcal{A}} - \mathcal{A}_\circ)U_j = O((\log T)^a T^{-1})$ in this case implying that the second order terms vanish. Thus we obtain the asymptotic distribution under the null hypothesis as the limiting distribution of

$$T\text{tr}[\langle \mathcal{K}_{\circ,j,\mathbb{C}} \varepsilon_t, x_{t,j,\mathbb{C}} \rangle \langle x_{t,j,\mathbb{C}}, x_{t,j,\mathbb{C}} \rangle^{-1}].$$

It is easy to verify that this test statistic is pivotal for complex and real unit roots. This proves Theorem 3.

Appendix C.4. Proof of Theorem 4

The result for $\tilde{C}_m$ can be shown using the results of [4]. As the eigenvalues are insensitive to changes in the basis we can assume without restriction of generality that the only unit root components in $\mathcal{T} X_t^{(m)}$ are contained in the first c_m rows:

$$c_t^{(m)} := \mathcal{T} X_t^{(m)} = \begin{bmatrix} c_{t,u}^{(m)} \\ c_{t,\bullet}^{(m)} \end{bmatrix}, \quad \tilde{D}_c = \begin{bmatrix} T^{-1} I_{c_m} & 0 \\ 0 & I_{n-c_m} \end{bmatrix}.$$

Due to the filtering, $c_{t,\bullet}^{(m)}$ is stationary while $c_{t,u}^{(m)}$ contains the unit root z_m. Then the relevant matrix $\hat{X}_m$ can be written as

$$\hat{X}_m := \langle c_{t-1}^\pi, p_t^\pi \rangle \langle p_t^\pi, p_t^\pi \rangle^{-1} \langle p_t^\pi, c_{t-1}^\pi \rangle \langle c_{t-1}^\pi, c_{t-1}^\pi \rangle^{-1}.$$

Since $p_t = \mathcal{K}\varepsilon_{t-1} + \sum_{j=1,j\neq m}^{S} \alpha_j \beta_j' X_{t-1}^{(j)} + [0, \tilde{\alpha}_m] c_{t-1}^{(m)}$, we consequently have $p_t^\pi = \mathcal{K}\varepsilon_{t-1}^\pi + [0, \tilde{\alpha}_m] c_{t-1}^\pi$. Therefore, for the three components of $\hat{X}_m$ we obtain with appropriate definitions of the random variables S_m, T_m and using standard asymptotics

$$\langle p_t^\pi, p_t^\pi \rangle = \langle \mathcal{K}\varepsilon_{t-1}^\pi + \tilde{\alpha}_m c_{t-1,\bullet}^\pi, \mathcal{K}\varepsilon_{t-1}^\pi + \tilde{\alpha}_m c_{t-1,\bullet}^\pi \rangle \to \mathcal{K}(\mathbb{E}\varepsilon_{t-1}\varepsilon_{t-1}')\mathcal{K}' + \tilde{\alpha}_m \mathbb{E} c_{t-1,\bullet}^\Pi (c_{t-1,\bullet}^\Pi)' \tilde{\alpha}_m' > 0,$$

$$\langle p_t^\pi, c_{t-1}^\pi \rangle = \langle \mathcal{K}\varepsilon_{t-1}^\pi + \tilde{\alpha}_m c_{t-1,\bullet}^\pi, c_{t-1}^\pi \rangle \xrightarrow{d} [S_m, \tilde{\alpha}_m \mathbb{E} c_{t-1,\bullet}^\Pi (c_{t-1,\bullet}^\Pi)'],$$

$$\langle p_t^\pi, c_{t-1}^\pi \rangle \langle c_{t-1}^\pi, c_{t-1}^\pi \rangle^{-1} \tilde{D}_c^{-1} = [0, \tilde{\alpha}_m] + \langle \mathcal{K}\varepsilon_{t-1}^\pi, c_{t-1}^\pi \rangle \langle c_{t-1}^\pi, c_{t-1}^\pi \rangle^{-1} \tilde{D}_c^{-1}$$

$$\langle \mathcal{K}\varepsilon_{t-1}^\pi, c_{t-1}^\pi \rangle \langle c_{t-1}^\pi, c_{t-1}^\pi \rangle^{-1} \tilde{D}_c^{-1} \xrightarrow{d} [T_m, 0].$$

Correspondingly the first block column $\hat{X}_{m,u}$ of $\hat{X}_m$ converges to zero such that $T\hat{X}_{m,u}$ converges in distribution while the second block column converges in probability without normalization. This shows that

$$T \sum_{i=1}^{c_m} \hat{\lambda}_i = T\text{tr}\left[U_m'(\hat{X}_m - X_m)[U_m - X_m^\dagger(\hat{X}_m - X_m)U_m]\right] + o_P(1) = \text{tr}[T\hat{X}_{m,uu}] + o_P(1)$$

converges in distribution. The limit is given in [4].

For the case of the estimated state note that the difference between the estimated and the true state is given as

$$\tilde{x}_t - x_t = \tilde{\Gamma}_p' \tilde{z}_{t,p} - \Gamma_p' \tilde{z}_{t,p} - \mathcal{A}^p x_{t-p} = (\tilde{\Gamma}_p - \Gamma_p)' \tilde{z}_{t,p} - \mathcal{A}^p x_{t-p}.$$

The strict minimum-phase assumption and the assumption on the increase of $p = p(T)$ implies that the second term can be neglected being $o_P(T^{-1})$. Furthermore

$$(\tilde{\Gamma}_p - \Gamma_p)'\tilde{z}_{t,p} = (\tilde{\Gamma}_p - \Gamma_p)'\tilde{D}_z^{-1}\tilde{D}_z\tilde{z}_{t,p}, \quad (\tilde{\Gamma}_p - \Gamma_p)'\tilde{D}_z^{-1} = O_P(T^{-1/2}).$$

Using this it can be concluded that

$$\langle \hat{p}_t, \hat{p}_t \rangle = \langle p_t, p_t \rangle + O_P(T^{-1/2}) \quad , \quad \langle \hat{p}_t, \hat{c}_{t-1,\bullet}^{(m)} \rangle = \langle p_t, c_{t-1,\bullet}^{(m)} \rangle + O_P(T^{-1/2}),$$
$$\langle \hat{p}_t, \hat{c}_{t-1,u}^{(m)} \rangle = \langle p_t, c_{t-1,u}^{(m)} \rangle + O_P(T^{-1/2}) \quad , \quad \langle \hat{c}_{t,u}^{(m)}, \hat{c}_{t,u}^{(k)} \rangle = \langle c_{t,u}^{(m)}, c_{t,u}^{(k)} \rangle + O_P(1).$$

These equations imply that the difference between the expression using the true state and the one using the estimated state converges to zero, implying that the two tests accept and reject jointly asymptotically under the null hypothesis.

References

1. Rodrigues, P.M.; Taylor, A. Alternative estimators and unit root tests for seasonal autoregressive processes. *J. Econom.* **2004**, *120*, 35–73. [CrossRef]
2. Johansen, S.; Schaumburg, E. Likelihood Analysis of Seasonal Cointegration. *J. Econom.* **1999**, *88*, 301–339. [CrossRef]
3. Hylleberg, S.; Engle, R.; Granger, C.; Yoo, B. Seasonal Integration and Cointegration. *J. Econom.* **1990**, *44*, 215–238. [CrossRef]
4. Cubadda, G. Complex Reduced Rank Models For Seasonally Cointegrated Time Series. *Oxf. Bull. Econ. Stat.* **2001**, *63*, 497–511. [CrossRef]
5. Cubadda, G.; Omtzigt, P. Small-sample improvements in the statistical analysis of seasonally cointegrated systems. *Comput. Stat. Data Anal.* **2005**, *49*, 333–348. [CrossRef]
6. Ahn, S.K.; Cho, S.; Seong, B. Inference of Seasonal Cointegration: Gaussian Reduced Rank Estimation and Tests for Various Types of Cointegration. *Oxford Bull. Econ. Stat.* **2004**, *66*, 261–284. [CrossRef]
7. Vivas, E.; Allende-Cid, H.; Salas, R. A Systematic Review of Statistical and Machine Learning Methods for Electrical Power Forecasting with Reported MAPE Score. *Entropy* **2020**, *22*, 1412. [CrossRef]
8. García-Martos, C.; Rodríguez, J.; Sánchez, M.J. Forecasting electricity prices and their volatilities using Unobserved Components. *Energy Econ.* **2011**, *33*, 1227–1239. [CrossRef]
9. Dufour, J.M.; Stevanović, D. Factor-augmented VARMA models with macroeconomic applications. *J. Bus. Econ. Stat.* **2013**, *31*, 491–506. [CrossRef]
10. Dias, G.; Kapetanios, G. Estimation and forecasting in vector autoregressive moving average models for rich datasets. *J. Econom.* **2018**, *202*, 75–91. [CrossRef]
11. Foroni, C.; Marcellino, M.; Stevanović, D. Mixed-frequency models with moving-average components. *J. Appl. Econom.* **2019**, *34*, 688–706. [CrossRef]
12. Kascha, C.; Trenkler, C. Simple Identification and specification of cointegrated VARMA models. *J. Appl. Econom.* **2015**, *30*, 675–702. [CrossRef]
13. Ravenna, F. Vector autoregressions and reduced form representations of DSGE models. *J. Monet. Econ.* **2007**, *54*, 2048–2064. [CrossRef]
14. Komunjer, I.; Zhu, Y. Likelihood ratio testing in linear state space models: An application to dynamic stochastic general equilibrium models. *J. Econom.* **2020**, *218*, 561–586. [CrossRef]
15. Bauer, D.; Wagner, M. A State Space Canonical Form for Unit Root Processes. *Econom. Theory* **2012**, *28*, 1313–1349. [CrossRef]
16. Larimore, W.E. System Identification, reduced order filters and modeling via canonical variate analysis. In Proceedings of the 1983 American Control Conference, San Francisco, CA, USA, 22–24 June 1983; pp. 445–451.
17. Bauer, D. Comparing the CCA subspace method to quasi maximum likelihood methods in the case of no exogenous inputs. *J. Time Ser. Anal.* **2006**, *26*, 631–668. [CrossRef]
18. Bauer, D. Using Subspace Methodes for Estimating ARMA models for multivariate time series with conditionally heteroskedastic innovations. *Econom. Theory* **2008**, *24*, 1063–1092.
19. Bauer, D. Using Subspace Methods to Model Long-Memory Processes. In *Theory and Applications of Time Series Analysis. ITISE 2018. Contributions to Statistics*; Valenzuela, O., Rojas, F., Pomares, H., Rojas, I., Eds.; Springer: Berlin/Heidelberg, Germany, 2019.
20. Bauer, D.; Wagner, M. Estimating Cointegrated Systems Using Subspace Algorithms. *J. Econom.* **2002**, *111*, 47–84. [CrossRef]
21. Bauer, D. Estimating linear dynamical systems using subspace methods. *Econom. Theory* **2005**, *21*, 181–211. [CrossRef]
22. Hannan, E.J.; Deistler, M. *The Statistical Theory of Linear Systems*; John Wiley: New York, NY, USA, 1998.
23. Chatelin, F. *Eigenvalues of Matrices*; John Wiley & Sons: Hoboken, NJ, USA, 1993.
24. Bauer, D. Asymptotic Distribution of Estimators in Reduced Rank Regression Settings When the Regressors Are Integrated. Technical Report. 2012. Available online: http://arxiv.org/abs/1211.1439 (accessed on 26 March 2021).
25. Phillips, P.C.B.; Durlauf, S.N. Multiple Time Series Regression with Integrated Processes. *Rev. Econ. Stud.* **1986**, *LIII*, 473–495. [CrossRef]

26. Carrasco, M.; Chen, X. Mixing and Moment Properties of Various GARCH and Stochastic Volatility Models. *Econom. Theory* **2002**, *18*, 17–39. [CrossRef]
27. Bauer, D.; Wagner, M. *Autoregressive Approximations to MFI(1) Processes*; Technical Report; Department for Mathematical Methods in Economics: TU Wien, Austria, 2004.
28. Bierens, H. Nonparametric cointegration analysis. *J. Econom.* **1997**, *77*, 379–404. [CrossRef]
29. Wagner, M. A Comparison of Johansen's, Bierens' and the Subspace Algorithm Method for Cointegration Analysis. *Oxf. Bull. Econ. Stat.* **2004**, *66*, 399–424. [CrossRef]
30. Johansen, S.; Nielsen, M. The cointegrated vector autoregressive model with general deterministic terms. *J. Econom.* **2018**, *202*, 214–229. [CrossRef]
31. Lee, H. Maximum Likelihood Inference on Cointegration and Seasonal Cointegration. *J. Econom.* **1992**, *54*, 1–47. [CrossRef]
32. Bauer, D.; Wagner, M. Using subspace algorithm cointegration analysis: Simulation performance and application to the term structure. *Comput. Stat. Data Anal.* **2009**, *53*, 1954–1973. [CrossRef]
33. Bauer, D. Order Estimation for Subspace Methods. *Automatica* **2001**, *37*, 1561–1573. [CrossRef]
34. Brockwell, P.J.; Davis, R.A. *Time Series: Theory and Methods*; Springer Series in Statistics, 2nd ed.; Springer: New York, NY, USA, 2006.
35. Qu, Z.; Perron, P. A Modified Information Criterion for Cointegration Tets Based on a VAR Approximation. *Econom. Theory* **2007**, *23*, 638–658. [CrossRef]
36. Mulla, R. Hourly Energy Consumption. Available online: www.kaggle.com/robikscube/hourly-energy-consumption/ (accessed on 22 January 2021).
37. del Barrio Castro, T.; Rodrigues, P.M.M.; Taylor, A.M.R. Temporal Aggregation of Seasonally Near-Integrated Processes. *J. Time Ser. Anal.* **2019**, *40*, 872–886. [CrossRef]
38. Bauer, D. Almost sure bounds on the estimation error for ols estimators when the regressors include certain MFI(1) processes. *Econom. Theory* **2009**, *25*, 571–582. [CrossRef]
39. Ahn, S.; Reinsel, G. Estimation of Partially Nonstationary Vector Autoregressive Models with Seasonal Behaviour. *J. Econom.* **1994**, *62*, 317–350. [CrossRef]
40. Bauer, D.; Deistler, M.; Scherrer, W. Consistency and Asymptotic Normality of some Subspace Algorithms for Systems Without Observed Inputs. *Automatica* **1999**, *35*, 1243–1254. [CrossRef]

Article

Ordinal Pattern Dependence in the Context of Long-Range Dependence

Ines Nüßgen * and Alexander Schnurr

Department of Mathematics, Siegen University, Walter-Flex-Straße 3, 57072 Siegen, Germany; schnurr@mathematik.uni-siegen.de
* Correspondence: nuessgen@mathematik.uni-siegen.de

Abstract: Ordinal pattern dependence is a multivariate dependence measure based on the co-movement of two time series. In strong connection to ordinal time series analysis, the ordinal information is taken into account to derive robust results on the dependence between the two processes. This article deals with ordinal pattern dependence for a long-range dependent time series including mixed cases of short- and long-range dependence. We investigate the limit distributions for estimators of ordinal pattern dependence. In doing so, we point out the differences that arise for the underlying time series having different dependence structures. Depending on these assumptions, central and non-central limit theorems are proven. The limit distributions for the latter ones can be included in the class of multivariate Rosenblatt processes. Finally, a simulation study is provided to illustrate our theoretical findings.

Keywords: ordinal patterns; time series; long-range dependence; multivariate data analysis; limit theorems

Citation: Nüßgen, I.; Schnurr, A. Ordinal Pattern Dependence in the Context of Long-Range Dependence. *Entropy* **2021**, *23*, 670. https://doi.org/10.3390/e23060670

Academic Editor: Christian H. Weiss

Received: 28 April 2021
Accepted: 19 May 2021
Published: 26 May 2021

Publisher's Note: MDPI stays neutral with regard to jurisdictional claims in published maps and institutional affiliations.

Copyright: © 2021 by the authors. Licensee MDPI, Basel, Switzerland. This article is an open access article distributed under the terms and conditions of the Creative Commons Attribution (CC BY) license (https://creativecommons.org/licenses/by/4.0/).

1. Introduction

The origin of the concept of ordinal patterns is in the theory of dynamical systems. The idea is to consider the order of the values within a data vector instead of the full metrical information. The ordinal information is encoded as a permutation (cf. Section 3). Already in the first papers on the subject, the authors considered entropy concepts related to this ordinal structure (cf. [1]). There is an interesting relationship between these concepts and the well-known Komogorov–Sinai entropy (cf. [2,3]). Additionally, an ordinal version of the Feigenbaum diagram has been dealt with e.g., in [4]. In [5], ordinal patterns were used in order to estimate the Hurst parameter in long-range dependent time series. Furthermore, Ref. [6] have proposed a test for independence between time series (cf. also [7]). Hence, the concept made its way into the area of statistics. Instead of long patterns (or even letting the pattern length tend to infinity), rather short patterns have been considered in this new framework. Furthermore, ordinal patterns have been used in the context of ARMA processes [8] and change-point detection within one time series [9]. In [10], ordinal patterns were used for the first time in order to analyze the dependence between two time series. Limit theorems for this new concept were proven in a short-range dependent framework in [11]. Ordinal pattern dependence is a promising tool, which has already been used in financial, biological and hydrological data sets, see in this context, also [12] for an analysis of the co-movement of time series focusing on symbols. In particular, in the context of hydrology, the data sets are known to be long-range dependent. Therefore, it is important to also have limit theorems available in this framework. We close this gap in the present article.

All of the results presented in this article have been established in the Ph.D. thesis of I. Nüßgen written under the supervision of A. Schnurr.

The article is structured as follows: in the subsequent section, we provide the reader with the mathematical framework. The focus is on (multivariate) long-range dependence. In Section 3, we recall the concept of ordinal pattern dependence and prove our main results. We present a simulation study in Section 4 and close the paper by a short outlook in Section 5.

2. Mathematical Framework

We consider a stationary d-dimensional Gaussian time series $(Y_j)_{j \in \mathbb{Z}}$ (for $d \in \mathbb{N}$), with:

$$Y_j := \left(Y_j^{(1)}, \ldots, Y_j^{(d)}\right)^t \tag{1}$$

such that $\mathbb{E}\left(Y_j^{(p)}\right) = 0$ and $\mathbb{E}\left(\left(Y_j^{(p)}\right)^2\right) = 1$ for all $j \in \mathbb{Z}$ and $p = 1, \ldots, d$. Furthermore, we require the cross-correlation function to fulfil $\left|r^{(p,q)}(k)\right| < 1$ for $p, q = 1, \ldots, d$ and $k \geq 1$, where the component-wise cross-correlation functions $r^{(p,q)}(k)$ are given by $r^{(p,q)}(k) = \mathbb{E}\left(Y_j^{(p)} Y_{j+k}^{(q)}\right)$ for each $p, q = 1, \ldots, d$ and $k \in \mathbb{Z}$. For each random vector Y_j, we denote the covariance matrix by Σ_d, since it is independent of j due to stationarity. Therefore, we have $\Sigma_d = \left(r^{(p,q)}(0)\right)_{p,q=1,\ldots,d}$.

We specify the dependence structure of $(Y_j)_{j \in \mathbb{Z}}$ and turn to long-range dependence: we assume that for the cross-correlation function $r^{(p,q)}(k)$ for each $p, q = 1, \ldots, d$, it holds that:

$$r^{(p,q)}(k) = L_{p,q}(k) k^{d_p + d_q - 1}, \tag{2}$$

with $L_{p,q}(k) \to L_{p,q}$ $(k \to \infty)$ for finite constants $L_{p,q} \in [0, \infty)$ with $L_{p,p} \neq 0$, where the matrix $L = (L_{p,q})_{p,q=1,\ldots,d}$ has full rank, is symmetric and positive definite. Furthermore, the parameters $d_p, d_q \in \left(0, \frac{1}{2}\right)$ are called long-range dependence parameters. Therefore, $(Y_j)_{j \in \mathbb{Z}}$ is multivariate long-range dependent in the sense of [13], Definition 2.1.

The processes we want to consider have a particular structure, namely for $h \in \mathbb{N}$, we obtain for fixed $j \in \mathbb{Z}$:

$$Y_{j,h} := \left(Y_j^{(1)}, \ldots, Y_{j+h-1}^{(1)}, Y_j^{(2)}, \ldots, Y_{j+h-1}^{(2)}, \ldots, Y_j^{(d)}, \ldots, Y_{j+h-1}^{(d)}\right)^t \in \mathbb{R}^{dh}. \tag{3}$$

The following relation holds between the *extendend process* $(Y_{j,h})_{j \in \mathbb{Z}}$ and the primarily regarded process $(Y_j)_{j \in \mathbb{Z}}$. For all $k = 1, \ldots, dh, j \in \mathbb{Z}$ we have:

$$Y_{j,h}^{(k)} = Y_{j+(k \mod h)-1}^{\lfloor \frac{k-1}{h} \rfloor + 1}, \tag{4}$$

where $\lfloor x \rfloor = \max\{k \in \mathbb{Z} : k \leq x\}$. Note that the process $(Y_{j,h})_{j \in \mathbb{Z}}$ is still a centered Gaussian process since all finite-dimensional marginals of $(Y_j)_{j \in \mathbb{Z}}$ follow a normal distribution. Stationarity is also preserved since for all $p, q = 1, \ldots, dh$, $p \leq q$ and $k \in \mathbb{Z}$, the cross-correlation function $r^{(p,q,h)}(k)$ of the process $(Y_{j,h})_{j \in \mathbb{Z}}$ is given by

$$r^{(p,q,h)}(k) = \mathbb{E}\left(Y_{j,h}^{(p)} Y_{j+k,h}^{(q)}\right)$$

$$= \mathbb{E}\left(Y_{j+(p \mod h)-1}^{\lfloor \frac{p-1}{h} \rfloor + 1} Y_{j+k+(q \mod h)-1}^{\lfloor \frac{q-1}{h} \rfloor + 1}\right)$$

$$= r^{\left(\lfloor \frac{p-1}{h} \rfloor + 1, \lfloor \frac{q-1}{h} \rfloor + 1\right)}(k + ((q-p) \mod h)) \tag{5}$$

and the last line does not depend on j. The covariance matrix $\Sigma_{d,h}$ of $Y_{j,h}$ has the following structure:

$$(\Sigma_{d,h})_{\substack{p,q=1,\ldots,d,\\ p\leq q}} = \left(r^{(p,q,h)}(0)\right)_{\substack{p,q=1,\ldots,dh,\\ p\leq q,}}$$

$$(\Sigma_{d,h})_{\substack{p,q=1,\ldots,d,\\ p>q}} = \left(r^{(q,p,h)}(0)\right)_{\substack{p,q=1,\ldots,dh,\\ q<p}}.$$

Hence, we arrive at:

$$\Sigma_{d,h} = \left(\Sigma_h^{(p,q)}\right)_{1\leq p,q\leq d'} \tag{6}$$

where $\Sigma_h^{(p,q)} = \mathbb{E}\left(\left(Y_1^{(p)},\ldots,Y_h^{(p)}\right)^t \left(Y_1^{(q)},\ldots,Y_h^{(q)}\right)\right) = \left(r^{(p,q)}(i-k)\right)_{1\leq i,k\leq h'}$ $p,q=1,\ldots,d$.
Note that $\Sigma_h^{(p,q)} \in \mathbb{R}^{h\times h}$ and $r^{(p,q)}(k) = r^{(q,p)}(-k)$, $k \in \mathbb{Z}$ since we are studying cross-correlation functions.

Therefore, we finally have to show that based on the assumptions on $(Y_j)_{j\in\mathbb{Z}'}$ the extended process is still long-range dependent.

Hence, we have to consider the cross-correlations again:

$$r^{(p,q,h)}(k) = r^{\left(\left\lfloor \frac{p-1}{h} \right\rfloor +1, \left\lfloor \frac{q-1}{h} \right\rfloor +1\right)}(k + ((q-p) \mod h))$$
$$= r^{(p^*,q^*)}(k+m^*)$$
$$\simeq r^{(p^*,q^*)}(k) \ (k \to \infty), \tag{7}$$

since $p^*, q^* \in \{1,\ldots,d\}$ and $m^* \in \{0,\ldots,h-1\}$, with $p^* := \left\lfloor \frac{p-1}{h} \right\rfloor + 1$, $q^* := \left\lfloor \frac{q-1}{h} \right\rfloor + 1$ and $m^* = (q-p) \mod h$.

Let us remark that $a_k \simeq b_k \Leftrightarrow \lim_{k\to\infty} \frac{a_k}{b_k} = 1$.

Therefore, we are still dealing with a multivariate long-range dependent Gaussian process. We see in the proofs of the following limit theorems that the crucial parameters that determine the asymptotic distribution are the long-range dependence parameters d_p, $p=1,\ldots,d$ of the original process $(Y_j)_{j\in\mathbb{Z}}$ and therefore, we omit the detailed description of the parameters d_{p^*} herein.

It is important to remark that the extended process $(Y_{j,h})_{j\in\mathbb{Z}}$ is also long-range dependent in the sense of [14], p. 2259, since:

$$\lim_{k\to\infty} \frac{k^D r^{(p,q,h)}(k)}{L(k)} = \lim_{k\to\infty} \frac{k^D r^{(p^*,q^*)}(k)}{L(k)}$$
$$= \lim_{k\to\infty} \frac{k^D L_{p^*,q^*} k^{d_{p^*}+d_{q^*}-1}}{L(k)}$$
$$=: b_{p^*,q^*}, \tag{8}$$

with:

$$D := \min_{p^*\in\{1,\ldots,d\}} \{1-2d_{p^*}\} \in (0,1) \tag{9}$$

and $L(k)$ can be chosen as any constant $L_{p,q}$ that is not equal to zero, so for simplicity, we assume without a loss of generality $L_{1,1} \neq 0$, and therefore, $L(k) = L_{1,1}$, since the condition in [14] only requires convergence to a finite constant b_{p^*,q^*}. Hence, we may apply the results in [14] in the subsequent results.

We define the following set, which is needed in the proofs of the theorems of this section.

$$P^* := \{p \in \{1, \ldots, d\} : d_p \geq d_q, \text{ for all } q \in \{1, \ldots, d\}\} \tag{10}$$

and denote the corresponding long-range dependence parameter to each $p \in P^*$ by

$$d^* := d_p, \quad p \in P^*.$$

We briefly recall the concept of Hermite polynomials as they play a crucial role in determining the limit distribution of functionals of multivariate Gaussian processes.

Definition 1. *(Hermite polynomial, [15], Definition 3.1)*
The j-th Hermite polynomial $H_j(x)$, $j = 0, 1, \ldots$, is defined as

$$H_j(x) := (-1)^j \exp\left(\frac{x^2}{2}\right) \frac{d^j}{dx^j} \exp\left(-\frac{x^2}{2}\right).$$

Their multivariate extension is given by the subsequent definition.

Definition 2. *(Multivariate Hermite polynomial, [15], p. 122)*
Let $d \in \mathbb{N}$. We define as d-dimensional Hermite polynomial:

$$H_k(x) := H_{k_1,\ldots,k_d}(x) := H_{k_1,\ldots,k_d}(x_1, \ldots, x_d) = \prod_{j=1}^d H_{k_j}(x_j),$$

with $k = (k_1, \ldots, k_d) \in \mathbb{N}_0^d \setminus \{(0, \ldots, 0)\}$.

Let us remark that the case $k = (0, \ldots, 0)$ is excluded here due to the assumption $\mathbb{E}(f(X)) = 0$.

Analogously to the univariate case, the family of multivariate Hermite polynomials $\{H_{k_1,\ldots,k_d}, k_1, \ldots, k_d \in \mathbb{N}\}$ forms an orthogonal basis of $L^2\left(\mathbb{R}^d, \varphi_{I_d}\right)$, which is defined as

$$L^2\left(\mathbb{R}^d, \varphi_{I_d}\right) := \left\{ f : \mathbb{R}^d \to \mathbb{R}, \int_{\mathbb{R}^d} f^2(x_1, \ldots, x_d) \varphi(x_1) \ldots \varphi(x_d) dx_d \ldots dx_1 < \infty \right\}.$$

The parameter φ_{I_d} denotes the density of the d-dimensional standard normal distribution, which is already divided into the product of the univariate densities φ in the formula above.

We denote the Hermite coefficients by

$$C(f, X, k) := C(f, I_d, k) := \langle f, H_k \rangle = \mathbb{E}(f(X) H_k(X)).$$

The Hermite rank $m(f, I_d)$ of f with respect to the distribution $\mathcal{N}(0, I_d)$ is defined as the largest integer m, such that:

$$\mathbb{E}\left(f(X) \prod_{j=1}^d H_{k_j}\left(X^{(j)}\right)\right) = 0 \text{ for all } 0 < k_1 + \ldots k_d < m.$$

Having these preparatory results in mind, we derive the multivariate Hermite expansion given by

$$f(X) - \mathbb{E}f(X) = \sum_{k_1+\ldots+k_d \geq m(f, I_d)} \frac{C(f, X, k)}{k_1! \ldots k_d!} \prod_{j=1}^d H_{k_j}\left(X^{(j)}\right). \tag{11}$$

We focus on the limit theorems for functionals with Hermite rank 2. First, we introduce the matrix-valued Rosenblatt process. This plays a crucial role in the asymptotics of functionals with Hermite rank 2 applied to multivariate long-range dependent Gaussian processes. We begin with the definition of a multivariate Hermitian–Gaussian random measure $\tilde{B}(d\lambda)$ with independent entries given by

$$\tilde{B}(d\lambda) = \left(\tilde{B}^{(1)}(d\lambda), \ldots, \tilde{B}^{(d)}(d\lambda)\right)^t, \tag{12}$$

where $\tilde{B}^{(p)}(d\lambda)$ is a univariate Hermitian–Gaussian random measure as defined in [16], Definition B.1.3. The multivariate Hermitian–Gaussian random measure $\tilde{B}(d\lambda)$ satisfies:

$$\mathbb{E}\left(\tilde{B}(d\lambda)\right) = 0,$$
$$\mathbb{E}\left(\tilde{B}(d\lambda)\tilde{B}(d\lambda)^*\right) = I_d \, d\lambda$$

and:

$$\mathbb{E}\left(\tilde{B}^{(p)}(d\lambda_1)\overline{\tilde{B}^{(q)}(d\lambda_2)}\right) = 0, \quad |\lambda_1| \neq |\lambda_2|, \quad p,q = 1, \ldots, d,$$

where $\tilde{B}(d\lambda)^* = \left(\overline{\tilde{B}^{(1)}(d\lambda)}, \ldots, \overline{\tilde{B}^{(d)}(d\lambda)}\right)$ denotes the Hermitian transpose of $\tilde{B}(d\lambda)$. Thus, following [14], Theorem 6, we can state the spectral representation of the matrix-valued Rosenblatt process $Z_{2,H}(t)$, $t \in [0,1]$ as

$$Z_{2,H}(t) = \left(Z_{2,H}^{(p,q)}(t)\right)_{p,q=1,\ldots,d}$$

where each entry of the matrix is given by

$$Z_{2,H}^{(p,q)}(t) = \int_{\mathbb{R}^2}'' \frac{\exp(it(\lambda_1 + \lambda_2)) - 1}{i(\lambda_1 + \lambda_2)} \tilde{B}^{(p)}(d\lambda_1)\tilde{B}^{(q)}(d\lambda_2).$$

The double prime in $\int_{\mathbb{R}^2}''$ excludes the diagonals $|\lambda_i| = |\lambda_j|$, $i \neq j$ in the integration. For details on multiple Wiener-Itô integrals, as can be seen in [17].

The following results were taken from [18], Section 3.2. The corresponding proofs were outsourced to the Appendix A.

Theorem 1. *Let $(Y_j)_{j \in \mathbb{Z}}$ be a stationary Gaussian process as defined in (1) that fulfils (2) for $d_p \in \left(\frac{1}{4}, \frac{1}{2}\right)$, $p = 1 \ldots, d$. For $h \in \mathbb{N}$ we fix:*

$$Y_{j,h} := \left(Y_j^{(1)}, \ldots, Y_{j+h-1}^{(1)}, \ldots, Y_j^{(d)}, \ldots, Y_{j+h-1}^{(d)}\right)^t \in \mathbb{R}^{dh}$$

with $Y_{j,h} \sim \mathcal{N}(0, \Sigma_{d,h})$ and $\Sigma_{d,h}$ as described in (6). Let $f : \mathbb{R}^{dh} \to \mathbb{R}$ be a function with Hermite rank 2 such that the set of discontinuity points D_f is a Null set with respect to the dh-dimensional Lebesgue measure. Furthermore, we assume f fulfills $\mathbb{E}\left(f^2\left(Y_{j,h}\right)\right) < \infty$. Then:

$$n^{-2d^*}(C_2)^{-\frac{1}{2}} \sum_{j=1}^n \left(f\left(Y_j^{(1)}, \ldots, Y_{j+h-1}^{(d)}\right) - \mathbb{E}\left(f\left(Y_j^{(1)}, \ldots, Y_{j+h-1}^{(d)}\right)\right)\right)$$
$$\xrightarrow{\mathcal{D}} \sum_{p,q \in P^*} \tilde{\alpha}^{(p,q)} Z_{2,d^*+1/2}^{(p,q)}(1), \tag{13}$$

where:

$$Z^{(p,q)}_{2,d^*+1/2}(1) = K_{p,q}(d^*) \int_{\mathbb{R}^2}'' \frac{\exp(i(\lambda_1+\lambda_2))-1}{i(\lambda_1+\lambda_2)} |\lambda_1 \lambda_2|^{-d^*} \tilde{B}^{(p)}_L(d\lambda_1) \tilde{B}^{(q)}_L(d\lambda_2).$$

The matrix $K(d^*)$ is a normalizing constant, as can be seen in [18], Corollary 3.6. Moreover, $\tilde{B}_L(d\lambda)$ is a multivariate Hermitian–Gaussian random measure with $\mathbb{E}(B_L(d\lambda)B_L(d\lambda)^*) = L\, d\lambda$ and L as defined in (2). Furthermore, $C_2 := \frac{1}{2d^*(4d^*-1)}$ is a normalizing constant and:

$$\tilde{\alpha}^{(p,q)} := \sum_{i,k=1}^{h} \alpha^{(p,q)}_{i,k}$$

where $\alpha^{(p,q)}_{i,k} = \alpha_{i+(p-1)h,k+(q-1)h}$ for each $p,q \in P^*$ and $i,k = 1,\ldots,h$ and:

$$(\alpha_{i,k})_{1 \le i,k \le dh} = \Sigma^{-1}_{d,h} C \Sigma^{-1}_{d,h}$$

where C denotes the matrix of second order Hermite coefficients, given by

$$C = (c_{i,k})_{1 \le i,k \le dh} = \mathbb{E}\Big(Y_{1,h}(f(Y_{1,h}) - \mathbb{E}(f(Y_{1,h}))) Y^t_{1,h}\Big).$$

It is possible to soften the assumptions in Theorem 1 to allow for mixed cases of short- and long-range dependence.

Corollary 1. *Instead of demanding in the assumptions of Theorem 1 that (2) holds for $(Y_j)_{j \in \mathbb{Z}}$ with the addition that for all $p = 1,\ldots,d$ we have $d_p \in \left(\frac{1}{4}, \frac{1}{2}\right)$, we may use the following condition. We assume that:*

$$r^{(p,q)}(k) = k^{d_p+d_q-1} L_{p,q}(k) \quad (k \to \infty)$$

with $L_{p,q}(k)$ as given in (2), but we do no longer assume $d_p \in \left(\frac{1}{4}, \frac{1}{2}\right)$ for all $p = 1,\ldots,d$ but soften the assumption to $d^* \in \left(\frac{1}{4}, \frac{1}{2}\right)$ and for $d_p \ne d^*$, $p = 1,\ldots,d$ we allow for $d_p \in (-\infty,0) \cup \left(0, \frac{1}{4}\right]$. Then, the statement of Theorem 1 remains valid.

However, with a mild technical assumption on the covariances of the one-dimensional marginal Gaussian processes that is often fulfilled in applications, there is another way of normalizing the partial sum on the right-hand side in Theorem 1, this time explicitly for the case $\#P^* = 2$ and $h \in \mathbb{N}$, such that the limit can be expressed in terms of two standard Rosenblatt random variables. This yields the possibility of further studying the dependence structure between these two random variables. In the following theorem, we assume $\#P^* = d = 2$ for the reader's convenience.

Theorem 2. *Under the same assumptions as in Theorem 1 with $\#P^* = d = 2$ and $d^* \in \left(\frac{1}{4}, \frac{1}{2}\right)$ and the additional condition that $r^{(1,1)}(l) = r^{(2,2)}(l)$, for $l = 0,\ldots,h-1$, and $L_{1,1} + L_{2,2} \ne L_{1,2} + L_{2,1}$, it holds that:*

$$n^{-2d^*}(C_2)^{-\frac{1}{2}} \sum_{j=1}^{n} \Big(f\big(Y^{(1)}_j,\ldots,Y^{(d)}_{j+h-1}\big) - \mathbb{E}f\big(Y^{(1)}_j,\ldots,Y^{(d)}_{j+h-1}\big) \Big)$$

$$\xrightarrow{\mathcal{D}} \big(\tilde{\alpha}^{(1,1)} - \tilde{\alpha}^{(1,2)}\big) \frac{L_{2,2} - L_{2,1} - L_{1,2} + L_{1,1}}{2} Z^*_{2,d^*+1/2}(1)$$

$$+ \big(\tilde{\alpha}^{(1,1)} + \tilde{\alpha}^{(1,2)}\big) \frac{L_{2,2} + L_{2,1} + L_{1,2} + L_{1,1}}{2} Z^{**}_{2,d^*+1/2}(1)$$

with $C_2 := \frac{1}{2d^*(4d^*-1)}$ being the same normalizing factor as in Theorem 1, $(\alpha_{i,k})_{1 \leq i,k \leq dh} = \Sigma_{d,h}^{-1} C \Sigma_{d,h}^{-1}$ and $C = (c_{i,k})_{1 \leq i,k \leq dh} = \mathbb{E}\left(Y_{1,h}(f(Y_{1,h}) - \mathbb{E}f(Y_{1,h}))Y_{1,h}^t\right)$. Note that $Z_{2,d^*+1/2}^*(1)$ and $Z_{2,d^*+1/2}^{**}(1)$ are both standard Rosenblatt random variables whose covariance is given by

$$\text{Cov}\left(Z_{2,d^*+1/2}^*(1), Z_{2,d^*+1/2}^{**}(1)\right) = \frac{(L_{2,2} - L_{1,1})^2}{(L_{1,1} + L_{2,2})^2 - (L_{1,2} + L_{2,1})^2}. \quad (14)$$

3. Ordinal Pattern Dependence

Ordinal pattern dependence is a multivariate dependence measure that compares the co-movement of two time series based on the ordinal information. First introduced in [10] to analyze financial time series, a mathematical framework including structural breaks and limit theorems for functionals of absolutely regular processes has been built in [11]. In [19], the authors have used the so-called symbolic correlation integral in order to detect the dependence between the components of a multivariate time series. Their considerations focusing on testing independence between two time series are also based on ordinal patterns. They provide limit theorems in the i.i.d.-case and otherwise use bootstrap methods. In contrast, in the mathematical model in the present article, we focus on asymptotic distributions of an estimator of ordinal pattern dependence having a bivariate Gaussian time series in the background but allowing for several dependence structures to arise. As it will turn out in the following, this yields central but also non-central limit theorems.

We start with the definition of an ordinal pattern and the basic mathematical framework that we need to build up the ordinal model.

Let S_h denote the set of permutations in $\{0, \ldots, h\}$, $h \in \mathbb{N}_0$ that we express as $(h+1)$-dimensional tuples, assuring that each tuple contains each of the numbers above exactly once. In mathematical terms, this yields:

$$S_h = \left\{\pi \in \mathbb{N}_0^{h+1} : 0 \leq \pi_i \leq h, \text{ and } \pi_i \neq \pi_k, \text{ whenever } i \neq k, \quad i, k = 0, \ldots, h\right\},$$

as can be seen in [11], Section 2.1.

The number of permutations in S_h is given by $\#S_h = (h+1)!$. In order to get a better intuitive understanding of the concept of ordinal patterns, we have a closer look at the following example, before turning to the formal definition.

Example 1. *Figure 1 provides an illustrative understanding of the extraction of an ordinal pattern from a data set. The data points of interest are colored in red and we consider a pattern of length $h = 3$, which means that we have to take $n = 4$ data points into consideration. We fix the points in time t_0, t_1, t_2 and t_3 and extract the data points from the time series. Then, we search for the point in time which exhibits the largest value in the resulting data and write down the corresponding time index. In this example, it was given by $t = t_1$. We order the data points by writing the time position of the largest value as the first entry, the time position of the second largest as the second entry, etc. Hence, the absolute values are ordered from largest to smallest and the ordinal pattern $(1, 0, 3, 2) \in S_3$ is obtained for the considered data points.*

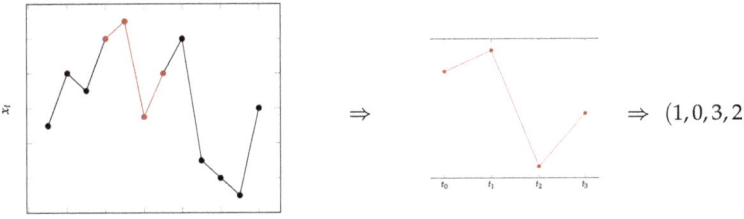

Figure 1. Example of the extraction of an ordinal pattern of a given data set.

Formally, the aforementioned procedure can be defined as follows, as can be seen in [11], Section 2.1.

Definition 3. *As the ordinal pattern of a vector $x = (x_0, \ldots, x_h) \in \mathbb{R}^{h+1}$, we define the unique permutation $\pi = (\pi_0, \ldots, \pi_h) \in S_h$:*

$$\Pi(x) = \Pi(x_0, \ldots, x_h) = (\pi_0, \ldots, \pi_h),$$

such that:

$$x_{\pi_0} \geq \ldots \geq x_{\pi_h},$$

with $\pi_{i-1} < \pi_i$ if $x_{\pi_{i-1}} = x_{\pi_i}$, $i = 1, \ldots, h$.

The last condition assures the uniqueness of π if there are ties in the data sets. In particular, this condition is necessary if real-world data are to be considered.

In Figure 2, all ordinal patterns of length $h = 2$ are shown. As already mentioned in the introduction, from the practical point of view, a highly desirable property of ordinal patterns is that they are not affected by monotone transformations, as can be seen in [5], p. 1783.

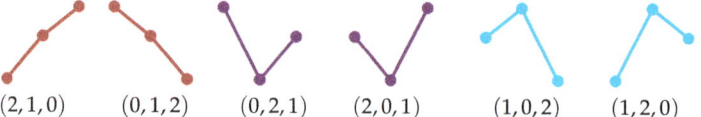

(2,1,0)   (0,1,2)   (0,2,1)   (2,0,1)   (1,0,2)   (1,2,0)

Figure 2. Ordinal patterns for $h = 2$.

Mathematically, this means that if $f : \mathbb{R} \to \mathbb{R}$ is strictly monotone, then:

$$\Pi(x_0, \ldots, x_h) = \Pi(f(x_0), \ldots, f(x_h)). \tag{15}$$

In particular, this includes linear transformations $f(x) = ax + b$, with $a \in \mathbb{R}^+$ and $b \in \mathbb{R}$.

Following [11], Section 1, the minimal requirement of the data sets we use for ordinal analysis in the time series context, i.e., for ordinal pattern probabilities as well as for ordinal pattern dependence later on, is *ordinal pattern stationarity (of order h)*. This property implies that the probability of observing a certain ordinal pattern of length h remains the same when shifting the moving window of length h through the entire time series and is not depending on the specific points in time. In the course of this work, the time series, in which the ordinal patterns occur, always have either stationary increments or are even stationary themselves. Note that both properties imply ordinal pattern stationarity. The reason why requiring stationary increments is a sufficient condition is given in the following explanation.

One fundamental property of ordinal patterns is that they are uniquely determined by the increments of the considered time series. As one can imagine in Example 1, the knowledge of the increments between the data points is sufficient to obtain the corresponding ordinal pattern. In mathematical terms, we can define another mapping $\tilde{\Pi}$, which assigns the corresponding ordinal pattern to each vector of increments, as can be seen in [5], p. 1783.

Definition 4. *We define for $y = (y_1, \ldots, y_h) \in \mathbb{R}^h$ the mapping $\tilde{\Pi} : \mathbb{R}^h \to S_h$:*

$$\tilde{\Pi}(y_1, \ldots, y_h) := \Pi(0, y_1, y_1 + y_2, \ldots, y_1 + \ldots + y_h),$$

such that for $y_i = x_i - x_{i-1}, i = 1, \ldots, h$, we obtain:

$$\tilde{\Pi}(y_1, \ldots, y_h) = \Pi(0, y_1, y_1 + y_2, \ldots, y_1 + \ldots + y_h)$$
$$= \Pi(0, x_1 - x_0, x_2 - x_0, \ldots, x_h - x_0)$$
$$= \Pi(x_0, x_1, x_2, \ldots, x_h).$$

We define the two mappings, following [5], p. 1784:

$$\mathcal{S} : S_h \to S_h, (\pi_0, \ldots, \pi_h) \to (\pi_h, \ldots, \pi_0),$$
$$\mathcal{T} : S_h \to S_h, (\pi_0, \ldots, \pi_h) \to (h - \pi_0, \ldots, h - \pi_h).$$

An illustrative understanding of these mappings is given as follows. The mapping $\mathcal{S}(\pi)$, which is the spatial reversion of the pattern π, is the reflection of π on a horizontal line, while $\mathcal{T}(\pi)$, the time reversal of π, is its reflection on a vertical line, as one can observe in Figure 3.

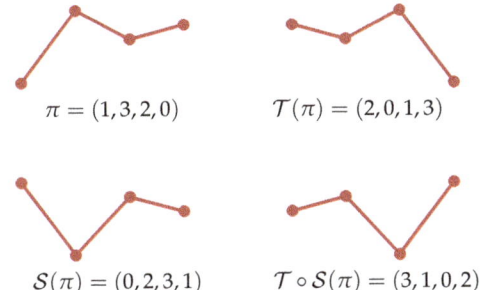

Figure 3. Space and time reversion of the pattern $\pi = (1, 3, 2, 0)$.

Based on the spatial reversion, we define a possibility to divide S_h into two disjoint sets.

Definition 5. *We define S_h^* as a subset of S_h with the property that for each $\pi \in S_h$, either π or $\mathcal{S}(\pi)$ are contained in the set, but not both of them.*

Note that this definition does not yield the uniqueness of S_h^*.

Example 2. *We consider the case $h = 2$ again and we want to divide S_2 into a possible choice of S_2^* and the corresponding spatial reversal. We choose $S_2^* = \{(2, 1, 0), (2, 0, 1), (1, 2, 0)\}$, and therefore, $S_2 \setminus S_2^* = \{(0, 1, 2), (1, 0, 2), (0, 2, 1)\}$. Remark that $S_2^* = \{(0, 1, 2), (2, 0, 1), (1, 2, 0)\}$ is also a possible choice. The only condition that has to be satisfied is that if one permutation is chosen for S_2^*, then its spatial reverse must not be an element of this set.*

We stick to the formal definition of ordinal pattern dependence, as it is proposed in [11], Section 2.1. The considered moving window consists of $h + 1$ data points, and hence, h increments. We define:

$$p := p_{X^{(1)}, X^{(2)}} := \mathbb{P}\left(\Pi\left(X_0^{(1)}, \ldots, X_h^{(1)}\right) = \Pi\left(X_0^{(2)}, \ldots, X_h^{(2)}\right)\right) \tag{16}$$

and:

$$q := q_{X^{(1)}, X^{(2)}} := \sum_{\pi \in S_h} \mathbb{P}\left(\Pi\left(X_0^{(1)}, \ldots, X_h^{(1)}\right) = \pi\right) \mathbb{P}\left(\Pi\left(X_0^{(2)}, \ldots, X_h^{(2)}\right) = \pi\right).$$

Then, we define ordinal pattern dependence OPD as

$$OPD := OPD_{X^{(1)}, X^{(2)}} := \frac{p - q}{1 - q}. \qquad (17)$$

The parameter q represents the hypothetical case of independence between the two time series. In this case, p and q would obtain equal values and therefore, OPD would equal zero. Regarding the other extreme, the case in which both processes coincide or one is a strictly monotone increasing transform of the other one, we obtain the value 1. However, in the following, we assume $p \in (0,1)$ and $q \in (0,1)$.

Note that the definition of ordinal pattern dependence in (17) only measures positive dependence. This is no restriction in practice, because negative dependence can be investigated in an analogous way, by considering $OPD_{X^{(1)}, -X^{(2)}}$. If one is interested in both types of dependence simultaneously, in [11], the authors propose to use $\left(OPD_{X^{(1)}, X^{(2)}}\right)_+ - \left(OPD_{X^{(1)}, -X^{(2)}}\right)_+$. To keep the notation simple, we focus on OPD as it is defined in (17).

We compare whether the ordinal patterns in $\left(X_j^{(1)}\right)_{j \in \mathbb{Z}}$ coincide with the ones in $\left(X_j^{(2)}\right)_{j \in \mathbb{Z}}$. Recall that it is an essential property of ordinal patterns that they are uniquely determined by the increment process. Therefore, we have to consider the increment processes $(Y_j)_{j \in \mathbb{Z}} = \left(\left(Y_j^{(1)}, Y_j^{(2)}\right)\right)_{j \in \mathbb{Z}}$ as defined in (1) for $d = 2$, where $Y_j^{(p)} = X_j^{(p)} - X_{j-1}^{(p)}$, $p = 1, 2$. Hence, we can also express p and q (and consequently OPD) as a probability that only depends on the increments of the considered vectors of the time series. Recall the definition of $\left(Y_{j,h}\right)_{j \in \mathbb{Z}}$ for $d = 2$, given by

$$Y_{j,h} = \left(Y_j^{(1)}, \ldots, Y_{j+h-1}^{(1)}, Y_j^{(2)}, \ldots, Y_{j+h-1}^{(2)}\right)^t,$$

such that $Y_{j,h} \sim \mathcal{N}(0, \Sigma_{2,h})$ with $\Sigma_{2,h}$ as given in (6).

In the course of this article, we focus on the estimation of p. For a detailed investigation of the limit theorems for estimators of OPD, we refer to [18]. We define the estimator of p, the probability of coincident patterns in both time series in a moving window of fixed length, by

$$\hat{p}_n = \frac{1}{n - h} \sum_{j=0}^{n-h-1} \mathbf{1}_{\left\{\Pi\left(X_j^{(1)}, \ldots, X_{j+h}^{(1)}\right) = \Pi\left(X_j^{(2)}, \ldots, X_{j+h}^{(2)}\right)\right\}}$$

$$= \frac{1}{n - h} \sum_{j=1}^{n-h} \mathbf{1}_{\left\{\tilde{\Pi}\left(Y_j^{(1)}, \ldots, Y_{j+h-1}^{(1)}\right) = \tilde{\Pi}\left(Y_j^{(2)}, \ldots, Y_{j+h-1}^{(2)}\right)\right\}},$$

where:

$$\tilde{\Pi}(Y_1, \ldots, Y_h) := \Pi(0, Y_1, Y_1 + Y_2, \ldots, Y_1 + \ldots + Y_h)$$
$$= \Pi(0, X_1 - X_0, \ldots, X_h - X_0)$$
$$= \Pi(X_0, X_1, \ldots, X_h).$$

Figure 4 illustrates the way ordinal pattern dependence is estimated by $\hat{p}_n$. The patterns of interest that are compared in each moving window are colored in red.

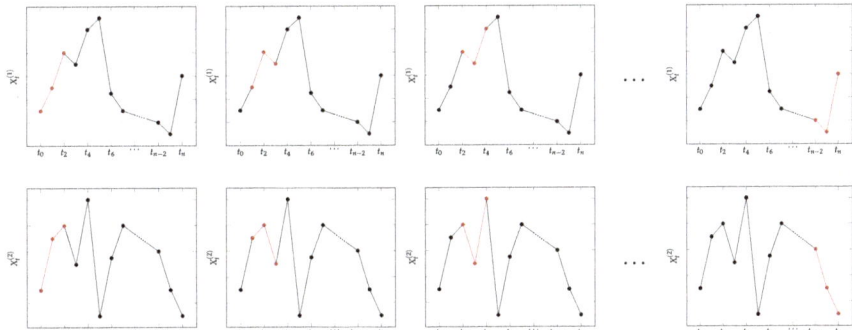

Figure 4. Illustration of estimation of ordinal pattern dependence.

Having emphasized the crucial importance of the increments, we define the following conditions on the increment process $(Y_j)_{j\in\mathbb{Z}}$: let $(Y_j)_{j\in\mathbb{Z}}$ be a bivariate, stationary Gaussian process with $Y_j^{(p)} \sim \mathcal{N}(0,1)$, $p=1,2$:

(L) We assume that $(Y_j)_{j\in\mathbb{Z}}$ fulfills (2) with d^* in $\left(\frac{1}{4},\frac{1}{2}\right)$. We allow for $\min\{d_1,d_2\}$ to be in the range $(-\infty,0)\cup\left(0,\frac{1}{4}\right]$.

(S) We assume $d_1,d_2\in(-\infty,0)\cup\left(0,\frac{1}{4}\right)$ such that the cross-correlation function of $(Y_j)_{j\in\mathbb{Z}}$ fulfills for $p,q=1,2$:

$$r^{(p,q)}(k) = k^{d_p+d_q-1}L_{p,q}(k) \quad (k\to\infty)$$

with $L_{p,q}(k)\to L_{p,q}$ and $L_{p,q}\in\mathbb{R}$ holds.

Furthermore, in both cases, it holds that $\left|r^{(p,q)}(k)\right|<1$ for $p,q=1,2$ and $k\geq 1$ to exclude ties.

We begin with the investigation of the asymptotics of $\hat{p}_n$. First, we calculate the Hermite rank of $\hat{p}_n$, since the Hermite rank determines for which ranges of d^* the estimator $\hat{p}_n$ is still long-range dependent. Depending on this range, different limit theorems may hold.

Lemma 1. *The Hermite rank of* $f(Y_{j,h}) = 1_{\left\{\tilde{\Pi}\left(Y_{j+1}^{(1)},\ldots,Y_{j+h}^{(1)}\right)=\tilde{\Pi}\left(Y_{j+1}^{(2)},\ldots,Y_{j+h}^{(2)}\right)\right\}}$ *with respect to* $\Sigma_{2,h}$ *is equal to 2.*

Proof. Following [20], Lemma 5.4 it is sufficient to show the following two properties:

(i) $m(f,\Sigma_{2,h})\geq 2$,
(ii) $m(f,I_{2,h})\leq 2$.

Note that the conclusion is not trivial, because $m(f,\Sigma_{2,h})\neq m(f,I_{2,h})$ in general, as can be seen in [15], Lemma 3.7. Lemma 5.4 in [20] can be applied due to the following reasoning. Ordinal patterns are not affected by scaling, therefore, the technical condition that $\Sigma_{2,h}^{-1} - I_{2,h}$ is positive semidefinite is fulfilled in our case. We can scale the standard deviation of the random vector $Y_{j,h}$ by any positive real number $\sigma>0$ since for all $j\in\mathbb{Z}$ we have:

$$\left\{\tilde{\Pi}\left(Y_j^{(1)},\ldots,Y_{j+h-1}^{(1)}\right) = \tilde{\Pi}\left(Y_j^{(2)},\ldots,Y_{j+h-1}^{(2)}\right)\right\}$$
$$=\left\{\tilde{\Pi}\left(\sigma Y_j^{(1)},\ldots,\sigma Y_{j+h-1}^{(1)}\right) = \tilde{\Pi}\left(\sigma Y_j^{(2)},\ldots,\sigma Y_{j+h-1}^{(2)}\right)\right\}.$$

To show property (i), we need to consider a multivariate random vector:

$$Y_{1,h} := \left(Y_1^{(1)}, \ldots, Y_h^{(1)}, Y_1^{(2)}, \ldots, Y_h^{(2)}\right)^t$$

with covariance matrix $\Sigma_{2,h}$. We fix $i = 1, \ldots, 2h$. We divide the set S_h into disjoint sets, namely into S_h^*, as defined in Definition 5 and the complimentary set $S_h \setminus S_h^*$. Note that:

$$-Y_{j,h} \stackrel{\mathcal{D}}{=} Y_{j,h}$$

holds. This implies:

$$\mathbb{E}\left(Y_{j,h}^{(i)} \mathbf{1}_{\left\{\tilde{\Pi}\left(Y_1^{(1)},\ldots,Y_h^{(1)}\right)=\tilde{\Pi}\left(Y_1^{(2)},\ldots,Y_h^{(2)}\right)=\pi\right\}}\right) = -\mathbb{E}\left(Y_{j,h}^{(i)} \mathbf{1}_{\left\{\tilde{\Pi}\left(Y_1^{(1)},\ldots,Y_h^{(1)}\right)=\tilde{\Pi}\left(Y_1^{(2)},\ldots,Y_h^{(2)}\right)=\mathcal{S}(\pi)\right\}}\right)$$

for $\pi \in S_h$. Hence, we arrive at:

$$\mathbb{E}\left(Y_{j,h}^{(i)} f(Y_{j,h})\right) = \mathbb{E}\left(Y_{j,h}^{(i)} \mathbf{1}_{\left\{\tilde{\Pi}\left(Y_1^{(1)},\ldots,Y_h^{(1)}\right)=\tilde{\Pi}\left(Y_1^{(2)},\ldots,Y_h^{(2)}\right)\right\}}\right)$$

$$= \sum_{\pi \in S_h} \mathbb{E}\left(Y_{j,h}^{(i)} \mathbf{1}_{\left\{\tilde{\Pi}\left(Y_1^{(1)},\ldots,Y_h^{(1)}\right)=\tilde{\Pi}\left(Y_1^{(2)},\ldots,Y_h^{(2)}\right)=\pi\right\}}\right)$$

$$= \sum_{\pi \in S_h^*} \mathbb{E}\left(Y_{j,h}^{(i)} \mathbf{1}_{\left\{\tilde{\Pi}\left(Y_1^{(1)},\ldots,Y_h^{(1)}\right)=\tilde{\Pi}\left(Y_1^{(2)},\ldots,Y_h^{(2)}\right)=\pi\right\}}\right)$$

$$- \sum_{\pi \in S_h \setminus S_h^*} \mathbb{E}\left(Y_{j,h}^{(i)} \mathbf{1}_{\left\{\tilde{\Pi}\left(Y_1^{(1)},\ldots,Y_h^{(1)}\right)=\tilde{\Pi}\left(Y_1^{(2)},\ldots,Y_h^{(2)}\right)=\mathcal{S}(\pi)\right\}}\right)$$

$$= 0$$

for $i = 1, \ldots, 2h$.

Consequently, $m(f, \Sigma_{2,h}) \geq 2$.

In order to prove (ii), we consider:

$$U_{1,h} := \left(U_1^{(1)}, \ldots, U_h^{(1)}, U_1^{(2)}, \ldots, U_h^{(2)}\right)^t$$

to be a random vector with independent $\mathcal{N}(0,1)$ distributed entries. For $i = 1, \ldots, h$ and $k = h+1, \ldots, 2h$ such that $k - h = i$, we obtain:

$$\mathbb{E}\left(U_{1,h}^{(i)} U_{1,h}^{(k)} f(U_{1,h})\right) = \mathbb{E}\left(U_i^{(1)} U_{k-h}^{(2)} \mathbf{1}_{\left\{\tilde{\Pi}\left(u_1^{(1)},\ldots,u_h^{(1)}\right)=\tilde{\Pi}\left(u_1^{(2)},\ldots,u_h^{(2)}\right)\right\}}\right)$$

$$= \sum_{\pi \in S_h} \mathbb{E}\left(U_i^{(1)} U_i^{(2)} \mathbf{1}_{\left\{\tilde{\Pi}\left(u_1^{(1)},\ldots,u_h^{(1)}\right)=\tilde{\Pi}\left(u_1^{(2)},\ldots,u_h^{(2)}\right)=\pi\right\}}\right)$$

$$= \sum_{\pi \in S_h} \left(\mathbb{E}\left(U_i^{(1)} \mathbf{1}_{\left\{\tilde{\Pi}\left(u_1^{(1)},\ldots,u_h^{(1)}\right)=\pi\right\}}\right)\right)^2$$

$$\neq 0,$$

since $\mathbb{E}\left(U_i^{(1)} \mathbf{1}_{\left\{\tilde{\Pi}\left(u_1^{(1)},\ldots,u_h^{(1)}\right)=\pi\right\}}\right) \neq 0$ for all $\pi \in S_h$. This was shown in the proof of Lemma 3.4 in [20].

All in all, we derive $m(f, \Sigma_{2,h}) = 2$ and hence, have proven the lemma. □

The case $m(f, \Sigma_{2,h}) = 2$ exhibits the property that the standard range of the long-range dependence parameter $d^* \in \left(0, \frac{1}{2}\right)$ has to be divided into two different sets. If $d^* \in \left(\frac{1}{4}, \frac{1}{2}\right)$, the transformed process $f\left(Y_{j,h}\right)_{j \in \mathbb{Z}}$ is still long-range dependent, as can be seen in [16],

Table 5.1. If $d^* \in \left(0, \frac{1}{4}\right)$, the transformed process is short-range dependent, which means by definition that the autocorrelations of the transformed process are summable, as can be seen in [13], Remark 2.3. Therefore, we have two different asymptotic distributions that have to be considered for the estimator $\hat{p}_n$ of coincident patterns.

3.1. Limit Theorem for the Estimator of p in Case of Long-Range Dependence

First, we restrict ourselves to the case that at least one of the two parameters d_1 and d_2 is in $\left(\frac{1}{4}, \frac{1}{2}\right)$. This assures $d^* \in \left(\frac{1}{4}, \frac{1}{2}\right)$. We explicitly include mixing cases where the process corresponding to $\min\{d_1, d_2\}$ is allowed to be long-range as well as short-range dependent.

Note that this setting includes the pure long-range dependence case, which means that for $p = 1, 2$, we have $d_p \in \left(\frac{1}{4}, \frac{1}{2}\right)$, or even $d_1 = d_2 = d^*$. However, in general, the assumptions are lower, such that we only require $d_p \in \left(\frac{1}{4}, \frac{1}{2}\right)$ for either $p = 1$ or $p = 2$ and the other parameter is also allowed to be in $(-\infty, 0)$ or $\left(0, \frac{1}{4}\right)$.

We can, therefore, apply the results of Corollary 1 and obtain the following asymptotic distribution for $\hat{p}_n$:

Theorem 3. *Under the assumption in (L), we obtain:*

$$n^{1-2d^*}(C_2)^{-\frac{1}{2}}(\hat{p}_n - p) \xrightarrow{\mathcal{D}} \sum_{p,q \in P^*} \tilde{\alpha}^{(p,q)} Z^{(p,q)}_{2,d^*+1/2}(1) \qquad (18)$$

with $Z^{(p,q)}_{2,d^*+1/2}(1)$ as given in Theorem 1 for $p, q \in P^*$ and $C_2 := \frac{1}{2d^*(4d^*-1)}$ being a normalizing constant. We have:

$$\tilde{\alpha}^{(p,q)} := \sum_{i,k=1}^{h} \alpha^{(p,q)}_{i,k}, \text{ where } \alpha^{(p,q)}_{i,k} = \alpha_{i+(p-1)h, k+(q-1)h},$$

for each $p, q \in P^*$ and $i, k = 1, \ldots, h$ and $(\alpha_{i,k})_{1 \leq i,k \leq dh} = \Sigma_{2,h}^{-1} C \Sigma_{2,h}^{-1}$, where the variable:

$$C = (c_{i,k})_{1 \leq i,k \leq 2h} = \mathbb{E}\left(Y_{1,h}\left(\mathbf{1}_{\{\tilde{\Pi}(Y_1^{(1)},\ldots,Y_h^{(1)}) = \tilde{\Pi}(Y_1^{(2)},\ldots,Y_h^{(2)})\}} - p\right) Y^t_{1,h}\right)$$

denotes the matrix of second order Hermite coefficients.

Proof. The proof of this theorem is an immediate application of the Corollary 1 and Lemma 1. Note that for $\hat{p}_n$ it holds that it is square integrable with respect to $Y_{j,h}$ and that the set of discontinuity points is a Null set with respect to the $2h$-dimensional Lebesgue measure. This is shown in [18], Equation (4.5). □

Following Theorem 2, we are also able to express the limit distribution above in terms of two standard Rosenblatt random variables by modifying the weighting factors in the limit distribution. Note that this requires slightly stronger assumptions as in Theorem 1.

Theorem 4. *Let (L) hold with $d_1 = d_2$. Additionally, we assume that $r^{(1,1)}(l) = r^{(2,2)}(l)$, for $l = 0, \ldots, h-1$, and $L_{1,1} + L_{2,2} \neq L_{1,2} + L_{2,1}$. Then, we obtain:*

$$n^{1-2d^*}(C_2)^{-\frac{1}{2}}(\hat{p}_n - p) \xrightarrow{\mathcal{D}} \left(\tilde{\alpha}^{(1,1)} - \tilde{\alpha}^{(1,2)}\right) \frac{L_{2,2} - L_{2,1} - L_{1,2} + L_{1,1}}{2} Z^*_{2,d^*+1/2}(1)$$

$$+ \left(\tilde{\alpha}^{(1,1)} + \tilde{\alpha}^{(1,2)}\right) \frac{L_{2,2} + L_{2,1} + L_{1,2} + L_{1,1}}{2} Z^{**}_{2,d^*+1/2}(1),$$

with C_2 and $\tilde{\alpha}^{(p,q)}$ as given in Theorem 3. Note that $Z^*_{2,d^*+1/2}(1)$ and $Z^{**}_{2,d^*+1/2}(1)$ are both standard Rosenblatt random variables, whose covariance is given by

$$\text{Cov}\left(Z^*_{2,d^*+1/2}(1), Z^{**}_{2,d^*+1/2}(1)\right) = \frac{(L_{2,2} - L_{1,1})^2}{(L_{1,1} + L_{2,2})^2 - (L_{1,2} + L_{2,1})^2}. \tag{19}$$

Remark 1. *Following [18], Corollary 3.14, if additionally $r^{(1,1)}(k) = r^{(2,2)}(k)$ and $r^{(1,2)}(k) = r^{(2,1)}(k)$ is fulfilled for all $k \in \mathbb{Z}$, then the two limit random variables following a standard Rosenblatt distribution in Theorem 4 are independent. Note that due to the considerations in [21], Equation (10), we know that the distribution of the sum of two independent standard Rosenblatt random variables is not standard Rosenblatt. However, this yields a computational benefit, as it is possible to efficiently simulate the standard Rosenblatt distribution, for details, as can be seen in [21].*

We turn to an example that deals with the asymptotic variance of the estimator of p in Theorem 3 in the case $h = 1$.

Example 3. *We focus on the case $h = 1$ and consider the underlying process $(Y_{j,1})_{j \in \mathbb{Z}} = \left(Y_j^{(1)}, Y_j^{(2)}\right)_{j \in \mathbb{Z}}$. It is possible to determine the asymptotic variance depending on the correlation $r^{(1,2)}(0)$ between these two increment variables.*

We start with the calculation of the second order Hermite coefficients in the case $\pi = (1,0)$. This corresponds to the event $\{Y_j^{(1)} \geq 0, Y_j^{(2)} \geq 0\}$, which yields:

$$c_{1,1}^{\pi,2} = \mathbb{E}\left(\left(\left(Y_j^{(1)}\right)^2 - 1\right)\mathbf{1}_{\{Y_j^{(1)} \geq 0, Y_j^{(2)} \geq 0\}}\right)$$

and:

$$c_{1,2}^{\pi,2} = \mathbb{E}\left(\left(Y_j^{(1)} Y_j^{(2)}\right)\mathbf{1}_{\{Y_j^{(1)} \geq 0, Y_j^{(2)} \geq 0\}}\right).$$

Due to $r^{(1,2)}(0) = r^{(2,1)}(0)$, we have $\left(Y_j^{(1)}, Y_j^{(2)}\right) \stackrel{\mathcal{D}}{=} \left(Y_j^{(2)}, Y_j^{(1)}\right)$ and therefore, $c_{1,1}^{\pi,2} = c_{2,2}^{\pi,2}$. We identify the second order Hermite coefficients as the ones already calculated in [20], Example 3.13, although we are considering two consecutive increments of a univariate Gaussian process there. However, since the corresponding values are only determined by the correlation between the Gaussian variables, we can simply replace the autocorrelation at lag 1 by the cross-correlation at lag 0. Hence, we obtain:

$$c_{1,1}^{\pi,2} = \varphi^2(0) r^{(1,2)}(0) \sqrt{1 - \left(r^{(1,2)}(0)\right)^2},$$
$$c_{1,2}^{\pi,2} = \varphi^2(0) \sqrt{1 - \left(r^{(1,2)}(0)\right)^2}.$$

Recall that the inverse $\Sigma_{2,1}^{-1} = (g_{i,j})_{i,j=1,2}$ of the correlation matrix of $\left(Y_j^{(1)}, Y_j^{(2)}\right)$ is given by

$$\Sigma_{2,1}^{-1} = \frac{1}{1 - \left(r^{(1,2)}(0)\right)^2}\begin{pmatrix} 1 & -r^{(1,2)}(0) \\ -r^{(1,2)}(0) & 1 \end{pmatrix}.$$

By using the formula for $\tilde{\alpha}^{(p,q)}$ obtained in [18], Equation (4.23), we derive:

$$\tilde{\alpha}_{\pi,2}^{(1,1)} = \alpha_{1,1}^{\pi,2} = \left(g_{1,1}^2 + g_{1,2}^2\right) c_{1,1}^{\pi,2} + 2 g_{1,1} g_{1,2} c_{1,2}^{\pi,2},$$
$$\tilde{\alpha}_{\pi,2}^{(1,2)} = \alpha_{1,2}^{\pi,2} = \left(g_{1,1}^2 + g_{1,2}^2\right) c_{1,2}^{\pi,2} + 2 g_{1,1} g_{1,2} c_{1,1}^{\pi,2}.$$

Plugging the second order Hermite coefficients and the entries of the inverse of the covariance matrix depending on $r^{(1,2)}(0)$ into the formulas, we arrive at:

$$\tilde{\alpha}_{\pi,2}^{(1,1)} = \frac{-\varphi^2(0)r^{(1,2)}(0)}{\left(1-\left(r^{(1,2)}(0)\right)^2\right)^{1/2}}$$

and:

$$\tilde{\alpha}_{\pi,2}^{(1,2)} = \frac{\varphi^2(0)}{\left(1-\left(r^{(1,2)}(0)\right)^2\right)^{1/2}}.$$

Therefore, in the case $h = 1$, we obtain the following factors in the limit variance in Theorem 3:

$$\tilde{\alpha}^{(1,1)} = \tilde{\alpha}^{(2,2)} = \frac{-2\varphi^2(0)r^{(1,2)}(0)}{\left(1-\left(r^{(1,2)}(0)\right)^2\right)^{1/2}}$$

$$\tilde{\alpha}^{(1,2)} = \tilde{\alpha}^{(2,1)} = \frac{2\varphi^2(0)}{\left(1-\left(r^{(1,2)}(0)\right)^2\right)^{1/2}}.$$

Remark 2. *It is not possible to analytically determine the limit variance for $h = 2$, as this includes orthant probabilities of a four-dimensional Gaussian distribution. Following [22], no closed formulas are available for these probabilities. However, there are fast algorithms at hand that calculate the limit variance efficiently. It is possible to take advantage of the symmetry properties of the multivariate Gaussian distribution to keep the computational cost of these algorithms low. For detail, as can be seen in [18], Section 4.3.1.*

3.2. Limit Theorem for the Estimator of p in Case of Short-Range Dependence

In this section, we focus on the case of $d^* \in (-\infty, 0) \cup \left(0, \frac{1}{4}\right)$. If $d^* \in \left(0, \frac{1}{4}\right)$, we are still dealing with a long-range dependent multivariate Gaussian process $\left(Y_{j,h}\right)_{j \in \mathbb{Z}}$. However, the transformed process $\hat{p}_n - p$ is no longer long-range dependent, since we are considering a function with Hermite rank 2, see also [16], Table 5.1. Otherwise, if $d^* \in (-\infty, 0)$, the process $\left(Y_{j,h}\right)_{j \in \mathbb{Z}}$ itself is already short-range dependent, since the cross-correlations are summable. Therefore, we obtain the following central limit theorem by applying Theorem 4 in [14].

Theorem 5. *Under the assumptions in (S), we obtain:*

$$n^{\frac{1}{2}}(\hat{p}_n - p) \xrightarrow{\mathcal{D}} \mathcal{N}(0, \sigma^2)$$

with:

$$\sigma^2 = \sum_{k=-\infty}^{\infty} \mathbb{E}\left[\left(\mathbf{1}_{\left\{\tilde{\Pi}\left(Y_1^{(1)},\ldots,Y_h^{(1)}\right)=\tilde{\Pi}\left(Y_1^{(2)},\ldots,Y_h^{(2)}\right)\right\}} - p\right)\right.$$
$$\left. \times \left(\mathbf{1}_{\left\{\tilde{\Pi}\left(Y_{1+k}^{(1)},\ldots,Y_{h+k}^{(1)}\right)=\tilde{\Pi}\left(Y_{1+k}^{(2)},\ldots,Y_{h+k}^{(2)}\right)\right\}} - p\right)\right].$$

We close this section with a brief retrospect of the results obtained. We established limit theorems for the estimator of p as probability of coincident pattern in both time series and hence, on the most important parameter in the context of ordinal pattern dependence. The long-range dependent case as well as the mixed case of short- and long-range dependence was considered. Finally, we provided a central limit theorem for a multivariate Gaussian time series that is short-range dependent if transformed by $\hat{p}_n$. In the subsequent section, we provide a simulation study that illustrates our theoretical findings. In doing so, we shed light on the Rosenblatt distribution and the distribution of the sum of Rosenblatt distributed random variables.

4. Simulation Study

We begin with the generation of a bivariate long-range dependent fractional Gaussian noise series $\left(Y_j^{(1)}, Y_j^{(2)}\right)_{j=1,\ldots,n}$.

First, we simulate two independent fractional Gaussian noise processes $\left(U_j^{(1)}\right)_{j=1,\ldots,n}$ and $\left(U_j^{(2)}\right)_{j=1,\ldots,n}$, derived by the R-package "longmemo", for a fixed parameter $H \in \left(\frac{1}{2}, 1\right)$ in both time series. For the reader's convenience, we denote the long-range dependence parameter d by $H = d + \frac{1}{2}$ as it is common, when dealing with fractional Gaussian noise and fractional Brownian motion. We refer to H as *Hurst parameter*, tracing back to the work of [23]. For $H = 0.7$ and $H = 0.8$ we generate $n = 10^6$ samples, for $H = 0.9$, we choose $n = 2 \cdot 10^6$. We denote the correlation function of univariate fractional Gaussian noise by $r_H^{(1,1)}(k), k \geq 0$. Then, we obtain $\left(Y_j^{(1)}, Y_j^{(2)}\right)_j$ for $j = 1, \ldots, n$:

$$Y_j^{(1)} = U_j^{(1)},$$
$$Y_j^{(2)} = \psi U_j^{(1)} + \phi U_j^{(2)}, \tag{20}$$

for $\psi, \phi \in \mathbb{R}$.

Note that this yields the following properties for the cross-correlations of the two processes for $k \geq 0$:

$$r_H^{(1,2)}(k) = \mathbb{E}\left(Y_j^{(1)} Y_{j+k}^{(2)}\right) = \psi r_H^{(1,1)}(k)$$
$$r_H^{(2,1)}(k) = r^{(1,2)}(-k) = \psi r_H^{(1,1)}(k)$$
$$r_H^{(2,2)}(k) = \mathbb{E}\left(Y_j^{(2)} Y_{j+k}^{(2)}\right) = \left(\psi^2 + \phi^2\right) r_H^{(1,1)}(k).$$

We use $\psi = 0.6$ and $\phi = 0.8$ to obtain unit variance in the second process.

Note that we chose the same Hurst parameter in both processes to get a better simulation result. The simulations of the processes $\left(Y_j^{(1)}\right)_{j \in \mathbb{Z}}$ and $\left(Y_j^{(2)}\right)_{j \in \mathbb{Z}}$ are visualized in Figure 5. On the left-hand side, the different fractional Gaussian noises depending on the Hurst parameter H are displayed. They represent the stationary long-range dependent Gaussian *increment* processes we need in the view of the limit theorems we derived in Section 3. The processes in which we are comparing the coincident ordinal patterns, namely $\left(X_j^{(1)}\right)_{j \in \mathbb{Z}}$ and $\left(X_j^{(2)}\right)_{j \in \mathbb{Z}}$, are shown on the right-hand side in Figure 5. The long-range dependent behavior of the increment processes is very illustrative in these processes: roughly speaking, they become smoother the larger the Hurst parameter gets.

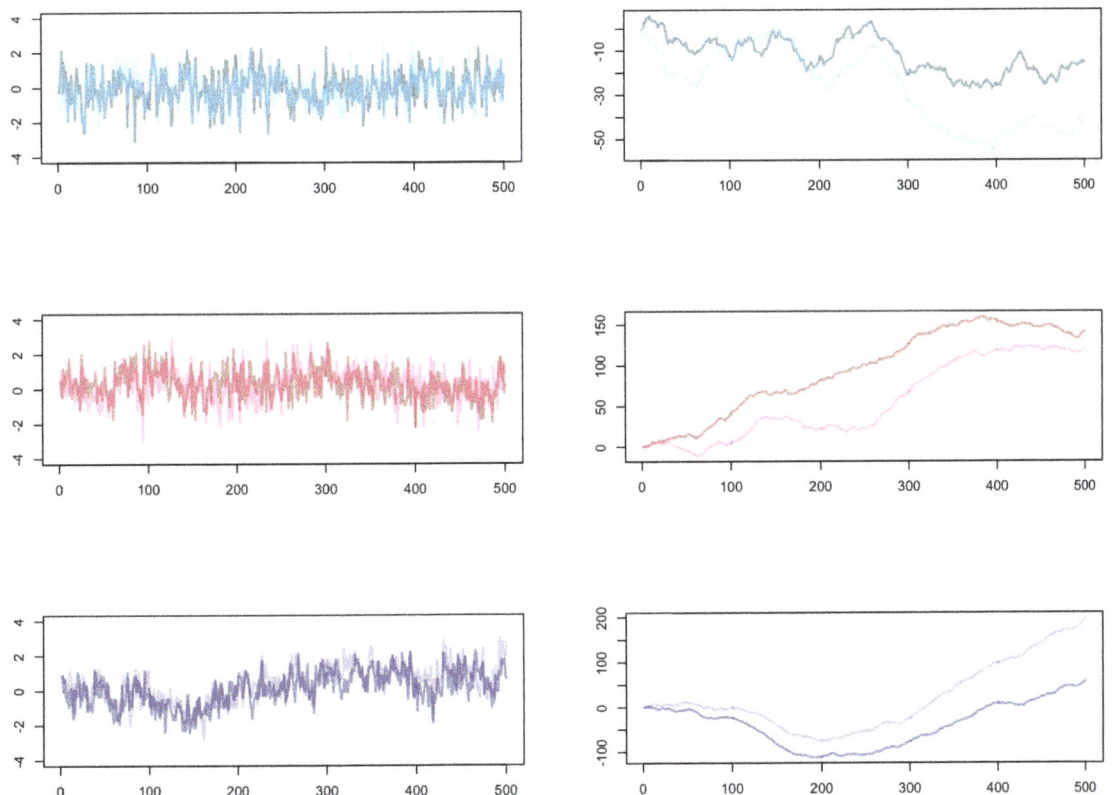

Figure 5. Plots of 500 data points of one path of two dependent fractional Gaussian noise processes (**left**) and the paths of the corresponding fractional Brownian motions (**right**) for different Hurst parameters: $H = 0.7$ (**top**), $H = 0.8$ (**middle**), $H = 0.9$ (**bottom**).

We turn to the simulation results for the asymptotic distribution of the estimator $\hat{p}_n$. The first limit theorem is given in Theorem 3 for $H = 0.8$ and $H = 0.9$. In the case of $H = 0.7$, a different limit theorem holds, see Theorem 5. Therefore, we turn to the simulation results of the asymptotic distribution of the estimator $\hat{p}_n$ of p, as shown in Figure 6 for pattern length $h = 2$. The asymptotic normality in case $H = 0.7$ can be clearly observed. We turn to the interpretation of the simulation results of the distribution of $\hat{p}_n - p$ for $H = 0.8$ and $H = 0.9$ as the weighted sum of the sample (cross-)correlations: we observe in the Q–Q plot for $H = 0.8$ that the samples in the upper and lower tail deviate from the reference line. For $H = 0.9$, a similar behavior in the Q–Q plot is observed.

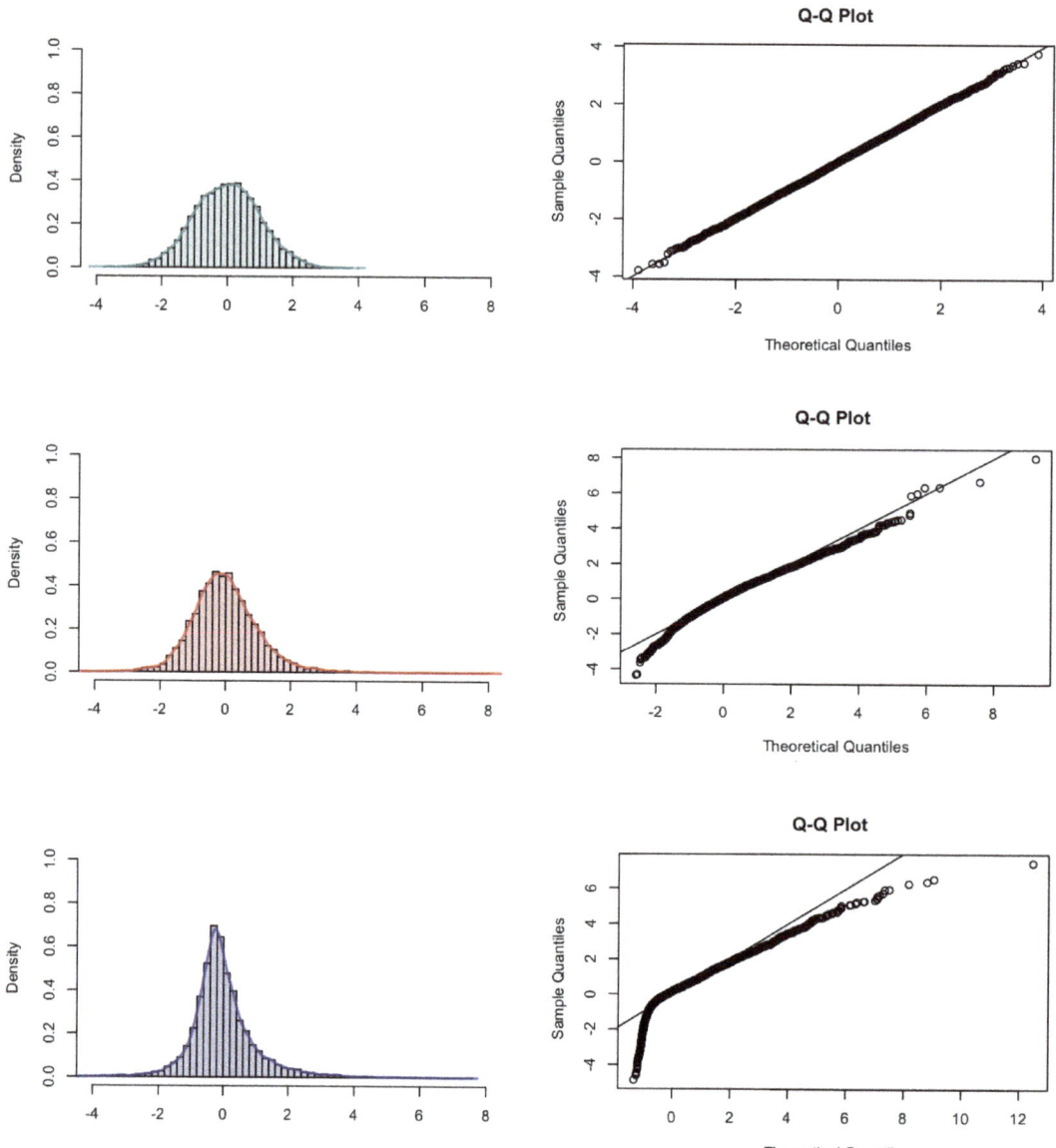

Figure 6. Histogram, kernel density estimation and Q–Q plot with respect to the normal distribution ($H = 0.7$) or to the Rosenblatt distribution of $\hat{p}_n - p$ with $h = 2$ for different Hurst parameters: $H = 0.7$ (**top**); $H = 0.8$ (**middle**); $H = 0.9$ (**bottom**).

We want to verify the result in Theorem 4 that it is possible, by a different weighting, to express the limit distribution of $\hat{p}_n - p$ as the distribution of the sum of two independent standard Rosenblatt random variables. The simulated convergence result is provided in Figure 7. We observed the standard Rosenblatt distribution.

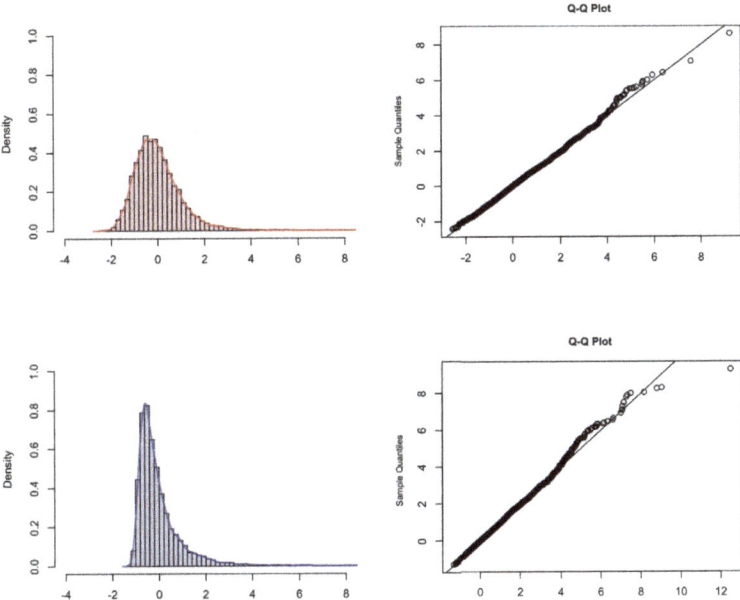

Figure 7. Histogram, kernel density estimation and Q–Q plot with respect to the Rosenblatt distribution of $\frac{1}{n}\sum_{j=1}^{n} H_2\left(Y_j^*\right)$ for different Hurst parameters: $H = 0.8$ (**top**); $H = 0.9$ (**bottom**).

5. Conclusions and Outlook

We considered limit theorems in the context of the estimation of ordinal pattern dependence in the long-range dependence setting. Pure long-range dependence, as well as mixed cases of short- and long-range dependence, were considered alongside the transformed short-range dependent case. Therefore, we complemented the asymptotic results in [11]. Hence, we made ordinal pattern dependence applicable for long-range dependent data sets as they arise in the context of neurology, as can be seen in [24] or artificial intelligence, as can be seen in [25]. As these kinds of data were already investigated using ordinal patterns, as can be seen, for example, in [26], this emphasizes the large practical impact of the ordinal approach in analyzing the dependence structure multivariate time series. This yields various research opportunities in these fields in the future.

Our results rely on the assumption of Gaussianity of the considered multivariate time series. If we focus on comparing the coincident ordinal patterns in a stationary long-range dependent bivariate time series, we highly benefit from the property of ordinal patterns not being affected by monotone transformations. It is possible to transform the data set to the Gaussian framework without losing the necessary ordinal information. In applications, this property is highly desirable. If we consider the more general setting, that is, stationary increments, the mathematical theory in the background gets a lot more complex leading to the limitations of our results. A crucial argument used in the proofs of the results in Section 2 is given in the *Reduction Theorem*, originally proven in Theorem 4.1 in [27] in the univariate case and extended to the multivariate setting in Theorem 6 in [14]. For further details, we refer the reader to the Appendix A. However, this result only holds in the Gaussian case. Limit theorems for the sample cross-correlation process of multivariate linear long-range dependent processes with Hermite rank 2 have recently been proven in Theorem 4 in [28]. This is possibly an interesting starting point to adapt the proofs in the Appendix A to this larger class of processes without requiring Gaussianity.

Considering the property of having a discrete bivariate time series in the background, an interesting extension is given in time continuous processes and the associated techniques of discretization to still regard the ordinal perspective. To think even further beyond our scope, a generalization to categorical data is conceivable and yields an interesting open research opportunity.

Author Contributions: Conceptualization, I.N. and A.S.; methodology and mathematical theory, I.N.; simulations, I.N.; validation, I.N. and A.S.; writing—original draft preparation, I.N.; writing—review and editing, A.S.; funding acquisition, A.S. All authors have read and agreed to the published version of the manuscript.

Funding: This research was funded by the German Research Foundation (DFG) grant number SCHN 1231/3-2.

Institutional Review Board Statement: Not applicable.

Informed Consent Statement: Not applicable.

Data Availability Statement: The data that support the findings of this study are available from the corresponding author, Ines Nüßgen, upon reasonable request.

Conflicts of Interest: The authors declare no conflict of interest.

Appendix A. Technical Appendix

All proofs in this appendix were taken from [18], Chapter 3.

Appendix A.1. Preliminary Results

Before turning to limit theorems, we introduce a possibility to decompose the d-dimensional Gaussian process $(Y_j)_{j \in \mathbb{Z}}$ using the Cholesky decomposition, as can be seen in [29]. Based on the definition of the multivariate normal distribution, as can be seen in [30], Definition 1.6.1, we find an upper triangular matrix $\tilde{A}$, such that $\tilde{A}\tilde{A}^t = \Sigma_d$. Then, it holds that:

$$Y_j \overset{\mathcal{D}}{=} \tilde{A} U_j^*, \tag{A1}$$

where U_j^* is a d-dimensional Gaussian process where each U_j^* has independent and identically $\mathcal{N}(0,1)$ distributed entries. We want to assure that $(U_j^*)_{j \in \mathbb{Z}}$ preserves the long-range dependent structure of $(Y_j)_{j \in \mathbb{Z}}$. Since we know from (2) that:

$$\mathbb{E}(Y_j Y_{j+k}) = \Gamma_Y(k) \simeq k^{D - \frac{1}{2} I_d} L k^{D - \frac{1}{2} I_d} \quad (k \to \infty),$$

the process (U_j^*) has to fulfill:

$$\mathbb{E}(U_j^* U_{j+k}^*) = \Gamma_{U^*}(k) \simeq k^{D - \frac{1}{2} I_d} L_U k^{D - \frac{1}{2} I_d} \quad (k \to \infty), \tag{A2}$$

with $L = \tilde{A} L_{U^*} \tilde{A}^t$.

Then, it holds for all $n \in \mathbb{N}$ that:

$$(Y_j, j = 1, \ldots, n) \overset{\mathcal{D}}{=} (\tilde{A} U_j^*, j = 1, \ldots, n). \tag{A3}$$

Note that the assumption in (A2) is only well-defined because we assumed $\left| r^{(p,q)}(k) \right| < 1$ for $k \geq 1$ and $p, q = 1, \ldots, d$ in (1). This becomes clear in the following considerations. In the proofs of the theorems in this chapter, we do not only need a decomposition of Y_j, but also of $Y_{j,h}$. As $Y_{j,h}$ is still a multivariate Gaussian process, the covariance matrix of $Y_{j,h}$

given by $\Sigma_{d,h}$ is positive definite. Hence, it is possible to find an upper triangular matrix A, such that $AA^t = \Sigma_{d,h}$. It holds thatL

$$Y_{j,h} \stackrel{\mathcal{D}}{=} AU_{j,h}$$

for:

$$U_{j,h} = \left(U^{(1)}_{(j-1)h+1}, \ldots, U^{(1)}_{jh}, \ldots, U^{(d)}_{(j-1)h+1}, \ldots, U^{(d)}_{jh}\right)^t.$$

The random vector $U_{j,h}$ consists of $(d \cdot h)$ independent and standard normally distributed random variables. We notice the different structure of $U_{j,h}$ compared to $Y_{j,h}$. We assure that for consecutive j, the entries in $U_{j,h}$ are all different while there are identical entries, for example in $Y_{1,h} = \left(Y^{(1)}_1, Y^{(1)}_2, \ldots, Y^{(d)}_h\right)^t$ and $Y_{2,h} = \left(Y^{(1)}_2, \ldots, Y^{(d)}_h, Y^{(d)}_{h+1}\right)^t$. This complicates our aim that:

$$\left(Y_{j,h}, j=1,\ldots,n\right)^t \stackrel{\mathcal{D}}{=} \left(AU_{j,h}, j=1,\ldots,n\right)^t \tag{A4}$$

holds.

The special structure of $\left(Y_{j,h}\right)_{j\in\mathbb{Z}}$, namely that consisting of h consecutive entries of each marginal process $\left(Y^{(p)}_j\right)$, $p = 1,\ldots,d$, alongside the dependence between two random vectors in the process $\left(Y_{j,h}\right)$, has to be reflected in the covariance matrix of $\left(U_{j,h}, j=1,\ldots,n\right)$. Hence, we need to check whether such a vector $\left(U_{j,h}, j=1,\ldots,n\right)$ exists, i.e., if there is a positive semi-definite matrix that fulfills these conditions. We define **A** as a block diagonal matrix with A as main-diagonal blocks and all off-diagonal blocks as $dh \times dh$-zero matrix.

We denote the covariance matrix of $\left(Y_{j,h}, j=1,\ldots,n\right)^t$ by $\Sigma_{Y,n}$ and define the following matrix:

$$\Sigma_{U,n} := \text{inv}(\mathbf{A})\Sigma_{Y,n}\text{inv}(\mathbf{A}^t). \tag{A5}$$

We know that $\Sigma_{Y,n}$ is a positive semi-definite for all $n \in \mathbb{N}$ because (Y_j) is a Gaussian process. Mathematically described, this means that:

$$x^t \Sigma_{Y,n} x \geq 0, \tag{A6}$$

for all $x = (x_1, \ldots, x_{nhd})^t \in \mathbb{R}^{nhd}$. We conclude:

$$\begin{aligned}x^t \Sigma_{U,n} x &= x^t \text{inv}(\mathbf{A})\Sigma_{Y,n}\text{inv}(\mathbf{A}^t) x \\ &= \left(\text{inv}(\mathbf{A}^t)x\right)^t \Sigma_{Y,n}\left(x^t\text{inv}(\mathbf{A})\right) \\ &\stackrel{(A6)}{\geq} 0.\end{aligned}$$

Therefore, $\Sigma_{U,n}$ is a positive semi-definite matrix for all $n \in \mathbb{N}$ and the random vector:

$$\left(U_{j,h}, j=1,\ldots,n\right)^t \mathcal{N} \sim (0, \Sigma_{U,n})$$

exists and (A4) holds. Note that we do not have any further information on the dependence structure within the process (U_j), in general, this process neither exhibits long-range dependence, nor independence, nor stationarity.

We continue with two preparatory results that are also necessary for proving Theorem 2.1.

Lemma A1. Let $(Y_j)_{j \in \mathbb{Z}}$ be a d-dimensional Gaussian process as defined in (1) that fulfills (2) with $d_1 = \ldots = d_d = d^*$, such that:

$$\Gamma_Y(k) = \mathbb{E}\left(Y_j Y_{j+k}^t\right) \simeq L k^{2d^* - 1}, \quad (k \to \infty).$$

Let C_2 be a normalization constant:

$$C_2 = \frac{1}{2d^*(4d^* - 1)}$$

and let B_Y be an upper triangular matrix, such that:

$$B_Y B_Y^t = L.$$

Furthermore, for $l \in \mathbb{N}$ we have:

$$\hat{\Gamma}_{Y,n}(l) = \frac{1}{n-l} \sum_{j=1}^{n-l} Y_j Y_{j+l}^t.$$

Then, for $h \in \mathbb{N}$ it holds that:

$$\left(n^{1-2d^*} (C_2)^{-1/2} (B_Y \otimes B_Y)^{-1} \mathrm{vec}(\hat{\Gamma}_n(l) - \Gamma(l)), \; l = 0, \ldots, h-1 \right)$$
$$\xrightarrow{D} \left(\mathrm{vec}\left(Z_{2,d^*+1/2}^{(p,q)}(1) \right)_{p,q=1,\ldots,d}, \; l = 0, \ldots, h-1 \right),$$

where $Z_{2,d^*+1/2}^{(p,q)}(1)$ has the spectral domain representation:

$$Z_{2,d^*+1/2}^{(p,q)}(1) = K_{p,q}(d^*) \int_{\mathbb{R}^2}'' \frac{\exp(i(\lambda_1 + \lambda_2)) - 1}{i(\lambda_1 + \lambda_2)} |\lambda_1 \lambda_2|^{-d^*} \tilde{B}^{(p)}(d\lambda_1) \tilde{B}^{(q)}(d\lambda_2)$$

where:

$$K_{p,q}^2(d^*) = \begin{cases} \frac{1}{2C_2(2\Gamma(1-2d^*)\sin(\pi d^*))^2}, & p = q \\ \frac{1}{C_2(2\Gamma(1-2d^*)\sin(\pi d^*))^2}, & p \neq q. \end{cases}$$

and $\tilde{B}(d\lambda) = \left(\tilde{B}^{(1)}(d\lambda), \ldots, \tilde{B}^{(d)}(d\lambda) \right)$ is a multivariate Hermitian–Gaussian random measure as defined in (12).

Proof. First, we can use (A1):

$$Y_j \stackrel{D}{=} \tilde{A} U_j^*,$$

such that $\left(U_j^* \right)$ is a multivariate Gaussian process with $U_j^* \sim \mathcal{N}(0, I_d)$ and $\left(U_j^* \right)$ is still long-range dependent, as can be seen in (A2). It is possible to decompose the sample cross-covariance matrix $\hat{\Gamma}_{Y,n}(l) - \Gamma_Y(l)$ with respect to (Y_j) at lag l given by

$$\hat{\Gamma}_{Y,n}(l) - \Gamma_Y(l) = \frac{1}{n-l} \sum_{j=1}^{n-l} Y_j Y_{j+l}^t - \mathbb{E}\left(Y_j Y_{j+l}^t \right)$$

to:

$$\hat{\Gamma}_{Y,n}(l) - \Gamma_Y(l) \stackrel{D}{=} \tilde{A} \left(\hat{\Gamma}_{U^*,n}(l) - \Gamma_{U^*}(l) \right) \tilde{A}^t,$$

where we define the sample cross-covariance matrix $\hat{\Gamma}_{U^*,n}(l) - \Gamma_{U^*}(l)$ with respect to $\left(U_j^*\right)$ at lag l by

$$\hat{\Gamma}_{U^*,n}(l) - \Gamma_{U^*}(l) = \frac{1}{n-l}\sum_{j=1}^{n-l} U_j^* U_{j+l}^* - \mathbb{E}\left(U_j^* U_{j+l}^*\right).$$

Each entry of:

$$\hat{\Gamma}_{U^*,n}(l) - \Gamma_{U^*}(l) = \left(\hat{r}_{n,U^*}^{(p,q)}(l) - r_{U^*}^{(p,q)}(l)\right)_{p,q=1,\ldots,d}$$

is given by

$$\hat{r}_{n,U^*}^{(p,q)}(l) - r_{U^*}^{(p,q)}(l) := \sum_{j=1}^{n} U_j^{*\,(p)} U_{j+l}^{*\,(q)} - \mathbb{E}\left(U_j^{*\,(p)} U_{j+l}^{*\,(q)}\right).$$

Following [31], proof of Lemma 7.4, the limit distribution of:

$$(\hat{\Gamma}_{U^*,n}(l) - \Gamma_{U^*}(l),\ l = 0,\ldots,h-1)$$

is equal to the limit distribution of:

$$(\hat{\Gamma}_{U^*,n}(0) - \Gamma_{U^*}(0),\ l = 0,\ldots,h-1).$$

We recall the assumption that $d^* = d_p$ for all $p = 1,\ldots,d$. We follow [14], Theorem 6 and use the Cramer–Wold device: Let $a_{1,1}, a_{1,2}, \ldots, a_{d,d} \in \mathbb{R}$. We are interested in the asymptotic behavior of:

$$n^{1-2d^*} \sum_{p,q=1}^{d} a_{p,q}\left(\hat{r}_{n,U}^{(p,q)}(0) - r_U^{(p,q)}(0)\right)$$

$$= n^{-2d^*} \sum_{j=1}^{n} \sum_{p,q=1}^{d} a_{p,q}\left(U_j^{*\,(p)} U_j^{*\,(q)} - \mathbb{E}\left(U_j^{*\,(p)} U_j^{*\,(q)}\right)\right).$$

We consider the function:

$$f\left(U_j^*\right) = \sum_{p,q=1}^{d} a_{p,q}\left(U_j^{*\,(p)} U_j^{*\,(q)} - \mathbb{E}\left(U_j^{*\,(p)} U_j^{*\,(q)}\right)\right) \tag{A7}$$

and may apply Theorem 6 in [14]. Using the Hermite decomposition of f as given in (11), we observe that f and therefore, $a_{p,q}$, $p,q = 1,\ldots,d$, only affects the Hermite coefficients. Indeed, using [15], Lemma 3.5, the Hermite coefficients reduce to $a_{p,q}$ for each summand on the right-hand side in (A7). Hence, we can state:

$$n^{-2d^*} \sum_{j=1}^{n} \sum_{p,q=1}^{d} a_{p,q}\left(U_j^{*\,(p)} U_j^{*\,(q)} - \mathbb{E}\left(U_j^{*\,(p)} U_j^{*\,(q)}\right)\right) \tag{A8}$$

$$\xrightarrow{\mathcal{D}} \sum_{p,q=1}^{d} a_{p,q} Z_{2,d^*+1/2}^{(p,q)}(1), \tag{A9}$$

where $Z_{2,d^*+1/2}^{(p,q)}(1)$ has the spectral domain representation:

$$Z_{2,d^*+1/2}^{(p,q)}(1) = K_{p,q}(d^*) \int_{\mathbb{R}^2}'' \frac{\exp(i(\lambda_1+\lambda_2))-1}{i(\lambda_1+\lambda_2)} |\lambda_1 \lambda_2|^{-d^*} \tilde{B}^{(p)}(d\lambda_1) \tilde{B}^{(q)}(d\lambda_2) \tag{A10}$$

where:

$$K_{p,q}^2(d^*) = \begin{cases} \frac{1}{2C_2(2\Gamma(1-2d^*)\sin(\pi d^*))^2}, & p = q \\ \frac{1}{C_2(2\Gamma(1-2d^*)\sin(\pi d^*))^2}, & p \neq q. \end{cases}$$

and $\tilde{B}(d\lambda) = \left(\tilde{B}^{(1)}(d\lambda), \ldots, \tilde{B}^{(d)}(d\lambda)\right)$ is an appropriate multivariate Hermitian–Gaussian random measure. Thus, we proved convergence in the distribution of the sample-cross correlation matrix:

$$n^{1-2d^*}\left(\hat{\Gamma}_{U^*,n}(0) - \Gamma_{U^*}(0)\right) \xrightarrow{D} \left(Z_{2,d^*+1/2}^{(p,q)}(1)\right)_{p,q=1,\ldots,d}.$$

We take a closer look at the covariance matrix of $\text{vec}(\hat{\Gamma}_{U^*,n}(0) - \Gamma_{U^*}(0))$. Following [28], Lemma 5.7, we observe:

$$n^{1-2d^*}(4d^*(4d^*-1))^{1/2}\text{Cov}\left(\text{vec}(\hat{\Gamma}_{U^*,n}(0) - \Gamma_{U^*}(0)), \text{vec}(\hat{\Gamma}_{U^*,n}(0) - \Gamma_{U^*}(0))\right)$$
$$= (I_{d^2} + K_{d^2})(L_{U^*} \otimes L_{U^*}),$$

with L_{U^*} as defined in (A3) and $\otimes$ denotes the Kronecker product. Furthermore, K_d denotes the commutation matrix that that transforms $\text{vec}(A)$ into $\text{vec}(A^t)$ for $A \in \mathbb{R}^{d \times d}$. Further details can be found in [32].

Hence, the covariance matrix of the vector of the sample cross-covariances is fully specified by the knowledge of L_{U^*} as it arises in the context of long-range dependence in (A3).

We obtain a relation between L and L_{U^*}, since:

$$\Gamma_Y(\cdot) = \tilde{A}\Gamma_U(\cdot)\tilde{A}^t.$$

Both:

$$\Gamma_Y(k) \simeq Lk^{2d^*-1} \ (k \to \infty)$$

and:

$$\Gamma_{U^*}(k) \simeq L_{U^*}k^{2d^*-1} \ (k \to \infty)$$

hold and we obtain:

$$L = \tilde{A}L_{U^*}\tilde{A}^t.$$

We study the covariance matrix of: $\text{vec}(\hat{\Gamma}_{Y,n}(0) - \Gamma_Y(0))$:

$$n^{1-2d^*}(4d^*(4d^*-1))^{1/2}\text{Cov}\left(\text{vec}(\hat{\Gamma}_{Y,n}(0) - \Gamma_Y(0)), \text{vec}(\hat{\Gamma}_{Y,n}(0) - \Gamma_Y(0))^t\right)$$
$$\to (I_{d^2} + K_{d^2})(L \otimes L) \quad (A11)$$
$$= (I_{d^2} + K_{d^2})(\tilde{A}L_{U^*}\tilde{A}^t) \otimes (\tilde{A}L_{U^*}\tilde{A}^t)$$
$$= (I_{d^2} + K_{d^2})(\tilde{A} \otimes \tilde{A}) \cdot (L_{U^*} \otimes L_{U^*}) \cdot (\tilde{A}^t \otimes \tilde{A}^t).$$

Let B_{U^*} be an upper triangular matrix, such that:

$$B_{U^*}B_{U^*}^t := L_{U^*}.$$

We know that such a matrix exists because L_{U^*} is positive definite. Analogously, we define B_Y:

$$B_Y := \tilde{A}B_{U^*}.$$

Then, it holds that:
$$B_Y B_Y^t = L.$$

We arrive at:
$$n^{1-2d^*}(C_2)^{-1/2}(B_Y \otimes B_Y)^{-1}\text{vec}(\hat{\Gamma}_{Y,n}(0) - \Gamma_Y(0))$$
$$\stackrel{\mathcal{D}}{=} n^{1-2d^*}(C_2)^{-1/2}(B_{U^*} \otimes B_{U^*})^{-1}(A \otimes A)^{-1}\text{vec}(\tilde{A}(\hat{\Gamma}_{U^*,n}(0) - \Gamma_{U^*}(0))\tilde{A}^t)$$
$$= n^{1-2d^*}(C_2)^{-1/2}(B_{U^*} \otimes B_{U^*})^{-1}\text{vec}(\hat{\Gamma}_{U^*,n}(0) - \Gamma_{U^*}(0))$$
$$\stackrel{D}{\to} \text{vec}\left(Z^{(p,q)}_{2,d^*+1/2}(1)\right)_{p,q=1,\ldots,d'}$$

where $Z^{(p,q)}_{2,d^*+1/2}(1)$ has the spectral domain representation:
$$Z^{(p,q)}_{2,d^*+1/2}(1) = K_{p,q}(d^*) \int_{\mathbb{R}^2}'' \frac{\exp(i(\lambda_1 + \lambda_2)) - 1}{i(\lambda_1 + \lambda_2)} |\lambda_1 \lambda_2|^{-d^*} \tilde{B}^{(p)}(d\lambda_1) \tilde{B}^{(q)}(d\lambda_2)$$

where:
$$K^2_{p,q}(d^*) = \begin{cases} \frac{1}{2C_2(2\Gamma(1-2d^*)\sin(\pi d^*))^2}, & p = q \\ \frac{1}{C_2(2\Gamma(1-2d^*)\sin(\pi d^*))^2}, & p \neq q. \end{cases}$$

and $\tilde{B}(d\lambda) = \left(\tilde{B}^{(1)}(d\lambda), \ldots, \tilde{B}^{(d)}(d\lambda)\right)$ is a multivariate Hermitian–Gaussian random measure as defined in (12). Note that the standardization on the left-hand side is appropriate since the covariance matrix of $\text{vec}(Z_{2,d^*+1/2}(1))$ is given by

$$\mathbb{E}\left(K^2(d^*) \int_{\mathbb{R}^2}'' \int_{\mathbb{R}^2}'' E_{\lambda_1,\lambda_2} \overline{E_{\lambda_3,\lambda_4}} \text{vec}\left(\tilde{B}(d\lambda_1)(\tilde{B}(d\lambda_2))^t\right)\right.$$
$$\left.\left(\text{vec}\left(\overline{\tilde{B}(d\lambda_3)(\tilde{B}(d\lambda_4))^t}\right)\right)^t\right). \quad (A12)$$

by denoting:
$$E_{\lambda_1,\lambda_2} := \frac{\exp(i(\lambda_1 + \lambda_2)) - 1}{i(\lambda_1 + \lambda_2)} |\lambda_1 \lambda_2|^{-d^*}.$$

We observe:
$$\mathbb{E}\left(\text{vec}\left(\tilde{B}(d\lambda_1)\tilde{B}(d\lambda_2)^t\right)\left(\text{vec}\left(\overline{\tilde{B}(d\lambda_3)(\tilde{B}(d\lambda_4))^t}\right)\right)^t\right)$$
$$= \begin{cases} I_{d^2} d\lambda_1 d\lambda_2, & |\lambda_1| = |\lambda_3| \neq |\lambda_2| = |\lambda_4|, \\ K_{d^2} d\lambda_1 d\lambda_2, & |\lambda_1| = |\lambda_4| \neq |\lambda_2| = |\lambda_3|, \end{cases} \quad (A13)$$

following [28], (27). Neither the case $|\lambda_1| = |\lambda_2|$ nor $|\lambda_3| = |\lambda_4|$ has to be incorporated as the diagonals are excluded in the integration in (A12). □

Corollary A1. *Under the assumptions of Lemma A1, there is a different representation of the limit random vector. For $h \in \mathbb{N}$, we obtain:*

$$\left(n^{1-2d^*}(C_2)^{-1/2}\text{vec}(\hat{\Gamma}_n(l) - \Gamma(l)), l = 0, \ldots, h-1\right) \stackrel{D}{\to} (\text{vec}(Z_{2,d^*+1/2}(1)) l = 0, \ldots, h-1),$$

where $\text{vec}(Z_{2,d^*+1/2}(1))$ has the spectral domain representation:

$$\text{vec}(Z_{2,d^*+1/2}(1)) = D_{K(d^*)} \int_{\mathbb{R}^2}^{\prime\prime} \frac{\exp(i(\lambda_1+\lambda_2))-1}{i(\lambda_1+\lambda_2)} |\lambda_1\lambda_2|^{-d^*} \text{vec}\left(\tilde{B}_L(d\lambda_1)\tilde{B}_L(d\lambda_2)^t\right).$$

The matrix $D_{K(d^*)}$ is a diagonal matrix:

$$D_{K(d^*)} = \text{diag}(\text{vec}(K(d^*))),$$

and $K(d^*) = \left(K_{p,q}(d^*)\right)_{p,q=1,\ldots,d}$ is such that:

$$K^2(d^*)_{p,q} = \begin{cases} \frac{1}{2C_2(2\Gamma(1-2d^*)\sin(\pi d^*))^2}, & p = q \\ \frac{1}{C_2(2\Gamma(1-2d^*)\sin(\pi d^*))^2}, & p \neq q. \end{cases}$$

Furthermore, $\tilde{B}_L(d\lambda)$ is a multivariate Hermitian–Gaussian random measure that fulfills:

$$\mathbb{E}\left(\tilde{B}_L(d\lambda)\tilde{B}_L(d\lambda)^*\right) = L\, d\lambda.$$

Proof. The proof is an immediate consequence of Lemma A1 using $\tilde{B}_L(d\lambda) = B_Y \tilde{B}(d\lambda)$ with $B_Y B_Y^t = L$ and $\tilde{B}(d\lambda)$ as defined in (12). □

Appendix A.2. Proof of Theorem 2.1

Proof. Without loss of generality, we assume $\mathbb{E}\left(f\left(Y_{j,h}\right)\right) = 0$. Following the argumentation in [20], Theorem 5.9, we first remark that $Y_{j,h} \stackrel{\mathcal{D}}{=} AU_{j,h}$ with $U_{j,h}$ and A as described in (A4) and (A5). We now want to study the asymptotic behavior of the partial sum $\sum_{j=1}^{n} f^*(U_j)$ where $f^*\left(U_{j,h}\right) := f\left(AU_{j,h}\right) \stackrel{\mathcal{D}}{=} f\left(Y_{j,h}\right)$. Since $m(f^*, I_{dh}) = m(f \circ A, I_{dh}) = m(f, \Sigma_{d,h}) = 2$, as can be seen in [15], Lemma 3.7, hence, we know by [14], Theorem 6, that these partial sums are dominated by the second order terms in the Hermite expansion of f^*:

$$\sum_{j=1}^{n} f^*\left(U_{j,h}\right) \sum_{j=1}^{n} \sum_{l_1+\ldots+l_{dh}=2} \mathbb{E}\left(f^*\left(U_{j,h}\right) H_{l_1,\ldots,l_{dh}}\left(U_{j,h}\right)\right) H_{l_1,\ldots,l_{dh}}\left(U_{j,h}\right) + o_{\mathbb{P}}\left(n^{2d^*}\right).$$

This follows from the multivariate extension of the Reduction Theorem as proven in [14]. We obtain:

$$\sum_{l_1+\ldots+l_{dh}=2} \mathbb{E}\left(f^*\left(U_{j,h}\right) H_{l_1,\ldots,l_{dh}}\left(U_{j,h}\right)\right) H_{l_1,\ldots,l_{dh}}\left(U_{j,h}\right)$$

$$= \sum_{i=1}^{dh} \mathbb{E}\left(f^*\left(U_{j,h}\right)\left(\left(U_{j,h}^{(i)}\right)^2 - 1\right)\right)\left(\left(U_{j,h}^{(i)}\right)^2 - 1\right) + \sum_{1\leq i,k\leq dh, i\neq k} \mathbb{E}\left(f^*\left(U_{j,h}\right) U_{j,h}^{(i)} U_{j,h}^{(k)}\right) U_{j,h}^{(i)} U_{j,h}^{(k)}$$

$$= \sum_{i=1}^{dh} \mathbb{E}\left(f^*\left(U_{j,h}\right)\left(U_{j,h}^{(i)}\right)^2\right)\left(\left(U_{j,h}^{(i)}\right)^2 - 1\right) + \sum_{1\leq i,k\leq dh, i\neq k} \mathbb{E}\left(f^*\left(U_{j,h}\right) U_{j,h}^{(i)} U_{j,h}^{(k)}\right) U_{j,h}^{(i)} U_{j,h}^{(k)},$$

since $\mathbb{E}\left(f^*\left(U_{j,h}\right)\right) = \mathbb{E}\left(f\left(Y_{j,h}\right)\right) = 0$. This results in:

$$\sum_{i=1}^{dh} \mathbb{E}\left(f^*\left(U_{j,h}\right)\left(U_{j,h}^{(i)}\right)^2\right)\left(\left(U_{j,h}^{(i)}\right)^2 - 1\right) + \sum_{1\leq i,k\leq dh, i\neq k} \mathbb{E}\left(f^*\left(U_{j,h}\right) U_{j,h}^{(i)} U_{j,h}^{(k)}\right) U_{j,h}^{(i)} U_{j,h}^{(k)}$$

$$= \sum_{1\leq i,k\leq dh} \mathbb{E}\left(f^*\left(U_{j,h}\right) U_{j,h}^{(i)} U_{j,h}^{(k)}\right) U_{j,h}^{(i)} U_{j,h}^{(k)} - \sum_{i=1}^{dh} \mathbb{E}\left(f^*\left(U_{j,h}\right)\left(U_{j,h}^{(i)}\right)^2\right). \tag{A14}$$

Note that:

$$\sum_{1\leq i,k\leq dh} \mathbb{E}\left(f^*\left(U_{j,h}\right)U_{j,h}^{(i)}U_{j,h}^{(k)}\right)\mathbb{E}\left(U_{j,h}^{(i)}U_{j,h}^{(k)}\right) = \sum_{i=1}^{dh}\mathbb{E}\left(f^*\left(U_{j,h}\right)\left(U_{j,h}^{(i)}\right)^2\right) \quad (A15)$$

since the entries of $U_{j,h}$ are independent for fixed j and identically $\mathcal{N}(0,1)$ distributed. Thus, the subtrahend in (A14) equals the expected value of the minuend.

Define $B := (b_{i,k})_{1\leq i,k\leq dh} \in \mathbb{R}^{(dh)\times(dh)}$ with:

$$b_{i,k} := \mathbb{E}\left(f^*\left(U_{j,h}\right)U_{j,h}^{(i)}U_{j,h}^{(k)}\right) = \mathbb{E}\left(f^*(U_1)U_1^{(i)}U_1^{(k)}\right)$$

since we are considering a stationary process. We obtain:

$$B = \mathbb{E}\left(U_{j,h}f^*\left(U_{j,h}\right)U_{j,h}^t\right) = \mathbb{E}\left(A^{-1}Y_{j,h}f\left(Y_{j,h}\right)Y_{j,h}^t\left(A^{-1}\right)^t\right).$$

Hence, we can state the following:

$$\sum_{1\leq i,k\leq dh} \mathbb{E}\left(f^*\left(U_{j,h}\right)U_{j,h}^{(i)}U_{j,h}^{(k)}\right)U_{j,h}^{(i)}U_{j,h}^{(k)}$$

$$= U_{j,h}^t B U_{j,h}$$

$$\stackrel{\mathcal{D}}{=} Y_{j,h}^t\left(A^{-1}\right)^t B A^{-1} Y_{j,h}$$

$$= Y_{j,h}^t\left(A^{-1}\right)^t A^{-1}\mathbb{E}\left(Y_{j,h}f\left(Y_{j,h}\right)Y_{j,h}^t\right)\left(A^{-1}\right)^t A^{-1} Y_{j,h}$$

$$= Y_{j,h}^t \Sigma_{d,h}^{-1}\mathbb{E}\left(Y_{j,h}f\left(Y_{j,h}\right)Y_{j,h}^t\right)\Sigma_{d,h}^{-1} Y_{j,h}$$

$$= Y_{j,h}^t \mathbb{A} Y_{j,h}$$

$$= \sum_{1\leq i,k\leq dh} Y_j^{(i)}Y_j^{(k)}\alpha_{ik}, \quad (A16)$$

where we define $\mathbb{A} := (\alpha_{ik})_{1\leq ik\leq dh} := \Sigma_{d,h}^{-1}C\Sigma_{d,h}^{-1}$, with $C := \mathbb{E}\left(Y_{j,h}f\left(Y_{j,h}\right)Y_{j,h}^t\right)$ as the matrix of second order Hermite coefficients, in contrast to B now with respect to the original considered process $\left(Y_{j,h}\right)_{j\in\mathbb{Z}}$.

Remembering the special structure of $Y_{j,h} = \left(Y_j^{(1)},\ldots,Y_{j+h-1}^{(1)},\ldots,Y_j^{(d)},Y_{j+h-1}^{(d)}\right)^t$, namely that $Y_{j,h}^{(k)} = Y_{j+(k \bmod h)-1}^{(\lfloor\frac{k-1}{h}\rfloor+1)}$, $k = 1,\ldots,dh$, we can see that:

$$\sum_{j=1}^n\sum_{1\leq ik\leq dh} Y_{j,h}^{(i)}Y_{j,h}^{(k)}\alpha_{ik} = \sum_{j=1}^n\sum_{1\leq ik\leq dh} Y_{j+(i \bmod h)-1}^{(\lfloor\frac{i-1}{h}\rfloor+1)} Y_{j+(k \bmod h)-1}^{(\lfloor\frac{k-1}{h}\rfloor+1)}\alpha_{ik}$$

$$= \sum_{j=1}^n\sum_{p,q=1}^d\sum_{i,k=1}^h Y_{j+i-1}^{(p)}Y_{j+k-1}^{(q)}\alpha_{ik}^{(p,q)}, \quad (A17)$$

where we divide:

$$\mathbb{A} = \begin{pmatrix} \mathbb{A}^{(1,1)} & \mathbb{A}^{(1,2)} & \cdots & \mathbb{A}^{(1,d)} \\ \mathbb{A}^{(2,1)} & \mathbb{A}^{(2,2)} & \cdots & \mathbb{A}^{(2,d)} \\ \vdots & \vdots & \vdots & \vdots \\ \mathbb{A}^{(d,1)} & \mathbb{A}^{(d,2)} & \cdots & \mathbb{A}^{(d,d)} \end{pmatrix},$$

with $\mathbb{A}^{(p,q)} = \left(\alpha_{i,k}^{(p,q)}\right)_{1 \leq i,k \leq h} \in \mathbb{R}^{h \times h}$ such that $\alpha_{i,k}^{(p,q)} = \alpha_{i+(p-1)h,k+(q-1)h}$ for each $p,q = 1,\ldots,d$ and $i,k = 1,\ldots,h$.

We can now split the considered sum in (A17) in a way such that we are afterwards able to express it in terms of sample cross-covariances. In order to do so, we define the sample cross-covariance at lag l by

$$\hat{r}_n^{(p,q)}(l) := \frac{1}{n} \sum_{j=1}^{n-l} X_j^{(p)} X_{j+l}^{(q)}$$

for $p,q = 1,\ldots,d$.

Note that in the case $h = 1$, it follows directly that:

$$\sum_{j=1}^{n} \sum_{p,q=1}^{d} \sum_{i,k=1}^{h} Y_{j+i-1}^{(p)} Y_{j+k-1}^{(q)} \alpha_{ik}^{(p,q)} = \sum_{p,q=1}^{d} \alpha_{1,1}^{(p,q)} \sum_{j=1}^{n} Y_j^{(p)} Y_j^{(q)} = n \sum_{p,q=1}^{d} \hat{r}_n^{(p,q)}(0).$$

The case $h = 2$ has to be regarded separately, too, and we obtain:

$$\sum_{j=1}^{n} \sum_{p,q=1}^{d} \sum_{i,k=1}^{2} Y_{j+i-1}^{(p)} Y_{j+k-1}^{(q)} \alpha_{ik}^{(p,q)}$$

$$= \sum_{p,q=1}^{d} \left(\alpha_{1,1}^{(p,q)} \sum_{j=1}^{n} Y_j^{(p)} Y_j^{(q)} + \alpha_{1,2}^{(p,q)} \sum_{j=1}^{n} Y_j^{(p)} Y_{j+1}^{(q)} + \alpha_{2,1}^{(p,q)} \sum_{j=1}^{n} Y_{j+1}^{(p)} Y_j^{(q)} + \alpha_{2,2}^{(p,q)} \sum_{j=1}^{n} Y_{j+1}^{(p)} Y_{j+1}^{(q)} \right)$$

$$= \sum_{p,q=1}^{d} \left(\alpha_{1,1}^{(p,q)} n \hat{r}_n^{(p,q)}(0) + \alpha_{1,2}^{(p,q)} \left(n \hat{r}_n^{(p,q)}(1) + \underbrace{Y_n^{(p)} Y_{n+1}^{(q)}}_{\star} \right) + \alpha_{2,1}^{(p,q)} \left(n \hat{r}_n^{(q,p)}(1) + \underbrace{Y_{n+1}^{(p)} Y_n^{(q)}}_{\star} \right) \right.$$

$$\left. + \alpha_{2,2}^{(p,q)} \left(n \hat{r}_n^{(p,q)}(0) + \underbrace{Y_{n+1}^{(p)} Y_{n+1}^{(q)}}_{\star} - \underbrace{Y_1^{(p)} Y_1^{(q)}}_{\star} \right) \right),$$

Note that for each of the terms labeled by $\star$, the following holds for $d^* \in \left(\frac{1}{4}, \frac{1}{2}\right)$:

$$n^{-2d^*} \star \xrightarrow{\mathbb{P}} 0, \ (n \to \infty).$$

We use this property later on when dealing with the asymptotics of the term in (A17). Finally, we consider the term in (A17) for $h \geq 3$ and arrive at:

$$\sum_{j=1}^{n} \sum_{p,q=1}^{d} \sum_{i,k=1}^{h} Y_{j+i-1}^{(p)} Y_{j+k-1}^{(q)} \alpha_{ik}^{(p,q)}$$

$$= \sum_{p,q=1}^{d} \sum_{i,k=1}^{h} \alpha_{ik}^{(p,q)} \sum_{j=i}^{n+i-1} Y_j^{(p)} Y_{j+k-i}^{(q)}$$

$$= \sum_{p,q=1}^{d} \sum_{l=0}^{h-1} \sum_{i=1}^{h-l} \alpha_{i,i+l}^{(p,q)} \sum_{j=i}^{n+i-1} Y_j^{(p)} Y_{j+l}^{(q)} + \sum_{p,q=1}^{d} \sum_{l=-(h-1)}^{-1} \sum_{i=1-l}^{h} \alpha_{i,i+l}^{(p,q)} \sum_{j=i}^{n+i-1} Y_j^{(p)} Y_{j+l}^{(q)}$$

$$= \sum_{p,q=1}^{d} \sum_{l=0}^{h-1} \sum_{i=1}^{h-l} \alpha_{i,i+l}^{(p,q)} \sum_{j=i}^{n+i-1} Y_j^{(p)} Y_{j+l}^{(q)}$$

$$+ \sum_{p,q=1}^{d} \sum_{l=1}^{h-1} \sum_{i=1}^{h-l} \alpha_{i+l,i}^{(p,q)} \sum_{j=i}^{n+i-1} Y_{j+l}^{(p)} Y_j^{(q)} \tag{A18}$$

$$
\begin{aligned}
&= \sum_{p,q=1}^{d} \sum_{i=1}^{h} \alpha_{i,i}^{(p,q)} \sum_{j=i}^{n+i-1} Y_j^{(p)} Y_j^{(q)} \\
&\quad + \sum_{p,q=1}^{d} \sum_{l=1}^{h-1} \sum_{i=1}^{h-l} \left(\alpha_{i,i+l}^{(p,q)} \sum_{j=i}^{n+i-1} Y_j^{(p)} Y_{j+l}^{(q)} + \alpha_{i+l,i}^{(p,q)} \sum_{j=i}^{n+i-1} Y_{j+l}^{(p)} Y_j^{(q)} \right) \quad \text{(A19)}
\end{aligned}
$$

$$
\begin{aligned}
&= \sum_{p,q=1}^{d} \left(\alpha_{1,1}^{(p,q)} \sum_{j=1}^{n} Y_j^{(p)} Y_j^{(q)} + \sum_{i=2}^{h} \alpha_{i,i}^{(p,q)} \sum_{j=i}^{n+i-1} Y_j^{(p)} Y_j^{(q)} \right) \\
&\quad + \sum_{p,q=1}^{d} \sum_{l=1}^{h-2} \left(\left(\alpha_{1,1+l}^{(p,q)} \sum_{j=1}^{n} Y_j^{(p)} Y_{j+l}^{(q)} + \alpha_{1+l,1}^{(p,q)} \sum_{j=1}^{n} Y_{j+l}^{(p)} Y_j^{(q)} \right) \right. \\
&\quad \left. + \sum_{i=2}^{h-l} \left(\alpha_{i,i+l}^{(p,q)} \sum_{j=i}^{n+i-1} Y_j^{(p)} Y_{j+l}^{(q)} + \alpha_{i+l,i}^{(p,q)} \sum_{j=i}^{n+i-1} Y_{j+l}^{(p)} Y_j^{(q)} \right) \right) \\
&\quad + \sum_{p,q=1}^{d} \left(\alpha_{1,h}^{(p,q)} \sum_{j=1}^{n} Y_j^{(p)} Y_{j+h-1}^{(q)} + \alpha_{h,1}^{(p,q)} Y_{j+h-1}^{(p)} Y_j^{(q)} \right) \quad \text{(A20)}
\end{aligned}
$$

$$
\begin{aligned}
&= \sum_{p,q=1}^{d} \left(\alpha_{1,1}^{(p,q)} n\hat{r}_n^{(p,q)}(0) + \sum_{i=2}^{h} \alpha_{i,i}^{(p,q)} \left(\underbrace{\sum_{j=n+1}^{n+i-1} Y_j^{(p)} Y_j^{(q)}}_{\star} + n\hat{r}_n^{(p,q)}(0) - \underbrace{\sum_{j=1}^{i-1} Y_j^{(p)} Y_j^{(q)}}_{\star} \right) \right) \\
&\quad + \sum_{p,q=1}^{d} \sum_{l=1}^{h-2} \left(\left(\alpha_{1,1+l}^{(p,q)} n\hat{r}_n^{(p,q)}(l) + \alpha_{1+l,1}^{(p,q)} n\hat{r}_n^{(q,p)}(l) \right) \right. \\
&\quad + \sum_{i=2}^{h-l} \left(\alpha_{i,i+l}^{(p,q)} \left(\underbrace{\sum_{j=n-l+1}^{n+i-1} Y_j^{(p)} Y_{j+l}^{(q)}}_{\star} + n\hat{r}_n^{(p,q)}(l) - \underbrace{\sum_{j=1}^{i-1} Y_j^{(p)} Y_{j+l}^{(q)}}_{\star} \right) \right. \\
&\quad \left. \left. + \alpha_{i+l,i}^{(p,q)} \left(\underbrace{\sum_{j=n-l+1}^{n+i-1} Y_{j+l}^{(p)} Y_j^{(q)}}_{\star} + n\hat{r}_n^{(q,p)}(l) - \underbrace{\sum_{j=1}^{i-1} Y_{j+l}^{(p)} Y_j^{(q)}}_{\star} \right) \right) \right) \\
&\quad + \sum_{p,q=1}^{d} \left(\alpha_{1,h}^{(p,q)} \left(\underbrace{\sum_{j=n-h+2}^{n} Y_j^{(p)} Y_{j+h-1}^{(q)}}_{\star} + n\hat{r}_n^{(p,q)}(h-1) \right) \right. \quad \text{(A21)} \\
&\quad \left. + \alpha_{h,1}^{(p,q)} \left(\underbrace{\sum_{j=n-h+2}^{n} Y_{j+h-1}^{(p)} Y_j^{(q)}}_{\star} + n\hat{r}_n^{(q,p)}(h-1) \right) \right). \quad \text{(A22)}
\end{aligned}
$$

Again for each of the terms labeled by ★ it holds for $d^* \in \left(\frac{1}{4}, \frac{1}{2}\right)$:

$$ n^{-2d^*} \star \xrightarrow{\mathbb{P}} 0, \ (n \to \infty), $$

since each ★ describes a sum with a finite number (independent of n) of summands. Therefore, we continue to express the terms denoted by ★ by $o_{\mathbb{P}}\left(n^{2d^*}\right)$.

With these calculations, we are able to re-express the partial sum, whose asymptotics we are interested in, in terms of the sample cross-correlations of the original long-range dependent process $(Y_j)_{j \in \mathbb{Z}}$.

Finally, the previous calculations lead to:

$$\sum_{j=1}^{n} f(Y_{j,h})$$

$$\stackrel{\mathcal{D}}{=} \sum_{j=1}^{n} f^{*}(U_{j,h})$$

$$\stackrel{(A14)}{=} \sum_{j=1}^{n} \left(\sum_{1 \leq i,k \leq dh} \mathbb{E}\left(f^{*}(U_{j,h}) U_{j,h}^{(i)} U_{j,h}^{(k)}\right) U_{j,h}^{(i)} U_{j,h}^{(k)} - \sum_{i=1}^{dh} \mathbb{E}\left(f^{*}(U_{j,h}) \left(U_{j,h}^{(i)}\right)^{2}\right) \right) + o_{\mathbb{P}}(n^{2d^{*}})$$

$$\stackrel{\mathcal{D}}{=} \sum_{j=1}^{n} \sum_{p,q=1}^{d} \sum_{i,k=1}^{h} \alpha_{ik}^{(p,q)} \left(Y_{j+i-1}^{(p)} Y_{j+k-1}^{(q)} - \mathbb{E}\left(Y_{j+i-1}^{(p)} Y_{j+k-1}^{(q)}\right) \right) + o_{\mathbb{P}}(n^{2d^{*}}), \quad (A23)$$

where (A23) follows, since (A15) yields:

$$\sum_{i=1}^{dh} \mathbb{E}\left(f^{*}(U_{j,h}) \left(U_{j,h}^{(i)}\right)^{2}\right) = \sum_{1 \leq i,k \leq dh} \mathbb{E}\left(f^{*}(U_{j,h}) U_{j,h}^{(i)} U_{j,h}^{(k)}\right) \mathbb{E}\left(U_{j,h}^{(i)} U_{j,h}^{(k)}\right)$$

$$\stackrel{(A17)}{=} \sum_{p,q=1}^{d} \sum_{i,k=1}^{h} \alpha_{ik}^{(p,q)} \mathbb{E}\left(Y_{j+i-1}^{(p)} Y_{j+k-1}^{(q)}\right).$$

Taking the parts containing the sample cross-correlations into account, we derive:

$$\sum_{j=1}^{n} \sum_{p,q=1}^{d} \sum_{i,k=1}^{h} \alpha_{ik}^{(p,q)} \left(Y_{j+i-1}^{(p)} Y_{j+k-1}^{(q)} - \mathbb{E}\left(Y_{j+i-1}^{(p)} Y_{j+k-1}^{(q)}\right) \right) + o_{\mathbb{P}}(n^{2d^{*}})$$

$$\stackrel{(A22)}{=} \sum_{p,q=1}^{d} \left(\alpha_{1,1}^{(p,q)} n \left(\hat{r}_{n}^{(p,q)}(0) - r^{(p,q)}(0) \right) + \sum_{i=2}^{h} \alpha_{i,i}^{(p,q)} n \left(\hat{r}_{n}^{(p,q)}(0) - r^{(p,q)}(0) \right) \right)$$

$$+ \sum_{p,q=1}^{d} \sum_{l=1}^{h-2} \left(\left(\alpha_{1,1+l}^{(p,q)} n \left(\hat{r}_{n}^{(p,q)}(l) - r^{(p,q)}(l) \right) + \alpha_{1+l,1}^{(p,q)} n \left(\hat{r}_{n}^{(q,p)}(l) - r^{(q,p)}(l) \right) \right) \right.$$

$$\left. + \sum_{i=2}^{h-l} \left(\alpha_{i,i+l}^{(p,q)} n \left(\hat{r}_{n}^{(p,q)}(l) - r^{(p,q)}(l) \right) + \alpha_{i+l,i}^{(p,q)} n \left(\hat{r}_{n}^{(q,p)}(l) - r^{(q,p)}(l) \right) \right) \right)$$

$$+ \sum_{p,q=1}^{d} \left(\alpha_{1,h}^{(p,q)} n \left(\hat{r}_{n}^{(p,q)}(h-1) - r^{(p,q)}(h-1) \right) + \alpha_{h,1}^{(p,q)} n \left(\hat{r}_{n}^{(q,p)}(h-1) - r^{(q,p)}(h-1) \right) \right)$$

$$(A24)$$

$$+ o_{\mathbb{P}}(n^{2d^{*}})$$

$$= n \sum_{p,q=1}^{d} \left(\sum_{l=0}^{h-1} \sum_{i=1}^{h-l} \alpha_{i,i+l}^{(p,q)} \left(\hat{r}_{n}^{(p,q)}(l) - r^{(p,q)}(l) \right) + \sum_{l=1}^{h-1} \sum_{i=1}^{h-l} \alpha_{i+l,i}^{(p,q)} \left(\hat{r}_{n}^{(q,p)}(l) - r^{(q,p)}(l) \right) \right)$$

$$+ o_{\mathbb{P}}(n^{2d^{*}}). \quad (A25)$$

We take a closer look at the impact of each long-range dependence parameter d_p, $p = 1, \ldots, d$ to the convergence of this sum. The setting we are considering does not allow for a normalization depending on p and q for each cross-correlation $\left(\hat{r}_{n}^{(p,q)}(l) - r^{(p,q)}(l) \right)$, $l = 0, \ldots, h-1$, but we need to find a normalization for all $p,q = 1, \ldots, d$. Hence, we need to remember the set $P^* := \{ p \in \{1, \ldots, d\} : d_p \geq d_q \forall q \in \{1, \ldots, d\} \}$ and the parameter

$d^* = \max_{p=1,\ldots,d} d_p$, such that for each $p \in P^*$, we have $d_p = d^*$. For each $p,q \in \{1,\ldots,d\}$ with $(p,q) \notin (P^* \times P^*)$ and $l = 0, \ldots, h-1$, we conclude that:

$$\mathbb{E}\left(\left(n^{1-2d^*}\left(\hat{r}_n^{(p,q)}(l) - r^{(p,q)}(l)\right)\right)^2\right) = n^{2(d_p+d_q-2d^*)}\mathbb{E}\left(\left(n^{1-d_p-d_q}\left(\hat{r}_n^{(p,q)}(l) - r^{(p,q)}(l)\right)\right)^2\right)$$
$$= n^{2d_p+2d_q-4d^*}C_2\left(L_{p,p}L_{q,q} + L_{p,q}L_{q,p}\right)$$
$$\xrightarrow{(n\to\infty)} 0, \qquad (A26)$$

since $d_p + d_q - 2d^* < 0$.

This implies that:

$$n^{1-2d^*}\left(\hat{r}_n^{(p,q)}(0) - r^{(p,q)}(0)\right) \xrightarrow{\mathbb{P}} 0$$

Hence, using Slutsky's theorem, the crucial parameters that determine the normalization and therefore, the limit distribution of (A27) are given in P^*. We have an equal long-range dependence parameter d^* to regard for all $p \in P^*$. Applying Lemma A1, we obtain the following, by using the symmetry in $l = 0$ of the cross-correlation function $r^{(p,q)}(0) = r^{(q,p)}(0)$ for $p,q \in P^*$:

$$\sum_{p,q=1}^{d}\left(\sum_{l=0}^{h-1}\sum_{i=1}^{h-l}\alpha_{i,i+l}^{(p,q)}\left(\hat{r}_n^{(p,q)}(l) - r^{(p,q)}(l)\right) + \sum_{l=1}^{h-1}\sum_{i=1}^{h-l}\alpha_{i+l,i}^{(p,q)}\left(\hat{r}_n^{(q,p)}(l) - r^{(q,p)}(l)\right)\right)$$
$$= \sum_{p,q \in P^*}\left(\sum_{l=0}^{h-1}\sum_{i=1}^{h-l}\alpha_{i,i+l}^{(p,q)}\left(\hat{r}_n^{(p,q)}(0) - r^{(p,q)}(0)\right) + \sum_{l=1}^{h-1}\sum_{i=1}^{h-l}\alpha_{i+l,i}^{(p,q)}\left(\hat{r}_n^{(q,p)}(0) - r^{(q,p)}(0)\right)\right)$$
$$+ o_{\mathbb{P}}(n^{2d^*-1})$$
$$= \sum_{p,q \in P^*}\left(\hat{r}_n^{(p,q)}(0) - r^{(p,q)}(0)\right)\left(\sum_{l=0}^{h-1}\sum_{i=1}^{h-l}\alpha_{i,i+l}^{(p,q)} + \sum_{l=1}^{h-1}\sum_{i=1}^{h-l}\alpha_{i+l,i}^{(p,q)}\right) + o_{\mathbb{P}}(n^{2d^*})$$
$$= \sum_{p,q \in P^*}\left(\hat{r}_n^{(p,q)}(0) - r^{(p,q)}(0)\right)\left(\sum_{i,k=1}^{h}\alpha_{i,k}^{(p,q)}\right) + o_{\mathbb{P}}(n^{2d^*})$$
$$= \sum_{p,q \in P^*}\tilde{\alpha}^{(p,q)}\left(\hat{r}_n^{(p,q)}(0) - r^{(p,q)}(0)\right) + o_{\mathbb{P}}(n^{2d^*}), \qquad (A27)$$

by defining $\tilde{\alpha}^{(p,q)} := \sum_{i,k=1}^{h}\alpha_{i,k}^{(p,q)}$. Applying the continuous mapping theorem given in [33], Theorem 2.3, to the result in Corollary A1, we arrive at:

$$n^{-2d^*}(C_2)^{-1/2}\sum_{j=1}^{n}f(Y_{j,h}) = n^{-2d^*}\left(n\sum_{p,q=1}^{d}\tilde{\alpha}^{(p,q)}\left(\hat{r}_n^{(p,q)}(0) - r^{(p,q)}(0)\right) + o_{\mathbb{P}}(n^{2d^*})\right)$$
$$= n^{1-2d^*}(C_2)^{-1/2}\sum_{p,q=1}^{d}\tilde{\alpha}^{(p,q)}\left(\hat{r}_n^{(p,q)}(0) - r^{(p,q)}(0)\right) + o_{\mathbb{P}}(1)$$
$$\xrightarrow{\mathcal{D}} \sum_{p,q \in P^*}\tilde{\alpha}^{(p,q)}Z_{2,d^*+1/2}^{(p,q)}(1),$$

where:

$$Z_{2,d^*+1/2}^{(p,q)}(1) = K_{p,q}(d^*)\int_{\mathbb{R}^2}\frac{\exp(i(\lambda_1+\lambda_2))-1}{i(\lambda_1+\lambda_2)}|\lambda_1\lambda_2|^{-d^*}\tilde{B}_L^{(p)}(d\lambda_1)\tilde{B}_L^{(q)}(d\lambda_2).$$

The matrix $K(d^*)$ is given in Corollary A1. Moreover, $\tilde{B}_L(d\lambda)$ is a multivariate Hermitian–Gaussian random measure with $\mathbb{E}(B_L(d\lambda)B_L(d\lambda)^*) = L\,d\lambda$ and L as defined in (2). □

Appendix A.3. Proof of Corollary 2.2

Proof. We assumed $d^* \in \left(\frac{1}{4}, \frac{1}{2}\right)$, because otherwise we leave the long-range dependent setting, since we are studying functionals with Hermite rank 2 and the transformed process would no longer be long-range dependent and limit theorems for functionals of short-range dependent processes would hold, as can be seen in Theorem 4 in [14]. This choice of d^* assures that the multivariate generalization of the Reduction Theorem as it is used in the proof of Theorem 2.1 still holds for these softened assumptions, as explained in (8).

We turn to the asymptotics of $g^{(p,q)}(Y_j)$. We obtain for all $p, q \in \{1, \ldots, d\} \setminus P^*$, i.e., excluding $d_p = d_q = d^*$ and for all $l = 0, \ldots, h-1$ as in (A26), that:

$$\mathbb{E}\left(\left(n^{1-2d^*}\left(\hat{r}_n^{(p,q)}(l) - r^{(p,q)}(l)\right)\right)^2\right) = n^{2(d_p+d_q-2d^*)}\mathbb{E}\left(\left(n^{1-d_p-d_q}\left(\hat{r}_n^{(p,q)}(l) - r^{(p,q)}(l)\right)\right)^2\right)$$

$$= n^{2d_p+2d_q-4d^*} C_2\left(L_{p,p}L_{q,q} + L_{p,q}L_{q,p}\right)$$

$$\xrightarrow{(n\to\infty)} 0, \qquad (A28)$$

since $d_p + d_q - 2d^* < 0$.

This implies that:

$$n^{1-2d^*}\left(\hat{r}_n^{(p,q)}(0) - r^{(p,q)}(0)\right) \xrightarrow{\mathbb{P}} 0.$$

Applying Slutsky's theorem, we observe that only $p, q \in P^*$ have an impact on the convergence behavior as it is given in (A27) and hence, the result in Theorem 2.1 holds. □

Appendix A.4. Proof of Theorem 2.3

Proof. We follow the proof of Theorem 2.1 until (A27), in order to obtain a limit distribution that can be expressed by the sum of two standard Rosenblatt random variables:

$$\sum_{p,q=1}^{2} \tilde{\alpha}^{(p,q)}\left(\hat{r}_n^{(p,q)}(0) - r^{(p,q)}(0)\right)$$

$$= \frac{1}{n}\sum_{j=1}^{n}\sum_{p,q=1}^{2} \tilde{\alpha}^{(p,q)}\left(Y_j^{(p)}Y_j^{(q)} - r^{(p,q)}(0)\right)$$

$$= \frac{1}{n}\sum_{j=1}^{n}\left(Y_j^{(1)}, Y_j^{(2)}\right)\begin{pmatrix} \tilde{\alpha}^{(1,1)} & \tilde{\alpha}^{(1,2)} \\ \tilde{\alpha}^{(2,1)} & \tilde{\alpha}^{(2,2)} \end{pmatrix}\left(Y_j^{(1)}, Y_j^{(2)}\right)^t$$

$$- \mathbb{E}\left(\left(Y_j^{(1)}, Y_j^{(2)}\right)\begin{pmatrix} \tilde{\alpha}^{(1,1)} & \tilde{\alpha}^{(1,2)} \\ \tilde{\alpha}^{(2,1)} & \tilde{\alpha}^{(2,2)} \end{pmatrix}\left(Y_j^{(1)}, Y_j^{(2)}\right)^t\right). \qquad (A29)$$

We remember that $\tilde{\alpha}^{(p,q)} = \sum_{i,k=1}^{h} \alpha_{i,k}^{(p,q)} = \sum_{i,k=1}^{h} \alpha_{i+(p-1)h, k+(q-1)h}$ for $p, q = 1, 2$ and $\mathbb{A} = (\alpha_{i,k})_{1 \le i,k \le 2h} = \Sigma_{2,h}^{-1} C \Sigma_{2,h}^{-1}$. Since $\Sigma_{2,h}^{-1}$ is the inverse of the covariance matrix $\Sigma_{2,h}$ of $Y_{1,h}$ it is a symmetric matrix. The matrix of second order Hermite coefficients C has the representation $C = \mathbb{E}\left(Y_{j,h} f(Y_{j,h}) Y_{j,h}^t\right)$ and therefore, $c_{i,k} = \mathbb{E}\left(Y_{j,h}^{(i)} Y_{j,h}^{(k)} f(Y_{j,h})\right) = c_{k,i}$ for each $i, k = 1, \ldots, 2h$. Then, $\mathbb{A}$ is also a symmetric matrix, since $\mathbb{A}^t = \left(\Sigma_{2,h}^{-1} C \Sigma_{2,h}^{-1}\right)^t = \left(\Sigma_{2,h}^{-1}\right)^t C^t \left(\Sigma_{2,h}^{-1}\right)^t = \mathbb{A}$. We can now show that $\begin{pmatrix} \tilde{\alpha}^{(1,1)} & \tilde{\alpha}^{(1,2)} \\ \tilde{\alpha}^{(2,1)} & \tilde{\alpha}^{(2,2)} \end{pmatrix}$ is a symmetric matrix, i.e., $\tilde{a}^{(1,2)} = \tilde{a}^{(2,1)}$. To this end, we define $\mathbb{I}_p = (0,0,\ldots,0,1,\ldots,1,0,\ldots,0)^t \in \mathbb{R}^{2h}$ such that $\mathbb{I}_p^{(i)} = 1$ only if $i = (p-1)h+1,\ldots,ph$, $p = 1,2$. Then, we obtain:

$$\tilde{a}^{(1,2)} = \sum_{i,k=1}^{h} \alpha_{i,k}^{(1,2)} = \left(\tilde{a}^{(1,2)}\right)^t = \left(\mathbb{I}_1^t A \mathbb{I}_2\right)^t = \mathbb{I}_2^t A \mathbb{I}_1 = \tilde{a}^{(2,1)}.$$

We now apply the new assumption that $r^{(1,1)}(l) = r^{(2,2)}(l)$, for $l = 0,\ldots,h-1$ and show $\tilde{a}^{(1,1)} = \tilde{a}^{(2,2)}$ with the symmetry features of the multivariate normal distribution discussed in (2.2) and in (2.3) in [18], since $c_{i,j} = c_{2h-i+1,2h-j+1}$, $i,j = 1,\ldots,2h$. We have to study:

$$\tilde{a}^{(2,2)} = \left(\mathbb{I}_2^t A \mathbb{I}_2\right)^t = \mathbb{I}_2^t \Sigma_{2,h}^{-1} C \Sigma_{2,h}^{-1} \mathbb{I}_2.$$

Since $\Sigma_{2,h}^{-1} = (g_{i,k})_{1 \leq i,k \leq 2h}$ is a symmetric and persymmetric matrix, we have $g_{i,k} = g_{k,i}$ and $g_{i,k} = g_{2h-i+1,2h-k+1}$ for $i,k = 1,\ldots,2h$. Then, we obtain:

$$\mathbb{I}_2^t \Sigma_{2,h}^{-1} = \left(\sum_{i=h+1}^{2h} g_{i,1},\ldots,\sum_{i=h+1}^{2h} g_{i,2h}\right)$$

$$= \left(\sum_{i=1}^{h} g_{i+h,1},\ldots,\sum_{i=1}^{h} g_{i+h,2h}\right)$$

$$= \left(\sum_{i=1}^{h} g_{h-i+1,2h},\ldots,\sum_{i=1}^{h} g_{h-i+1,1}\right)$$

$$= \left(\sum_{i=1}^{h} g_{i,2h},\ldots,\sum_{i=1}^{h} g_{i,1}\right)$$

$$= \left(\sum_{i=1}^{h} g_{2h,i},\ldots,\sum_{i=1}^{h} g_{1,i}\right)$$

$$=: (\tilde{g}_{2h},\ldots,\tilde{g}_1).$$

Note that:

$$\Sigma_{2,h}^{-1} \mathbb{I}_1 = \left(\sum_{i=1}^{h} g_{1,i},\ldots,\sum_{i=1}^{h} g_{2h,i}\right)^t = (\tilde{g}_1,\ldots,\tilde{g}_{2h})^t.$$

Then, we arrive at:

$$\tilde{a}^{(2,2)} = \left(\mathbb{I}_2^t A \mathbb{I}_2\right)^t = \mathbb{I}_2^t \Sigma_{2,h}^{-1} C \Sigma_{2,h}^{-1} \mathbb{I}_2$$

$$= \sum_{i,k=1}^{2h} \tilde{g}_{2h-i+1} \tilde{g}_{2h-k+1} c_{i,k}$$

$$= \sum_{i,k=1}^{2h} \tilde{g}_{2h-i+1} \tilde{g}_{2h-k+1} c_{2h-i+1,2h-k+1}$$

$$= \sum_{i,k=1}^{2h} \tilde{g}_i \tilde{g}_k c_{i,k}$$

$$= \mathbb{I}_1^t \Sigma_{2,h}^{-1} C \Sigma_{2,h}^{-1} \mathbb{I}_1$$

$$= \tilde{a}^{(1,1)}.$$

Therefore, we have to deal with a special type of 2×2 matrix, since the original matrix in the formula (A27), namely $\begin{pmatrix} \tilde{\alpha}^{(1,1)} & \tilde{\alpha}^{(1,2)} \\ \tilde{\alpha}^{(2,1)} & \tilde{\alpha}^{(2,2)} \end{pmatrix}$, has now reduced to $\begin{pmatrix} \tilde{\alpha}^{(1,1)} & \tilde{\alpha}^{(1,2)} \\ \tilde{\alpha}^{(1,2)} & \tilde{\alpha}^{(1,1)} \end{pmatrix}$.

Finally, we know that any real-valued symmetric matrix A can be decomposed via diagonalization into an orthogonal matrix V and a diagonal matrix D, where the entries of the latter one are determined via the eigenvalues of A, for details, as can be seen in [34], p. 327.

We can explicitly give formulas for the entries of these matrices here:

$$V = \begin{pmatrix} -2^{-1/2} & 2^{-1/2} \\ 2^{-1/2} & 2^{-1/2} \end{pmatrix}, \quad D = \begin{pmatrix} \lambda_1 = \tilde{\alpha}^{(1,1)} - \tilde{\alpha}^{(1,2)} & 0 \\ 0 & \lambda_2 = \tilde{\alpha}^{(1,1)} + \tilde{\alpha}^{(1,2)} \end{pmatrix},$$

such that:

$$VDV = \begin{pmatrix} \frac{\lambda_1+\lambda_2}{2} & \frac{\lambda_2-\lambda_1}{2} \\ \frac{\lambda_2-\lambda_1}{2} & \frac{\lambda_1+\lambda_2}{2} \end{pmatrix} = \begin{pmatrix} \tilde{\alpha}^{(1,1)} & \tilde{\alpha}^{(1,2)} \\ \tilde{\alpha}^{(1,2)} & \tilde{\alpha}^{(1,1)} \end{pmatrix}.$$

Therefore, continuing with (A29), we now have the representation:

$$\frac{1}{n} \sum_{j=1}^{n} \left(Y_j^{(1)}, Y_j^{(2)}\right) \begin{pmatrix} \tilde{\alpha}^{(1,1)} & \tilde{\alpha}^{(1,2)} \\ \tilde{\alpha}^{(2,1)} & \tilde{\alpha}^{(2,2)} \end{pmatrix} \left(Y_j^{(1)}, Y_j^{(2)}\right)^t$$

$$- \mathbb{E}\left(\left(Y_j^{(1)}, Y_j^{(2)}\right) \begin{pmatrix} \tilde{\alpha}^{(1,1)} & \tilde{\alpha}^{(1,2)} \\ \tilde{\alpha}^{(2,1)} & \tilde{\alpha}^{(2,2)} \end{pmatrix} \left(Y_j^{(1)}, Y_j^{(2)}\right)^t\right)$$

$$= \frac{1}{n} \sum_{j=1}^{n} \left(Y_j^{(1)}, Y_j^{(2)}\right) VDV \left(Y_j^{(1)}, Y_j^{(2)}\right)^t - \mathbb{E}\left(\left(Y_j^{(1)}, Y_j^{(2)}\right) VDV \left(Y_j^{(1)}, Y_j^{(2)}\right)^t\right)$$

$$= \frac{1}{n} \sum_{j=1}^{n} \frac{\tilde{\alpha}^{(1,1)} - \tilde{\alpha}^{(1,2)}}{2} \left(\left(Y_j^{(2)} - Y_j^{(1)}\right)^2 - \mathbb{E}\left(Y_j^{(2)} - Y_j^{(1)}\right)^2\right)$$

$$+ \frac{1}{n} \sum_{j=1}^{n} \frac{\tilde{\alpha}^{(1,1)} + \tilde{\alpha}^{(1,2)}}{2} \left(\left(Y_j^{(1)} + Y_j^{(2)}\right)^2 - \mathbb{E}\left(Y_j^{(1)} + Y_j^{(2)}\right)^2\right)$$

$$= \frac{1}{n} \sum_{j=1}^{n} \left(\tilde{\alpha}^{(1,1)} - \tilde{\alpha}^{(1,2)}\right)\left(1 - r^{(1,2)}(0)\right) \left(\left(\frac{Y_j^{(2)} - Y_j^{(1)}}{\sqrt{2 - 2r^{(1,2)}(0)}}\right)^2 - 1\right)$$

$$+ \frac{1}{n} \sum_{j=1}^{n} \left(\tilde{\alpha}^{(1,1)} + \tilde{\alpha}^{(1,2)}\right)\left(1 + r^{(1,2)}(0)\right) \left(\left(\frac{Y_j^{(1)} + Y_j^{(2)}}{\sqrt{2 + 2r^{(1,2)}(0)}}\right)^2 - 1\right)$$

$$= \frac{1}{n} \left(\tilde{\alpha}^{(1,1)} - \tilde{\alpha}^{(1,2)}\right)\left(1 - r^{(1,2)}(0)\right) \sum_{j=1}^{n} H_2\left(Y_j^*\right)$$

$$+ \frac{1}{n} \left(\tilde{\alpha}^{(1,1)} + \tilde{\alpha}^{(1,2)}\right)\left(1 + r^{(1,2)}(0)\right) \sum_{j=1}^{n} H_2\left(Y_j^{**}\right), \quad (A30)$$

with $Y_j^* := \frac{Y_j^{(2)} - Y_j^{(1)}}{\sqrt{2 - 2r^{(1,2)}(0)}}$ and $Y_j^{**} := \frac{Y_j^{(1)} + Y_j^{(2)}}{\sqrt{2 + 2r^{(1,2)}(0)}}$.

Now, note that:

$$\mathbb{E}\left(Y_j^* Y_j^{**}\right) = \mathbb{E}\left(\frac{Y_j^{(2)} - Y_j^{(1)}}{\sqrt{2 - 2r^{(1,2)}(0)}} \frac{Y_j^{(1)} + Y_j^{(2)}}{\sqrt{2 + 2r^{(1,2)}(0)}}\right) = 0.$$

Therefore, we created a bivariate long-range dependent Gaussian process, since:

$$\begin{pmatrix} -1 & 1 \\ 1 & 1 \end{pmatrix} \left(Y_j^{(1)}, Y_j^{(2)} \right)^t = \left(Y_j^*, Y_j^{**} \right)^t \sim \mathcal{N}(0, I_2)$$

with cross-covariance function:

$$\begin{aligned} r_*^{(1,2)}(k) := \mathbb{E}\left(Y_j^* Y_{j+k}^{**} \right) &= \mathbb{E}\left(\frac{Y_j^{(2)} - Y_j^{(1)}}{\sqrt{2 - 2r^{(1,2)}(0)}} \frac{Y_{j+k}^{(1)} + Y_{j+k}^{(2)}}{\sqrt{2 + 2r^{(1,2)}(0)}} \right) \\ &= \frac{r^{(2,1)}(k) + r^{(2,2)}(k) - r^{(1,1)}(k) - r^{(1,2)}(k)}{2\sqrt{\left(1 - r^{(1,2)}(0)\right)\left(1 + r^{(1,2)}(0)\right)}} \\ &\simeq \frac{L_{2,2} + L_{2,1} - L_{1,2} - L_{1,1}}{2\sqrt{\left(1 - r^{(1,2)}(0)\right)\left(1 + r^{(1,2)}(0)\right)}} k^{2d^* - 1}. \end{aligned} \quad (A31)$$

Note that the covariance functions have the following asymptotic behavior:

$$\begin{aligned} r_*^{(1,1)}(k) := \mathbb{E}\left(Y_j^* Y_{j+k}^* \right) &= \mathbb{E}\left(\frac{Y_j^{(2)} - Y_j^{(1)}}{\sqrt{2 - 2r^{(1,2)}(0)}} \frac{Y_{j+k}^{(2)} - Y_{j+k}^{(1)}}{\sqrt{2 - 2r^{(1,2)}(0)}} \right) \\ &= \frac{r^{(2,2)}(k) - r^{(2,1)}(k) - r^{(1,2)}(k) + r^{(1,1)}(k)}{2 - 2r^{(1,2)}(0)} \\ &\simeq \underbrace{\frac{L_{2,2} - L_{2,1} - L_{1,2} + L_{1,1}}{2 - 2r^{(1,2)}(0)}}_{=: L_{1,1}^*} k^{2d^* - 1} \end{aligned}$$

and analogously:

$$r_*^{(2,2)}(k) := \mathbb{E}\left(Y_j^{**} Y_{j+k}^{**} \right) \simeq \underbrace{\frac{L_{2,2} + L_{2,1} + L_{1,2} + L_{1,1}}{2 + 2r^{(1,2)}(0)}}_{=: L_{2,2}^*} k^{2d^* - 1}.$$

We can now apply the result of [14], Theorem 6, since we created a bivariate Gaussian process with independent entries for fixed j. Note that for the function we apply here, namely $\tilde{f}\left(Y_j^*, Y_j^{**} \right) = H_2\left(Y_j^* \right) + H_2\left(Y_j^{**} \right)$ the weighting factors in [14], Theorem 6, reduce to $e_{1,1} = e_{2,2} = 1$ and $e_{1,2} = e_{2,1} = 0$. These weighting factors fit into the result in [14], (3.6) and (3.7), which even yields the joint convergence of the vector of both univariate summands, $\left(H_2\left(Y_j^* \right), H_2\left(Y_j^{**} \right) \right)$, suitably normalized to a vector of two (dependent) Rosenblatt random variables. Since the long-range dependence property in [13], Definition 2.1 is more specific than in [14], p. 2259, (3.1) (see the considerations in (8)), we are able to scale the variances of each Rosenblatt random variable to 1 and give the covariance between them, by using the normalization given in [15], Theorem 4.3. We obtain:

$$n^{-2d^*}(2C_2)^{-1/2}\left(\tilde{\alpha}^{(1,1)} - \tilde{\alpha}^{(1,2)}\right)\left(1 - r^{(1,2)}(0)\right)\sum_{j=1}^{n}H_2\left(Y_j^*\right)$$

$$+ n^{-2d^*}(2C_2)^{-1/2}\left(\tilde{\alpha}^{(1,1)} + \tilde{\alpha}^{(1,2)}\right)\left(1 + r^{(1,2)}(0)\right)\sum_{j=1}^{n}H_2\left(Y_j^{**}\right)$$

$$\xrightarrow{\mathcal{D}} \left(\tilde{\alpha}^{(1,1)} - \tilde{\alpha}^{(1,2)}\right)\left(1 - r^{(1,2)}(0)\right)L_{1,1}^* Z_{2,d^*+1/2}^*(1)$$

$$+ \left(\tilde{\alpha}^{(1,1)} + \tilde{\alpha}^{(1,2)}\right)\left(1 + r^{(1,2)}(0)\right)L_{2,2}^* Z_{2,d^*+1/2}^{**}$$

$$= \left(\tilde{\alpha}^{(1,1)} - \tilde{\alpha}^{(1,2)}\right)\frac{L_{2,2} - L_{2,1} - L_{1,2} + L_{1,1}}{2} Z_{2,d^*+1/2}^*(1)$$

$$+ \left(\tilde{\alpha}^{(1,1)} + \tilde{\alpha}^{(1,2)}\right)\frac{L_{2,2} + L_{2,1} + L_{1,2} + L_{1,1}}{2} Z_{2,d^*+1/2}^{**}(1)$$

with $C_2 := \frac{1}{2d^*(4d^*-1)}$ being the same normalizing factor as in Theorem 2.1.

We observe that $Z_{2,d^*+1/2}^*(1)$ and $Z_{2,d^*+1/2}^{**}(1)$ are both standard Rosenblatt random variables. Following Corollary A1, their covariance is given by

$$\text{Cov}\left(Z_{2,d^*+1/2}^*(1), Z_{2,d^*+1/2}^{**}(1)\right) = \frac{\left(L_{1,2}^* + L_{2,1}^*\right)^2}{L_{1,1}^* L_{2,2}^*}$$

$$= \frac{2\left((L_{2,2} - L_{1,1})^2\right)}{4\left(1 - r^{(1,2)}(0)\right)\left(1 + r^{(1,2)}(0)\right)}\left(L_{1,1}^* L_{2,2}^*\right)^{-1}$$

$$= \frac{(L_{2,2} - L_{1,1})^2}{(L_{1,1} + L_{2,2})^2 - (L_{1,2} + L_{2,1})^2}.$$

Note that $(L_{1,1} + L_{2,2})^2 - (L_{1,2} + L_{2,1})^2 \neq 0$ is fulfilled since $L_{1,1} + L_{2,2} \neq L_{1,2} + L_{2,1}$.

□

References

1. Bandt, C.; Pompe, B. Permutation entropy: A natural complexity measure for time series. *Phys. Rev. Lett.* **2002**, *88*, 174102. [CrossRef] [PubMed]
2. Keller, K.; Sinn, M. Kolmogorov–Sinai entropy from the ordinal viewpoint. *Phys. D Nonlinear Phenom.* **2010**, *239*, 997–1000. [CrossRef]
3. Gutjahr, T.; Keller, K. Ordinal Pattern Based Entropies and the Kolmogorov–Sinai Entropy: An Update. *Entropy* **2020**, *22*. [CrossRef] [PubMed]
4. Keller, K.; Sinn, M. Ordinal analysis of time series. *Phys. A Stat. Mech. Its Appl.* **2005**, *356*, 114–120. [CrossRef]
5. Sinn, M.; Keller, K. Estimation of ordinal pattern probabilities in Gaussian processes with stationary increments. *Comput. Stat. Data Anal.* **2011**, *55*, 1781–1790. [CrossRef]
6. Matilla-García, M.; Ruiz Marín, M. A non-parametric independence test using permutation entropy. *J. Econom.* **2008**, *144*, 139–155. [CrossRef]
7. Caballero-Pintado, M.V.; Matilla-García, M.; Marín, M.R. Symbolic correlation integral. *Econom. Rev.* **2019**, *38*, 533–556. [CrossRef]
8. Bandt, C.; Shiha, F. Order patterns in time series. *J. Time Ser. Anal.* **2007**, *28*, 646–665. [CrossRef]
9. Unakafov, A.M.; Keller, K. Change-Point Detection Using the Conditional Entropy of Ordinal Patterns. *Entropy* **2018**, *20*. [CrossRef]
10. Schnurr, A. An ordinal pattern approach to detect and to model leverage effects and dependence structures between financial time series. *Stat. Pap.* **2014**, *55*, 919–931. [CrossRef]
11. Schnurr, A.; Dehling, H. Testing for Structural Breaks via Ordinal Pattern Dependence. *J. Am. Stat. Assoc.* **2017**, *112*, 706–720. [CrossRef]
12. López-García, M.N.; Sánchez-Granero, M.A.; Trinidad-Segovia, J.E.; Puertas, A.M.; Nieves, F.J.D.l. Volatility Co-Movement in Stock Markets. *Mathematics* **2021**, *9*, 598. [CrossRef]

13. Kechagias, S.; Pipiras, V. Definitions and representations of multivariate long-range dependent time series. *J. Time Ser. Anal.* **2015**, *36*, 1–25. [CrossRef]
14. Arcones, M.A. Limit theorems for nonlinear functionals of a stationary Gaussian sequence of vectors. *Ann. Probab.* **1994**, *22*, 2242–2274. [CrossRef]
15. Beran, J.; Feng, Y.; Ghosh, S.; Kulik, R. *Long-Memory Processes*; Springer: Berlin/Heidelberg, Germany, 2013.
16. Pipiras, V.; Taqqu, M.S. *Long-Range Dependence and Self-Similarity*; Cambridge University Press: Cambridge, UK, 2017; Volume 45.
17. Major, P. Non-central limit theorem for non-linear functionals of vector valued Gaussian stationary random fields. *arXiv* **2019**, arXiv:1901.04086.
18. Nüßgen, I. Ordinal Pattern Analysis: Limit Theorems for Multivariate Long-Range Dependent Gaussian Time Series and a Comparison to Multivariate Dependence Measures. Ph.D. Thesis, University of Siegen, Siegen, Germany, 2021.
19. Caballero-Pintado, M.V.; Matilla-García, M.; Rodríguez, J.M.; Ruiz Marín, M. Two Tests for Dependence (of Unknown Form) between Time Series. *Entropy* **2019**, *21*. [CrossRef]
20. Betken, A.; Buchsteiner, J.; Dehling, H.; Münker, I.; Schnurr, A.; Woerner, J.H. Ordinal Patterns in Long-Range Dependent Time Series. *arXiv* **2019**, arXiv:1905.11033.
21. Veillette, M.S.; Taqqu, M.S. Properties and numerical evaluation of the Rosenblatt distribution. *Bernoulli* **2013**, *19*, 982–1005. [CrossRef]
22. Abrahamson, I. Orthant probabilities for the quadrivariate normal distribution. *Ann. Math. Stat.* **1964**, *35*, 1685–1703. [CrossRef]
23. Hurst, H.E. Long-term storage capacity of reservoirs. *Trans. Amer. Soc. Civil Eng.* **1951**, *116*, 770–808. [CrossRef]
24. Karlekar, M.; Gupta, A. Stochastic modeling of EEG rhythms with fractional Gaussian noise. In Proceedings of the 2014 22nd European Signal Processing Conference (EUSIPCO), Lisbon, Portugal, 1–5 September 2014; pp. 2520–2524.
25. Shu, Y.; Jin, Z.; Zhang, L.; Wang, L.; Yang, O.W. Traffic prediction using FARIMA models. In Proceedings of the 1999 IEEE International Conference on Communications (Cat. No.99CH36311), Vancouver, BC, Canada, 6–10 June 1999; Volume 2, pp. 891–895.
26. Keller, K.; Lauffer, H.; Sinn, M. Ordinal analysis of EEG time series. *Chaos Complex. Lett.* **2007**, *2*, 247–258.
27. Taqqu, M.S. Weak convergence to fractional Brownian motion and to the Rosenblatt process. *Probab. Theory Relat. Fields* **1975**, *31*, 287–302.
28. Düker, M.C. Limit theorems in the context of multivariate long-range dependence. *Stoch. Process. Appl.* **2020**, *130*, 5394–5425. [CrossRef]
29. Golub, G.H.; Van Loan, C.F. *Matrix Computations*; JHU press: Baltimore, MD, USA, 2013; Volume 3.
30. Brockwell, P.J.; Davis, R.A. *Time Series: Theory and Methods*; Springer-Verlag New York, Inc.: New York, NY, USA, 1991.
31. Düker, M.C. Limit Theorems for Multivariate Long-Range Dependent Processes. *arXiv* **2017**, arXiv:1704.08609v3.
32. Magnus, J.R.; Neudecker, H. The commutation matrix: Some properties and applications. *Ann. Stat.* **1979**, *7*, 381–394. [CrossRef]
33. Van der Vaart, A.W. *Asymptotic Statistics*; Cambridge University Press: Cambridge, UK, 2000; Volume 3.
34. Beutelspacher, A. *Lineare Algebra*; Springer: Berlin/Heidelberg, Germany, 2003.

Article

A New First-Order Integer-Valued Autoregressive Model with Bell Innovations

Jie Huang and Fukang Zhu *

School of Mathematics, Jilin University, 2699 Qianjin Street, Changchun 130012, China; jiehuang19@mails.jlu.edu.cn
* Correspondence: fzhu@jlu.edu.cn

Abstract: A Poisson distribution is commonly used as the innovation distribution for integer-valued autoregressive models, but its mean is equal to its variance, which limits flexibility, so a flexible, one-parameter, infinitely divisible Bell distribution may be a good alternative. In addition, for a parameter with a small value, the Bell distribution approaches the Poisson distribution. In this paper, we introduce a new first-order, non-negative, integer-valued autoregressive model with Bell innovations based on the binomial thinning operator. Compared with other models, the new model is not only simple but also particularly suitable for time series of counts exhibiting overdispersion. Some properties of the model are established here, such as the mean, variance, joint distribution functions, and multi-step-ahead conditional measures. Conditional least squares, Yule–Walker, and conditional maximum likelihood are used for estimating the parameters. Some simulation results are presented to access these estimates' performances. Real data examples are provided.

Keywords: Bell distribution; count time series; estimation; INAR; overdispersion

Citation: Huang, J.; Zhu, F. A New First-Order Integer-Valued Autoregressive Model with Bell Innovations. *Entropy* **2021**, *23*, 713. https://doi.org/10.3390/e23060713

Received: 5 April 2021
Accepted: 1 June 2021
Published: 4 June 2021

Publisher's Note: MDPI stays neutral with regard to jurisdictional claims in published maps and institutional affiliations.

Copyright: © 2021 by the authors. Licensee MDPI, Basel, Switzerland. This article is an open access article distributed under the terms and conditions of the Creative Commons Attribution (CC BY) license (https://creativecommons.org/licenses/by/4.0/).

1. Introduction

In recent years, studying count time series has attracted a lot of attention in different fields, such as finance, medical science, and insurance. There are many models for count data that have been proposed by scholars. The most famous model was first introduced by McKenzie (1985) [1] and Al-Osh and Alzaid (1987) [2] based on the binomial thinning ∘ operator (Steutel and van Harn 1979 [3]) called the first-order integer-valued autoregressive (INAR(1)) process. Given a non-negative integer-valued random variable (r.v.) X and a constant $\alpha \in (0,1)$, the binomial thinning operator ∘ is defined as $\alpha \circ X = \sum_{i=1}^{X} \xi_i$, where the counting series ξ_i is a sequence of independent identically distributed (i.i.d.) Bernoulli r.v.s with $P(\xi_i = 1) = 1 - P(\xi_i = 0) = \alpha$. Then, the form of the INAR(1) model is

$$X_t = \alpha \circ X_{t-1} + \epsilon_t, \ t = 0, 1, 2, \ldots, \qquad (1)$$

where ϵ_t is a sequence of i.i.d. discrete r.v.s, with the mean μ_ϵ and finite variance σ_ϵ^2. ϵ_t is independent of ξ_i and X_{t-s} for $s \geq 1$. According to Alzaid and Al-Osh (1988) [4], we know that the mean and variance of the INAR(1) model are

$$\mu := \mu_X = \frac{\mu_\epsilon}{1-\alpha} \text{ and } \sigma^2 := \sigma_X^2 = \frac{\sigma_\epsilon^2 + \alpha\mu_\epsilon}{1-\alpha^2}, \text{ respectively.}$$

For innovation ϵ_t, the Poisson distribution is often assumed as the distribution of ϵ_t in the INAR(1) model. A natural characteristic of the Poisson distribution is equidispersion; i.e., its mean and variance are equal to each other. In practice, however, many data examples are overdispersed (variance is greater than mean) relative to the Poisson distribution. For this reason, the INAR(1) model with Poisson innovations is not always suitable for modeling integer-valued time series. Therefore, several models which describe the overdispersion phenomena have been discussed in the statistical literature.

One common approach is to change the thinning operation in the INAR(1) model. Weiß (2018) [5] summarized several alternative thinning operators, such as random coefficient thinning, iterated thinning and quasi-binomial thinning. Ristić et al. (2009) [6] proposed the negative binomial thinning operator and defined the corresponding INAR(1) process with geometric marginal distributions. Liu and Zhu (2021) [7] generalized the binomial thinning operator to the extended binomial one.

Changing the distribution of innovations is also used to modify the INAR(1) model. Jung et al. (2005) [8] indicated that the INAR(1) model with negative binomial innovations (NB-INAR(1)) is appropriate for generating overdispersion. Jazi et al. (2012) [9] defined a zero-inflated Poisson ZIP(ρ, λ) for innovations (ZIP-INAR(1)), because a frequent occurrence in overdispersion is that the incidence of zero counts is greater than expected from the Poisson distribution. Jazi et al. (2012) [10] proposed a modification of the INAR(1) model with geometric innovations (G-INAR(1)) for modeling overdispersed count data. Schweer and Weiß (2014) [11] investigated the compound Poisson INAR(1) (CP-INAR(1)) model, which is suitable for fitting datasets with overdispersion. According to Schweer and Weiß (2014) [11], we can also know that the negative binomial distribution and the geometric distribution both belong to the compound Poisson distribution. Livio et al. (2018) [12] presented the INAR(1) model with the Poisson–Lindley innovations, i.e., PL-INAR(1). Bourguignon et al. (2019) [13] introduced the INAR(1) model with the double Poisson (DP-INAR(1)) and generalized Poisson innovations (GP-INAR(1)). Qi et al. (2019) [14] considered zero-and-one inflated INAR(1)-type models, and Cunha et al. (2021) [15] introduced an INAR(1) model with Borel innovations to model zero truncated count time series.

This paper applies the second approach to dealing with overdispersion. Although several models have been proposed in recent years, most of the considered distributions are based on some generalizations of the Poisson distribution and have more than one parameter, such as the zero-inflated Poisson, compound Poisson, double Poisson, and generalized Poisson distributions. Here we use a relatively simple distribution introduced by Castellares et al. (2018) [16] for the innovations, i.e., the Bell distribution. It has the advantages of having only one parameter, belonging to the exponential family, having a simple probability mass function, and having infinite divisibility. Infinite divisibility is significant for constructing the binomial thinning INAR(1) model. Further, the Bell distribution is suitable for modeling some overdispersed count data. Therefore, we introduce a new INAR(1) model with Bell innovations (BL-INAR(1)), which can account for overdispersion in an INAR(1) framework.

In order to observe whether the BL-INAR(1) model has advantages, we compare it with the INAR(1) model with Poisson innovations (P-INAR(1)), G-INAR(1), PL-INAR(1), NB-INAR(1), ZIP-INAR(1), DP-INAR(1), and GP-INAR(1) models. Different information criteria, such as Akaike's information criterion (AIC) [17], the Bayesian information criterion (BIC) [18], the consistent Akaike information criterion (CAIC) [19], and the Hannan–Quinn information criterion (HQIC) [20], are used to compare the above eight models. By comparing the results of different information criteria, it can be seen that the BL-INAR(1) model is competitive when modeling the overdispersed integer-valued time series data, which shows that the proposed BL-INAR(1) model is meaningful; see Section 5 for more details.

We organize the remaining parts of this paper as follows. In Section 2, we briefly review the Bell distribution, including its definition and some properties. Then we propose the BL-INAR(1) model, and its basic properties are constructed; conditional mean and variance are obtained. Section 3 discusses estimates of the model parameters by using the conditional least squares (CLS), Yule–Walker (YW), and conditional maximum likelihood (CML) methods. In Section 4, a numerical simulation of the estimates is presented with some discussions. In Section 5, we compare the proposed model with the other seven INAR(1)-type models when fitting two real data examples, which show the superior performances of the proposed model. The paper concludes in Section 6.

2. The BL-INAR(1) Model

In this section, we present a brief review of the Bell distribution (Castellares et al., 2018 [16]). Its definition and some properties are presented. Later we introduce the BL-INAR(1) model and derive some basic properties of it.

2.1. The Bell Distribution

At first, we introduce the Bell numbers. Bell (1934) [21] has provided the following expansion:

$$\exp(e^x - 1) = \sum_{n=0}^{\infty} \frac{B_n}{n!} x^n, \quad x \in \mathbb{R},$$

where B_n is the Bell number defined by

$$B_n = \frac{1}{e} \sum_{k=0}^{\infty} \frac{k^n}{k!}. \tag{2}$$

The Bell number B_n is the n-th moment of the Poisson distribution with parameter equal to 1. Some Bell numbers are listed as follows. $B_0 = B_1 = 1, B_2 = 2, B_3 = 5, B_4 = 15, B_5 = 52, B_6 = 203, B_7 = 877, B_8 = 4140, B_9 = 21{,}147, B_{10} = 115{,}975, B_{11} = 678{,}570, B_{12} = 4{,}213{,}597$ and $B_{13} = 27{,}644{,}437$.

For the convenience of the reader, we introduce the following definition and properties of the Bell distribution described in Castellares et al. (2018) [16]:

Definition 1. *A discrete r.v. Z taking values in $\mathbb{N}_0 = \{0, 1, 2, \ldots\}$ has a Bell distribution with parameter $\theta > 0$, denoted as $Z \sim \text{Bell}(\theta)$, if its probability mass function is given by*

$$\Pr(Z = z) = \frac{\theta^z e^{-e^\theta + 1} B_z}{z!}, \quad z \in \mathbb{N}_0, \tag{3}$$

where B_z is the Bell number in (2).

We can see that the Bell distribution has only one parameter, and it belongs to the one-parameter exponential family of distributions. If $Z \sim \text{Bell}(\theta)$, the probability generating function is

$$G_Z(s) = E\left(s^Z\right) = \exp\left(e^{s\theta} - e^\theta\right), \quad |s| < 1.$$

The mean and variance of Z are

$$E(Z) = \theta e^\theta \text{ and } \text{Var}(Z) = \theta(1+\theta)e^\theta, \text{ respectively.} \tag{4}$$

Note that $\text{Var}(Z)/E(Z) = 1 + \theta > 1$; hence, the Bell distribution is overdispersed, which means the Bell distribution may be suitable for count data with overdispersion in certain situations.

There are some other interesting properties of the Bell distribution, including the following: (i) the Poisson distribution is not nested in the Bell family, but for small values of the parameter, the Bell distribution approaches the Poisson distribution; (ii) it is identifiable, strongly unimodal and infinitely divisible; (iii) a r.v. $Z \sim \text{Bell}(\theta)$ has the same distribution as $Y_1 + Y_2 + \cdots + Y_N$, where Y_n has zero-truncated Poisson distribution with parameter θ, and $N \sim \text{Poisson}(e^\theta - 1)$. See Castellares et al. (2018) [16] for more properties.

Additionally, there are some papers based on the Bell distribution, and the following are a few related references: Batsidis et al. (2020) [22] proposed and studied a goodness-of-fit test for the Bell distribution, which is consistent against fixed alternatives; Castellares et al. (2020) [23] presented a new two-parameter Bell–Touchard discrete distribution; Lemonte et al. (2020) [24] introduced a zero-inflated Bell regression model for count data; Muhammad et al. (2021) [25] proposed a Bell ridge regression as a solution to the multicollinearity problems.

2.2. Definition and Properties of the BL-INAR(1) Process

In this section, we give the definition of the BL-INAR(1) process, and its basic statistical properties are derived.

Definition 2. *Let $\{X_t\}_{t \in \mathbb{N}_0}$ be an INAR(1) process according to (1). It refers to a BL-INAR(1) model if the innovations $\{\epsilon_t\}_{t \in \mathbb{N}_0}$ are a sequence of i.i.d. Bell(θ) r.v.s given by (3); i.e.,*

$$\begin{cases} X_t = \alpha \circ X_{t-1} + \epsilon_t, & t \geq 1, \\ \epsilon_t \sim \text{Bell}(\theta), \end{cases} \quad (5)$$

where $0 < \alpha < 1$ and $\theta > 0$, and ϵ_t is independent of ξ_i and X_{t-1} for $t \geq 1$.

According to Equation (4), we know the mean and variance of ϵ_t are finite; therefore, the process of $\{X_t\}_{t \in \mathbb{N}_0}$ in (5) is an ergodic stationary Markov chain (Du and Li, (1991) [26]) with transition probabilities

$$P_{ij} = P(X_t = i | X_{t-1} = j) = P(\alpha \circ X_{t-1} + \epsilon_t = i | X_{t-1} = j)$$

$$= \sum_{m=0}^{\min(i,j)} P(\alpha \circ X_{t-1} = m | X_{t-1} = j) P(\epsilon_t = i - m)$$

$$= \sum_{m=0}^{\min(i,j)} \binom{j}{m} \alpha^m (1-\alpha)^{j-m} \frac{\theta^{i-m} e^{-e^\theta+1} B_{i-m}}{(i-m)!}, \quad i,j = 0,1 \ldots$$

Further, we can obtain the joint probability function as follows:

$$f(i_1, i_2 \ldots i_T) = P(X_1 = i_1, X_2 = i_2, \ldots, X_T = i_T)$$

$$= P(X_1 = i_1) P(X_2 = i_2 | X_1 = i_1) \ldots P(X_T = i_T | X_{T-1} = i_{T-1})$$

$$= P(X_1 = i_1) \prod_{k=1}^{T-1} \left[\sum_{m=0}^{\min(i_k, i_{k+1})} \binom{i_k}{m} \alpha^m (1-\alpha)^{i_k - m} P(\epsilon_{k+1} = i_{k+1} - m) \right]. \quad (6)$$

The conditional mean, conditional variance, mean, variance, covariance and autocorrelation function of the BL-INAR(1) process are given in the following lemma.

Lemma 1. *Let X_t be the process in Definition 2. Then it has the following properties:*
(i) $E[X_t | X_{t-1}] = \alpha X_{t-1} + \mu_\epsilon = \alpha X_{t-1} + \theta e^\theta$;
(ii) $\text{Var}[X_t | X_{t-1}] = \alpha(1-\alpha) X_{t-1} + \sigma_\epsilon^2 = \alpha(1-\alpha) X_{t-1} + \theta(1+\theta) e^\theta$;
(iii) $\mu := E[X_t] = \frac{\theta e^\theta}{1-\alpha}$;
(iv) $\sigma^2 := \text{Var}[X_t] = \frac{\theta e^\theta (1+\alpha+\theta)}{1-\alpha^2}$;
(v) $\gamma_k := \text{Cov}(X_k, X_{k+1}) = \alpha^k \sigma^2$;
(vi) $\rho_k := \text{Corr}(X_k, X_{k+1}) = \alpha^k$.

The proof of Lemma 1 is similar to that of Theorem 1 of Qi et al. (2019) [14], so it is omitted.

According to Lemma 1, the dispersion index (Fisher, 1950 [27]) of X_t is derived as follows:

$$I_x := \frac{\sigma^2}{\mu} = 1 + \frac{\theta}{1+\alpha} > 1;$$

thus, the BL-INAR(1) process is suited for overdispersed integer-valued time series.

Additionally, we can obtain the k-step ahead conditional mean and k-step ahead conditional variance of the BL-INAR(1) process in the following theorem.

Theorem 1. *The k-step ahead conditional mean and k-step ahead conditional variance of the BL-INAR(1) process are given, respectively, by:*

$$E(X_{t+k}|X_t) = \alpha^k X_t + \mu_\epsilon \frac{1-\alpha^k}{1-\alpha},$$

and

$$\text{Var}(X_{t+k}|X_t) = \alpha^k\left(1-\alpha^k\right)X_t + \mu_\epsilon \frac{(\alpha-\alpha^k)(1-\alpha^k)}{1-\alpha^2} + \sigma_\epsilon^2 \frac{1-\alpha^{2k}}{1-\alpha^2}.$$

For more details about the proof of this theorem, see Qi et al. (2019) [14] and Ristić, Bakouch, and Nastić (2009) [6]. It is easy to see that if $k \to \infty$, $E(X_{t+k}|X_t) \to \frac{\mu_\epsilon}{1-\alpha} = \frac{\theta e^\theta}{1-\alpha}$ and $\text{Var}(X_{t+k}|X_t) \to \frac{\alpha \mu_\epsilon + \sigma_\epsilon^2}{1-\alpha^2} = \frac{\theta e^\theta(1+\alpha+\theta)}{1-\alpha^2}$, which are the unconditional mean and unconditional variance of X_t, respectively.

3. Estimation of Parameters

The true values of parameters α and θ are unknown in practice; therefore, we need to estimate the value of (α, θ). Sometimes we have to give an estimate of (α, μ) first to get the estimate of (α, θ). In this section, we consider three methods for estimating parameters, namely, CLS, YW and CML.

3.1. Conditional Least Squares Estimation

The CLS estimates of the parameters α and θ are obtained by

$$(\hat{\alpha}, \hat{\theta}) = \arg\min \sum_{t=2}^{T} [X_t - E(X_t|X_{t-1})]^2,$$

and the CLS estimates of (α, μ) are given by

$$\hat{\alpha}_{CLS} = \frac{(T-1)\sum_{t=2}^{T} X_t X_{t-1} - \sum_{t=2}^{T} X_t \sum_{t=2}^{T} X_{t-1}}{(T-1)\sum_{t=2}^{T} X_{t-1}^2 - \left(\sum_{t=2}^{T} X_{t-1}\right)^2},$$

and

$$\hat{\mu}_{CLS} = \frac{\sum_{t=2}^{T} X_t - \hat{\alpha}_{CLS}\sum_{t=2}^{T} X_{t-1}}{(T-1)(1-\hat{\alpha}_{CLS})}.$$

Then, the CLS estimate of θ can be obtained by solving the equation $\hat{\theta}_{CLS} e^{\hat{\theta}_{CLS}} = \hat{\mu}_{CLS}(1-\hat{\alpha}_{CLS})$.

According to Theorems 3.1 and 3.2 in Tjøstheim (1986) [28], we can establish the consistency and asymptotic normality of the CLS estimates $\hat{\alpha}_{CLS}$ and $\hat{\mu}_{CLS}$ in the following theorem. The proofs of Theorem 2 and the following theorem are given in Appendix A.

Theorem 2. *Let $\hat{\alpha}_{CLS}$ and $\hat{\mu}_{CLS}$ be the CLS estimates of the BL-INAR(1) process; then $(\hat{\alpha}_{CLS}, \hat{\mu}_{CLS})'$ is strongly consistent for (α, μ); and the asymptotic distribution follows as:*

$$\sqrt{T}(\hat{\alpha}_{CLS} - \alpha, \hat{\mu}_{CLS} - \mu)' \xrightarrow{d} \mathbf{N}(0, \Sigma),$$

where

$$\Sigma = \begin{bmatrix} \frac{\alpha(1+\alpha)\mu+\sigma_\epsilon^2}{(1-\alpha)^2} & \frac{\alpha\sigma^2}{\mu(1+\mu)} \\ \frac{\alpha\sigma^2}{\mu(1+\mu)} & \frac{\alpha(1-\alpha)(\mu_3-2\mu\sigma^2-\mu^3)+\sigma_\epsilon^2\sigma^2}{\mu^2(1+\mu)^2} \end{bmatrix},$$

and $\mu_3 = E(X_t^3) = \frac{(1-\alpha^3)\mu^3 + (1+2\alpha^2-3\alpha^3)\mu\sigma^2 + \alpha^2(1-\alpha)\sigma^2}{1-\alpha^3}$.

Using the delta method, we can obtain the limit distribution of $(\hat{\alpha}, \hat{\theta})$, and we can also know that $\hat{\theta}$ is consistent.

3.2. Yule–Walker Estimation

Let $X_1, \ldots, X_T$ come from the process $\{X_t\}$ in Definition 2. The sample mean is $\bar{X} = \frac{1}{T}\sum_{t=1}^{T} X_t$, and the sample autocorrelation function is

$$\hat{\rho}_k = \frac{\sum_{t=1}^{T-k}(X_t - \bar{X})(X_{t+k} - \bar{X})}{\sum_{t=1}^{n}(X_t - \bar{X})^2}.$$

From Lemma 1, we know $\rho_k = \alpha^k$, thus the Yule–Walker (YW) estimate of α is given by

$$\hat{\alpha}_{YW} = \hat{\rho}(1) = \frac{\sum_{t=1}^{T-1}(X_t - \bar{X})(X_{t+1} - \bar{X})}{\sum_{t=1}^{T}(X_t - \bar{X})^2},$$

and

$$\hat{\mu}_{YW} = \bar{X},$$

with $\mu = \dfrac{\theta e^\theta}{1-\alpha}$; then the estimate of θ can be obtained.

For asymptotic properties of the YW estimates, Freeland and McCabe (2005) [29] showed that the YW and CLS estimates are asymptotically equivalent for a Poisson INAR(1) process. The next theorem shows that the conclusion holds for our BL-INAR(1) process.

Theorem 3. *In the BL-INAR(1) process, CLS and YW estimates are asymptotically equivalent, i.e.,*

$$\hat{\alpha}_{CLS} - \hat{\alpha}_{YW} = o_p(T^{-\frac{1}{2}}) \quad \text{and} \quad \hat{\theta}_{CLS} - \hat{\theta}_{YW} = o_p(T^{-\frac{1}{2}}).$$

3.3. Conditional Maximum Likelihood Estimation

According to the joint probability function (6), the likelihood function can be obtained as:

$$f(x_1, x_2, \ldots, x_T) = P(X_1 = x_1) \prod_{t=1}^{T-1} P(X_{t+1} = x_{t+1} | X_t = x_t)$$

$$= f(x_1) \prod_{t=1}^{T-1} \left[\sum_{m=0}^{\min(x_t, x_{t+1})} \binom{x_t}{m} \alpha^m (1-\alpha)^{x_t - m} P(\epsilon_{t+1} = x_{t+1} - m) \right].$$

To condition on variable X_1, we can obtain the conditional log likelihood function as:

$$L(\alpha, \theta) = \sum_{t=1}^{T-1} \log P(X_{t+1} = x_{t+1} | X_t = x_t),$$

the CML estimates of (α, θ) are the values of $(\hat{\alpha}_{CML}, \hat{\theta}_{CML})$ obtained by maximizing the conditional log likelihood function $L(\alpha, \theta)$. It is easy to check that the BL-INAR(1) process satisfies conditions (C1)–(C6) of Franke and Seligmann (1993) [30]; thus, the CML estimates $(\hat{\alpha}_{CML}, \hat{\theta}_{CML})$ are consistent and asymptotically normal. The proof is similar to those of Theorems 22.4 and 22.5 of Franke and Seligmann (1993) [30], so it is omitted.

4. Simulation

A Monte Carlo simulation was conducted to study the performances of the CLS, YW, and CML estimates of the BL-INAR(1) model. The CML estimates were obtained by using the BFGS quasi-Newton nonlinear optimization algorithm with numerical derivatives. We considered YW estimates as initial values for the algorithm. The simulation was conducted using R programming language, and the size of the sample was 100, 250, 500, or 1000.

The number of replicates was 1000. For the true values of parameters, we considered $\alpha = 0.25, 0.5,$ and 0.75 and $\theta = 0.5,$ and 1.5.

First, we give the Q–Q plots of the CLS, YW, and CML estimates for the BL-INAR(1) model with sample size $T = 1000$, $\alpha = 0.5$, and $\theta = 1.5$ in Figure 1. From the six Q–Q plots, we can see that they contain roughly straight lines; i.e., the estimates of the parameters are normally distributed. Then, the numerical simulation results are presented in Tables 1 and 2. By comparing the two tables, we can find that with the same θ and T, the mean squared error (MSE) for the estimate of θ increased with the increase of α, but the MSE for the estimate of α decreased. Additionally, the MSE for the estimate of θ increased with the increase of θ with the same α and T, but the MSE for the estimate of α decreased. Furthermore, we can observe that the estimates of CLS and YW are similar, and the bias tended toward zero for all estimates as the sample size increased. The estimates of CML converged faster to the true parameter values. We conclude that the CML estimates produced the smallest mean square errors, and CML performed better than CLS and YW.

Table 1. Empirical means and mean squared errors (in parentheses) of the estimates of the parameters for some values of α and θ of the BL-INAR(1) model.

T	$\hat{\alpha}_{CLS}$	$\hat{\theta}_{CLS}$	$\hat{\alpha}_{YW}$	$\hat{\theta}_{YW}$	$\hat{\alpha}_{CML}$	$\hat{\theta}_{CML}$
			$(\alpha, \theta) = (0.25, 0.5)$			
100	0.220445	0.507901	0.218464	0.508838	0.238497	0.500617
	(0.011615)	(0.003871)	(0.011564)	(0.003814)	(0.007120)	(0.003049)
250	0.240011	0.502610	0.239107	0.503039	0.246433	0.500099
	(0.004665)	(0.001520)	(0.004653)	(0.001501)	(0.002658)	(0.001146)
500	0.245254	0.500707	0.244710	0.500965	0.247766	0.499777
	(0.002286)	(0.000758)	(0.002284)	(0.000759)	(0.001300)	(0.000603)
1000	0.246458	0.500868	0.246197	0.500974	0.249120	0.499768
	(0.001195)	(0.000379)	(0.001197)	(0.000380)	(0.000714)	(0.000290)
			$(\alpha, \theta) = (0.5, 0.5)$			
100	0.475430	0.508229	0.469977	0.512782	0.495566	0.497046
	(0.010198)	(0.005396)	(0.010256)	(0.005296)	(0.004046)	(0.003083)
250	0.488517	0.504259	0.486491	0.505890	0.497636	0.499045
	(0.003895)	(0.002128)	(0.003911)	(0.002123)	(0.001723)	(0.001332)
500	0.493426	0.502160	0.492388	0.503029	0.498222	0.499355
	(0.001857)	(0.001026)	(0.001868)	(0.001025)	(0.000866)	(0.000643)
1000	0.496426	0.501635	0.495922	0.502043	0.499262	0.499936
	(0.000914)	(0.000530)	(0.000916)	(0.000529)	(0.000412)	(0.000322)
			$(\alpha, \theta) = (0.75, 0.5)$			
100	0.714977	0.535092	0.707308	0.543643	0.745993	0.500460
	(0.006966)	(0.011276)	(0.007639)	(0.011838)	(0.001321)	(0.003355)
250	0.736256	0.513360	0.733057	0.517222	0.748974	0.498915
	(0.002354)	(0.004357)	(0.002456)	(0.004432)	(0.000494)	(0.001352)
500	0.743674	0.505799	0.742084	0.507695	0.749245	0.499568
	(0.001052)	(0.001967)	(0.001079)	(0.001983)	(0.000243)	(0.000681)
1000	0.746006	0.504828	0.745283	0.505726	0.749925	0.500221
	(0.000546)	(0.001001)	(0.000554)	(0.001011)	(0.000132)	(0.000309)

Table 2. Empirical means and mean squared errors (in parentheses) of the estimates of the parameters for some values of α and θ of the BL-INAR(1) model.

T	$\hat{\alpha}_{CLS}$	$\hat{\theta}_{CLS}$	$\hat{\alpha}_{YW}$	$\hat{\theta}_{YW}$	$\hat{\alpha}_{CML}$	$\hat{\theta}_{CML}$
			$(\alpha, \theta) = (0.25, 1.5)$			
100	0.230059	1.508538	0.227601	1.510489	0.252786	1.492707
	(0.010409)	(0.007500)	(0.010294)	(0.007323)	(0.004877)	(0.004375)
250	0.243290	1.503077	0.242313	1.503896	0.250278	1.498450
	(0.003994)	(0.002928)	(0.003976)	(0.002898)	(0.001810)	(0.001660)
500	0.244804	1.503143	0.244310	1.503531	0.249992	1.499429
	(0.001917)	(0.001459)	(0.001914)	(0.001451)	(0.000913)	(0.000829)
1000	0.248715	1.500420	0.248470	1.500628	0.251744	1.498222
	(0.000984)	(0.000745)	(0.000983)	(0.000745)	(0.000477)	(0.000422)
			$(\alpha, \theta) = (0.5, 1.5)$			
100	0.472192	1.522593	0.467254	1.528069	0.497401	1.497773
	(0.008714)	(0.011950)	(0.008913)	(0.011884)	(0.002653)	(0.004999)
250	0.489125	1.509054	0.487244	1.511225	0.499745	1.498361
	(0.003127)	(0.004609)	(0.003148)	(0.004594)	(0.000991)	(0.001856)
500	0.496116	1.502407	0.495032	1.503670	0.501865	1.496562
	(0.001650)	(0.002493)	(0.001660)	(0.002487)	(0.000584)	(0.001078)
1000	0.497904	1.501246	0.497432	1.501800	0.500976	1.498100
	(0.000826)	(0.001314)	(0.000827)	(0.001314)	(0.000274)	(0.000502)
			$(\alpha, \theta) = (0.75, 1.5)$			
100	0.721350	1.547389	0.713291	1.565523	0.749555	1.495159
	(0.005627)	(0.025790)	(0.006188)	(0.026764)	(0.000827)	(0.005581)
250	0.736975	1.522062	0.733930	1.529286	0.749880	1.497782
	(0.002181)	(0.011299)	(0.002278)	(0.011488)	(0.000343)	(0.002363)
500	0.742692	1.512717	0.741144	1.516338	0.749888	1.498329
	(0.000919)	(0.005007)	(0.000947)	(0.005076)	(0.000158)	(0.001101)
1000	0.747785	1.503485	0.747046	1.505296	0.750224	1.499187
	(0.000476)	(0.002670)	(0.000479)	(0.002670)	(0.000083)	(0.000541)

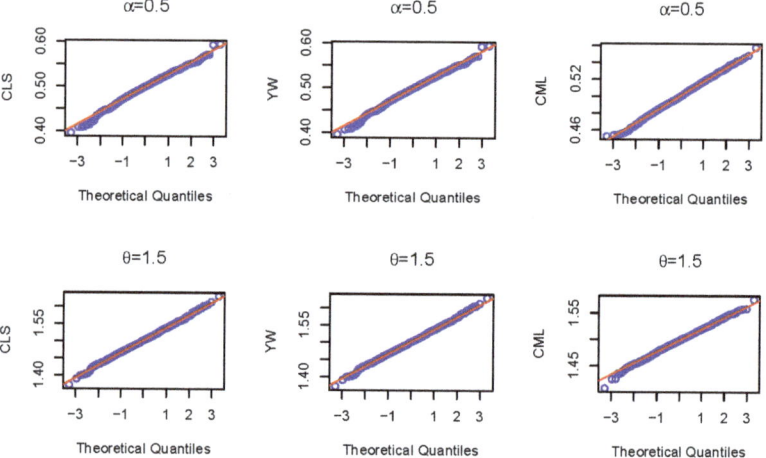

Figure 1. The Q–Q plots of the CLS, YW, and CML estimates for the BL-INAR(1) model with sample size $T = 1000$.

5. Real Data Examples

In this section, we present two applications of the BL-INAR(1) model to real datasets, and compare it with the P-INAR(1), G-INAR(1), PL-INAR(1), NB-INAR(1), ZIP-INAR(1),

DP-INAR(1), and GP-INAR(1) models. Results of the comparison are discussed here as well.

5.1. Disconduct Data

The first dataset is a monthly count of disconduct in the first census tract in Rochester, which can be obtained from Available online: http://www.forecastingprinciples.com (accessed on 8 May 2012). The data comprise 132 observations ($T = 132$) starting from January 1991 and ending in December 2001.

The time plot, histogram, autocorrelation function (ACF), and partial autocorrelation function (PACF) are provided in Figure 2. We applied the Ljung–Box test (Ljung and Box (1978) [31]) to check whether this time series dataset has any autocorrelation. The p-value of the Ljung–Box test is 1.317×10^{-5}, which is less than 0.05. This means that the time series data have some autocorrelation, and according to the PACF diagram, the data are first-order autocorrelated, which shows that the AR(1)-type process is appropriate for modeling this dataset.

The sample mean and variance of the data are $\bar{X} = 1.6288$ and $S_X^2 = 2.4455$, respectively. Thus, we got the dispersion index $\hat{I}_x = S_X^2/\bar{X} = 1.5014$. According to the overdispersion test of Schweer and Weiß (2014) [11], the critical value of the data is 1.1994. The dispersion index $\hat{I}_x$ exceeds the critical value, which means that the equidispersed P-INAR(1) model is not a good choice for the data.

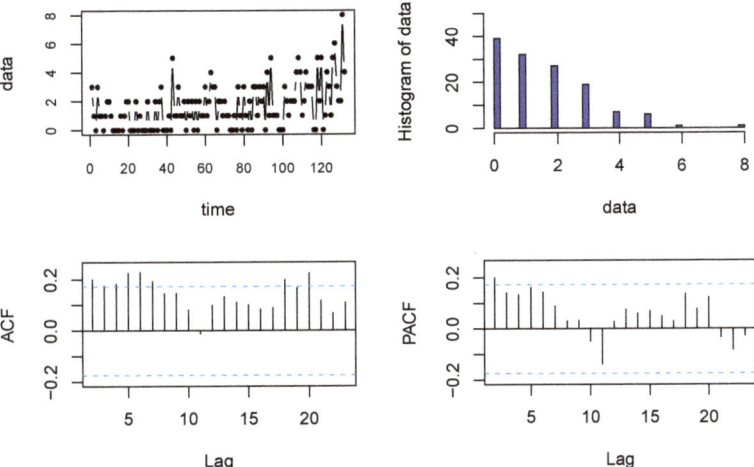

Figure 2. The time plot, histogram, ACF, and PACF of disconduct data.

For comparison, we calculated the CML estimates of parameters, and the AIC, BIC, CAIC, HQIC, fitted mean, and fitted variance of the BL-INAR(1) model, the P-INAR(1) model, the G-INAR (1) model, the PL-INAR(1) model, the ZIP-INAR(1) model, the NB-INAR(1) model, the DP-INAR(1) model, and the GP-INAR(1) model. Among the eight models, the first four are two-parameter models and the last four are three-parameter models. The results are presented in Table 3. We found that the AIC, BIC, CAIC, and HQIC of the BL-INAR(1) model were smaller than those of others. We also found that the fitted means of all eight models were near to the sample mean, and the fitted mean of the PL-INAR(1) model was the closest to the sample mean. In terms of fitted variance, Table 3 shows that the fitted variance of the BL-INAR(1) model performed better than those of the other seven models.

Table 3. CML estimates, AIC, BIC, CAIC, HQIC, fitted mean, fitted variance, and RMSE for eight INAR(1) models of disconduct data.

Model	Parameters	AIC	BIC	CAIC	HQIC	Mean	Variance	RMSE
BL-INAR	$\hat{\alpha} = 0.1882$ $\hat{\theta} = 0.6718$	**441.7380**[1]	**447.5036**[1]	**449.5036**[1]	**444.0809**[1]	1.6201	**2.5361**[2]	3.2205
P-INAR	$\hat{\alpha} = 0.1496$ $\hat{\lambda} = 1.3773$	456.4653	462.2309	464.2309	458.8082	1.6197	2.1512	3.2497
G-INAR	$\hat{\alpha} = 0.2405$ $\hat{\pi} = 0.4482$	446.1416	451.9072	453.9072	448.4845	1.6207	3.2290	**3.1803**[1]
PL-INAR	$\hat{\alpha} = 0.2197$ $\hat{\theta} = 1.1545$	444.3542	450.1198	452.1198	446.6971	**1.6254**[2]	0.2928	3.1929
NB-INAR	$\hat{\alpha} = 0.1845$ $\hat{n} = 1.9345$ $\hat{\pi} = 0.5942$	445.4351	454.0835	457.0835	448.9494	1.6201	2.5542	3.2233
ZIP-INAR	$\hat{\alpha} = 0.1992$ $\hat{\lambda} = 1.8674$ $\hat{\rho} = 0.3052$	442.7224	451.3708	454.3708	446.2367	1.6202	2.3903	3.2121
DP-INAR	$\hat{\alpha} = 0.1865$ $\hat{\mu} = 1.2121$ $\hat{\phi} = 0.5122$	443.7622	452.4106	455.4106	447.2765	1.4900	2.6859	3.3155
GP-INAR	$\hat{\alpha} = 0.1820$ $\hat{\mu} = 1.0254$ $\hat{\phi} = 0.2262$	445.8156	454.4640	457.4640	449.3300	1.6200	2.5386	3.2252

[1] Bold text means the smallest value in the column. [2] Bold text means that this value is the closest in the column to the sample value described in the text.

For the prediction, we used the first 126 observations to estimate the parameters, and then predicted the last six observations. The predicted values of the disconduct data could be given by $E(X_{t+k} \mid X_t) = \alpha^k X_t + \mu_\epsilon \dfrac{1 - \alpha^k}{1 - \alpha}$. For a further comparison of models, we calculated the root mean square values of the prediction errors (RMSEs) for the last 6 months of the data, and the RMSE is defined as RMSE $= \sqrt{\dfrac{1}{6} \sum_{k=1}^{6} (X_{t+k} - \hat{X}_{t+k})^2}$. We present the RMSE results of eight models in the last column of Table 3. From the table, we can see that the RMSE of the G-INAR(1) model was best. The RMSE of the BL-INAR(1) model is smaller than those of the P-INAR(1) model, the NB-INAR(1) model, the DP-INAR(1) model, and the GP-INAR(1) model; and a little larger than those of the G-INAR(1) model, the PL-INAR(1) model, and the ZIP-INAR(1) model. Although the fitted mean and RMSE of the BL-INAR(1) model are not the best, it is the best choice under the other five criteria. Further, we analyze the Pearson residuals, and Figure 3 plots the ACF, PACF, and Q–Q plots of residuals. The ACF and PACF graphs show no correlation between residuals, which is supported by the result of the Ljung–Box test with a p-value of $0.05251 > 0.05$. The Q–Q plots appear to be roughly normally distributed, as we expected. Hence, we can conclude that the BL-INAR(1) model is the most suitable among those available for this dataset.

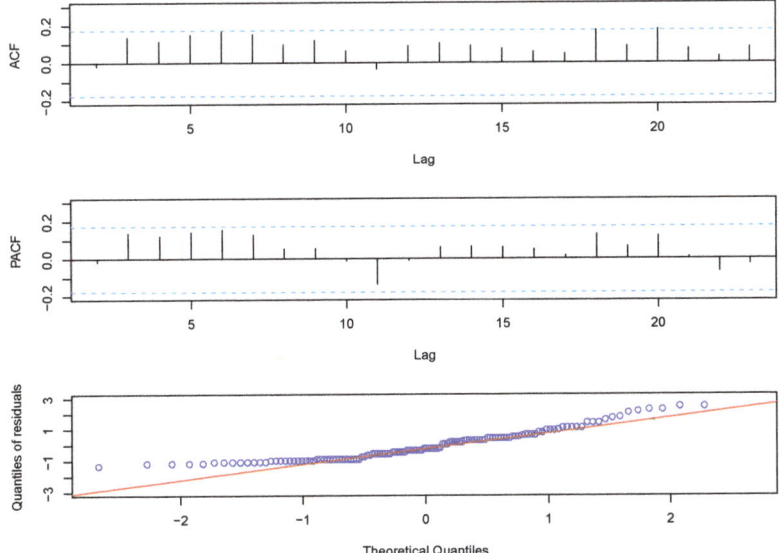

Figure 3. The ACF, PACF, and Q–Q plots of the Pearson residual for disconduct data using the BL-INAR(1) model.

5.2. Strikes Data

The second dataset, which was analyzed by Weiß (2010) [32], is the monthly number of work stoppages (strikes and lock-outs) of 1000 or more workers for the period 1994–2002. It was published by the US Bureau of Labor Statistics and can be obtained by online at the address Available online: http://www.bls.gov/wsp/ (accessed on 8 May 2012). The data contain 108 observations, and the time plot, histogram, ACF, and PACF are provided in Figure 4. As with the previous example, the Ljung–Box test was used to check whether the strike data have any autocorrelation. The p-value of the Ljung–Box test was 2.372×10^{-8}, which shows that the time series data have some autocorrelation, and according to the PACF diagram, it is also first-order autocorrelated, so an AR(1)-type process is appropriate for modeling this dataset.

The sample mean, variance, and dispersion index were calculated to be 4.9444, 7.8488, and 1.5874, respectively. According to the overdispersion test, the critical value of the data is 1.2808, and we observe that it was inappropriate to use the P-INAR(1) model to fit the data. The CML estimates, AIC, BIC, CAIC, HQIC, fitted mean, and fitted variance for the BL-INAR(1), P-INAR(1), G-INAR(1), PL-INAR(1), NB-INAR(1), ZIP-INAR(1), DP-INAR(1), and GP-INAR(1) models were obtained and are shown in Table 4. We see that the AIC, BIC, CAIC, and HQIC of the BL-INAR(1) model are smaller than those of others, and the fitted mean of the BL-INAR(1) model is not much different from those of the other seven models. Further, we can see that the BL-INAR(1) model performed better than others when calculating the fitted variance. Similarly to the previous example, the first 102 observations were used to estimate the parameters and predict the last six observations. The RMSE of the predictions is also presented in Table 4. We can observe that the RMSE of the G-INAR(1) model is the smallest; however, it is only 0.05 less than the RMSE of the BL-INAR(1) model. As in the previous example, although the BL-INAR(1) model was not the best under the fitted mean and RMSE criteria, it performed best under the other five criteria. Additionally, we show the Pearson residuals analysis. Figure 5 gives the ACF, PACF, and Q–Q plots of the residuals. We found that there is no evidence of any significant correlation within the residuals, a finding also supported by the Ljung–Box test with a p-value of 0.9522, which is greater than 0.05. The Q–Q plot also appears to be roughly normally distributed. Thus, according to above discussions and its simplicity, we can conclude that the BL-INAR(1) model was the most appropriate.

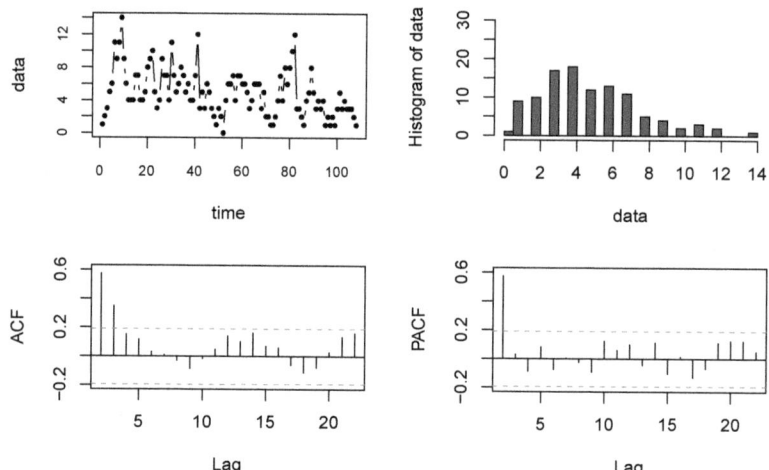

Figure 4. The time plot, histogram, ACF, and PACF of data on strikes.

Table 4. CML estimates, AIC, BIC, CAIC, HQIC, fitted mean, fitted variance, and RMSE from eight INAR(1) models of strike data.

Model	Parameters	AIC	BIC	CAIC	HQIC	Mean	Variance	RMSE
BL-INAR	$\hat{\alpha} = 0.5789$ $\hat{\theta} = 0.8747$	**468.1557**[1]	**473.5199**[1]	**475.5199**[1]	**470.3307**[1]	4.9813	**7.7408**[2]	2.2659
P-INAR	$\hat{\alpha} = 0.5061$ $\hat{\lambda} = 2.4603$	473.0936	478.4578	480.4578	475.2686	4.9813	9.8110	2.3331
G-INAR	$\hat{\alpha} = 0.6235$ $\hat{\pi} = 0.3478$	475.3209	480.6852	482.6852	477.4960	4.9813	10.7361	**2.2121**[1]
PL-INAR	$\hat{\alpha} = 0.6062$ $\hat{\theta} = 0.7911$	471.9345	477.2987	479.2987	474.1095	5.0016	1.8876	2.2489
NB-INAR	$\hat{\alpha} = 0.5483$ $\hat{n} = 3.8582$ $\hat{\pi} = 0.6317$	469.6850	477.7314	480.7314	472.9476	4.9813	6.8573	2.2969
ZIP-INAR	$\hat{\alpha} = 0.5785$ $\hat{\lambda} = 2.6343$ $\hat{\rho} = 0.2030$	470.9985	479.0449	482.0449	474.2610	4.9813	6.6692	2.2663
DP-INAR	$\hat{\alpha} = 0.5617$ $\hat{\mu} = 2.1727$ $\hat{\phi} = 0.5924$	469.5585	477.6048	480.6048	472.8210	**4.9576**[2]	7.1420	2.2659
GP-INAR	$\hat{\alpha} = 0.5464$ $\hat{\mu} = 1.8003$ $\hat{\phi} = 0.2032$	469.7467	477.7930	480.7930	473.0092	4.9813	6.8335	2.2986

[1] Bold means the smallest value in the column. [2] Bold means that this value is the closest in the column to the sample value described in the text.

Combined with the above two examples and the advantages of the Bell distribution with one parameter and a simple form, the BL-INAR(1) model is competitive with the other seven models.

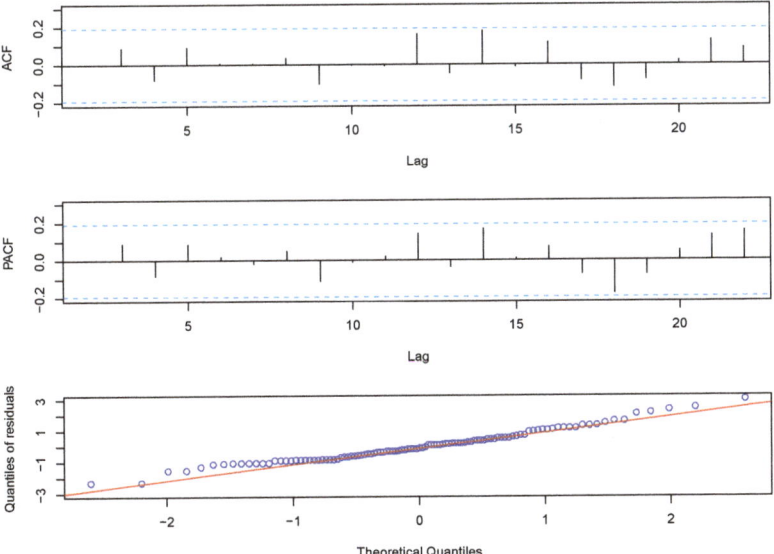

Figure 5. The ACF, PACF, and Q–Q plots of the Pearson residual for strike data using the BL-INAR(1) model.

6. Conclusions

A new INAR(1) model with Bell innovations based on the binomial thinning operator was introduced in this paper. Based on the overdispersed property of the Bell distribution, we found that the BL-INAR(1) model is suitable for overdispersed data. Some basic properties of the model were obtained, such as transition probabilities, conditional mean, conditional variance, mean, variance, covariance, autocorrelation function, and k-step ahead conditional mean and variance. For unknown parameters, CLS, YW, and CML methods are used to estimate them. The Q–Q plots showed that the estimates of the parameters are normally distributed. The simulated results revealed that the CML estimates of parameters of the BL-INAR(1) model were better than the CLS and YW estimates. Finally, by comparing the AIC values, BIC values, CAIC values, HQIC values, fitted means, fitted variances, and RMSE values of the predictions among eight INAR(1) models, two real datasets both showed that the BL-INAR(1) model fits better than other INAR(1) models. The analysis of residuals also shows that the BL-INAR(1) model provided adequate fits to those datasets.

Although there are many overdispersed INAR(1) models, some interesting properties of the Bell distribution, such as having one parameter, infinitely divisibility, having a simple probability mass function, belonging to the one-parameter exponential family of distributions, and for a parameter with a small value, having the Bell distribution approach the Poisson distribution, make the BL-INAR(1) model competitive. Some extended distributions of the Bell distribution, such as the zero-inflated Bell distribution and the Bell–Touchard distribution, provide ideas for us to study related INAR models in the future.

Author Contributions: Conceptualization, F.Z.; methodology, F.Z.; software, J.H.; validation, J.H. and F.Z.; formal analysis, J.H.; investigation, J.H. and F.Z.; resources, F.Z.; data curation, J.H.; writing—original draft preparation, J.H.; writing—review and editing, F.Z.; visualization, J.H.; supervision, F.Z.; project administration, F.Z.; funding acquisition, F.Z. All authors have read and agreed to the published version of the manuscript.

Funding: Zhu's work is supported by National Natural Science Foundation of China, grant numbers 11871027 and 11731015.

Institutional Review Board Statement: Not applicable.

Informed Consent Statement: Not applicable.

Data Availability Statement: The disconduct data and the strike data are available at http://www.forecastingprinciples.com (accessed on 1 June 2021) and http://www.bls.gov/wsp/ (accessed on 1 June 2021), respectively .

Conflicts of Interest: The authors declare no conflict of interest.

Appendix A

Appendix A.1. Proof of Theorem 2

To prove this theorem, we need to show that the conditions given in Theorems 3.1 and 3.2 of Tjøstheim (1986) [28] are satisfied.

Define $\phi = (\alpha, \mu)$, and the true value of the unknown parameter $\phi_0 = (\alpha_0, \mu_0)$. According to Lemma 1, we know that $E[X_t^2] < \infty$ and that $E[X_t|X_{t-1}]$ is almost surely three times differentiable in an open set Φ containing ϕ_0.

Condition 1:

$$E\left\{\left|\frac{\partial E(X_t \mid X_{t-1})}{\partial \phi_i}(\boldsymbol{\phi}_0)\right|^2\right\} < \infty \text{ and } E\left\{\left|\frac{\partial^2 E(X_t \mid X_{t-1})}{\partial \phi_i \partial \phi_j}(\boldsymbol{\phi}_0)\right|^2\right\} < \infty, \text{ for } i,j = 1,2.$$

According to $E(X_t \mid X_{t-1}) = \alpha X_{t-1} + (1-\alpha)\mu$, we have

$$E\left\{\left|\frac{\partial E(X_t \mid X_{t-1})}{\partial \alpha}(\boldsymbol{\phi}_0)\right|^2\right\} = E\left\{|X_{t-1} - \mu_0|^2\right\} = \text{Var}(X_{t-1}) < \infty \text{ and}$$

$$E\left\{\left|\frac{\partial E(X_t \mid X_{t-1})}{\partial \mu}(\boldsymbol{\phi}_0)\right|^2\right\} = E\left\{|1 - \alpha_0|^2\right\} = (1-\alpha_0)^2 < \infty.$$

For the second derivative of $E(X_t \mid X_{t-1})$, we have

$$E\left\{\left|\frac{\partial^2 E(X_t \mid X_{t-1})}{\partial \alpha^2}(\boldsymbol{\phi}_0)\right|^2\right\} = E\left\{\left|\frac{\partial^2 E(X_t \mid X_{t-1})}{\partial \mu^2}(\boldsymbol{\phi}_0)\right|^2\right\} = 0 < \infty \text{ and}$$

$$E\left\{\left|\frac{\partial^2 E(X_t \mid X_{t-1})}{\partial \alpha \partial \mu}(\boldsymbol{\phi}_0)\right|^2\right\} = 1 < \infty.$$

Condition 2:

The vectors $\partial E(X_t \mid X_{t-1})(\theta_0)/\partial \phi_i, i,j = 1,2$ are linearly independent in the sense that if a_1 and a_2 are arbitrary real numbers such that

$$E\left\{\left|\sum_{i=1}^{2} a_i \frac{\partial E(X_t \mid X_{t-1})}{\partial \phi_i}(\boldsymbol{\phi}_0)\right|^2\right\} = 0,$$

then $a_1 = a_2 = 0$. Note that

$$E\left\{\left|a_1 \frac{\partial E(X_t \mid X_{t-1})}{\partial \alpha}(\boldsymbol{\phi}_0) + a_2 \frac{\partial E(X_t \mid X_{t-1})}{\partial \mu}(\boldsymbol{\phi}_0)\right|^2\right\} = 0 \Rightarrow$$

$$E\left\{|a_1(X_{t-1} - \mu_0) + a_2(1 - \alpha_0)|^2\right\} = 0 \Rightarrow$$

$$\underbrace{a_1^2 \text{Var}(X_{t-1})}_{\geq 0} + \underbrace{a_2^2(1-\alpha_0)^2}_{\geq 0} = 0 \Rightarrow$$

$$a_1^2 \underbrace{\text{Var}(X_{t-1})}_{>0} = 0 \text{ and } a_2^2 \underbrace{(1-\alpha_0)^2}_{>0} = 0 \Rightarrow$$

$$a_1^2 = 0 \text{ and } a_2^2 = 0.$$

Then $a_1 = a_2 = 0$.

Condition 3:

For $\phi \in \Phi$, there exist functions $G_{t-1}^{ijk}(X_1, \ldots, X_{t-1})$ and $H_t^{ijk}(X_1, \ldots, X_t)$ for $i, j = 1, 2$ such that

$$M_{t-1}^{ijk}(\phi) = \left| \frac{\partial E(X_t \mid X_{t-1})}{\partial \phi_i}(\phi) \frac{\partial^2 E(X_t \mid X_{t-1})}{\partial \phi_j \partial \phi_k}(\phi) \right| \leq G_{t-1}^{ijk}, \ E\left(G_{t-1}^{ijk}\right) < \infty,$$

$$N_t^{ijk}(\phi) = \left| \{X_t - E(X_t \mid X_{t-1})(\phi)\} \frac{\partial^3 E(X_t \mid X_{t-1})}{\partial \phi_i \partial \phi_j \partial \phi_k}(\phi) \right| \leq H_t^{ijk}, \ E\left(H_t^{ijk}\right) < \infty.$$

Note that $M_{t-1}^{111}(\phi) = M_{t-1}^{122}(\phi) = M_{t-1}^{211}(\phi) = M_{t-1}^{222}(\phi) = 0$ and

$$M_{t-1}^{112}(\phi) = M_{t-1}^{121}(\phi) = |X_{t-1} - \mu|,$$
$$M_{t-1}^{212}(\phi) = M_{t-1}^{221}(\phi) = |\alpha - 1| < 1;$$

then we can choose $G_{t-1}^{ijk}(\phi) = (X_{t-1} - \mu)^2 + 1, \forall i, j, k = 1, 2$, which guarantees that $M_{t-1}^{ijk}(\phi) < G_{t-1}^{ijk}(\phi)$ and $E(G_{t-1}^{ijk}) = \text{Var}(X_{t-1} + 1) < \infty$.

For $N_t^{ijk}(\phi)$, it is easy to know that $N_t^{ijk}(\phi) = 0, \forall i, j, k = 1, 2$. So we choose $H_t^{ijk}(\phi) = 0$, $\forall i, j, k = 1, 2$ to satisfy $N_t^{ijk}(\phi) < H_t^{ijk}(\phi)$ and $E(H_t^{ijk}) = 0 < \infty$.

The above three conditions ensure that $(\hat{\alpha}_{CLS}, \hat{\mu}_{CLS})$ is a strongly consistent estimator for (α, μ). According to Theorem 3.2 in Tjøstheim (1986) [28], the asymptotic distribution of $(\hat{\alpha}_{CLS}, \hat{\mu}_{CLS})$ is

$$\sqrt{T}(\hat{\alpha}_{CLS} - \alpha, \hat{\mu}_{CLS} - \mu)' \xrightarrow{d} N(0, \Sigma),$$

where $\Sigma = U^{-1} R U^{-1}$,

$$U = E\left\{ \frac{\partial E(X_t \mid X_{t-1})}{\partial \phi}^T(\phi) \cdot \frac{\partial E(X_t \mid X_{t-1})}{\partial \phi}(\phi) \right\},$$

$$R = E\left\{ \frac{\partial E(X_t \mid X_{t-1})}{\partial \phi}^T(\phi) f_{t|t-1}(\phi) \frac{\partial E(X_t \mid X_{t-1})}{\partial \phi}(\phi) \right\},$$

and

$$f_{t|t-1}(\phi) = E\left\{ (X_t - E(X_t \mid X_{t-1}))(X_t - E(X_t \mid X_{t-1}))^T \mid X_{t-1} \right\}.$$

We can then find that

$$\Sigma = \begin{bmatrix} \dfrac{\alpha(1+\alpha)\mu + \sigma_\epsilon^2}{(1-\alpha)^2} & \dfrac{\alpha\sigma^2}{\mu(1+\mu)} \\ \dfrac{\alpha\sigma^2}{\mu(1+\mu)} & \dfrac{\alpha(1-\alpha)(\mu_3 - 2\mu\sigma^2 - \mu^3) + \sigma_\epsilon^2 \sigma^2}{\mu^2(1+\mu)^2} \end{bmatrix},$$

where $\mu_3 = E(X_t^3) = \dfrac{(1-\alpha^3)\mu^3 + (1 + 2\alpha^2 - 3\alpha^3)\mu\sigma^2 + \alpha^2(1-\alpha)\sigma^2}{1-\alpha^3}$, which follows from the following derivation:

$$\mu_3 = E[X_t^3] = E[X_t^2(\alpha \circ X_{t-1} + \epsilon_t)] = E[E[X_t^2(\alpha \circ X_{t-1} + \epsilon_t) \mid X_{t-1}]]$$
$$= E[\alpha X_{t-1} E[X_t^2 \mid X_{t-1}] + \mu_\epsilon E[X_t^2 \mid X_{t-1}]]$$
$$= E[\alpha X_{t-1} E[X_t^2 \mid X_{t-1}]] + \mu_\epsilon E[X_t^2],$$

according to Lemma 1,

$$E[X_t^2|X_{t-1}] = \text{Var}[X_t|X_{t-1}] + (E[X_t|X_{t-1}])^2$$
$$= \alpha^2 X_t^2 + (\alpha(1-\alpha) + 2\alpha\mu_\epsilon)X_t + \mu_\epsilon^2 + \sigma_\epsilon^2;$$

then, we have

$$\mu_3 = E[\alpha X_{t-1} E[X_t^2|X_{t-1}]] + \mu_\epsilon E[X_t^2]$$
$$= \alpha^3 E[X_{t-1}^3] + \alpha^2(1-\alpha)E[X_{t-1}^2] + 2\alpha^2\mu_\epsilon E[X_{t-1}^2] + \alpha\mu(\mu_\epsilon^2 + \sigma_\epsilon^2) + \mu_\epsilon E[X_t^2]$$
(where $\mu_\epsilon = (1-\alpha)\mu$ and $\sigma_\epsilon^2 = (1-\alpha^2)\sigma^2 - \alpha(1-\alpha)\mu$)
$$= \alpha^3 \mu_3 + (1-\alpha^3)\mu^3 + (1 + 2\alpha^2 - 3\alpha^3)\mu\sigma^2 + \alpha^2(1-\alpha)\sigma^2.$$

Thus, we obtain that

$$\mu_3 = \frac{(1-\alpha^3)\mu^3 + (1+2\alpha^2-3\alpha^3)\mu\sigma^2 + \alpha^2(1-\alpha)\sigma^2}{1-\alpha^3}.$$

Appendix A.2. Proof of Theorem 3

The proof is similar to that of Theorem 4.2 in Cunha et al. (2021) [15]. For estimator $\hat{\alpha}$, we have

$$\sqrt{T}(\hat{\alpha}_{YW} - \hat{\alpha}_{CLS}) = \frac{(D_{CLS} - D_{YW})}{D_{CLS}D_{YW}} \cdot \frac{\sum_{t=2}^T X_t X_{t-1}}{\sqrt{T}}$$
$$- \frac{\left(D_{CLS}\sum_{t=2}^T \frac{X_t}{T} - D_{YW}\sum_{t=2}^T \frac{X_t}{T-1}\right)}{D_{CLS}D_{YW}} \cdot \frac{\sum_{t=1}^T X_t}{\sqrt{T}}$$
$$- \frac{D_{YW}}{D_{CLS}D_{YW}} \cdot \sum_{t=2}^T \frac{X_t}{T-1} \frac{X_T}{\sqrt{T}} + \frac{D_{CLS}}{D_{CLS}D_{YW}} \cdot \bar{X} \frac{(X_T - \bar{X})}{\sqrt{T}}$$
$$= o_p(1)O_p(1) - o_p(1)O_p(1) - O_p(1)O_p(1)o_p(1) + O_p(1)O_p(1)o_p(1)$$
$$= o_p(1),$$

where $D_{CLS} = \frac{1}{T}\left[\sum_{t=2}^T X_{t-1}^2 - \frac{1}{T-1}\left(\sum_{t=2}^T X_{t-1}\right)^2\right]$ and $D_{YW} = \frac{1}{T}\sum_{t=1}^T (X_t - \bar{X})^2 = \frac{1}{T}\sum_{t=1}^T X_t^2 - \bar{X}^2$. For estimator $\hat{\theta}$, we only need to prove $\sqrt{T-1}(\hat{\mu}_{\epsilon,CLS} - \hat{\mu}_{\epsilon,YW})$ is $o_p(1)$.

$$\sqrt{T-1}(\hat{\mu}_{\epsilon,CLS} - \hat{\mu}_{\epsilon,YW})$$
$$= \sqrt{T-1}\left(\frac{\sum_{t=2}^T X_t - \hat{\alpha}_{CLS}\sum_{t=2}^T X_{t-1}}{T-1} - \bar{X}(1-\hat{\alpha}_{YW})\right)$$
$$= \frac{1}{\sqrt{T-1}}\left(\sum_{t=2}^T X_t - \hat{\alpha}_{CLS}\sum_{t=2}^T X_{t-1} - T\bar{X}(1-\hat{\alpha}_{YW}) + \bar{X}(1-\hat{\alpha}_{YW})\right)$$
$$= \frac{\hat{\alpha}_{CLS}X_T - X_1}{\sqrt{T-1}} - \sqrt{\frac{T}{T-1}} \cdot \sqrt{T}(\hat{\alpha}_{CLS} - \hat{\alpha}_{YW})\bar{X} + \frac{\bar{X}}{\sqrt{T-1}}(1-\hat{\alpha}_{YW})$$
$$= o_p(1) - o_p(1) + o_p(1) = o_p(1).$$

References

1. McKenzie, E. Some simple models for discrete variate time series. *Water Resour. Bull.* **1985**, *21*, 645–650. [CrossRef]
2. Al-Osh, M.A.; Alzaid, A.A. First-order integer-valued autoregressive (INAR(1)) process. *J. Time Ser. Anal.* **1987**, *8*, 261–275. [CrossRef]
3. Steutel, F.W.; van Harn, K. Discrete analogues of self-decomposability and stability. *Ann. Probab.* **1979**, *7*, 893–899. [CrossRef]
4. Alzaid, A.A.; Al-Osh, M.A. First-order integer-valued autoregressive process: Distributional and regression properties. *Stat. Neerl.* **1988**, *42*, 53–61. [CrossRef]

5. Weiß, C.H. *An Introduction to Discrete-Valued Time Series*; John Wiley & Sons: Hoboken, NJ, USA, 2018.
6. Ristić, M.M.; Bakouch, H.S.; Nastić, A.S. A new geometric first-order integer-valued autoregressive (NGINAR(1)) process. *J. Stat. Plan. Inference* 2009, *139*, 2218–2226. [CrossRef]
7. Liu, Z.; Zhu, F. A new extension of thinning-based integer-valued autoregressive models for count data. *Entropy* 2021, *23*, 62. [CrossRef] [PubMed]
8. Jung, R.C.; Ronning, G.; Tremayne, A.R. Estimation in conditional first order autoregression with discrete support. *Stat. Pap.* 2005, *46*, 195–224. [CrossRef]
9. Jazi, M.A.; Jones, G.; Lai, C.-D. First-order integer valued AR processes with zero inflated Poisson innovations. *J. Time Ser. Anal.* 2012, *33*, 954–963. [CrossRef]
10. Jazi, M.A.; Jones, G.; Lai, C.-D. Integer valued AR(1) with geometric innovations. *J. Iran. Stat. Soc.* 2012, *11*, 173–190.
11. Schweer, S.; Weiß, C.H. Compound Poisson INAR(1) processes: Stochastic properties and testing for overdispersion. *Comput. Stat. Data Anal.* 2014, *77*, 267–284. [CrossRef]
12. Livio, T.; Mamode Khan, N.; Bourguignon, M.; Bakouch, H.S. An INAR(1) model with Poisson–Lindley innovations. *Econ. Bull.* 2018, *38*, 1505–1513.
13. Bourguignon, M.; Rodrigues, J.; Santos-Neto, M. Extended Poisson INAR(1) processes with equidispersion, underdispersion and overdispersion. *J. Appl. Stat.* 2019, *46*, 101–118. [CrossRef]
14. Qi, X.; Li, Q.; Zhu, F. Modeling time series of count with excess zeros and ones based on INAR(1) model with zero-and-one inflated Poisson innovations. *J. Comput. Appl. Math.* 2019, *346*, 572–590. [CrossRef]
15. Cunha, E.T.D.; Bourguignon, M.; Vasconcellos, K.L.P. On shifted integer-valued autoregressive model for count time series showing equidispersion, underdispersion or overdispersion. *Commun. Stat.-Theory Methods* 2021. [CrossRef]
16. Castellares, F.; Ferrari, S.L.P.; Lemonte, A.J. On the Bell distribution and its associated regression model for count data. *Appl. Math. Model.* 2018, *56*, 172–185. [CrossRef]
17. Akaike, H. Information theory as an extension of the maximum likelihood principle. In *Proceedings of the Second International Symposium on Information Theory*; Petrov, B.N., Csaki, F., Eds.; Akadémiai Kiado: Budapest, Hungary, 1973; pp. 267–281.
18. Schwarz G. Estimating the Dimension of a Model. *Ann. Stat.* 1978, *6*, 461–464. [CrossRef]
19. Bozdogan, H. Model selection and Akaike's Information Criterion (AIC): The general theory and its analytical extensions. *Psychometrika* 1978, *52*, 345–370. [CrossRef]
20. Hannan, E.J.; Quinn, B.G. The Determination of the Order of an Autoregression. *J. R. Stat. Soc. Ser. B* 1979, *41*, 190–195. [CrossRef]
21. Bell, E.T. Exponential polynomials. *Ann. Math.* 1934, *35*, 258–277. [CrossRef]
22. Batsidis, A.; Jiménez-Gamero, M.D.; Lemonte, A.J. On goodness-of-fit tests for the Bell distribution. *Metrika* 2020, *83*, 297–319. [CrossRef]
23. Castellares, F.; Lemonte, A.J.; Moreno–Arenas, G. On the two-parameter Bell–Touchard discrete distribution. *Commun. Stat.-Theory Methods* 2020, *49*, 4834–4852. [CrossRef]
24. Lemonte, A.J.; Moreno-Arenas, G.; Castellares, F. Zero-inflated Bell regression models for count data. *J. Appl. Stat.* 2020, *47*, 265–286. [CrossRef]
25. Muhammad, A.; Muhammad, N.A.; Abdul, M. On the estimation of Bell regression model using ridge estimator. *Commun. Stat.-Simul. Comput.* 2021. [CrossRef]
26. Du, J.G.; Li, Y. The integer valued autoregressive (INAR(p)) model. *J. Times Ser. Anal.* 1991, *12*, 129–142.
27. Fisher, R.A. The significance of deviations from expectation in a Poisson series. *Biometrics* 1950, *6*, 17–24. [CrossRef]
28. Tjøstheim, D. Estimation in nonlinear time series models. *Stoch. Process. Their Appl.* 1986, *21*, 251–273. [CrossRef]
29. Freeland, R.K.; McCabe, B. Asymptotic properties of CLS estimates in the Poisson AR(1) model. *Stat. Probab. Lett.* 2005, *73*, 147–153. [CrossRef]
30. Franke, J.; Seligmann, T. Conditional maximum likelihood estimates for INAR(1) processes and their application to modelling epileptic seizure counts. In *Developments in Time Series Analysis*; Rao, T.S., Ed.; Chapman and Hall/CRC: Boca Raton, FL, USA, 1993; pp. 310–330.
31. Ljung, G.M.; Box, G.E.P. On a measure of lack of fit in time series models. *Biometrika* 1978, *65*, 297–303. [CrossRef]
32. Weiß, C.H. The INARCH(1) model for overdispersed time series of Counts. *Commun. Stat.-Simul. Comput.* 2010, *39*, 1269–1291. [CrossRef]

Article

A New Extension of Thinning-Based Integer-Valued Autoregressive Models for Count Data

Zhengwei Liu and Fukang Zhu *

School of Mathematics, Jilin University, 2699 Qianjin Street, Changchun 130012, China; zhengweil18@mails.jlu.edu.cn
* Correspondence: fzhu@jlu.edu.cn

Abstract: The thinning operators play an important role in the analysis of integer-valued autoregressive models, and the most widely used is the binomial thinning. Inspired by the theory about extended Pascal triangles, a new thinning operator named extended binomial is introduced, which is a general case of the binomial thinning. Compared to the binomial thinning operator, the extended binomial thinning operator has two parameters and is more flexible in modeling. Based on the proposed operator, a new integer-valued autoregressive model is introduced, which can accurately and flexibly capture the dispersed features of counting time series. Two-step conditional least squares (CLS) estimation is investigated for the innovation-free case and the conditional maximum likelihood estimation is also discussed. We have also obtained the asymptotic property of the two-step CLS estimator. Finally, three overdispersed or underdispersed real data sets are considered to illustrate a superior performance of the proposed model.

Keywords: extended binomial distribution; INAR; thinning operator; time series of counts

1. Introduction

Counting time series naturally occur in many contexts, including actuarial science, epidemiology, finance, economics, etc. The last few years have witnessed the rapid development of modeling time series of counts. One of the most common approaches for modeling integer-valued autoregressive (INAR) time series is based on thinning operators. In order to fit different kinds of situations, many corresponding operators have been developed; see [1] for a detailed discussion on thinning-based INAR models.

The most popular thinning operator is the binomial thinning operator introduced by [2]. Let X be a non-negative integer-valued random variable and $\alpha \in (0,1)$, the binomial thinning operator is defined as

$$\alpha \circ X = \sum_{i=1}^{X} B_i, \quad X > 0, \tag{1}$$

and 0 otherwise, where $\{B_i\}$ is a sequence of independent identically distributed (i.i.d.) Bernoulli random variables with fixed success probability α, and B_i is independent of X. Based on the binomial thinning operator, [3,4] independently proposed an INAR(1) model as follows

$$X_t = \alpha \circ X_{t-1} + \epsilon_t, \quad t \in \mathbb{Z}, \tag{2}$$

where $\{\epsilon_t\}$ is a sequence of i.i.d. integer-valued random variables with finite mean and variance. Since this seminal work, the INAR-type models have received considerable attention. For recent literature on this topic, see [5,6], among others.

Note that B_i in (1) follows a Bernoulli distribution, so $\alpha \circ X$ is always less than or equal to X; in other words, the first part of the right side in (2) cannot be greater than X_{t-1}, which limits the flexibility of the model. Although it has such a shortcoming, the simple form makes it easy to estimate the parameter, and it also has many similar properties to the multiplication operator in the continuous case. For this reason, there have still been many extensions of the binomial thinning operator since its emergence. Zhu and Joe [7] proposed the expectation thinning operator, which is the generalization of binomial thinning from the perspective of a probability generating function (pgf). Although this extension is very successful, the estimation procedure is a little complicated. Compared with this extension, the thinning operator we proposed is simpler and more intuitive. For recent developments, Yang et al. [8] proposed the generalized Poisson (GP) thinning operator, which is defined by replacing B_i with a GP counting series. Although the GP thinning operator is flexible and adaptable, we argue that it has a potential drawback: the GP distribution is not a strict probability distribution in the conventional sense. Recently, Aly and Bouzar [9] introduced a two-parameter expectation thinning operator based on a linear fractional probability generating function, which can be regarded as a general case of at least nine thinning operators. Kang et al [10] proposed a new flexible thinning operator, which is named GSC because of three initiators of the counting series: Gómez-Déniza, Sarabia and Calderín-Ojeda.

Although the binomial thinning operator is very popular, it may not perform very well in large numerical value counting time series. This is because under such circumstances, the predicted data are often volatile, and the data are more likely to be non-stationary when the numerical value is large. We intend to establish a new thinning operator which meets the following requirements: (i) it is an extension of the binomial thinning operator; (ii) it contains two parameters to achieve flexibility, (iii) it has a simple structure and is easy to implement.

Based on the above considerations, we propose a new thinning operator based on the extended binomial (EB) distribution. The operator has two parameters: real-valued α and integer-valued m ($0 \leq \alpha \leq 1$, $m \geq 2$), which is more flexible compared to some single parameter thinning, and the binomial thinning operator (1) can be regarded as a special case of $m = 2$ in the EB thinning. The case of $m > 2$ in the EB thinning usually performs better than $m = 2$ in some large value data sets. In other words, the EB thinning alleviates the main defect of the binomial thinning to some extent. Since the EB thinning is not a special case of the expectation thinning in [9], we have further extended the framework of thinning-based INAR models to provide a new way in practical application. Therefore, an INAR(1) model is proposed based on the EB thinning operator, which is an extension of the model (2) and can more accurately and flexibly capture the dispersed features in real data.

This paper is organized as follows. In Section 2, we review the properties of the EB distribution and then introduce the EB thinning operator. Based on the new thinning operator, we propose a new INAR(1) model. In Section 3, two-step conditional least squares estimation is investigated for the innovation-free case of the model and the asymptotic property of the estimator is obtained. The conditional maximum likelihood estimation is discussed and the numerical simulations. In Section 4, we focus on forecasting and introduce two criteria to compare the prediction results for three overdispersed or underdispersed real data sets, which are considered to illustrate a better performance of the proposed model. In Section 5, we give some conclusions and related discussions.

2. A New INAR(1) Model

The EB distribution comes from the theory about Pascal's triangles, which can be regarded as a multivariate case of the binomial distribution; see [11] for more details. Based on this distribution, we introduce the EB thinning operator and propose a corresponding INAR(1) model.

2.1. EB Distribution

The EB random variable $X_n(m, \alpha)$, denoted as $EB(m, n, \alpha)$, which is defined as follows:

$$P(X_n(m, \alpha) = r) = C_m(n, r) \alpha^r \beta^{(m-1)n-r}, \quad 0 \leq r \leq (m-1)n, \tag{3}$$

where m and n are both integers satisfying $m \geq 2$ and $n \geq 1$; $C_m(n, r)$ can be calculated as

$$C_m(n, r) = \sum_{s=0}^{s_1} (-1)^s \binom{n}{s} \binom{r + n - sm - 1}{n - 1},$$

where $s_1 = \min\{n, \text{integer part in } r/m\}$; and α and β in (3) satisfy the following restriction:

$$\beta^{m-1} + \alpha \beta^{m-2} + \alpha^2 \beta^{m-3} + \ldots + \alpha^{m-1} = 1, \quad 0 \leq \alpha \leq 1, \quad 0 \leq \beta \leq 1. \tag{4}$$

The above restriction is equivalent to $\beta^m - \alpha^m = \beta - \alpha$. The mean and variance of EB random variables $X_n(m, \alpha)$ are

$$E(X_n(m, \alpha)) = n\alpha \frac{1 - m\alpha^{m-1}}{\beta - \alpha}, \quad \text{Var}(X_n(m, \alpha)) = n\alpha\beta \frac{1 - m^2(\alpha\beta)^{m-1}}{(\beta - \alpha)^2},$$

respectively. The pgf of $X_n(m, \alpha)$ can be written as $G(t) = E(t^{X_n(m,\alpha)}) = \left(\frac{\beta^m - \alpha^m t^m}{\beta - \alpha t}\right)^n$.

As $X_n(m, \alpha)$ can be expressed as a convolution, the EB distribution has the reproductive property. Specifically, if $Y_1, Y_2, \ldots, Y_k$ are independent random variables with $Y_i \sim EB(m, n_i, \alpha)$ in (3), then $\sum_{i=1}^k Y_i \sim EB(m, \sum_{i=1}^k n_i, \alpha)$. Notice that random variable $Y_i \sim EB(2, 1, \alpha)$ is equivalent to a Bernoulli random variable satisfying $P(Y_i = 1) = 1 - P(Y_i = 0) = 1 - \alpha$.

2.2. EB Thinning Operator

According to discussions in 2.1, we construct the EB thinning operator based on the configuration of $n = 1$. Let $\{U_i(m, \alpha)\}$ be a sequence of i.i.d. random variables with common distribution $EB(m, 1, \alpha)$, i.e., $P(U_i(m, \alpha) = z) = \alpha^z \beta^{(m-1)-z}, z = 0, \ldots, m-1$, where α and β satisfy (4). Note that the mean and variance of U_i are

$$\mu = E(U_i) := \alpha \frac{1 - m\alpha^{m-1}}{\beta - \alpha}, \quad \sigma^2 = \text{Var}(U_i) := \alpha\beta \frac{1 - m^2(\alpha\beta)^{m-1}}{(\beta - \alpha)^2}. \tag{5}$$

One can easily see that $\mu < \sigma^2$ if and only if

$$m\alpha^{m-2}(m\beta^m - \beta + \alpha) < 1. \tag{6}$$

For any $3 \leq m < \infty$, the left-hand side of (6) approaches 0 as $\alpha \to 0$ and m as $\alpha \to 1$, respectively. Hence, $\mu < \sigma^2$ or $\mu \geq \sigma^2$ is possible. When $m = 2$, which corresponds to the binomial distribution, and (5) gives $\mu = \alpha$ and $\sigma^2 = \alpha\beta$ with $\alpha + \beta = 1$, then we have $\mu > \sigma^2$ for all $0 < \alpha < 1$. When $\alpha \geq \beta$ and $m \to \infty$, one can easily show that $\mu = \theta/(1-\theta)$, $\sigma^2 = \theta/(1-\theta)^2$, where $\theta = \alpha/\beta$. Therefore, $\mu \leq \sigma^2$ for all $0 < \theta < 1$ in this case.

Based on the reproductive property of the EB distribution, we define the EB thinning operator "⊚" as follows: for a non-negative integer-valued random variable X,

$$(m, \alpha) \odot X = \sum_{i=1}^{X} U_i(m, \alpha), \quad X > 0,$$

where $(m, \alpha) \odot X = 0$ if $X = 0$. Note that the EB thinning operator reduces to the binomial operator (1) when $m = 2$. It is easy to know that $(m, \alpha) \odot X \leq X$ or $> X$, so the EB thinning operator is quite flexible when dealing with the overdispersed or underdispersed data sets.

Remark 1. *The computation of (α, m) for given (μ, σ^2) is based on (4) and (5). The solution can be obtained by solving these nonlinear equations. When $m = 3$, $\beta = (-\alpha + \sqrt{4 - 3\alpha^2})/2$ and when $m = 4$,*

$$\beta = \sqrt[3]{-\left(\frac{10\alpha^3}{27} - \frac{1}{2}\right) + \sqrt{\left(\frac{10\alpha^3}{27} - \frac{1}{2}\right)^2 + \left(\frac{2\alpha^2}{9}\right)^3}}$$
$$+ \sqrt[3]{-\left(\frac{10\alpha^3}{27} - \frac{1}{2}\right) - \sqrt{\left(\frac{10\alpha^3}{27} - \frac{1}{2}\right)^2 + \left(\frac{2\alpha^2}{9}\right)^3}} - \frac{\alpha}{3}.$$

For more complex cases ($m \geq 5$), we can derive the solution (α, β, m) by solving these large-scale nonlinear systems, and a more detailed calculation procedure is given in Section 3.3.

2.3. EB-INAR(1) Model

Based on the EB thinning operator, we define the EB-INAR(1) model as follows:

$$X_t = (m, \alpha) \odot X_{t-1} + \epsilon_t, \quad t = 1, 2, \ldots \tag{7}$$

where $\alpha \in (0, 1)$, $\{X_t\}$ is a sequence of non-negative integer-valued random variables; the innovation process $\{\epsilon_t\}$ is a sequence of i.i.d. integer-valued random variables with finite mean and variance; and ϵ_t is independent of $\{X_s, s < t\}$.

In order to obtain the estimation equations, we give some conditional or unconditional moments of the EB-INAR(1) model in the following proposition.

Proposition 1. *Suppose $\{X_t\}$ is a stationary process defined by (7) and let $\mu < 1$; then for $t \geq 1$,*

1. $E(X_t | X_{t-1}) = \mu X_{t-1} + \mu_\epsilon$;
2. $E(X_t) = \frac{\mu_\epsilon}{1-\mu}$;
3. $Var(X_t | X_{t-1}) = \sigma^2 X_{t-1} + \sigma_\epsilon^2$;
4. $Var(X_t) = \frac{\sigma^2 \mu_\epsilon + \sigma_\epsilon^2(1-\mu)}{(1-\mu)^2(1+\sigma^2)}$;
5. $Cov(X_t, X_{t-h}) = \mu^h Var(X_{t-h})$ and $Corr(X_t, X_{t-h}) = \mu^h$, for $h = 0, 1, 2, \ldots$

where μ_ϵ and σ_ϵ^2 are the expectation and variance of the innovation ϵ_t, respectively.

The proof of some of these properties mentioned above is given in Appendix A.

Remark 2. *Inspired by the INAR(p) model in [12], we can further extend this model to INAR(p); the EB-INAR(p) model is defined as follows:*

$$X_t = (m, \alpha_1) \odot X_{t-1} + \ldots + (m, \alpha_p) \odot X_{t-p} + \epsilon_t, \quad t = 2, 3, \ldots$$

where $\alpha_1, \ldots, \alpha_p \in (0, 1)$, m is an integer satisfying $m \geq 2$, $\{X_t\}$ is a sequence of non-negative integer-valued random variables, the innovation process $\{\epsilon_t\}$ is a sequence of i.i.d. integer-valued random variables with finite mean and variance, and ϵ_t is independent with $\{X_s, s < t\}$.

We will show that the new model can accurately and flexibly capture the dispersion features of real data in Section 4.

3. Estimation

We use the two-step conditional least squares estimation proposed by [13] to investigate the innovation-free case and the asymptotic properties of the estimators are obtained. Conditional maximum likelihood estimation for the parametric cases are also discussed. Finally, we demonstrate the finite sample performance via simulation studies.

3.1. Two-Step Conditional Least Squares Estimation

Denote $\theta_1 = (\mu, \mu_\epsilon)^\top$, $\theta_2 = (\sigma^2, \sigma_\epsilon^2)^\top$ and $\theta = (\theta_1^\top, \theta_2^\top)^\top$. The two-step CLS estimation will be conducted by the following two steps.

Step 1.1. The estimator for θ_1.

Let $g_1(\theta_1, X_{t-1}) = E(X_t|X_{t-1}) = \mu X_{t-1} + \mu_\epsilon$, $q_{1t}(\theta_1) = (X_t - g_1(\theta_1, X_{t-1}))^2$. Let

$$Q_1(\theta_1) = \sum_{t=1}^n q_{1t}(\theta_1)$$

be the CLS criterion function. Then the CLS estimator $\hat{\theta}_{1,CLS} := (\hat{\mu}_{CLS}, \hat{\mu}_{\epsilon,CLS})^\top$ of θ_1 can be obtained by solving the score equation $\frac{\partial Q_1(\theta_1)}{\partial \theta_1} = 0$, which implies a closed-form solution:

$$\hat{\theta}_{1,CLS} = \begin{pmatrix} \sum_{t=1}^n X_{t-1}^2 & \sum_{t=1}^n X_{t-1} \\ \sum_{t=1}^n X_{t-1} & n \end{pmatrix}^{-1} \begin{pmatrix} \sum_{t=1}^n X_t X_{t-1} \\ \sum_{t=1}^n X_t \end{pmatrix}.$$

Step 1.2. The estimator for θ_2.

Let $Y_t = X_t - E(X_t|X_{t-1})$, $g_2(\theta_2, X_{t-1}) = \text{Var}(X_t|X_{t-1}) = \sigma^2 X_{t-1} + \sigma_\epsilon^2$. Then

$$E(Y_t^2|X_{t-1}) = E((X_t - E(X_t|X_{t-1})^2)|X_{t-1})$$
$$= \text{Var}(X_t|X_{t-1}) = g_2(\theta_2, X_{t-1}).$$

Let $q_{2t}(\theta_2) = (Y_t^2 - g_2(\theta_2, X_{t-1}))^2$; then the CLS criterion function for θ_2 can be written as

$$Q_2(\theta_2) = \sum_{t=1}^n q_{2t}(\theta_2).$$

By solving the score equation $\frac{\partial Q_2(\theta_2)}{\partial \theta_2} = 0$, we can obtain the CLS estimator $\hat{\theta}_{2,CLS} := (\hat{\sigma}_{CLS}^2, \hat{\sigma}_{\epsilon,CLS}^2)^\top$ of θ_2, which also is a closed-form solution:

$$\hat{\theta}_{2,CLS} = \begin{pmatrix} \sum_{t=1}^n X_{t-1}^2 & \sum_{t=1}^n X_{t-1} \\ \sum_{t=1}^n X_{t-1} & n \end{pmatrix}^{-1} \begin{pmatrix} \sum_{t=1}^n Y_t^2 X_{t-1} \\ \sum_{t=1}^n Y_t^2 \end{pmatrix}.$$

Step 2. Estimating parameters (m, α) via the method of moments.

The estimator $(\hat{m}, \hat{\alpha})$ of (m, α), which is called a two-step CLS estimator, can be obtained by solving the following estimation equations:

$$\begin{cases} \hat{\mu}_{CLS} = \alpha \frac{1 - m\alpha^{m-1}}{\beta - \alpha}, \\ \hat{\sigma}_{CLS}^2 = \alpha\beta \frac{1 - m^2(\alpha\beta)^{m-1}}{(\beta - \alpha)^2}, \end{cases} \quad (8)$$

where α and β satisfy (4).

Therefore, the resulting CLS estimator is $\hat{\Theta}_{CLS} = (\hat{m}_{CLS}, \hat{\alpha}_{CLS}, \hat{\mu}_{\epsilon,CLS}, \hat{\sigma}_{\epsilon,CLS}^2)^\top$. To study the asymptotic behaviour of the estimator, we make the following assumptions:

Assumption 1. $\{X_t\}$ *is a stationary and ergodic process;*

Assumption 2. $EX_t^4 < \infty$.

Proposition 2. *Under assumptions 1 and 2, the CLS estimator $\hat{\theta}_{1,CLS}$ is strongly consistent and asymptotically normal:*

$$\sqrt{n}(\hat{\theta}_{1,CLS} - \theta_{1,0}) \xrightarrow{L} N(0, V_1^{-1} W_1 V_1^{-1}),$$

where $V_1 := E\left(\frac{\partial}{\partial \theta_1} g_1(\theta_{1,0}, X_0) \frac{\partial}{\partial \theta_1^\top} g_1(\theta_{1,0}, X_0)\right)$, $W_1 := E\left(q_{11}(\theta_{1,0}) \frac{\partial}{\partial \theta_1} g_1(\theta_{1,0}, X_0) \frac{\partial}{\partial \theta_1^\top} g_1(\theta_{1,0}, X_0)\right)$, and $\theta_{1,0} = (\mu_0, \mu_{\epsilon_0})$ denotes the true value of θ_1.

To obtain the asymptotic normality of $\hat{\theta}_{2,CLS}$, we make a further assumption:

Assumption 3. $EX_t^6 < \infty$.

Then we have the following proposition.

Proposition 3. *Under assumptions 1 and 3, the CLS estimator $\hat{\theta}_{2,CLS}$ is strongly consistent and asymptotically normal:*

$$\sqrt{n}(\hat{\theta}_{2,CLS} - \theta_{2,0}) \xrightarrow{L} N(0, V_2^{-1} W_2 V_2^{-1}),$$

where $V_2 := E\left(\frac{\partial}{\partial \theta_2} g_2(\theta_{2,0}, X_0) \frac{\partial}{\partial \theta_2^\top} g_2(\theta_{2,0}, X_0)\right)$, $W_2 := E\left(q_{21}(\theta_{2,0}) \frac{\partial}{\partial \theta_2} g_2(\theta_{2,0}, X_0) \frac{\partial}{\partial \theta_2^\top} g_2(\theta_{2,0}, X_0)\right)$, *and* $\theta_{2,0} = (\sigma_0^2, \sigma_{\epsilon_0}^2)$ *denotes the true value of θ_2.*

Based on Propositions 2 and 3 and Theorem 3.2 in [14], we have the following proposition.

Proposition 4. *Under assumptions 1 and 3, the CLS estimator $\hat{\theta}_{CLS} = (\hat{\theta}_{1,CLS}, \hat{\theta}_{2,CLS})^\top$ is strongly consistent and asymptotically normal:*

$$\sqrt{n}(\hat{\theta}_{CLS} - \theta_0) \xrightarrow{L} N(0, \Omega),$$

where

$$\Omega = \begin{pmatrix} V_1^{-1} W_1 V_1^{-1} & V_1^{-1} M V_2^{-1} \\ V_2^{-1} M^\top V_1^{-1} & V_2^{-1} W_2 V_2^{-1} \end{pmatrix},$$

$M = E\left(\sqrt{q_{11}(\theta_{1,0}) q_{21}(\theta_{2,0})} \frac{\partial}{\partial \theta_1} g_1(\theta_{1,0}, X_0) \frac{\partial}{\partial \theta_2^\top} g_2(\theta_{2,0}, X_0)\right)$, *and* $\theta_0 = (\theta_{1,0}, \theta_{2,0})^\top$ *denotes the true value of θ.*

We do the following preparation to establish Proposition 5. Based on (5), solve the equation about (m, α), and denote the solution as $(h_1(\mu, \sigma^2), h_2(\mu, \sigma^2))$. Let

$$D = D(\mu, \sigma^2) = \begin{pmatrix} \partial h_1/\partial \mu & \partial h_1/\partial \sigma^2 \\ \partial h_2/\partial \mu & \partial h_2/\partial \sigma^2 \end{pmatrix}. \quad (9)$$

Based on Proposition 4, we state the strong consistency and asymptotic normality of $(\hat{m}, \hat{\alpha})^\top$ in the following proposition.

Proposition 5. *Under assumptions 1 and 3, the CLS estimator $(\hat{m}_{CLS}, \hat{\alpha}_{CLS})^\top$ is strongly consistent and asymptotically normal:*

$$\sqrt{n}\begin{pmatrix} \hat{m}_{CLS} - m_0 \\ \hat{\alpha}_{CLS} - \alpha_0 \end{pmatrix} \xrightarrow{L} N(0, D\Sigma D^\top),$$

where D is given in (9); $\Sigma = diag(IV_1^{-1} W_1 V_1^{-1} I^\top, IV_2^{-1} W_2 V_2^{-1} I^\top)$ with $I = (1, 0)$; m_0 and α_0 denote the true values of m and α, respectively.

The brief proofs of Propositions 2–5 are given in Appendix A.

3.2. Conditional Maximum Likelihood Estimation

We maximize the likelihood function with respect to the model parameters $\theta = (m, \alpha, \delta)$ to get the conditional maximum likelihood (CML) estimate of the parametric case

$$L(X_1 = x_1, \ldots, X_N = x_N | \theta) = P_\theta(X_1 = x_1) \prod_{i=1}^{N} P_\theta(X_i = x_i | X_{i-1} = x_{i-1}, \ldots, X_1 = x_1)$$

$$= P_\theta(X_1 = x_1) \prod_{i=1}^{N} P_\theta(X_i = x_i | X_{i-1} = x_{i-1}),$$

where δ is the parameter of ϵ_i, P_{X_1} is the pmf for X_1 and $P_\theta(X_{i+1}|X_i)$ is the conditional pmf. Since the marginal distribution is difficult to obtain in general, a simple approach is conditional on the observed X_1. By essentially ignoring the dependency on the initial value and considering the CML estimate given X_1 as an estimate for θ by maximizing the conditional log-likelihood

$$l(X_1 = x_1, \ldots, X_N = x_N | \theta) = \sum_{i=2}^{N} \log P_\theta(X_i | X_{i-1})$$

over Θ, we denote the CML estimate by $\hat{\theta} = (\hat{m}, \hat{\alpha}, \hat{\delta})$. The log-likelihood function is as follows:

$$l(X_1 = x_1, \ldots, X_N = x_N | \theta) = \sum_{i=2}^{N} \log \left\{ \sum_{w=0}^{\min\{(m-1)x_{i-1}, x_i\}} C_m(x_{i-1}, w) \alpha^w \beta^{(m-1)x_{i-1}-w} \cdot P(\epsilon_i = x_i - w) \right\},$$

where α and β satisfy (4); ϵ_i follows a non-negative discrete distribution with a parameter δ. In what follows, we consider two cases: $m = 3, 4$.

Case 1: For $m=3$ with Poisson innovation, i.e., $\epsilon_t \sim P(\delta)$.

$$l(X_1 = x_1, \ldots, X_N = x_N | \theta) = \sum_{i=2}^{N} \log \left\{ \sum_{w=0}^{\min\{2x_{i-1}, x_i\}} \sum_{t_1=\max\{0, x_{i-1}-w\}}^{x_{i-1}-\frac{w}{2}} \binom{x_{i-1}}{t_1} \binom{x_{i-1}-t_1}{2x_{i-1}-2t_1-w} \right.$$

$$\left. \cdot (\beta^2)^{t_1} (\alpha\beta)^{2x_{i-1}-2t_1-w} (\alpha^2)^{t_1-x_{i-1}+w} \frac{\delta^{(x_i-w)}}{(x_i-w)!} e^{-\delta} \right\},$$

where β is given in Remark 1.

Case 2: For $m=4$ with geometric innovation, i.e. $\epsilon_t \sim Ge(\delta) = (1-\delta)^k \delta$ for $k = 0, 1, 2, \ldots$

$$l(X_1 = x_1, \ldots, X_N = x_N | \theta) = \sum_{i=2}^{N} \log \left\{ \sum_{w=0}^{\min\{3x_{i-1}, x_i\}} \sum_{t_1=x_{i-1}-w}^{x_{i-1}-\frac{w}{3}} \sum_{t_2=\max\{0, 2x_{i-1}-2t_1-w\}}^{\frac{3x_{i-1}-3t_1-w}{2}} \binom{x_{i-1}}{t_1} \binom{x_{i-1}-t_1}{t_2} \right.$$

$$\left. \cdot \binom{x_{i-1}-t_1-t_2}{3x_{i-1}-3t_1-2t_2-w} (\beta^3)^{t_1} (\alpha\beta^2)^{t_2} (\alpha^2\beta)^{3x_{i-1}-3t_1-2t_2-w} \right.$$

$$\left. \cdot (\alpha^3)^{2t_1+t_2-2x_{i-1}+w} (1-\delta)^{(x_i-w)} \delta \right\},$$

where β is given in Remark 1. For higher order m, the formula is a little tedious, which is omitted here. For the estimate of EB-INAR(p), the CML estimation is too complicated, but the two-step CLS estimation is quite feasible, the procedure is similar to the case of $p = 1$. For this reason, we only consider the case of EB-INAR(1) in simulation studies.

3.3. Simulation

A Monte Carlo simulation study was conducted to evaluate the finite sample performance of the estimator. For CLS estimation, we used the package BB in R for solving and optimizing large-scale nonlinear systems to solve Equations (4) and (8). For CML estimation, we used the package maxLik in R to maximize the log-likelihood function.

We considered the following configurations of the parameters:

- Poisson INAR(1) models with $\theta = (m, \alpha, \delta)^\top$:
  (A1) $= (3, 0.2, 1)^\top$; (A2) $= (3, 0.1, 0.5)^\top$; (A3) $= (4, 0.2, 1)^\top$; (A4) $= (4, 0.1, 0.5)^\top$;
- Geometric INAR(1) models with $\theta = (m, \alpha, \delta)^\top$:
  (B1) $= (3, 0.3, 0.5)^\top$; (B2) $= (3, 0.4, 2/3)^\top$; (B3) $= (4, 0.3, 0.5)^\top$; (B4) $= (4, 0.4, 2/3)^\top$.

In simulations, we chose sample sizes $n = 100$, 200 and 400 with $M = 500$ replications for each choice of parameters. The root mean squared error (RMSE) was calculated to evaluate the performance of the estimator according to the following formula: RMSE $= \sqrt{\frac{1}{M-1} \sum_{j=1}^{M} (\hat{\zeta}_j - \zeta_0)^2}$, where $\hat{\zeta}_j$ is the estimator of ζ_0 in the jth replication.

For the CLS estimate, the solutions of (4) and (8) are sensitive to $\hat{\mu}$ and $\hat{\sigma}^2$, so we adopted the following estimation procedure. First, calculate 500 groups of $\hat{\mu}$ and $\hat{\sigma}^2$ estimates, then use the mean values of $\hat{\mu}$ and $\hat{\sigma}^2$ to solve the Equations (4) and (8). The simulation results of CLS are summarized in Table 1. We found that the estimation values are closer to the true value and the values of RMSE gradually decrease as the sample size increases.

Table 1. Means of estimates, RMSEs (within parentheses) by CLS.

Case	n	$\hat{m}$	$\hat{\alpha}$	$\hat{\delta}$
A1	100	2.9100	0.1758	1.0220(0.1595)
	200	2.9219	0.1931	1.0142(0.1148)
	400	2.9995	0.1984	1.0090(0.0771)
A2	100	2.7628	0.0886	0.5014(0.0893)
	200	2.8169	0.0873	0.5049(0.0668)
	400	2.9132	0.0942	0.5004(0.0449)
A3	100	3.7885	0.1713	1.0265(0.1594)
	200	3.8450	0.1902	1.0165(0.1127)
	400	3.8957	0.1952	1.0113(0.0789)
A4	100	3.8421	0.0912	0.5059(0.0912)
	200	3.8483	0.0912	0.5031(0.0603)
	400	3.9590	0.0981	0.5012(0.0439)
B1	100	2.9074	0.3122	0.4981(0.0511)
	200	2.9588	0.3115	0.5008(0.0365)
	400	2.9858	0.3087	0.4984(0.0270)
B2	100	3.0719	0.3841	0.6578(0.0553)
	200	3.0523	0.3877	0.6569(0.0400)
	400	3.0986	0.3924	0.6600(0.0307)
B3	100	3.6127	0.2703	0.4962(0.0536)
	200	3.7937	0.2961	0.4980(0.0404)
	400	3.9217	0.2984	0.4970(0.0269)
B4	100	3.9575	0.3935	0.6417(0.0653)
	200	4.0388	0.3929	0.6556(0.0478)
	400	4.0027	0.3953	0.6606(0.0353)

Note: RMSE, root mean squared error.

As it is a little difficult to estimate the parameter m in CML estimation, we considered m as known. The simulation results of CML estimators are given in Table 2. For all cases, all estimates generally show small values of RMSE, and the values of RMSE gradually decrease as the sample size increases.

Table 2. Means of estimates, RMSEs (within parentheses) by CML.

Case	n	100	200	400
A1	$\hat{\alpha}$	0.1858(0.0656)	0.1927(0.0476)	0.1970(0.0347)
	$\hat{\delta}$	1.0070(0.1420)	1.0043(0.1057)	1.0057(0.0806)
A2	$\hat{\alpha}$	0.1041(0.0575)	0.1027(0.0453)	0.0973(0.0357)
	$\hat{\delta}$	0.4939(0.0804)	0.4964(0.0593)	0.4973(0.0422)
A3	$\hat{\alpha}$	0.1827(0.0592)	0.1923(0.0406)	0.1966(0.0306)
	$\hat{\delta}$	1.0186(0.1358)	1.0030(0.1059)	1.0002(0.0789)
A4	$\hat{\alpha}$	0.0938(0.0500)	0.0940(0.0414)	0.0987(0.0342)
	$\hat{\delta}$	0.4993(0.0790)	0.5052(0.0583)	0.4967(0.0420)
B1	$\hat{\alpha}$	0.2982(0.0456)	0.2980(0.0329)	0.2992(0.0237)
	$\hat{\delta}$	0.5086(0.0469)	0.5013(0.0323)	0.5016(0.0224)
B2	$\hat{\alpha}$	0.3888(0.0455)	0.3949(0.0331)	0.3980(0.0222)
	$\hat{\delta}$	0.6685(0.0501)	0.6681(0.0374)	0.6664(0.0256)
B3	$\hat{\alpha}$	0.2904(0.0441)	0.2958(0.0276)	0.2990(0.0219)
	$\hat{\delta}$	0.5039(0.0480)	0.5003(0.0338)	0.5006(0.0264)
B4	$\hat{\alpha}$	0.3867(0.0348)	0.3940(0.0240)	0.3965(0.0163)
	$\hat{\delta}$	0.6650(0.0539)	0.6674(0.0382)	0.6662(0.0271)

Note: RMSE, root mean squared error.

4. Real Data Examples

In this section, three real data sets, including overdispersed and underdispersed settings, are considered to illustrate the better performance of the proposed model. The first example is overdispersed crime data in Pittsburgh; the second is overdispersed stock data in New York Stock Exchange (NYSE); and the third is underdispersed crime data in Pittsburgh, which was also analyzed by [15]. As is well known, in time series analysis, forecasting is very important in model evaluation. We first introduce two criteria on forecasting, and other preparations.

4.1. Forecasting

Before introducing the evaluation criterion, we briefly introduce the basic procedure as follows: First, we divide the $n_1 + n_2$ data into two parts, the training set with the first n_1 data and the prediction set with the last n_2 data. The training set is used to estimate the parameters and evaluate the fitness of the model. Then we can evaluate the efficiency of each model by comparing the following criteria between prediction data and the real data in the prediction set.

Similar to the procedure in [16], which performs an out-of-sample experiment to compare forecasting performances of two model-based bootstrap approaches, we introduce the forecasting procedure as follows: For each $t = (n_1 + 1), \ldots, (n_1 + n_2 - 5)$ we estimate an INAR(1) model for the data $x_1, \ldots, x_t$, then we use the fitted result based on $x_1, \ldots, x_t$ to generate the next five forecasts, which is called the 5-step ahead forecast $x_{t+1}^F, \ldots, x_{t+5}^F$ for each t in $\{(n_1 + 1), \ldots, (n_1 + n_2 - 5)\}$, where x_t^F is the forecast at time t. In this way we obtain many sequences of $1, 2, \ldots, 5$ step-ahead forecasts, finally we replicate the whole procedure P times. Then we can evaluate the point forecast accuracy by the forecast mean square error (FMSE) defined as

$$\text{FMSE} = \frac{1}{P} \sum_{i=(n_1+1)}^{n_2} (x_i - \overline{x}_i^F)^2,$$

and forecast mean absolute error (FMAE) defined as

$$\text{FMAE} = \frac{1}{P} \sum_{i=(n_1+1)}^{n_2} |x_i - \overline{x}_i^F|,$$

where x_i is the true value of the data, $\overline{x}_i^F$ is the mean of all the forecasts at i and P is the number of replicates.

4.2. Overdispersed Cases

We consider two overdispersed data sets, the first one contains 144 observations and represents monthly tallies of crime data from the Forecasting Principles website http://www.forecastingprinciples.com, and these crimes are reported in the police car beats in Pittsburgh from January 1990 to December 2001; the second one is Empire District Electric Company (EDE) data set from the Trades and Quotes (TAQ) set in NYSE, which contains 300 observations, and it was also analyzed by [17].

4.2.1. P1V Data

The 45th P1V (Part 1 Violent Crimes) data set contains crimes of murder, rape, robbery and other kinds; see more details in the data dictionary on the Forecasting Principles website. Figure 1 plots the time series plot, the autocorrelation function (ACF) and the partial autocorrelation function (PACF) of 45th data of P1V series, respectively. The maximum value of the data is 15 and the minimum is 0; the mean is 4.3333; the variance is 7.4685. From the ACF plot, we found that the data are dependent. From the PACF plots, we can see that only the first sample is significant, which strongly suggests an INAR(1) model.

First, we divided the data set into two parts–the training set with the first $n_1 = 134$ counting data and the prediction set with the last $n_2 = 10$ data. We fit the training set by the following models: expectation thinning INAR(1) (ETINAR(1)) model in [9], GSC thinning INAR(1) (GSCINAR(1)) model in [10], the binomial thinning INAR(1) model and EB thinning EB-INAR(1) models with $m = 3, 4$. According to the mean and variance of P1V data, we used one of the most common settings–geometric distribution–as the distribution of the innovation in above models.

In order to compare the effectiveness of the models, we consider the following evaluation criteria: (1) AIC. (2) The mean and standard error of Pearson residual r_t and its related Ljung–Box statistics, where the Pearson residuals are defined as

$$r_t = \frac{X_t - \hat{\mu} X_{t-1} - \hat{\mu}_\epsilon}{[\hat{\sigma}^2 X_{t-1} + \hat{\sigma}_\epsilon^2]^{\frac{1}{2}}}, \quad t = 1, 2, \ldots,$$

where $\hat{\mu}$ and $\hat{\sigma}^2$ are the estimated expectation and variance for related thinning operators, respectively. (3) Three goodness-of-fit statistics: RMS (root mean square error), MAE (mean absolute error) and MdAE (median absolute error), where the error is defined by $X_t - E(X_t|X_{t-1})$, $t = 1, \ldots, n_1$. (4) The mean of the data $\hat{\bar{x}}$ on the training set calculated by the estimated results.

Next, focusing on forecasting, we generated $P = 100$ replicates based on the training set for each model. Then we calculated the FMSE and FMAE for each model.

All results of the fitted models are given in Table 3. There is no evidence of any correlation within the residuals of all five models, which is also supported by the Ljung–Box statistic based on 15 lags (because $\chi^2_{0.05}(14) = 23.6847$). There were no significant differences for the RMS, MAE, MdAE and $\hat{\bar{x}}$ values (the true mean of the 134 training set was 4.3880) of the models. In other words, no model performed the best in terms of these four criteria, so we also considered AIC. Since the CML estimator cannot be adopted in GSCINAR(1), one can only compare other criteria.

Considering the fitness on the training set, the EB-INAR(1) with $m = 3$ has the smallest AIC, EB-INAR(1) with $m = 4$ has almost the same AIC as $m = 3$. For the results on forecasting, EB-INAR(1) with $m = 4$ has the smallest FMSE and the second smallest FMAE among all models. EB-INAR(1) with $m = 3$ has the second smallest FMSE and the smallest FMAE. Based on these results, we conclude that EB-INAR(1) with $m = 3, 4$ performs better than INAR(1), ETINAR(1) and GSCINAR(1).

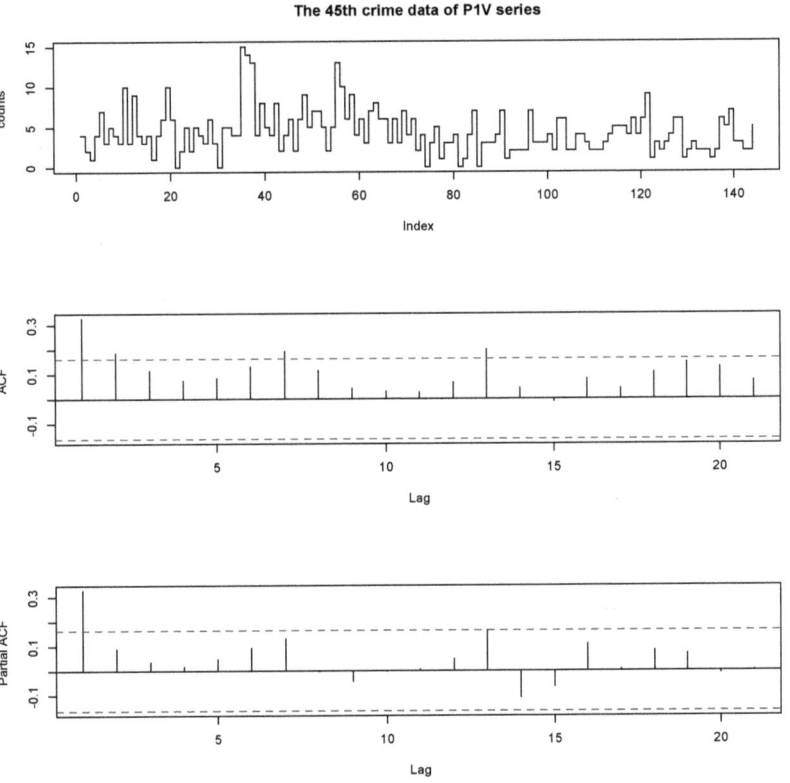

Figure 1. The data, autocorrelation function (ACF) and partial autocorrelation function (PACF) of 45th P1V series.

4.2.2. Stock Data

We analyzed another overdispersed data set of Empire District Electric Company (EDE) from the Trades and Quotes (TAQ) data set in NYSE. The data are about the number of trades in 5 min intervals between 9:45 a.m. and 4:00 p.m. in the first quarter of 2005 (3 January–31 March 2005, 61 trading days). Here we analyze a portion of the data between first to fourth trading days. As there are 75 5 min intervals per day, the sample size was $T = 300$.

Figure 2 plots the time series plot, the ACF and the PACF of the EDE series. The maximum value of the data is 25 and the minimum is 0; the mean is 4.6933; and the variance is 14.1665. It seems that the series is not completely stationary with several outliers or influential observations based on the time series plot. Zhu et al. [18] analyzed the Poisson autoregression for the stock transaction data with extreme values, which can be considered in the current setting. From the ACF plot, we found that the data are dependent. From the PACF plots, we can see that only the first sample is significant, which strongly suggests an INAR(1) model. We used the same procedures and criteria as before. We used the geometric distribution as the distribution of the innovation in above models.

First divide the data set into two parts–the training set with the first $n_1 = 270$ data and the prediction set with the last $n_2 = 30$ data. All results of the fitted models are given in Table 4. Among all models, EB-INAR(1) with $m = 4$ has the smallest AIC, and there is no evidence of any correlation within the residuals of all five models, which is also supported by the Ljung–Box statistic based on 15 lags. There are no significant differences for the RMS, MAE, MdAE and $\hat{\bar{x}}$ values (the true mean of the 270 training set was 4.3407) of all considered models. For the results of prediction, EB-INAR(1) with $m = 4$ has the smallest FMSE and FMAE among all models. Based on the above results, we conclude that EB-INAR(1) with $m = 4$ performs best for this data set.

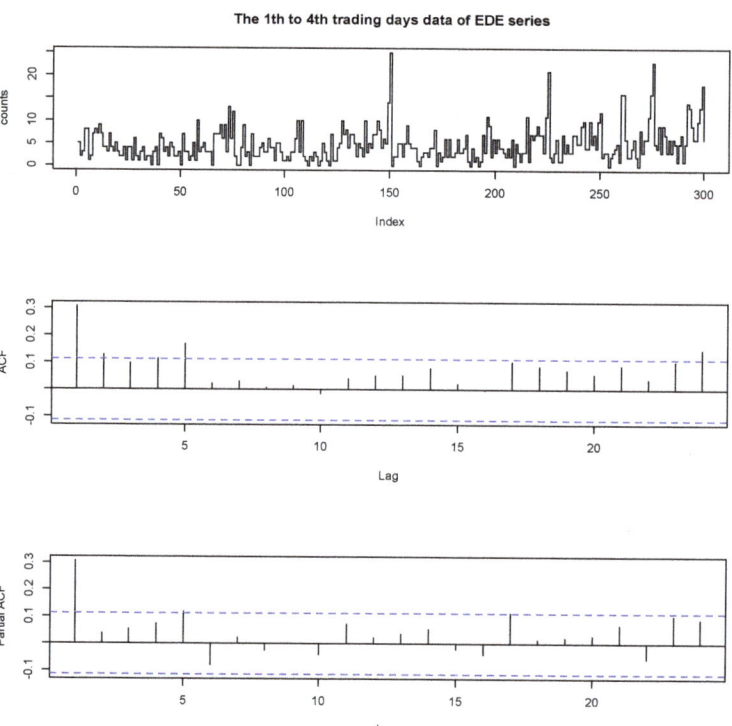

Figure 2. The data, ACF and PACF of first to fourth trading days of EDE series.

Table 3. Fitting results, AIC and some characteristics of P1V data.

Model	Estimates	AIC	$\bar{r}_t$	std(r_t)	Ljung-Box	RMS	MAE	MdAE	$\hat{\bar{x}}$	FMSE	FMAE
INAR(1)	$\hat{\alpha}=0.3955$ $\hat{\delta}=0.2741$	641.9136	0.0020	0.8008	12.5545	2.6384	2.0236	1.5828	4.3811	0.3988	0.1809
ETINAR(1)	$\hat{q}=0.4942$ $\hat{r}=0.3739$ $\hat{\delta}=0.3112$	639.3243	0.0134	0.8392	16.6092	2.6736	2.0628	1.6958	4.3758	0.3847	0.1821
GSCINAR(1)	$\hat{\gamma}=0.8944$ $\hat{\delta}=0.2515$	—	0.0005	0.6805	12.7469	2.6304	2.0027	1.7395	4.3838	0.4115	0.1935
EB-INAR(1) with $m=3$	$\hat{\alpha}=0.3299$ $\hat{\delta}=0.3050$	636.2386	0.0097	0.8525	15.7722	2.6666	2.0563	1.7167	4.3771	0.3770	0.1748
EB-INAR(1) with $m=4$	$\hat{\alpha}=0.3164$ $\hat{\delta}=0.3156$	636.2887	0.0145	0.8526	17.0785	2.6789	2.0673	1.6815	4.3756	0.3589	0.1791

Table 4. Fitting results, AIC and some characteristics of EDE data.

Model	Estimates	AIC	$\bar{r}_t$	std(r_t)	Ljung-Box	RMS	MAE	MdAE	$\hat{\bar{x}}$	FMSE	FMAE
INAR(1)	$\hat{\alpha}=0.2457$ $\hat{\delta}=0.2341$	1341.648	0.0003	0.8590	13.0696	3.3407	2.4817	2.0075	4.3358	10.2602	1.2368
ETINAR(1)	$\hat{q}=0.3204$ $\hat{r}=0.4125$ $\hat{\delta}=0.2533$	1339.397	0.0041	0.8811	16.4070	3.3570	2.4910	2.0534	4.3357	10.6103	1.2688
GSCINAR(1)	$\hat{\gamma}=0.9578$ $\hat{\delta}=0.2284$	—	−0.0003	0.8027	13.1596	3.3397	2.4812	2.0409	4.3361	10.4577	1.2362
EB-INAR(1) with $m=3$	$\hat{\alpha}=0.2200$ $\hat{\delta}=0.2448$	1338.071	0.0018	0.8801	14.5878	3.3477	2.4863	2.0299	4.3352	10.1038	1.2317
EB-INAR(1) with $m=4$	$\hat{\alpha}=0.2194$ $\hat{\delta}=0.2486$	1337.554	0.0027	0.8838	15.3238	3.3514	2.4880	2.0224	4.3350	10.0463	1.2279

4.3. Underdispersed Case

The 11th FAMVIOL data set contains the crimes of family violence, which can also be obtained from the Forecasting Principles website. Figure 3 plots the time series plot, the ACF and the PACF of the 11th data set of FAMVIOL series. The maximum value of the data is 3 and the minimum is 0; the mean is 0.4027; and the variance is 0.3820. We use the procedures and criteria in Section 4.2.1 to compare different models. According to the mean and the variance of FAMVIOL data, we use one of the most common settings-Poisson distribution as the distribution of the innovation in above models.

All results of the fitted models are given in Table 5. There is no evidence of any correlation within the residuals of all five models, which is also supported by the Ljung–Box statistic based on 15 lags. There are no significant differences about the criteria on the fitness and forecasting of all models. ETINAR(1) with the biggest AIC, performed the worst in these models.

Now let us have a brief summary. For the P1V data and stock data, which are overdispersed with slightly high-count data, the EB-INAR(1) of $m > 2$ is obviously better than $m = 2$. For the FAMVIOL data, which is underdispersed with small-count data, the EB-INAR(1) with $m > 2$ is also competitive.

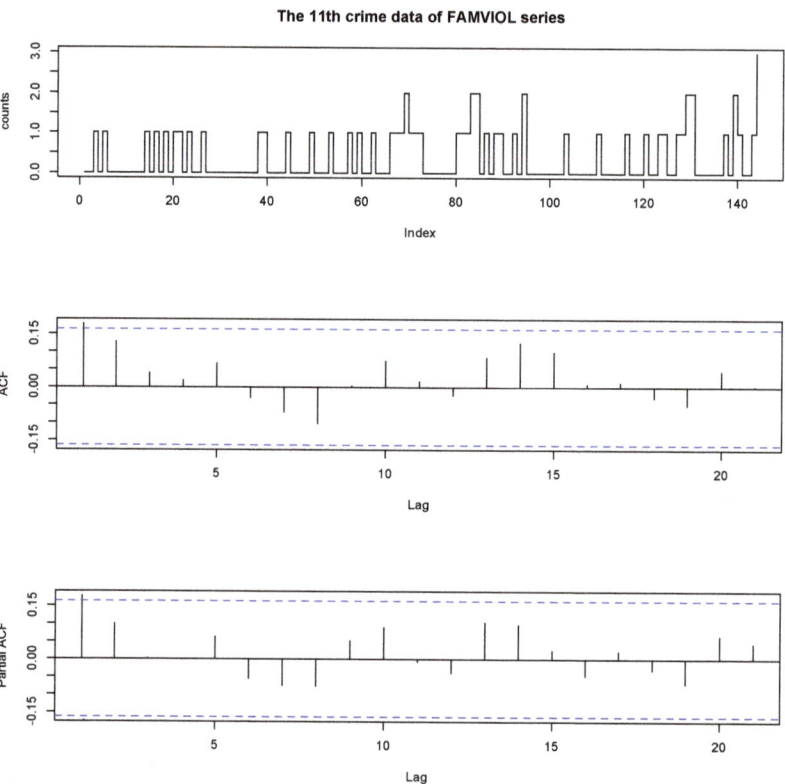

Figure 3. The data, ACF and PACF of the 11th data set of the FAMVIOL series.

Table 5. Fitting results, AIC and some characteristics of FAMVIOL data.

Model	Estimates	AIC	$\bar{r}_t$	$\text{std}(r_t)$	Ljung-Box	RMS	MAE	MdAE	$\hat{\bar{x}}$	FMSE	FMAE
INAR(1)	$\hat{\alpha} = 0.1750$ $\hat{\delta} = 0.3101$	206.5983	−0.0005	0.9157	8.6672	0.5596	0.4940	0.4851	0.3759	0.1111	0.0790
ETINAR(1)	$\hat{q} = 0.1774$ $\hat{p} = 0.0267$ $\hat{\delta} = 0.3092$	208.5826	−0.0003	0.9122	8.6344	0.5596	0.4939	0.4866	0.3759	0.1205	0.0837
GSCINAR(1)	$\hat{\gamma} = 0.9741$ $\hat{\delta} = 0.3045$	——	0.0023	0.8609	8.3735	0.5595	0.4932	0.4944	0.3759	0.1020	0.0814
EB-INAR(1) with $m = 3$	$\hat{\alpha} = 0.1505$ $\hat{\delta} = 0.3068$	206.5335	0.0007	0.9005	8.5485	0.5595	0.4935	0.4900	0.3757	0.1135	0.0792
EB-INAR(1) with $m = 4$	$\hat{\alpha} = 0.1371$ $\hat{\delta} = 0.3131$	206.8414	−0.0014	0.8943	8.4928	0.5597	0.4944	0.4802	0.3759	0.1149	0.0808

5. Conclusions

This paper proposes an EB-INAR(1) model based on the newly constructed EB thinning operator, which is an extension of the thinning-based INAR models. We gave the estimation method for parameters and established the asymptotic properties of the estimators for the innovation-free case. Based on the simulations and real data analysis, the EB-INAR(1) model can accurately and flexibly capture the dispersion features of the data, which shows its effectiveness and practicality. Compared with other models, such as ETINAR(1) and GSCINAR(1), our model is competitive.

We point out that many existing integer-valued models can be generalized by replacing the binomial thinning operator with the EB thinning operator, such as those models in [19–23]. In addition, we can extend the considered first-order INAR model to the higher-order one. More research will be studied in the future.

Author Contributions: Conceptualization, Z.L. and F.Z.; methodology, F.Z.; software, Z.L.; validation, Z.L. and F.Z.; formal analysis, Z.L.; investigation, Z.L. and F.Z.; resources, F.Z.; data curation, Z.L.; writing—original draft preparation, Z.L.; writing—review and editing, F.Z.; visualization, Z.L.; supervision, F.Z.; project administration, F.Z.; funding acquisition, F.Z. All authors have read and agreed to the published version of the manuscript.

Funding: Zhu's work is supported by National Natural Science Foundation of China grant numbers 11871027 and 11731015, and Cultivation Plan for Excellent Young Scholar Candidates of Jilin University.

Conflicts of Interest: The authors declare no conflict of interest.

Appendix A

Proof of Proposition 1. Since 1–4 are easy to verify, we only prove 5. By the law of total covariance, we have

$$
\begin{aligned}
Cov(X_t, X_{t-h}) &= Cov\big(E(X_t|X_{t-1},\dots), E(X_{t-h}|X_{t-1},\dots)\big) + E\big(Cov(X_t, X_{t-h}|X_{t-1},\dots)\big) \\
&= Cov(\mu X_{t-1} + \mu_\epsilon, X_{t-h}) \\
&\quad + E\big(E(X_t - E(X_t|X_{t-1},\dots))(X_{t-h} - E(X_{t-h}|X_{t-1},\dots))|X_{t-1},\dots\big) \\
&= \mu \cdot Cov(X_{t-1}, X_{t-h}) + 0 \\
&= \cdots \\
&= \mu^h \cdot Var(X_{t-h}).
\end{aligned}
$$

Thus, the autocorrelation function $Corr(X_t, X_{t-h}) = \mu^h$. □

Proof of Propositions 2 and 3. Propositions 2 and 3 are similar to Theorems 1 and 2 in [8], which can be proved by verifying the regularity conditions of Theorems 3.1 and 3.2 in [24]. For instance, in the proof of Proposition 2, the partial derivatives $\partial g(\alpha, F_{m-1})/\partial \alpha_i$ have finite fourth moments in [24], which correspond to Assumption 2 in Section 3.1, $u_m^2(\alpha)$ in [24] is corresponds to $q_{1t}(\theta_1)$ in Step 1.1. Hence, Proposition 2 can be regarded as a direct conclusion of Theorem 3.2.

Besides, the proof of Proposition 3 is similar to Proposition 2; the procedure is almost the same as Theorem 3.2 in [24]. □

Proof of Proposition 4. Similarly to the Theorem in [25], based on Theorem 3.2 in [14], we have

$$\sqrt{n}(\hat{\theta}_{CLS} - \theta_0) \xrightarrow{L} N(0, \Omega),$$

where

$$\Omega = \begin{pmatrix} V_1^{-1} W_1 V_1^{-1} & V_1^{-1} M V_2^{-1} \\ V_2^{-1} M^\top V_1^{-1} & V_2^{-1} W_2 V_2^{-1} \end{pmatrix}.$$

Based on the proof of the Theorem in [25], $\sqrt{q_{1t}(\theta_1)}$ corresponds to u_t and $\sqrt{q_{2t}(\theta_2)}$ corresponds to U_t in [25]. Based on the result of V_1, W_1 in Proposition 2 and V_2, W_2 in Proposition 3, we can obtain $M = E\left(\sqrt{q_{11}(\theta_{1,0})q_{21}(\theta_{2,0})}\frac{\partial}{\partial \theta_1}g_1(\theta_{1,0}, X_0)\frac{\partial}{\partial \theta_2^\top}g_2(\theta_{2,0}, X_0)\right)$. □

Proof of Proposition 5. Since the solutions $(h_1(\mu, \sigma^2), h_2(\mu, \sigma^2))$ about (m, α) in (3.1) are real-valued and have a nonzero differential, Proposition 5 is an application of the δ-method, for example, which can be found in Theorem A on p.122 of [26]. □

References

1. Weiß, C.H. *An Introduction to Discrete-Valued Time Series*; John Wiley & Sons, Inc.: Hoboken, NJ, USA, 2018.
2. Steutel, F.W.; van Harn, K. Discrete analogues of self-decomposability and stability. *Ann. Probab.* **1979**, *7*, 893–899. [CrossRef]
3. McKenzie, E. Some simple models for discrete variate time series. *Water Resour. Bull.* **1985**, *21*, 645–650. [CrossRef]
4. Al-Osh, M.; Alzaid, A. First-order integer-valued autoregressive (INAR(1)) processes. *J. Time Ser. Anal.* **1987**, *8*, 261–275. [CrossRef]
5. Schweer, S. A goodness-of-fit test for integer-valued autoregressive processes. *J. Time Series Anal.* **2016**, *37*, 77–98. [CrossRef]
6. Chen, C.W.S.; Lee, S. Bayesian causality test for integer-valued time series models with applications to climate and crime data. *J. R. Stat. Soc. Ser. C. Appl. Stat.* **2017**, *66*, 797–814. [CrossRef]
7. Zhu, R.; Joe, H. Negative binomial time series models based on expectation thinning operators. *J. Statist. Plann. Inference* **2010**, *140*, 1874–1888. [CrossRef]
8. Yang, K.; Kang, Y.; Wang, D.; Li, H.; Diao, Y. Modeling overdispersed or underdispersed count data with generalized Poisson integer-valued autoregressive processes. *Metrika* **2019**, *82*, 863–889. [CrossRef]
9. Aly, E.; Bouzar, N. Expectation thinning operators based on linear fractional probability generating functions. *J. Indian Soc. Probab. Stat.* **2019**, *20*, 89–107. [CrossRef]
10. Kang, Y.; Wang, D.; Yang, K.; Zhang, Y. A new thinning-based INAR(1) process for underdispersed or overdispersed counts. *J. Korean Statist. Soc.* **2020**, *49*, 324–349. [CrossRef]
11. Balasubramanian, K.; Viveros, R.; Balakrishnan, N. Some discrete distributions related to extended Pascal triangles. *Fibonacci Quart.* **1995**, *33*, 415–425.
12. Du, J.; Li, Y. The integer-valued autoregressive (INAR(p)) model. *J. Time Ser. Anal.* **1991**, *12*, 129–142.
13. Li, Q.; Zhu, F. Mean targeting estimator for the integer-valued GARCH(1,1) model. *Statist. Pap.* **2020**, *61*, 659–679. [CrossRef]
14. Nicholls, D.F.; Quinn, B.G. *Random Coefficient Autoregressive Models: An Introduction*; Springer: New York, NY, USA, 1982.
15. Bakouch, H.S.; Ristić, M.M. A mixed thinning based geometric INAR(1) model. *Metrika* **2010**, *72*, 265–280. [CrossRef]
16. Bisaglia, L.; Gerolimetto, M. Model-based INAR bootstrap for forecasting INAR(p) models. *Comput. Statist.* **2019**, *34*, 1815–1848. [CrossRef]
17. Jung, R.C.; Liesenfeld, R.; Richard, J.-F. Dynamic factor models for multivariate count data: An application to stock-market trading activity. *J. Bus. Econom. Statist.* **2011**, *29*, 73–85. [CrossRef]
18. Zhu, F.; Liu, S.; Shi, L. Local influence analysis for Poisson autoregression with an application to stock transaction data. *Stat. Neerl.* **2016**, *70*, 4–25. [CrossRef]
19. Zhang, H.; Wang, D.; Zhu, F. Inference for INAR(p) processes with signed generalized power series thinning operator. *J. Statist. Plann. Inference* **2010**, *140*, 667–683. [CrossRef]
20. Zhang, H.; Wang, D.; Zhu, F. Generalized RCINAR(1) process with signed thinning operator. *Comm. Statist. Theory Methods* **2012**, *41*, 1750–1770. [CrossRef]
21. Qi, X.; Li, Q.; Zhu, F. Modeling time series of count with excess zeros and ones based on INAR(1) model with zero-and-one inflated Poisson innovations. *J. Comput. Appl. Math.* **2019**, *346*, 572–590. [CrossRef]
22. Liu, Z.; Li, Q.; Zhu, F. Random environment binomial thinning integer-valued autoregressive process with Poisson or geometric marginal. *Braz. J. Probab. Stat.* **2020**, *34*, 251–272. [CrossRef]
23. Qian, L.; Li, Q.; Zhu, F. Modelling heavy-tailedness in count time series. *Appl. Math. Model.* **2020**, *82*, 766–784. [CrossRef]
24. Klimko, L.A.; Nelson, P.I. On conditional least squares estimation for stochastic processes. *Ann. Statist.* **1978**, *6*, 629–642. [CrossRef]
25. Zhu, F.; Wang, D. Estimation of parameters in the NLAR(p) model. *J. Time Ser. Anal.* **2008**, *29*, 619–628. [CrossRef]
26. Serfling, R.J. *Approximation Theorems of Mathematical Statistics*; John Wiley & Sons, Inc.: New York, NY, USA, 1980.

Article

A New Overdispersed Integer-Valued Moving Average Model with Dependent Counting Series

Kaizhi Yu and Huiqiao Wang *

School of Statistics, Southwestern University of Finance and Economics, Chengdu 611130, China; yukz@swufe.edu.cn
* Correspondence: 118020209004@smail.swufe.edu.cn

Abstract: A new integer-valued moving average model is introduced. The assumption of independent counting series in the model is relaxed to allow dependence between them, leading to the overdispersion in the model. Statistical properties were established for this new integer-valued moving average model with dependent counting series. The Yule–Walker method was applied to estimate the model parameters. The estimator's performance was evaluated using simulations, and the overdispersion test of the INMA(1) process was applied to examine the dependence between counting series.

Keywords: integer-valued moving average model; counting series; dispersion test

Citation: Yu, K.; Wang, H. A New Overdispersed Integer-Valued Moving Average Model with Dependent Counting Series. *Entropy* **2021**, *23*, 706. https://doi.org/10.3390/e23060706

Academic Editor: Christian H. Weiss

Received: 30 April 2021
Accepted: 31 May 2021
Published: 2 June 2021

Publisher's Note: MDPI stays neutral with regard to jurisdictional claims in published maps and institutional affiliations.

Copyright: © 2021 by the authors. Licensee MDPI, Basel, Switzerland. This article is an open access article distributed under the terms and conditions of the Creative Commons Attribution (CC BY) license (https://creativecommons.org/licenses/by/4.0/).

1. Introduction

Integer-valued time series can be encountered in numerous fields, such as epidemiology, insurance, and intraday stock transitions. The most widely used model is the integer-valued autoregressive (INAR) model, a recursive model first introduced by Alzaid and Al-Osh [1] and is similar to the traditional autoregressive (AR) model. Du and Li [2] generalized the model to the p-th order (which was called INAR(p) model) and proved the ergodic and Markov properties of the model. Similar to the continuous-valued moving average model, the q-th order integer-valued moving average model INMA(q) was introduced by Al-Osh and Alzaid [3], which is a slightly different form proposed by McKenzie [4].

Many researchers generalize the INAR model to deal with different real-life situations. Weiß [5] presented a new INAR(p) model showing possible marginal distributions of the DSD family. This model overcomes the difficulty of choosing the appropriate marginal distribution. Monteiro and Scotto [6] defined the periodic integer-valued autoregressive model, driven by a periodic sequence of independent Poisson-distributed random variables. Weiß [7] proposed the extended Poisson INAR(1) model, where the innovations are assumed to be serially dependent. Zhu [8] introduced a negative binomial INGARCH model to handle integer-valued time series with overdispersion and potential extreme observations. The study by Weiß [9] discussed threshold models for integer-valued time series with infinite range and briefly discussed new models for counting data time series with a finite range. Kang and Wang [10] generalized the mixture INAR(1) model based on mixing Pegram and binomial thinning operator. Li and Wang [11] proposed the first-order mixed integer-valued autoregressive process with zero-inflated generalized power series innovations, which contains the commonly used zero-inflated Poisson and geometric distributions. To handle the non-stationary integer-valued time series with a large dispersion, Kim and Park [12] introduced an integer-valued autoregressive process with a signed binomial thinning operator (INARS(p)).

Various modified thinning operators have been proposed to capture the specificity of real data, and many new INAR-type models have been defined. For example, Zheng and Basawa [13] introduced the random coefficient thinning operator while Ristić and

Bakouch [14] proposed the negative binomial thinning operator. The pth-order integer-valued autoregressive process with signed generalized power series thinning operator was proposed by Zhang et al. [15].

The most significant generalization of the thinning operator was made by Ristić et al. [16] which they called the dependent Bernoulli thinning operator. They constructed the new sequence of the Bernoulli random variable allowing correlation between counting series. Based on this, Miletić Ilić et al. [17] proposed the new model, based on the mix of regular binomial thinning and dependent thinning operator. For more details of thinning operators, refer to Weiß [18].

MA-type models are very important in time series analysis. The method of moving average is generally popular in statistical and mathematical analyses. Some researchers study forecasting using the moving average method. Winters [19] analyzed the exponentially weighted moving average method for forecasting sales. Cox [20] proposed the weighted moving average method to predict the Markov series. Landauskas et al. [21] introduced an algebraic approach to select the appropriate weight coefficients for weight moving averages performing better than classical moving average predictors. Wind speed prediction using combined time series model and neural network prediction was studied by Nan et al. [22]. Other applied weight moving average methods in control charts. Alevizakos et al. [23] proposed the triple exponentially weighted moving average control chart (TEWMA), which improves the detection ability of the classical control chart. Capizzi and Masarotto [24] proposed the adaptive exponentially weighted moving average control chart (AEWMA), which weights past observations of the monitored process using a suitable function of the current error. Adegoke et al. [25] studied the multivariate homogeneously weighted moving average (MHWMA) control chart for monitoring a process mean vector.

Some researchers constructed MA-type models from the perspective of count time series. Brännäs and Quoreshi [26] used the INMA process to model the number of transactions for intraday stocks and extended the model to include explanatory variables. Brännäs and Hall [27] mainly focused on the estimation in the INMA model. The construction of the thinning operation from these studies is based on independent counting series, which is a strong assumption. Thus, to make the model more flexible in capturing the specificity of different data types, this assumption should be relaxed.

For small counts of intraday transactions in stocks per minute, the decision of buyer or seller could be affected by the public news, which means decisions from different individuals may not be uncorrelated. The change in inventories can sometimes be described as an INMA process. However, during a certain period, the change can be influenced by the same external factor. For the INMA model applying a discrete risk model, the number of claims for certain insurance will be affected by the same factor, such as natural disasters. Thus, these claims are no longer independent. The independence between counting series should be relaxed. Allowing the correlation between counting series is natural, and therefore the INMA model based on dependent counting series can be derived to handle different real data situations.

The rest paper is organized as follows. Section 2 presents the model construction and discussions on some relevant statistical properties, and Section 3 discussed the estimation of unknown parameters. Section 4 shows the numerical simulation results and give dispersion test for dependence between counting series, while the conclusions are given in Section 5.

2. The Model and Basic Properties

2.1. The Model Construction

The counting series $\{U\}_{i\in\mathbb{N}}$ of the integer-valued model is defined as:

$$U_i = (1 - V_i)W_i + V_i Z$$

$\{W\}_{i\in\mathbb{N}}$ is a sequence of i.i.d random variable with $W_i \sim B(1,\beta)$, $\beta \in [0,1]$. $\{V\}_{i\in\mathbb{N}}$ is a sequence of i.i.d random variable with $V_i \sim B(1,\theta)$, $\theta \in [0,1]$. Z is a random variable with $Z \sim B(1,\beta)$. The operator $\circ_\theta$ is defined by $\beta \circ_\theta \varepsilon_t = \sum_{i=1}^{\varepsilon_t} U_i$, it is the dependent Bernoulli thinning operator, where ε_t is a non-negative integer-valued random variable. Based on this construction, we can easily verify that $E(U_i) = \beta$, $Var(U_i) = \beta(1-\beta)$, $corr(U_i, U_j) = \theta^2$, which has promising dependence between the counting series. Now we generalize the dependent count series to the INMA(q) process, For convenience, we use $\circ$ instead of $\circ_\theta$ to simplify the notation.

Definition 1 (Dependent Counting series Integer-valued Moving Average Model (DCINMA)). *The DCINMA(q) model is defined as:*

$$X_t = \varepsilon_t + \beta_1 \circ \varepsilon_{t-1} + \ldots + \beta_q \circ \varepsilon_{t-q}$$

$\beta_j \circ \varepsilon_{t-j}, j = 1, 2, \ldots, q$ *is the dependent Bernoulli thinning operator, and the following conditions should be satisfied.*

A1. $\{\varepsilon_t\}_{t \in \mathbb{N}}$ *is a sequence of i.i.d non-negative random variables.*
A2. *The counting variable* $\{U_i\}_{i \in \mathbb{N}}$ *is independent of* ε_t *for any* i, t.
A3. $\beta_j \circ \varepsilon_{t-j}$, *for any* $j = 1, 2, \ldots, q$ *are mutually independent.*

2.2. The Numerical Properties for DCINMA(q) Model

We denote μ_ε and σ_ε^2 as the mean and variance of term ε_t.

Theorem 1. *The numerical characteristics of $\{X_t\}$ in Definition 1 are as follows:*

(i) $E(X_t) = \mu_\varepsilon(1 + \sum_{i=1}^{q} \beta_i)$

(ii) $Var(X_t) = \sigma_\varepsilon^2 + \sum_{i=1}^{q}[\mu_\varepsilon \beta_i + \mu_\varepsilon^2 \theta^2 \beta_i(1-\beta_i) - \mu_\varepsilon(\theta^2 \beta_i - \theta^2 \beta_i^2 + \beta_i^2) + \beta_i^2 \sigma_\varepsilon^2]$

(iii)
$$cov(X_t, X_{t-k}) = \begin{cases} \sigma_\varepsilon^2 \sum_{i=0}^{q-k} \beta_i \beta_{i+k} & k = 1, \ldots, q \\ 0 & k \geq q+1 \end{cases}$$

Proof. See Appendix A. □

Theorem 2. $\{X_t\}$ *is the process defined in Definition 1, then* $\{X_t\}$ *is a covariance stationary process.*

Proof. It can be seen from the Theorem 1 that the unconditional mean and unconditional variance of X_t is a finite constant given the distribution of ε_t. Thus, X_t is a stationary process. □

Theorem 3. $\{X_t\}$ *is the process defined in Definition 1, then* $\{X_t\}$ *is ergodic in mean and autocovariance function.*

Proof. See Appendix A. □

2.3. The Probability Generating Functions for DCINMA Model

Ristić et al. [16] derived the probability generating function of the $\sum_{i=1}^{n} U_i$ as follows:

$$\Phi_U = E[s^{(U_1 + U_2 + \ldots + U_n)}]$$
$$= (1-\beta)(1 - \beta(1-\theta)(1-s))^n + \beta(1 - (\beta + \theta - \beta\theta)(1-s))^n$$

The above equation implies that the term $\sum_{i=1}^{n} U_i$ has a distribution of:

$$U_1 + U_2 + \ldots + U_n = \begin{cases} Bin(n, \beta(1-\theta)) & w.p \quad 1-\beta \\ Bin(n, \beta + \theta - \beta\theta) & w.p \quad \beta \end{cases}$$

Then the probability generating function (PGF) of $\{X_t\}$ is:

$$\phi_{X_n}(s) = [(1-\beta) \cdot \phi_\varepsilon(1 - \beta(1-\theta)(1-s)) + \beta \cdot \phi_\varepsilon(1 - (\beta + \theta - \beta\theta)(1-s))] \cdot \phi_\varepsilon(s)$$

$\phi_\varepsilon(s)$ is the probability generating function of the ε_t. Given the distribution of ε_t, the explicit expression of the probability generating function can be derived. Suppose Poisson distribution of ε_t, then

$$\phi_{X_n}(s) = [(1-\beta) \cdot e^{-\lambda\beta(1-\theta)(1-s)} + \beta e^{-\lambda(\beta+\theta-\beta\theta)(1-s)}] \cdot e^{(\lambda(s-1))}$$

The probability generating function is defined by the probabilities. The uniqueness of a power series expansion implies that the probability generating function in turn defines probabilities.

Therefore, we can derive the probability of $\{X_t\}$. For $X_t = j$, the probability mass function of X_t is:

$$P(X_t = j) = [\frac{1}{j!} \frac{d^j \phi(s)}{ds^j}]_{s=0}$$

The bivariate probability generating function of $\{X_t\}$ is $\Phi_{X_t, X_{t-1}}(s_1, s_2)$. Thus, deriving the explicit expression of bivariate probability generating function with Poisson innovation is easy for DCINMA(1) process.

$$E(s_1^{X_t} s_2^{X_{t-1}}) = E(s_1^{\beta \circ \varepsilon_{t-1} + \varepsilon_t} \cdot s_2^{\beta \circ \varepsilon_{t-2} + \varepsilon_{t-1}})$$
$$= E(s_1^{\beta \circ \varepsilon_{t-1}} \cdot s_1^{\varepsilon_t} \cdot s_2^{\beta \circ \varepsilon_{t-2}} \cdot s_2^{\varepsilon_{t-1}})$$
$$= E(s_1^{\varepsilon_t}) \cdot E(s_1^{\beta \circ \varepsilon_{t-2}}) \cdot E(s_1^{\beta \circ \varepsilon_{t-2}} \cdot s_2^{\varepsilon_{t-1}})$$

Given the Poisson distribution of the innovation term, we can obtain:

$$E(s_1^{\varepsilon_t}) = e^{\lambda(s_1-1)}, E(s_2^{\beta \circ \varepsilon_{t-1}}) = (1-\beta) \cdot e^{-\lambda\beta(1-\theta)(1-s_2)} + \beta \cdot e^{-\lambda(\beta+\theta-\beta\theta)(1-s_2)}$$

$$E(s_1^{\beta \circ \varepsilon_{t-1}} \cdot s_2^{\varepsilon_{t-1}}) = (1-\beta) \cdot e^{\lambda s_2[1-\beta(1-\theta)(1-s_2)]} + \beta \cdot e^{\lambda s_2[(1-\beta-\theta-\beta\theta)(1-s_1)]}.$$

2.4. Compare with the INMA(q) Model

The mean and the covariance of the q-th order integer-valued moving average model has the same expression, and the variance of the INMA(q) process is:

$$V_{inma} = \sigma_\varepsilon^2 + \sum_{i=1}^{q}[\sigma_\varepsilon^2 \beta_i + \mu_\varepsilon \beta_i(1-\beta_i)]$$

From Theorem 1, the variance of the DCINMA process presents a more complicated expression than the INMA(q) process due to the correlation between the counting series of parameter θ. For the Poisson innovation, the overdispersion index of the INMA and DCINMA model is as follows:

$$I_{inma} = 1, I_{dinma} = \frac{1 + \beta + \beta\lambda\theta^2(1-\beta)}{1+\beta}$$

Since the value of λ, β and θ are all non-negative, the term $1 + \beta\lambda\theta^2(1-\beta) > 1$. When two models (INMA(1) and DCINMA(1)) share the same λ and β, the θ will determine whether there is dependence between counting series ($\theta \neq 0$). Thus, if we want to test $\theta = 0$, it is equivalent to evaluating whether the model is overdispersed. If the value of θ is 0, the model degenerates to INMA model.

2.5. Compare the Entropy with INMA(1) Model

Entropy is an important concept in physics, but it can also be applied to other disciplines, including cosmology and economics. Entropy is a measure of the randomness or disorder of a system. In our case, entropy can be seen as a dispersion measure for the model. Thus, we evaluate the model from the perspective of entropy. The definition of Shannon entropy as follows:

$$H(Y) = -\sum_{i=1}^{n} p(y_i) \cdot \ln p(y_i)$$

where Y is a discrete random variable with probability mass function taking values on $y_1, \ldots, y_n$. We denote the $\phi_{dinma}(s)$ and $\phi_{inma}(s)$ as probability generating functions of the DCINMA(1) and INMA(1) process, respectively. Suppose the same innovation term for the two models follows the Poisson distribution. We can rewrite the probability generating function of them as:

$$\phi_{dinma}(s) = (1-\beta) \cdot e^{(s-1) \cdot [\lambda\beta(1-\theta)+\lambda]} + \beta \cdot e^{(s-1) \cdot [\lambda(\beta+\theta-\beta\theta+\lambda)]}$$
$$\phi_{inma}(s) = e^{(s-1) \cdot (\lambda\beta+\lambda)}$$

Thus, we can conclude the distribution of DCINMA(1) and INMA(1) process based on the definition of the probability generating function. $X_t^{inma(1)}$ and $X_t^{dinma(1)}$ denote the sample from INMA(1) model and DCINMA(1) model.

$$X_t^{inma(1)} \sim Poi(\lambda\beta + \lambda)$$
$$X_t^{dinma(1)} \sim \begin{cases} Poi(\lambda\beta(1-\theta) + \lambda) & w.p \quad 1-\beta \\ Poi(\lambda(\beta+\theta-\beta\theta) + \lambda) & w.p \quad \beta \end{cases}$$

The Shannon entropy for both models can be derived as follows:

$$H(X_t^{inma(1)}) = (\lambda\beta + \lambda)[1 - \log((\lambda\beta + \lambda))] + e^{(\lambda\beta+\lambda)} \cdot \sum_{k=0}^{\infty} \frac{(\lambda\beta + \lambda)^k \log(k!)}{k!}$$

$$H(X_t^{dinma(1)}) = (1-\beta) \cdot \{(\lambda\beta(1-\theta) + \lambda)[1 - \log((\lambda\beta(1-\theta) + \lambda))]$$
$$+ e^{(\lambda\beta(1-\theta)+\lambda)} \cdot \sum_{k=0}^{\infty} \frac{(\lambda\beta(1-\theta) + \lambda)^k \log(k!)}{k!}\}$$
$$+ \beta \cdot \{(\lambda(\beta+\theta-\beta\theta) + \lambda)[1 - \log((\lambda(\beta+\theta-\beta\theta) + \lambda))]$$
$$+ e^{(\lambda(\beta+\theta-\beta\theta)+\lambda)} \cdot \sum_{k=0}^{\infty} \frac{(\lambda(\beta+\theta-\beta\theta) + \lambda)^k \log(k!)}{k!}\}$$

The expression of entropy for the DCINMA(1) model is more complicated than the INMA(1) model due to the additional parameter θ.

3. Parameter Estimation

The estimation of the INMA model is complicated. Brännäs and Hall [27] discussed the Yule–Walker estimator, generalized moment method (GMM) based on the probability generating function (PGF) function, and the conditional least square method. Here, we did not attempt to use maximum likelihood estimation, which requires density functions that are generally not easily obtained in the INMA model, especially for this dependent situation. The results of generalized moments method-based probability generating function (PGF) function estimator are highly correlated with the values of z_1 and z_2 in $\Phi_k(z_1, z_2)$, which are not stable. On the other hand, for the conditional least square method, the number of estimation equations are less than the number of parameters. This means that there is an additional parameter θ to be estimated.

Therefore, we derive the Yule–Walker estimator to obtain the unknown parameters for the Poisson DCINMA(q) model. We denote the following symbols: μ_X is the sample mean of $\{X_t\}$, μ_ε is the mean of innovation term ε_t. γ_0 is the sample variance of $\{X_t\}$, $\gamma_1, \gamma_2, \ldots, \gamma_q$ are the sample covariance of 1-th, 2-th,..., q-th order. The β_0 for the equations below is 1.

Then, the unknown parameters in the DCINMA(q) model can be solved by equations through the sample moments function, which is as follows:

$$\begin{cases} \mu_X = \mu_\varepsilon(1 + \beta_1 + \beta_2 + \ldots + \beta_q) \\ \gamma_0 = \sigma_\varepsilon^2 + \sum_{i=1}^{q}[\mu_\varepsilon \beta_i + \mu_\varepsilon^2 \theta^2 \beta_i(1-\beta_i) - \mu_\varepsilon(\theta^2 \beta_i - \theta^2 \beta_i^2 + \beta_i^2) + \beta_i^2 \sigma_\varepsilon^2] \\ \gamma_1 = \lambda(\beta_0 \beta_1 + \beta_1 \beta_2 + \ldots + \beta_{q-1}\beta_q) \\ \gamma_2 = \lambda(\beta_0 \beta_2 + \beta_1 \beta_3 + \ldots + \beta_{q-2}\beta_q) \\ \vdots \\ \gamma_{q-1} = \lambda(\beta_0 \beta_{q-1} + \beta_1 \beta_q) \\ \gamma_q = \lambda(\beta_0 \beta_q) \end{cases}$$

4. Simulation Study

4.1. Estimation of the Model Parameters

In this section, we present some simulation results to show the performance of the estimator using different sample sizes. ($n = 100, 300, 700, 1000$). We focused on the Poisson DCINMA(1) model. The three group parameter values considered in this model are as follows:

Model A: ($\lambda = 1$, $\theta = 0.6$, $\beta = 0.2$)
Model B: ($\lambda = 4$, $\theta = 0.7$, $\beta = 0.1$)
Model C: ($\lambda = 5$, $\theta = 0.5$, $\beta = 0.1$)

We used the above parameter groups to generate the data and applied Yule–Walker method, then computed the bias and standard error based on 10,000 replications for each parameter group. The estimation results and their performance are reported in Table 1.

As shown in Table 1, we obtained nearly convergent estimators in all cases. In all cases, as the sample size increased, the bias decreased.

Table 1. Some numerical results of the estimates for true values of the parameters λ, θ and β.

Sample Size	$\hat{\lambda}$	$\hat{\theta}$	$\hat{\beta}$
(a) True values: $\lambda = 1$, $\theta = 0.6$, $\beta = 0.2$			
100	0.9875	0.5999	0.2641
Bias	0.0124	0.0001	−0.0641
Standard Error	0.1755	0.2065	0.1705
300	0.9858	0.6192	0.2627
Bias	0.0141	−0.0192	−0.0627
Standard Error	0.1082	0.2049	0.1085
700	0.9831	0.6489	0.2663
Bias	0.0168	−0.0489	−0.0663
Standard Error	0.0699	0.1922	0.0717
1000	0.9817	0.6718	0.2684
Bias	0.0182	−0.0718	−0.0684
Standard Error	0.0597	0.1829	0.0600

Table 1. *Cont.*

Sample Size	$\hat{\lambda}$	$\hat{\theta}$	$\hat{\beta}$
(b) True values: $\lambda = 4$, $\theta = 0.7$, $\beta = 0.1$			
100	3.7816	0.6173	0.1690
Bias	0.2183	0.0826	−0.0690
Standard Error	0.5296	0.2031	0.1353
300	3.8872	0.6611	0.1394
Bias	0.1127	0.0388	−0.0394
Standard Error	0.3364	0.1767	0.0855
700	3.9123	0.7208	0.1324
Bias	0.0876	−0.0208	−0.0324
Standard Error	0.2337	0.1353	0.0599
1000	3.9144	0.7430	0.1322
Bias	0.0855	−0.0430	−0.0322
Standard Error	0.2012	0.1142	0.0514
(c) True values: $\lambda = 5$, $\theta = 0.5$, $\beta = 0.1$			
100	4.7799	0.5709	0.1556
Bias	0.2200	−0.0709	−0.0556
Standard Error	0.5777	0.2059	0.1142
300	4.9129	0.5703	0.1279
Bias	0.0870	−0.0703	−0.0279
Standard Error	0.3715	0.1852	0.0730
700	4.9343	0.5688	0.1243
Bias	0.0656	−0.0688	−0.0243
Standard Error	0.2615	0.1529	0.0520
1000	4.9340	0.5738	0.1245
Bias	0.0659	−0.0738	−0.0245
Standard Error	0.2211	0.1330	0.0442

4.2. Testing for Dependence between Counting Series

From Section 2.5, when the two processes have the same λ and β, the DCINMA model always presents overdispersion due to θ. We tested whether the value of θ is 0, which is equivalent to assessing whether the DCINMA process presents overdispersion. Aleksandrov and Weiß [28] proposed the diagnostic test for the INMA process. In this section, only the overdispersion test was applied. Under the null hypothesis $H_0 : \theta = 0$ (the process does not present overdispersion), the distribution of index dispersion has the following form:

$$\hat{I}_{disp} \xrightarrow{d} N(1 - \frac{1}{T}\frac{1+3\beta}{1+\beta}, \frac{1}{T}(2 + 4(\frac{\beta}{1+\beta})^2))$$

We then analyzed the simulation results to assess the performance of the overdispersion test. The nominal level $\alpha = 0.05$ was employed, for sample sizes, $n = 100, 150, 250$. We did 1000 replications to calculate the size and power of the test. The following parameter groups were considered:

Model D: ($\lambda = 3$, $\theta = 0.4$, $\beta = 0.5$)
Model E: ($\lambda = 5$, $\theta = 0.4$, $\beta = 0.3$)
Model F: ($\lambda = 6$, $\theta = 0.3$, $\beta = 0.4$)
Model G: ($\lambda = 7$, $\theta = 0.7$, $\beta = 0.8$)

From Table 2, given the same values for λ and β, under the null hypothesis, $\theta = 0$, the size of $\hat{I}_{disp}$ are close to nominal level. Under the alternative situation, $\theta \neq 0$, for all cases, as n increase the power quickly increases to 1.

Table 2. Size and power of DCINMA(1) and INMA(1) model.

λ	β	θ	n		
			100	250	500
3	0.4	0	0.054	0.057	0.058
		0.5	0.431	0.877	0.937
5	0.4	0	0.053	0.057	0.047
		0.3	0.447	0.824	0.901
6	0.3	0	0.052	0.055	0.047
		0.4	0.63	0.8	0.99
7	0.7	0	0.052	0.046	0.053
		0.8	0.998	1	1

5. Real Data Example

We then applied the proposed model to a real dataset. Unlike the integer-valued autoregressive model, since the density function of this DCINMA process is hard to obtain, the Akaike information criterion (AIC) is difficult to use in measuring the fitness of such an INMA-type model. To assess the performance of the model, we adapted the parametric resampling method by Jung et al. [29].

We used crime data available from the Forecasting Principles site. The data consist of 144 observations of monthly larceny counts for the City of Pittsburgh from January of 1990 to December of 2001. It would be highly improbable for criminals to remain in the same place for a long time. They probably would flee in various directions to commit offenses, so that the INMA-type model is appropriate in this case.

The calculated of sample mean (4.73) and variance (6.40) suggest that the dataset is overdispersed. In addition, the sample autocorrelation function (ACF) in Figure 1 shows that the order of the model should be set to one. We fitted the data set into the DCINMA(1) model and INMA(1) model to evaluate the model performance for overdispersed data. The estimation results for the two models are presented in Table 3.

Table 3. Estimation results of DCINMA(1) and INMA(1) model.

	λ	β	θ
DCINMA(1)	3.21	0.47	0.80
INMA(1)	3.82	0.24	0

The parametric resampling method can be performed in several steps:

Step 1: Generate samples with length equal to the original dataset from the fitted model for R times.

Step 2: Compute the autocorrelation function (ACF) for each sample series. Then derive the empirical sample autocorrelation function (ACF) for the different lag orders.

Step 3: Using the results from **Step 2**, compute the $100(1 - \alpha/2)$ and $100(\alpha/2)$ quantiles.

For the given example, R was set to 5000, and α was set to 0.05. We plotted the acceptance envelope for DCINMA(1) model and INMA(1) model, and the results are shown in Figures 2 and 3. The sample autocorrelations for the real dataset lie within the acceptance envelopes of the DCINMA(1) model except for only one point, while the important second order of the autocorrelation function (ACF) exceed the upper bound of the acceptance envelopes of the INMA(1) model. It is clear that the acceptance envelopes of the DCINMA(1) model performs better than the acceptance envelopes of the INMA(1) model. Thus, the proposed model is suitable for this dataset, adequately representing the autocorrelation for real dataset.

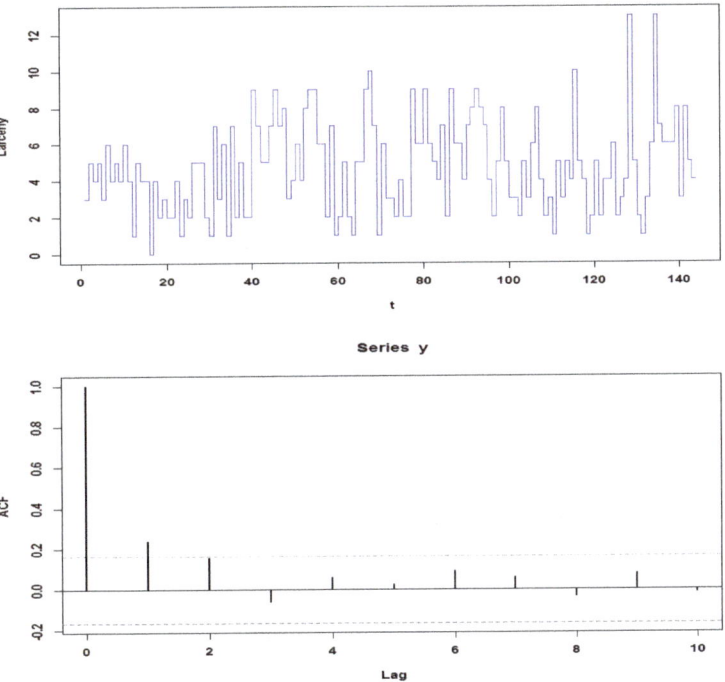

Figure 1. Time series plot (top figure) and autocorrelation function (below figure panel) for actual Larceny dataset.

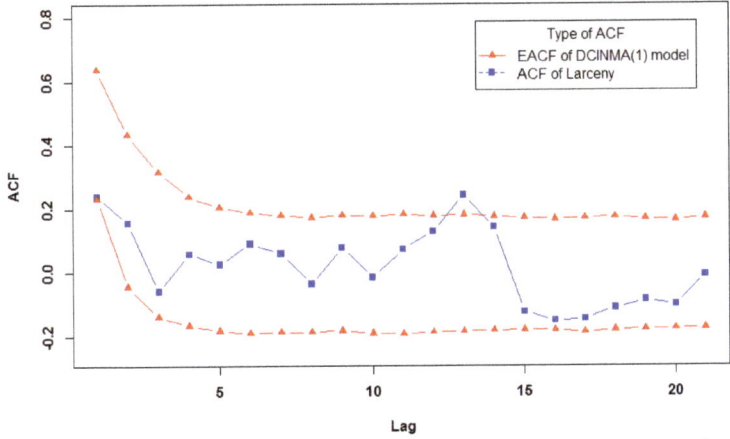

Figure 2. Acceptance envelope of the DCINMA(1) model for the autocorrelation function for the larceny dataset.

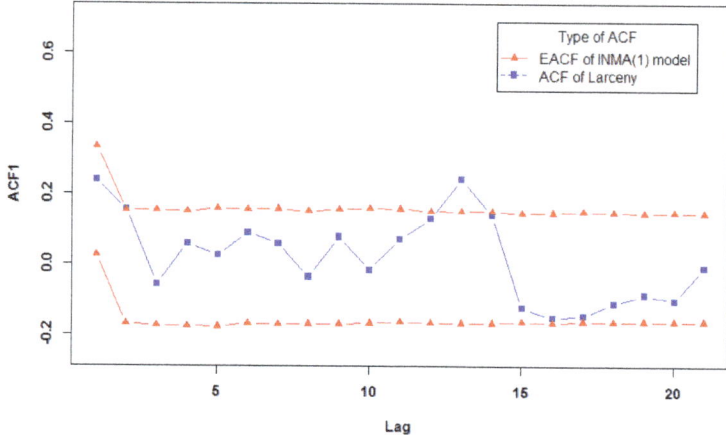

Figure 3. Acceptance envelope of the INMA(1) model for the autocorrelation function for the larceny dataset.

6. Conclusions

In this paper, we constructed a new integer-valued moving average model with dependent counting series. The statistical properties of the proposed model were discussed and evaluated. The parameter estimation of the proposed model is based on the Yule–Walker method. The new model presents overdispersion due to the dependence parameter θ, which means the dependence between counting series can be verified by an overdispersion test. Numerical simulation results were used to evaluate the performance of the estimation and overdispersion test.

Future extensions for this study are as follows. First, we only focused on the stationary DCINMA(1) model in this paper. However, cases with non-stationary series are more common. The switched system is an important model in studying hybrid systems, particularly from the perspective of control science and engineering. Refer to ([30–32]) for more detailed discussion. The mechanism conducts the transformation between different subsystems, providing an approach to generalize the DCINMA process into a non-stationary case. A function can be introduced to control the change for different parameter values of the stationary case, which characterize non-stationary in series. Second, the parameter θ in our model provides a probability for overdispersion. In a switched system, the switching signal is a piecewise constant function, depending on time, external signal, output, and its own past value. Thus, the weighted switching signal function with weights θ and $(1-\theta)$ can be considered, where the weight θ gives the probability for switching to different subsystems.

Author Contributions: Conceptualization, K.Y. and H.W.; methodology, K.Y.; software, H.W.; validation, K.Y. and H.W.; formal analysis, K.Y.; investigation, H.W.; resources, K.Y.; data curation, K.Y.; writing—original draft preparation, H.W.; writing—review and editing, K.Y.; visualization, H.W.; supervision, K.Y.; project administration, K.Y.; funding acquisition, K.Y. All authors have read and agreed to the published version of the manuscript.

Funding: This research received no external funding.

Data Availability Statement: The data used to support the findings of this study are included within the article.

Acknowledgments: This research was supported by the National Social Science Fund of China (No. 18BTJ039) and Fundamental Research Funds for the Central Universities (No. JBK2102021). In addition, the authors will be grateful for the editor and reviewers' comments and suggestions, which led to improvements of the paper.

Conflicts of Interest: The authors declare no conflict of interest.

Appendix A

Proof of Theorem 1. we give the detailed derivation process for 1-th order, the q-th order can be derived in the same manner. term (i) and (ii) of the Theorem 1 are easy to verify, so we focus on the term (iii).

$$Var(X_t) = Var(\beta \circ \varepsilon_{t-1}) + Var(\varepsilon_t)$$
$$= V[E[\beta \circ \varepsilon_{t-1}|\varepsilon_{t-1}]] + E[V[\beta \circ \varepsilon_{t-1}|\varepsilon_{t-1}]] + Var(\varepsilon_t)$$

Then, we focus on the term $V[\beta \circ \varepsilon_{t-1}|\varepsilon_{t-1}]$.

$$V[\beta \circ \varepsilon_{t-1}|\varepsilon_{t-1}] = E[(\sum_{i=1}^{\varepsilon_{t-1}} U_i)^2|\varepsilon_{t-1}] - [E[\sum_{i=1}^{\varepsilon_{t-1}} U_i|\varepsilon_{t-1}]]^2$$
$$= \varepsilon_{t-1} \cdot E[U_1^2] + (\varepsilon_{t-1} - 1) \cdot \varepsilon_{t-1} \cdot E[U_1 \cdot U_2] - [\varepsilon_{t-1} \cdot \beta]^2$$
$$= \varepsilon_{t-1} \cdot (\beta \cdot (1-\beta) + \beta^2) + \varepsilon_{t-1} \cdot (\varepsilon_{t-1} - 1) \cdot (\theta^2 \cdot \beta \cdot (1-\beta) + \beta^2)$$
$$- \beta^2 \cdot \varepsilon_{t-1}^2$$

$$E[V[\beta \circ \varepsilon_{t-1}|\varepsilon_{t-1}]] = (\beta \cdot (1-\beta) + \beta^2) \cdot E[\varepsilon_{t-1}] - E[\beta^2 \cdot \varepsilon_{t-1}^2]$$
$$+ E[\varepsilon_{t-1} \cdot (\varepsilon_{t-1} - 1)] \cdot (\theta^2 \cdot \beta \cdot (1-\beta) + \beta^2)$$
$$= \lambda \cdot (\beta \cdot (1-\beta) - \beta^2) - \beta^2 \cdot (\lambda + \lambda^2)$$
$$+ (\lambda + \lambda^2 - \lambda) \cdot (\theta^2 \cdot \beta \cdot (1-\beta) + \beta^2)$$
$$= \lambda^2 \cdot \beta \cdot (1-\beta) \cdot \theta^2 + \beta \cdot \lambda - \beta^2 \cdot \lambda$$

$$Var(X_t) = V[\beta \cdot \varepsilon_{t-1}]\lambda^2 \cdot \beta \cdot (1-\beta) \cdot \theta^2 + \beta \cdot \lambda - \beta^2 \cdot \lambda + \lambda$$
$$= \beta \cdot \lambda + \lambda^2 \cdot \beta \cdot (1-\beta) \cdot \theta^2 + \lambda$$

□

Proof of Theorem 3. let $Y_t = X_t - \mu_X$, $\{X_t\}$ is the process defined in Section 2. To prove the Y_t is ergodic in autocovariance function, it is sufficient to show that:

$$lim_{T \to \infty} \frac{1}{T} \sum_{l=0}^{T} \{E(Y_t Y_{t+v} Y_{t+l} Y_{t+v+l}) - R^2(v)\} = 0$$

It is obvious $E(Y_t^i) < \infty$. For $i = 1, 2, 3, 4$, let $E(Y_t^2) = c_2, E(Y_t^4) = c_4$.
The case $v = 0$:

$$E(Y_t Y_{t+v} Y_{t+l} Y_{t+v+l}) = E(Y_t^2 Y_{t+l}^2)$$

If $l = 1$,

$$E(Y_t Y_{t+v} Y_{t+l} Y_{t+v+l}) = E(Y_t^2 Y_{t+l}^2) \leq (E(|Y_t^2|^2 |Y_{t+l}^2|^2))^{1/2} = c_4$$

If $l > 1$, Y_t and Y_{t+l} are irrelevant, then

$$E(Y_t Y_{t+v} Y_{t+l} Y_{t+v+l}) = E(Y_t^2 Y_{t+l}^2) = c_2^2 = \gamma_Y^2(0)$$

$$\frac{1}{T}\sum_{l=0}^{T}\{E(Y_t Y_{t+v} Y_{t+l} Y_{t+v+l}) - R^2(v)\} \leq \frac{1}{T}\{c_4 - \gamma_Y^2(0)\} \to 0$$

The case $v \geq 1$:
If $l \leq v+1$:

$$E(Y_t Y_{t+v} Y_{t+l} Y_{t+v+l}) \leq [E(|Y_t Y_{t+v}|^2 |Y_{t+l} Y_{t+l+v}|^2)]^{\frac{1}{2}} \leq [E(|Y_t|^4 |Y_{t+v}|^4 |Y_{t+l}|^4 |Y_{t+l+v}|^4)]^{\frac{1}{4}} = c_4$$

If $l > v+1$, $Y_t Y_{t+l}$ and $Y_{t+l} Y_{t+l+v}$ are irrelevant, then

$$E(Y_t Y_{t+v} Y_{t+l} Y_{t+v+l}) = E(Y_t Y_{t+v}) E(Y_{t+l} Y_{t+l+v}) = R^2(v)$$

$$\frac{1}{T}\sum_{l=0}^{T}\{E(Y_t Y_{t+v} Y_{t+l} Y_{t+v+l}) - R_v^2\} \leq \frac{1}{T}\{(c_4 - R^2(v))(v+1)\} \to 0$$

We can obtain:

$$lim_{T \to \infty} E[(\hat{R}_n^2(v) - R^2(v))^2] = 0$$

Which implies:

$$\hat{R}^2(v) \xrightarrow{p} R^2(v)$$

The above equation is holding for variable Y_t:

$$Y_t = X_t - \mu_X, \hat{R}_Y^2(v) \xrightarrow{p} R_Y^2(v)$$

We need to prove that

$$P(|\hat{R}_X^2(v) - R_X^2(v)| \geq \epsilon) \to 0$$

Because of $R_X^2(v) = R_Y^2(v)$, rewriting the above equation as

$$P(|\hat{R}_X^2(v) - R_X^2(v)| \geq \epsilon) = P(|\hat{R}_X^2(v) - \hat{R}_Y^2(v) + \hat{R}_Y^2(v) - R_X^2(v)| \geq \epsilon)$$
$$\leq P(|\hat{R}_X^2(v) - \hat{R}_Y^2(v)| \geq \epsilon/2) + P(|\hat{R}_Y^2(v) - R_Y^2(v)| \geq \epsilon/2)$$

$\hat{R}_Y^2(v)$ and $\hat{R}_X^2(v)$ expressing as

$$\hat{R}_X^2(v) = \frac{1}{T}\sum_{t=1}^{T-k}(X_{t+k} - \bar{X})(X_t - \bar{X}), \hat{R}_Y^2(v) = \frac{1}{T}\sum_{t=1}^{T-k}Y_{t+k}Y_t = \frac{1}{T}\sum_{t=1}^{T-k}(X_{t+k} - \mu_X)(X_t - \mu_X)$$

Then the following expression can be derived

$$P(|\hat{R}_X^2(v) - \hat{R}_Y^2(v)| \geq \epsilon/2) = P((\bar{X}^2 - \mu_X^2) + (\mu_X - \bar{X})(\bar{X}_{t+k} + \bar{X}_T) \geq \epsilon/2)$$
$$\leq P(\bar{X}^2 - \mu_X^2 \geq \epsilon/4) + P((\mu_X - \bar{X})(\bar{X}_{t+k} + \bar{X}_T) \geq \epsilon/4)$$

Since $\bar{X} \xrightarrow{p} \mu_X$, by Slutsky's theorem

$$\bar{X}^2 - \mu_X^2 \xrightarrow{p} 0, (\mu_X - \bar{X})(\bar{X}_{t+k} + \bar{X}_T) \xrightarrow{p} 0$$

Consequently,
$$P(|\hat{R}_X^2(v) - R_X^2(v)| \geq \epsilon) \to 0, T \to \infty$$

□

References

1. Alzaid, A.A.; Al-Osh, M.A. An integer-valued pth-order autoregressive structure (INAR(p)) process. *J. Appl. Probab.* **1990**, *27*, 314–324. [CrossRef]
2. Du, J.; Li, Y. The integer-valued autoregressive (INAR(p)) model. *J. Time Ser. Anal.* **1991**, *12*, 129–142.
3. Al-Osh, M.A.; Alzaid, A.A. Integer-valued moving average (INMA) process. *Stat. Pap.* **1988**, *29*, 281–300. [CrossRef]
4. McKenzie, E. Some ARMA models for dependent sequences of Poisson count. *Adv. Appl. Probab.* **1988**, *20*, 822–835. [CrossRef]
5. Weiß, C.H. The combined INAR(p) models for time series of counts. *Stat. Probab. Lett.* **2008**, *78*, 1817–1822. [CrossRef]
6. Monteiro, M.; Scotto, M.G.; Pereira, I. Integer-valued autoregressive processes with periodic structure. *J. Stat. Plan. Infer.* **2010**, *140*, 1529–1541. [CrossRef]
7. Weiß, C.H. A Poisson INAR(1) model with serially dependent innovations. *Metrika* **2015**, *78*, 829–851. [CrossRef]
8. Zhu, F. Modeling time series of counts with com-poisson INGARCH models. *Math. Comput. Model.* **2012**, *56*, 191–203. [CrossRef]
9. Möller, T.; Weiß, C.H. Threshold models for integer-valued time series with infinite or finite range. In *Stochastic Models, Statistics and Their Applications*; Steland, A., Rafajłowicz, E., Szajowski, K., Eds.; Springer: Geneva, Switzerland, 2015; pp. 327–334.
10. Yao, K.; Wang, D. A new INAR(1) process with bounded support for counts showing equidispersion, underdispersion and overdispersion. *Stat. Pap.* **2019**, *62*, 745–767.
11. Li, C.; Wang, D. First-order mixed integer-valued autoregressive processes with zero-inflated generalized power series innovations. *J. Korean Stat. Soc.* **2015**, *44*, 232–246. [CrossRef]
12. Kim, H.Y.; Park, Y. A non-stationary integer-valued autoregressive model. *Stat. Pap.* **2008**, *49*, 485. [CrossRef]
13. Zheng, H.; Basawa, I.V. Inference for pth-order random coefficient integer-valued autoregressive processes. *J. Time Ser. Anal.* **2006**, *27*, 411–440. [CrossRef]
14. Ristić, M.M.; Bakouch, H.S. A new geometric first-order integer-valued autoregressive (NGINAR(1)) process. *J. Stat. Plan. Infer.* **2009**, *139*, 2218–2226. [CrossRef]
15. Zhang, H.; Wang, D.; Zhu, F. Inference for INAR(p) processes with signed generalized power series thinning operator. *J. Stat. Plan. Infer.* **2010**, *140*, 667–683. [CrossRef]
16. Ristić, M.M.; Nastić, A.S.; Miletić Ilić, A.V. A geometric time series model with dependent Bernoulli counting series. *J. Time Ser. Anal.* **2013**, *34*, 466–476. [CrossRef]
17. Miletić Ilić, A.V.; Ristić, M.M.; Nastić, A.S.; Bakouch, H.S. An INAR(1) model based on a mixed dependent and independent counting series. *J. Stat. Comput. Simul.* **2018**, *88*, 290–304. [CrossRef]
18. Weiß, C.H. Thinning operations for modeling time series of counts—A survey. *Adv. Appl. Probab.* **2008**, *92*, 319. [CrossRef]
19. Winters, P.R. Forecasting sales by exponentially weighted moving averages. *Manag. Sci.* **1960**, *6*, 231–362. [CrossRef]
20. Cox, D.R. Prediction by exponentially weighted moving averages and related methods. *J. R. Stat. Soc. Ser. B-Stat. Methodol.* **1961**, *23*, 414–422. [CrossRef]
21. Landauskas, M.; Navickas, Z.; Vainoras, A.; Ragulskis, M. Weighted moving averaging revisited—An algebraic approach. *Comput. Appl. Math.* **2017**, *36*, 1545–1558. [CrossRef]
22. Nan, X.; Li, Q.; Qiu, D.; Zhao, Y.; Guo, X. Short-term wind speed syntheses correcting forecasting model and its application. *Int. J. Electr. Power Energy Syst.* **2013**, *49*, 264–268. [CrossRef]
23. Alevizakos, V.; Chatterjee, K.; Koukouvinos, C. The triple exponentially weighted moving average control chart. *Qual. Technol. Quant. Manag.* **2021**, *18*, 326–354. [CrossRef]
24. Capizzi, G.; Masarotto, G. An adaptive exponentially weighted moving average control chart. *Technometrics* **2003**, *45*, 199–207. [CrossRef]
25. Adegoke, N.A.; Abbasi, S.A.; Smith, A.N.H.; Anderson, M.J.; Pawley, M.D.M. A multivariate homogeneously weighted moving average control chart. *IEEE Access* **2019**, *7*, 9586–9597. [CrossRef]
26. Brännäs, K.; Quoreshi, A.M.M.S. Integer-valued moving average modelling of the number of transactions in stocks. *Appl. Financ. Econ.* **2010**, *20*, 1429–1440. [CrossRef]
27. Brännäs, K.; Hall, A. Estimation in integer-valued moving average models. *Appl. Stoch. Models. Bus. Ind.* **2001**, *17*, 277–291. [CrossRef]
28. Aleksandrov, B.; Weiß, C.H. Parameter estimation and diagnostic tests for INMA(1) processes. *Test* **2019**, *29*, 196–232. [CrossRef]
29. Jung, R.C.; McCabe, P.M.; Tremayne, A.R. Model validation and diagnostics. In *Handbook of Discrete-Valued Time*; Davis, R.A., Holan, S.H., Lund, R., Ravishanker, N., Eds.; Chapman & Hall/CRC Press: Boca Raton, FL, USA, 2015; pp. 189–218.
30. Liberzon, D.; Morse, A.S. Basic problems in stability and design of switched systems. *IEEE Control Syst. Mag.* **1999**, *19*, 59–70.
31. Blanchini, F.; Miani, S. *Set-Theoretic Methods in Control*, 2nd ed.; Birkhäuser: Basle, Switzerland, 2015.
32. Jin, C.; Wang, R.; Wang, Q. Stabilization of switched systems with time-dependent switching signal. *J. Frankl. Inst.-Eng. Appl. Math.* **2020**, *357*, 13552–13568. [CrossRef]

Statistical Inference for Periodic Self-Exciting Threshold Integer-Valued Autoregressive Processes

Congmin Liu [1], Jianhua Cheng [1] and Dehui Wang [2,*]

1. School of Mathematics, Jilin University, 2699 Qianjin Street, Changchun 130012, China; lcm18@mails.jlu.edu.cn (C.L.); chengjh@jlu.edu.cn (J.C.)
2. School of Economics, Liaoning University, Shenyang 110036, China
* Correspondence: wangdh@jlu.edu.cn

Abstract: This paper considers the periodic self-exciting threshold integer-valued autoregressive processes under a weaker condition in which the second moment is finite instead of the innovation distribution being given. The basic statistical properties of the model are discussed, the quasi-likelihood inference of the parameters is investigated, and the asymptotic behaviors of the estimators are obtained. Threshold estimates based on quasi-likelihood and least squares methods are given. Simulation studies evidence that the quasi-likelihood methods perform well with realistic sample sizes and may be superior to least squares and maximum likelihood methods. The practical application of the processes is illustrated by a time series dataset concerning the monthly counts of claimants collecting short-term disability benefits from the Workers' Compensation Board (WCB). In addition, the forecasting problem of this dataset is addressed.

Keywords: periodic autoregression; integer-valued threshold models; parameter estimation

1. Introduction

There has been considerable interest in integer-valued time series because of their wide range of applications, including epidemiology, finance, and disease modeling. Examples of such data are as follows: the number of major global earthquakes per year, monthly crimes in a particular country or region, and patient numbers in a hospital per month over a period of time, etc. Following the first-order integer-valued autoregressive (INAR(1)) models introduced by Al-Osh and Alzaid [1], INAR models have been widely used, see Du and Li [2], Jung et al. [3], Weiß [4], Ristić et al. [5], Zhang et al. [6], Li et al. [7], Kang et al. [8] and Yu et al. [9], among others. However, for so-called piecewise phenomenon such as high thresholds, sudden bursts of large values, and time volatility, the INAR model will not work well. The threshold models (Tong [10]; Tong and Lim [11]) have attracted much attention and have been widely used to model nonlinear phenomena. To capture the piecewise phenomenon of integer-valued time series, Monteiro et al. [12] introduced a class of self-exciting threshold integer-valued autoregressive (SETINAR) models driven by independent Poisson-distributed random variables. Wang et al. [13] proposed a self-excited threshold Poisson autoregressive (SETPAR) model. Yang et al. [14] considered a class of SETINAR processes that properly capture flexible asymmetric and nonlinear responses without assuming the distributions for the errors. Yang et al. [15] introduced an integer-valued threshold autoregressive process based on a negative binomial thinning operator (NBTINAR(1)).

In addition, there are many sources of business, economic and meteorology time series data showing a periodically varying phenomenon that repeats itself after a regular period of time. It may be affected by seasonal factors and human activities. For dealing with the processes exhibiting periodic patterns, Bennett [16] and Gladyshev [17] proposed periodically correlated random processes. Then, Bentarzi and Hallin [18], Lund and Basawa [19], Basawa and Lund [20], and Shao [21], among other authors, studied the periodic autoregressive moving-average (PARMA) models in some detail. To capture the periodic phenomenon of integer-valued time series, Monteiro et al. [22] proposed the periodic integer-valued autoregressive models of order one (PINAR(1)) with period T, driven by a periodic sequence of independent Poisson-distributed random variables. Hall et al. [23] considered the extremal behavior of periodic integer-valued moving-average sequences. Santos et al. [24] introduced a multivariate PINAR model with time-varying parameters. The analysis of periodic self-exciting threshold integer-valued autoregressive (PSETINAR$(2;1,1)_T$) processes was introduced by Pereira et al. [25]. Manaa and Bentarzi [26] established the existence of high moment and the strict periodic stationarity for the PSETINAR$(2;1,1)_T$ processes. The CLS and CML methods are applied to estimate the parameters while using the nested sub-sample search (NeSS) algorithm proposed by Li and Tong [27] to estimate the periodic threshold parameters. A drawback of this PSETINAR$(2;1,1)_T$ model is that the mean and variance of Poisson distribution are equal, which is not always true in the real data. Therefore, in this paper, we remove the assumption of Poisson distribution, only specify the relationship between mean and variance of observations, develop quasi-likelihood inference for the PSETINAR$(2;1,1)_T$ processes, and consider the estimation of thresholds.

Quasi-likelihood is a non-parametric inference method proposed by Wedderburn [28]. It is very useful in cases where the exact distributional information is not available, while only the relation between mean and variance of the observation is given, and it enjoys a certain robustness of validity. Quasi-likelihood has been widely applied. For example, Azrak and Mélard [29] proposed a simple and efficient algorithm to evaluate the exact quasi-likelihood of ARMA models with time-dependent coefficients; Christou and Fokianos [30] studied probabilistic properties and quasi-likelihood estimation for negative binomial time series models; Li et al. [31] studied the quasi-likelihood inference for the self-exciting threshold integer-valued autoregressive (SETINAR(2,1)) processes under a weaker condition; Yang et al. [32] modeled overdispersed or underdispersed count data with generalized Poisson integer-valued autoregressive (GPINAR(1)) processes and investigated the maximum quasi-likelihood estimators.

The remainder of this paper is organized as follows. In Section 2, we redefine the PSETINAR$(2;1,1)_T$ processes under weak conditions and discuss their basic properties. In Section 3, we consider the quasi-likelihood inference for the unknown parameters. Thresholds estimation is also discussed. Section 4 presents some simulation results for the estimates. In Section 5, we give an application of the proposed processes to a real dataset. The forecasting problem of this dataset is addressed. Concluding remarks are given in Section 6. All proofs are postponed to the Appendix A.

2. The Model and Its Properties

The periodic self-exciting threshold integer-valued autoregressive model of order one with two regimes (PSETINAR$(2;1,1)_T$) (originally proposed by Pereira et al. [25], and further studied by Manaa and Bentarzi [26]) is defined by the recursive equation:

$$X_t = \begin{cases} \alpha_t^{(1)} \circ X_{t-1} + Z_t, & X_{t-1} \leq r_t, \\ \alpha_t^{(2)} \circ X_{t-1} + Z_t, & X_{t-1} > r_t, \end{cases} \quad t \in \mathbb{Z} \qquad (1)$$

with threshold parameters $r_t = r_j$, autoregressive coefficients $\alpha_t^{(k)} = \alpha_j^{(k)} \in (0,1)$, for $k = 1, 2, t = j + sT, j = 1, 2, \ldots, T, s \in \mathbb{Z}$, and $T \in \mathbb{N}_0$. Note that Equation (1) admits the representation

$$X_{j+sT} = \left(\alpha_j^{(1)} \circ X_{j+sT-1}\right) I_{j+sT-1}^{(1)} + \left(\alpha_j^{(2)} \circ X_{j+sT-1}\right) I_{j+sT-1}^{(2)} + Z_{j+sT}, \qquad (2)$$

where

(i) $I^{(1)}_{j+sT-1} := I\{X_{j+sT-1} \leq r_j\}, I^{(2)}_{j+sT-1} := 1 - I^{(1)}_{j+sT-1} = I\{X_{j+sT-1} > r_j\}$, in which $\{r_j, j = 1, 2, \ldots, T\}$ is a set of thresholds value;

(ii) The thinning operator "∘" is defined as

$$\alpha_j^{(k)} \circ X_{j+sT-1} = \sum_{i=1}^{X_{j+sT-1}} U_{i,j+sT}\left(\alpha_j^{(k)}\right), \quad (3)$$

in which $\{U_{i,j+sT}\left(\alpha_j^{(k)}\right), j = 1, 2, \ldots, T, s \in \mathbb{Z}\}$ is a sequence of independent periodic Bernoulli random variables with $P\left(U_{i,j+sT}\left(\alpha_j^{(k)}\right) = 1\right) = 1 - P\left(U_{i,j+sT}\left(\alpha_j^{(k)}\right) = 0\right) = \alpha_j^{(k)}, k = 1, 2$;

(iii) $\{Z_{j+sT}, j = 1, 2, \ldots, T, s \in \mathbb{Z}\}$ constitutes a sequence of independent periodic random variables with $E(Z_{j+sT}) = \lambda_j$, $Var(Z_{j+sT}) = \sigma^2_{z,j}$, which is assumed to be independent of $\{X_{j+sT-1}\}$ and $\{\alpha_j^{(k)} \circ X_{j+sT-1}\}$.

Remark 1. *The innovation of PSETINAR$(2; 1, 1)_T$ process defined by Pereira et al. [25] and Manaa and Bentarzi [26] is a sequence of independent periodic Poisson-distributed random variables with mean λ_j, that is $\{Z_t\} \sim P(\lambda_j)$, where $t = j + sT$, $j = 1, 2, \ldots, T$, $s \in \mathbb{Z}$. In this paper, we use $E(Z_{j+sT}) = \lambda_j$, $Var(Z_{j+sT}) = \sigma^2_{z,j}$ instead of the assumption of periodic Poisson distribution for $\{Z_{j+sT}\}$, so that the model is more flexible.*

The following proposition establishes the conditional mean and the conditional variance of the PSETINAR$(2; 1, 1)_T$ process, which plays an important role in the study of the process properties and parameter estimations.

Proposition 1. *For any fixed $j = 1, 2, \ldots, T$, with $T \in \mathbb{N}_0$, the conditional mean and the conditional variance of the process $\{X_t\}$ for $t = j + sT$ and $s \in \mathbb{Z}$ defined in (2) are given by*

(i) $E(X_{j+sT}|X_{j+sT-1}) = \alpha_j^{(1)} X_{j+sT-1} I^{(1)}_{j+sT-1} + \alpha_j^{(2)} X_{j+sT-1} I^{(2)}_{j+sT-1} + \lambda_j$,

(ii) $Var(X_{j+sT}|X_{j+sT-1}) = \sum_{k=1}^{2} \alpha_j^{(k)}\left(1 - \alpha_j^{(k)}\right) X_{j+sT-1} I^{(k)}_{j+sT-1} + \sigma^2_{z,j}$.

The following theorem states the ergodicity of the PSETINAR$(2; 1, 1)_T$ process (2). This property is useful in deriving the asymptotic properties of the parameter estimators.

Theorem 1. *For any fixed $j = 1, 2, \ldots, T$, with $T \in \mathbb{N}_0$, the process $\{X_t\}$ for $t = j + sT$ and $s \in \mathbb{Z}$ defined in (2) is an ergodic Markov chain.*

3. Parameters Estimation

Suppose we have a series of observations $\{X_{j+sT}, j = 1, 2, \ldots, T, s \in \mathbb{N}_0\}$ generated from the PSETINAR$(2; 1, 1)_T$ process. The goal of this section is to estimate the unknown parameters vector $\beta = (\beta_1, \ldots, \beta_{3T})' \triangleq (\alpha_1^{(1)}, \alpha_1^{(2)}, \lambda_1, \alpha_2^{(1)}, \alpha_2^{(2)}, \lambda_2, \ldots, \alpha_T^{(1)}, \alpha_T^{(2)}, \lambda_T)'$ and threshold parameters vector $r = (r_1, r_2, \ldots, r_T)'$. This section is divided into two subsections. In Section 3.1, we estimate the parameters vector β by using the maximum quasi-likelihood (MQL) method when the thresholds value is known. We consider the maximum quasi-likelihood (MQL) and conditional least square (CLS) estimators of thresholds r in Section 3.2.

3.1. Estimation of Parameters β

As described in Proposition 1 (ii), we have the variance of X_t conditional on X_{t-1}, let $\theta_j \triangleq \left(\theta_j^{(1)}, \theta_j^{(2)}, \sigma^2_{z,j}\right)'$ with $\theta_j^{(k)} = \alpha_j^{(k)}\left(1 - \alpha_j^{(k)}\right)$, $k = 1, 2$, $j = 1, 2, \ldots, T$, then the $Var(X_{j+sT}|X_{j+sT-1})$ admits the representation

$$V_{\theta_j}(X_{j+sT}|X_{j+sT-1}) \triangleq Var(X_{j+sT}|X_{j+sT-1}) = \theta_j^{(1)} X_{j+sT-1} I_{j+sT-1}^{(1)} + \theta_j^{(2)} X_{j+sT-1} I_{j+sT-1}^{(2)} + \sigma_{z,j}^2,$$

for $\forall j = 1, 2, \ldots, T, s \in \mathbb{N}_0$.

As discussed in Wedderburn [28], we have the set of standard quasi-likelihood estimating equations:

$$L(\beta) = \sum_{s=0}^{N-1} \sum_{j=1}^{T} \frac{X_{j+sT} - E(X_{j+sT}|X_{j+sT-1})}{V_{\theta_j}(X_{j+sT}|X_{j+sT-1})} \frac{\partial E(X_{j+sT}|X_{j+sT-1})}{\partial \beta_i} = 0, \quad (4)$$

for $i = 1, \ldots, 3T$, where N is the total number of cycles. By solving (4), the quasi-likelihood estimator can be obtained.

This method is essentially a two-step estimation, if θ_j is unknown, we propose substituting a suitable consistent estimator of θ_j obtained by other means, getting modified quasi-likelihood estimating equations and then solving them for the primary parameters of interest. In the modified quasi-likelihood estimating equations, we replace θ_j with a suitable consistent estimator $\hat{\theta}_j$. For simplicity in notation, we define $V_{\hat{\theta}_j}^{-1} \triangleq V_{\hat{\theta}_j}^{-1}(X_{j+sT}|X_{j+sT-1})$. This approach leads to the modified quasi-likelihood estimator $\hat{\beta}_{MQL}$ of β (see Zheng, Basawa and Datta [33]):

$$\hat{\beta}_{MQL} = Q_N^{-1} q_N, \quad (5)$$

where

$$Q_N = \begin{bmatrix} Q_{1,N} & 0 & \cdots & 0 \\ 0 & Q_{2,N} & \cdots & 0 \\ \vdots & \vdots & \ddots & \vdots \\ 0 & 0 & \cdots & Q_{T,N} \end{bmatrix},$$

and

$$q_N = (q_{1,N}, q_{2,N}, \ldots, q_{T,N})',$$

moreover, the 0's are (3×3)-null matrices, $Q_{j,N}$ and $q_{j,N}$ ($j = 1, 2, \ldots, T$) given by

$$Q_{j,N} = \begin{bmatrix} \sum_{s=0}^{N-1} V_{\hat{\theta}_j}^{-1} X_{j+sT-1}^2 I_{j+sT-1}^{(1)} & 0 & \sum_{s=0}^{N-1} V_{\hat{\theta}_j}^{-1} X_{j+sT-1} I_{j+sT-1}^{(1)} \\ 0 & \sum_{s=0}^{N-1} V_{\hat{\theta}_j}^{-1} X_{j+sT-1}^2 I_{j+sT-1}^{(2)} & \sum_{s=0}^{N-1} V_{\hat{\theta}_j}^{-1} X_{j+sT-1} I_{j+sT-1}^{(2)} \\ \sum_{s=0}^{N-1} V_{\hat{\theta}_j}^{-1} X_{j+sT-1} I_{j+sT-1}^{(1)} & \sum_{s=0}^{N-1} V_{\hat{\theta}_j}^{-1} X_{j+sT-1} I_{j+sT-1}^{(2)} & \sum_{s=0}^{N-1} V_{\hat{\theta}_j}^{-1} \end{bmatrix},$$

$$q_{j,N} = \left(\sum_{s=0}^{N-1} V_{\hat{\theta}_j}^{-1} X_{j+sT} X_{j+sT-1} I_{j+sT-1}^{(1)}, \sum_{s=0}^{N-1} V_{\hat{\theta}_j}^{-1} X_{j+sT} X_{j+sT-1} I_{j+sT-1}^{(2)}, \sum_{s=0}^{N-1} V_{\hat{\theta}_j}^{-1} X_{j+sT} \right)'.$$

Note that we use consistent estimator $\hat{\theta}_j = \left(\hat{\alpha}_j^{(1)} \left(1 - \hat{\alpha}_j^{(1)} \right), \hat{\alpha}_j^{(2)} \left(1 - \hat{\alpha}_j^{(2)} \right), \hat{\sigma}_{z,j}^2 \right)'$ instead of θ_j.

Next, the proposition gives consistent estimators $\hat{\sigma}_{z,j}^2$ of $\sigma_{z,j}^2$, which depends on some consistent estimators $\hat{\alpha}_j^{(k)}$ and $\hat{\lambda}_j$ with $k = 1, 2, j = 1, 2, \ldots, T$.

Proposition 2. The following variance estimators for $\{Z_{j+sT}\}$ with $j = 1, 2, \ldots, T, s \in \mathbb{N}_0$ are consistent:

(i) $\hat{\sigma}_{1,z,j}^2 = \dfrac{1}{N} \sum_{s=0}^{N-1} \left(X_{j+sT} - \sum_{k=1}^{2} \hat{\alpha}_j^{(k)} X_{j+sT-1} I_{j+sT-1}^{(k)} - \hat{\lambda}_j \right)^2$

$\qquad - \dfrac{1}{N} \sum_{k=1}^{2} \sum_{s=0}^{N-1} \hat{\alpha}_j^{(k)} \left(1 - \hat{\alpha}_j^{(k)}\right) X_{j+sT-1} I_{j+sT-1}^{(k)}, \qquad (6)$

(ii) $\hat{\sigma}_{2,z,j}^2 = \hat{\sigma}_{x,j}^2 - \hat{p}_j \left[\hat{\alpha}_j^{(1)^2} \hat{\sigma}_j^{2(1)} + \hat{\alpha}_j^{(1)} \left(1 - \hat{\alpha}_j^{(1)}\right) \hat{\mu}_j^{(1)} \right]$

$\qquad - (1 - \hat{p}_j) \left[\hat{\alpha}_j^{(2)^2} \hat{\sigma}_j^{2(2)} + \hat{\alpha}_j^{(2)} \left(1 - \hat{\alpha}_j^{(2)}\right) \hat{\mu}_j^{(2)} \right]$

$\qquad - \hat{p}_j (1 - \hat{p}_j) \left(\hat{\alpha}_j^{(1)} \hat{\mu}_j^{(1)} - \hat{\alpha}_j^{(2)} \hat{\mu}_j^{(2)} \right)^2, \qquad (7)$

for $k = 1, 2, j = 1, 2, \ldots, T, s \in \mathbb{N}_0$, in which $\hat{\alpha}_j^{(k)}$ and $\hat{\lambda}_j$ are consistent estimators of $\alpha_j^{(k)}$ and λ_j (for example, we can use the CLS estimators given in Theorem 3.1 of Pereira et al. [25]), furthermore

$$\bar{X}_j = \dfrac{1}{N} \sum_{s=0}^{N-1} X_{j+sT}, \quad \hat{\sigma}_{x,j}^2 = \dfrac{1}{N} \sum_{s=0}^{N-1} \left(X_{j+sT} - \bar{X}_j \right)^2,$$

$$N_j^{(k)} = \sum_{s=0}^{N-1} I_{j+sT-1}^{(k)}, \quad \hat{\mu}_j^{(k)} = \dfrac{1}{N_j^{(k)}} \sum_{s \in \{I_{j+sT-1}^{(k)} = 1\}} X_{j+sT},$$

$$\hat{p}_j = \dfrac{1}{N} \sum_{s=0}^{N-1} I_{j+sT-1}^{(1)}, \quad \hat{\sigma}_j^{2(k)} = \dfrac{1}{N_j^{(k)}} \sum_{s \in \{I_{j+sT-1}^{(k)} = 1\}} \left(X_{j+sT} - \hat{\mu}_j^{(k)} \right)^2.$$

The two estimations are based on conditional variance $\mathrm{Var}(X_{j+sT} | X_{j+sT-1})$ and variance $\mathrm{Var}(X_{j+sT})$, respectively. The details can be found in the Appendix A.

To study the asymptotic behavior of the estimator $\hat{\beta}_{MQL}$, we make the following assumptions about the process of $\{X_t\}$:

(C1) By Proposition 1 in Pereira et al. [25], we assume the $\{X_t\}$ is a strictly ciclostationary process;
(C2) $E|X_t|^4 < \infty$.

Now for the asymptotic properties of the quasi-likelihood estimator $\hat{\beta}_{MQL}$ given by (5), we have the following asymptotic distribution.

Theorem 2. Let $\{X_t\}$ be a $\mathrm{PSETINAR}(2; 1, 1)_T$ process defined in (2), then under the assumptions (C1)-(C2), the estimator $\hat{\beta}_{MQL}$ given by (5) is asymptotically normal,

$$\sqrt{N} \left(\hat{\beta}_{MQL} - \beta \right) \to \mathcal{N} \left(0, H^{-1}(\theta) \right),$$

where

$$H(\theta) = \begin{bmatrix} H_1(\theta) & 0 & \cdots & 0 \\ 0 & H_2(\theta) & \cdots & 0 \\ \vdots & \vdots & \ddots & \vdots \\ 0 & 0 & \cdots & H_T(\theta) \end{bmatrix},$$

with matrices H_j ($j = 1, 2, \ldots, T$) given by

$$H_j(\theta) = \begin{bmatrix} E\left(V_{\theta_j}^{-1}(X_j|X_{j-1})X_{j-1}^2 I_{j-1}^{(1)}\right) & 0 & E\left(V_{\theta_j}^{-1}(X_j|X_{j-1})X_{j-1} I_{j-1}^{(1)}\right) \\ 0 & E\left(V_{\theta_j}^{-1}(X_j|X_{j-1})X_{j-1}^2 I_{j-1}^{(2)}\right) & E\left(V_{\theta_j}^{-1}(X_j|X_{j-1})X_{j-1} I_{j-1}^{(2)}\right) \\ E\left(V_{\theta_j}^{-1}(X_j|X_{j-1})X_{j-1} I_{j-1}^{(1)}\right) & E\left(V_{\theta_j}^{-1}(X_j|X_{j-1})X_{j-1} I_{j-1}^{(2)}\right) & E\left(V_{\theta_j}^{-1}(X_j|X_{j-1})\right) \end{bmatrix}.$$

It is worth mentioning that this theorem reflects the consistency of the estimator $\hat{\beta}_{MQL}$.

3.2. Estimation of Thresholds Vector r

Note that in the real data application, the threshold values are also unknown. In this subsection, we estimate the thresholds vector $r = (r_1, r_2, \ldots, r_T)'$. Here, we further promote the nested sub-sample search (NeSS) algorithm (see, e.g., Yang et al. [15], Li and Tong [27], and Li et al. [31]) and use conditional least squares (CLS) and modified quasi-likelihood (MQL) principles to estimate r.

For some fixed $\lambda = (\lambda_1, \lambda_2, \ldots, \lambda_T)'$, the application of the conditional least squares principle yields the sum of squared errors:

$$S_N(r, \lambda) = \sum_{s=0}^{N-1} \sum_{j=1}^{T} \left(X_{j+sT} - \sum_{k=1}^{2} \frac{\sum_{s=0}^{N-1}\left(X_{j+sT}X_{j+sT-1}I_{j+sT-1}^{(k)} - \lambda_j X_{j+sT-1}I_{j+sT-1}^{(k)}\right)}{\sum_{s=0}^{N-1} X_{j+sT-1}^2 I_{j+sT-1}^{(k)}} X_{j+sT-1}I_{j+sT-1}^{(k)} - \lambda_j \right)^2,$$

and then the thresholds vector r can be estimated by minimizing $S_N(r, \lambda)$,

$$\hat{r} = \arg\min_{r \in [\underline{r}, \bar{r}]} S_N(r, \lambda), \tag{8}$$

where $\underline{r}$ and $\bar{r}$ are some known lower and upper bounds of r. In practice, they can be selected as the minimum and maximum values in each cycle of the sample. For convenience, we consider an alternative objective function

$$J_N(r, \lambda) = S_N - S_N(r, \lambda),$$

where

$$S_N = \sum_{s=0}^{N-1} \sum_{j=1}^{T} \left(X_{j+sT} - \frac{\sum_{s=0}^{N-1}\left(X_{j+sT}X_{j+sT-1} - \lambda_j X_{j+sT-1}\right)}{\sum_{s=0}^{N-1} X_{j+sT-1}^2} X_{j+sT-1} - \lambda_j \right)^2.$$

Now, the optimization in (8) is equivalent to

$$\hat{r}_{CLS} = \arg\max_{r \in [\underline{r}, \bar{r}]} J_N(r, \lambda), \tag{9}$$

where $\hat{r}_{CLS}$ is the conditional least squares estimator of the thresholds vector r.

Inspired by the method of conditional least squares, we investigate the performances of r by using the quasi-likelihood principle. The modified quasi-likelihood estimator $\hat{r}_{MQL}$ of r is obtained by maximizing the expression

$$\tilde{J}_N(r, \lambda) = \tilde{S}_N - \tilde{S}_N(r, \lambda),$$

which yields

$$\hat{r}_{MQL} = \arg\max_{r \in [\underline{r}, \bar{r}]} \tilde{J}_N(r, \lambda), \tag{10}$$

where

$\breve{S}_N(r, \lambda)$

$$= \sum_{s=0}^{N-1} \sum_{j=1}^{T} V_{\hat{\theta}_j}^{-1} \left(X_{j+sT} - \sum_{k=1}^{2} \frac{\sum_{s=0}^{N-1} V_{\hat{\theta}_j}^{-1} \cdot \left(X_{j+sT} X_{j+sT-1} I_{j+sT-1}^{(k)} - \lambda_j X_{j+sT-1} I_{j+sT-1}^{(k)} \right)}{\sum_{s=0}^{N-1} V_{\hat{\theta}_j}^{-1} \cdot X_{j+sT-1}^2 I_{j+sT-1}^{(k)}} X_{j+sT-1} I_{j+sT-1}^{(k)} - \lambda_j \right)^2,$$

and

$$\tilde{S}_N = \sum_{s=0}^{N-1} \sum_{j=1}^{T} V_{\hat{\theta}_j}^{-1} \left(X_{j+sT} - \frac{\sum_{s=0}^{N-1} V_{\hat{\theta}_j}^{-1} \cdot \left(X_{j+sT} X_{j+sT-1} - \lambda_j X_{j+sT-1} \right)}{\sum_{s=0}^{N-1} V_{\hat{\theta}_j}^{-1} \cdot X_{j+sT-1}^2} X_{j+sT-1} - \lambda_j \right)^2.$$

It is worth mentioning that there are unknown parameters λ_j with $j = 1, \ldots, T$ when we use (9) and (10) to estimate thresholds vector r. As argued in Li and Tong [27], Yang et al. [14], and Yang et al. [15], when λ and r are one-dimensional parameters, we can choose any positive number as the value of λ without worrying about getting a wrong result of $\hat{r}$. Fortunately, we also find out by simulations that the estimations of r by maximizing $\breve{J}_N(r, \lambda)$ and $\tilde{J}_N(r, \lambda)$ do not depend on the value of λ. In order to give an intuitive impression of $\breve{J}_N(r, \lambda)/N$, we generate a set of data with Model I (given in Section 4, i.e., $T = 2, N = 50, \beta = (0.2, 0.1, 3, 0.8, 0.1, 7)', r = (8, 4)'$), and plot the shapes of $\breve{J}_N(r, \lambda)/N$. From Figure 1, we can see that for different values of λ, the shape of $\breve{J}_N(r, \lambda)/N$ changes, but the maximum value in each subfigure is obtained at the true thresholds vector $r = (8, 4)'$. In practice, we can choose the mean in each cycle of the samples for $\lambda_j, j = 1, 2, \ldots, T$.

Actually, using the quasi-likelihood method to estimate the thresholds is a three-step estimation procedure, and we now present the algorithm to implement our estimation procedure as follows:

Step 1: *Choose the upper bound $\bar{r}$ and lower bound $\underline{r}$ of r, solve (9) to get the $\hat{r}_{CLS}$ with $\lambda_j = \bar{X}_j = \frac{1}{N} \sum_{s=0}^{N-1} X_{j+sT}, j = 1, 2, \ldots, T$;*

Step 2: *Fix $\hat{r}_{CLS}$ at the current value, solve (6) or (7) to get the $\hat{\sigma}_{z,j}^2, j = 1, 2, \ldots, T$, where $\alpha_j^{(k)}$ and λ_j with $k = 1, 2$ can be estimated by other methods, then solve (5) to get $\hat{\beta}_{MQL}$.*

Step 3: *Fix $\hat{\theta}_j = \left(\hat{\alpha}_{j,MQL}^{(1)} \left(1 - \hat{\alpha}_{j,MQL}^{(1)}\right), \hat{\alpha}_{j,MQL}^{(2)} \left(1 - \hat{\alpha}_{j,MQL}^{(2)}\right), \hat{\sigma}_{z,j}^2 \right)', j = 1, 2, \ldots, T$ at its estimated value from Step 2, choose the same upper bound $\bar{r}$ and lower bound $\underline{r}$ as in Step 1, solve (10) to get $\hat{r}_{MQL}$.*

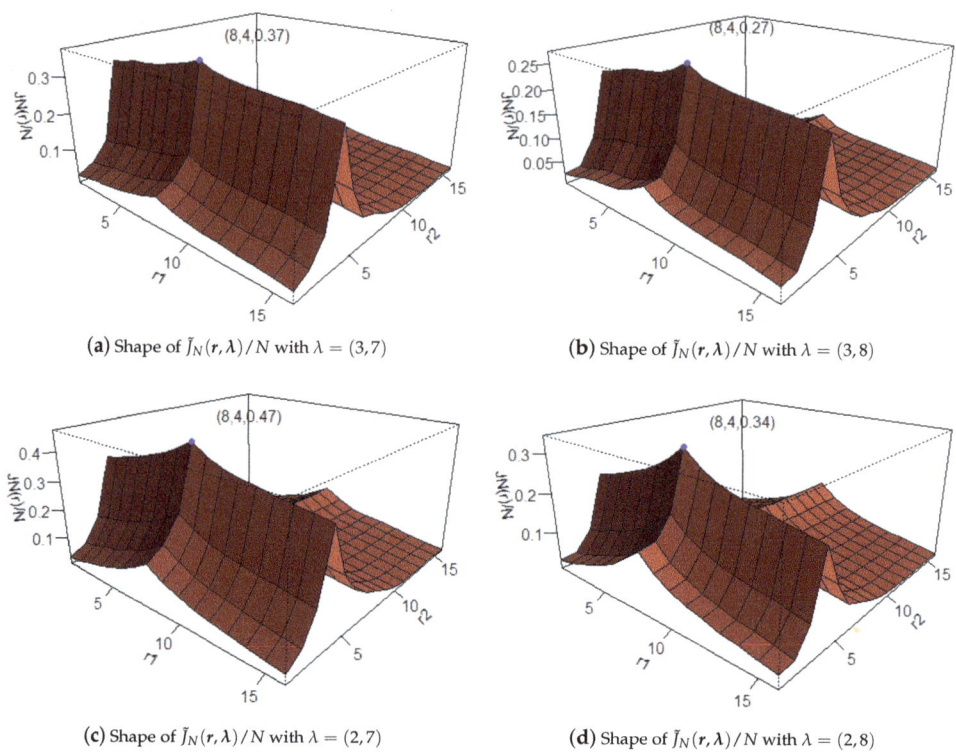

(a) Shape of $\tilde{J}_N(r,\lambda)/N$ with $\lambda=(3,7)$
(b) Shape of $\tilde{J}_N(r,\lambda)/N$ with $\lambda=(3,8)$
(c) Shape of $\tilde{J}_N(r,\lambda)/N$ with $\lambda=(2,7)$
(d) Shape of $\tilde{J}_N(r,\lambda)/N$ with $\lambda=(2,8)$

Figure 1. The shapes of $\tilde{J}_N(r,\lambda)/N$.

4. Simulation Study

In this section, we conduct simulation studies to illustrate the finite sample performances of the estimates. The initial value X_0 is fixed at 0. In order to capture the characteristics of the data from the PSETINAR$(2;1,1)_T$ process, we first generate a set of data with the distribution of innovations $\{Z_t\}$ given by Model I (mentioned below in this section) and parameters $\beta = (0.2, 0.45, 1, 0.2, 0.45, 2, 0.65, 0.45, 1, 0.65, 0.45, 2, 0.2, 0.45, 3, 0.2, 0.45, 7, 0.8, 0.45, 7, 0.2, 0.1, 3, 0.8, 0.1, 7, 0.2, 0.1, 7, 0.8, 0.45, 2)'$, $r = (3, 3, 3, 1, 3, 3, 5, 9, 3, 6, 7)'$, $T = 11$, $N = 50$. The parameter vectors we choose here are randomly selected, and there are slight differences between the parameters of each cycle, the thresholds vector of r was chosen such that there are enough data in each regime. We give the sample path in the first six cycles in Figure 2, of which $N = 6$. We can see that even if there are slight differences between the parameters of each cycle, the dataset still exhibits periodic characteristics.

To report the performances of the estimates, we conduct simulation studies under the following three models:

Model I. Assume that $\{Z_t\}$ is a sequence of i.i.d periodic Poisson distributed random variables with mean $E(Z_t) = Var(Z_t) = \lambda_j$ for $t = j + sT, j = 1, 2, \ldots, T, s \in \mathbb{N}_0$.

Model II. Assume that $\{Z_t\}$ is a sequence of i.i.d. periodic Geometric distributed random variables with p.m.f. given by

$$p(Z_{j+sT} = z) = \frac{\lambda_j^z}{(1+\lambda_j)^{1+z}}, z = 0, 1, 2, \ldots$$

with $E(Z_t) = \lambda_j$, $Var(Z_t) = \lambda_j(1+\lambda_j)$ for $t = j + sT, j = 1, 2, \ldots, T, s \in \mathbb{N}_0$.

Model III. Assume that $\{Z_t\}$ is a sequence of i.i.d mixed distributed random variables,

$$Z_t = \Delta_t Z_{1t} + (1-\Delta_t)Z_{2t},$$

where $\{\Delta_t\}$ is a sequence of i.i.d periodic Bernoulli distributed random variables with $P(\Delta_t = 1) = 1 - P(\Delta_t = 0) = \rho_j, \rho = (\rho_1, \rho_2, \ldots, \rho_T)$ for $t = j + sT, j = 1, 2, \ldots, T, s \in \mathbb{N}_0$, which is independent of $\{Z_{it}\}, i = 1, 2$.

For $\{Z_{1t}\}$ given in Model I and $\{Z_{2t}\}$ given in Model II, we can easily see that $E(Z_t) = \lambda_j, Var(Z_t) = \lambda_j^2(1-\rho_j) + \lambda_j$.

For each model, we generate the data with $X_0 = 0$, set $T = 3$ and the sample sizes $n = NT = 150, 300, 900$. All the calculations are performed under the $\mathcal{R}3.6.2$ software with 1000 replications. We use the command *constrOptim* to optimize the objective function of the maximum likelihood estimation. The threshold vector is calculated by the algorithms discussed in Section 3.2. Other algorithms are based on the explicit expressions.

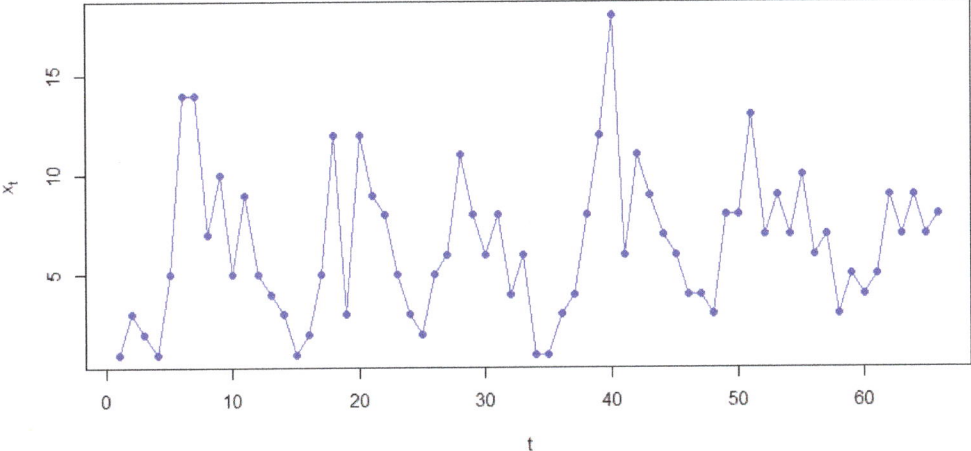

Figure 2. Sample path of the first six cycles.

4.1. Performances of the $\hat{\beta}_{CLS}$, $\hat{\beta}_{MQL}$ and $\hat{\beta}_{CML}$

Pereira et al. [25] provided a theoretical basis for the conditional least squares (CLS) and conditional maximum likelihood (CML) estimators of the parameters vector β in the PSETINAR$(2;1,1)_T$ process but did not conduct simulation research. Manaa and Bentarzi [26] provided the asymptotic properties of the estimators and compared their performance through a simulation study. To compare the performance of the three estimators $\hat{\beta}_{CLS}, \hat{\beta}_{CML}$ and $\hat{\beta}_{MQL}$ (given in Section 3), we conduct simulation studies for these three estimators under Models I to III. The parameters are selected as follows:

Series A. $\beta = (0.2, 0.45, 1, 0.2, 0.45, 2, 0.8, 0.45, 2)', r = (3, 2, 2)'$.
Series B. $\beta = (0.65, 0.45, 1, 0.65, 0.45, 2, 0.35, 0.45, 2)', r = (2, 2, 3)'$.
Series C. $\beta = (0.2, 0.45, 3, 0.2, 0.45, 7, 0.8, 0.45, 7)', r = (12, 7, 9)'$.

To eliminate the influence of the change of parameters on estimates, we choose the series randomly and change the parameters with fixed $\alpha^{(k)}, k = 1, 2$ or λ separately. The selection of these thresholds ensures there are enough data in each regime.

Spectral analysis starts from finding hidden periodicity, and it is an important subject of time series frequency domain analysis. The approach for studying hidden periods based on frequency domain analysis is the periodogram method, proposed by Schuster [34]; the rigorous examination is shown in Fisher [35]. For a series of observations $\{X_t\}$, $t = 1, 2, \ldots, n$, the periodogram is defined as

$$I_n(f_k) = \frac{1}{n}|\sum_{t=1}^{n} X_t e^{-i2\pi f_k t}|^2 = a_k^2 + b_k^2, \quad (11)$$

where

$$a_k = \begin{cases} \frac{1}{\sqrt{n}}(\sum_{t=1}^{n} X_t \cos(2\pi f_k t))^2, & k = 1, 2, \ldots, \left[\frac{n-1}{2}\right], \\ \frac{1}{\sqrt{n}} \sum_{t=1}^{n} (-1)^t X_t, & k = \frac{n}{2}, \end{cases}$$

$$b_k = \begin{cases} \frac{1}{\sqrt{n}}(\sum_{t=1}^{n} X_t \sin(2\pi f_k t))^2, & k = 1, 2, \ldots, \left[\frac{n-1}{2}\right], \\ 0, & k = \frac{n}{2}, \end{cases}$$

and the period $T = [1/\text{argmax}_f I_n(f_k)]$, where $[\cdot]$ denotes the integer part of a number.

The sample path and periodogram of the Series A, B and C under Model I are plotted in Figure 3 to show the periodic characteristics. Because the period is three and short, it is difficult to see the period from the sample path, but the periodogram can show the period very well. In addition, the simulation results are summarized in Tables 1–9.

As expected, biases and MSE of the estimators decrease as the sample size N increases, which is in agreement with the asymptotic properties of the estimators: asymptotic unbiasedness and consistency. Most of the biases and MSE in Model II are larger than those in Model I. Maybe this is because that the variance of $\{Z_t\}$ in Model II is larger than that in Model I, which leads to the fluctuation of data.

Tables 1–6 summarize the simulation results for different series under Model I and Model II. From these tables, we can see that most of the biases and MSE of $\hat{\beta}_{MQL}$ are smaller than $\hat{\beta}_{CLS}$. Perhaps it is because that the MQL method uses more information about the data than the CLS method. Therefore, the MQL method can obtain the optimal value more accurately. In addition, most of the biases of $\hat{\beta}_{MQL}$ are smaller than $\hat{\beta}_{CML}$, while the MSE is larger, which is because the CML uses the distribution. If the distribution is correct, it is indeed better than the MQL. It is worth mentioning that the CML method is more complicated and time-consuming than the MQL method in the simulation procedure. We can conclude that the MQL estimators are better than CLS estimators, and the CML estimators are not unanimously better than MQL estimators.

Table 1. Bias and MSE for Series A of Model I (MSE in parentheses): CLS, MQL and CML.

N	Method	$\alpha_1^{(1)}$	$\alpha_1^{(2)}$	λ_1	$\alpha_2^{(1)}$	$\alpha_2^{(2)}$	λ_2	$\alpha_3^{(1)}$	$\alpha_3^{(2)}$	λ_3
50	CLS	0.001	−0.001	0.001	−0.018	−0.004	0.006	0.008	0.005	−0.025
		(0.052)	(0.014)	(0.253)	(0.131)	(0.024)	(0.230)	(0.160)	(0.024)	(0.326)
	MQL	0.000	−0.002	0.006	−0.015	−0.004	0.002	0.011	0.006	−0.030
		(0.054)	(0.014)	(0.266)	(0.126)	(0.023)	(0.220)	(0.156)	(0.024)	(0.316)
	CML	0.024	0.010	−0.047	0.054	0.019	−0.079	0.003	0.007	−0.027
		(0.024)	(0.008)	(0.117)	(0.062)	(0.016)	(0.126)	(0.047)	(0.013)	(0.134)
100	CLS	0.004	0.000	−0.006	0.013	−0.001	−0.005	0.002	−0.003	0.008
		(0.026)	(0.007)	(0.132)	(0.058)	(0.011)	(0.108)	(0.085)	(0.012)	(0.168)
	MQL	0.004	0.000	−0.006	0.013	−0.001	−0.006	−0.001	−0.004	0.012
		(0.024)	(0.007)	(0.120)	(0.057)	(0.011)	(0.105)	(0.082)	(0.011)	(0.162)
	CML	0.012	0.004	−0.023	0.036	0.007	−0.034	0.003	0.000	−0.001
		(0.014)	(0.004)	(0.067)	(0.036)	(0.008)	(0.073)	(0.024)	(0.006)	(0.066)
300	CLS	−0.003	−0.002	0.009	0.002	0.000	−0.005	−0.002	0.000	−0.001
		(0.010)	(0.003)	(0.051)	(0.020)	(0.004)	(0.034)	(0.028)	(0.004)	(0.055)
	MQL	−0.002	−0.001	0.007	0.001	0.000	−0.004	−0.003	0.000	0.000
		(0.009)	(0.002)	(0.045)	(0.019)	(0.004)	(0.033)	(0.027)	(0.003)	(0.053)
	CML	0.000	0.000	0.000	0.003	0.001	−0.007	0.001	0.002	−0.006
		(0.005)	(0.001)	(0.025)	(0.014)	(0.003)	(0.024)	(0.007)	(0.002)	(0.020)

Table 2. Bias and MSE for Series B of Model I (MSE in parentheses): CLS, MQL and CML.

N	Method	$\alpha_1^{(1)}$	$\alpha_1^{(2)}$	λ_1	$\alpha_2^{(1)}$	$\alpha_2^{(2)}$	λ_2	$\alpha_3^{(1)}$	$\alpha_3^{(2)}$	λ_3
50	CLS	0.009	0.001	0.003	0.014	0.003	−0.015	−0.013	−0.009	0.032
		(0.119)	(0.015)	(0.238)	(0.166)	(0.031)	(0.365)	(0.105)	(0.026)	(0.525)
	MQL	0.010	0.001	0.003	0.013	0.003	−0.014	−0.012	−0.009	0.031
		(0.129)	(0.015)	(0.241)	(0.161)	(0.030)	(0.354)	(0.104)	(0.026)	(0.516)
	CML	0.006	0.003	0.001	0.014	0.006	−0.020	0.008	0.003	−0.019
		(0.043)	(0.007)	(0.090)	(0.062)	(0.016)	(0.150)	(0.045)	(0.014)	(0.229)
100	CLS	0.007	0.000	−0.001	−0.022	−0.009	0.042	−0.004	−0.002	0.003
		(0.061)	(0.008)	(0.133)	(0.076)	(0.014)	(0.173)	(0.046)	(0.012)	(0.222)
	MQL	0.008	0.000	−0.003	−0.023	−0.010	0.044	−0.004	−0.002	0.003
		(0.055)	(0.007)	(0.116)	(0.076)	(0.014)	(0.172)	(0.045)	(0.012)	(0.216)
	CML	0.002	0.000	−0.001	−0.004	−0.001	0.013	0.000	0.001	−0.008
		(0.018)	(0.003)	(0.040)	(0.031)	(0.008)	(0.078)	(0.027)	(0.007)	(0.127)
300	CLS	0.003	0.000	−0.003	0.002	0.000	0.002	−0.003	−0.001	−0.001
		(0.020)	(0.003)	(0.043)	(0.026)	(0.005)	(0.060)	(0.017)	(0.004)	(0.081)
	MQL	0.003	−0.001	−0.002	0.001	0.000	0.004	−0.002	0.000	−0.004
		(0.019)	(0.002)	(0.039)	(0.025)	(0.005)	(0.058)	(0.016)	(0.004)	(0.077)
	CML	−0.002	−0.002	0.003	0.003	0.001	−0.001	−0.003	0.000	−0.002
		(0.006)	(0.001)	(0.014)	(0.009)	(0.002)	(0.025)	(0.009)	(0.003)	(0.043)

Table 3. Bias and MSE for Series C of Model I (MSE in parentheses): CLS, MQL and CML.

N	Method	$\alpha_1^{(1)}$	$\alpha_1^{(2)}$	λ_1	$\alpha_2^{(1)}$	$\alpha_2^{(2)}$	λ_2	$\alpha_3^{(1)}$	$\alpha_3^{(2)}$	λ_3
50	CLS	−0.013	−0.010	0.146	−0.010	−0.003	0.053	−0.010	−0.007	0.054
		(0.022)	(0.011)	(2.088)	(0.082)	(0.022)	(1.915)	(0.078)	(0.026)	(3.823)
	MQL	−0.010	−0.008	0.117	−0.010	−0.003	0.052	−0.014	−0.009	0.079
		(0.022)	(0.010)	(2.000)	(0.082)	(0.021)	(1.913)	(0.075)	(0.025)	(3.709)
	CML	0.003	0.001	−0.015	0.044	0.021	−0.201	0.003	0.000	−0.033
		(0.012)	(0.006)	(1.119)	(0.044)	(0.013)	(1.054)	(0.025)	(0.010)	(1.286)
100	CLS	0.001	−0.002	0.015	−0.003	0.001	0.013	0.002	−0.003	0.022
		(0.014)	(0.006)	(1.323)	(0.043)	(0.011)	(1.046)	(0.038)	(0.012)	(1.772)
	MQL	0.000	−0.003	0.034	−0.002	0.001	0.008	0.001	−0.003	0.027
		(0.012)	(0.006)	(1.203)	(0.042)	(0.011)	(1.027)	(0.037)	(0.012)	(1.726)
	CML	0.006	0.001	−0.029	0.018	0.010	−0.085	0.011	0.003	−0.043
		(0.007)	(0.003)	(0.672)	(0.026)	(0.007)	(0.657)	(0.012)	(0.005)	(0.620)
300	CLS	0.000	0.000	0.006	0.002	0.002	−0.014	0.006	0.003	−0.040
		(0.006)	(0.003)	(0.586)	(0.014)	(0.004)	(0.350)	(0.013)	(0.004)	(0.606)
	MQL	0.001	0.000	0.002	0.001	0.001	−0.010	0.005	0.002	−0.032
		(0.005)	(0.002)	(0.527)	(0.014)	(0.004)	(0.341)	(0.012)	(0.004)	(0.589)
	CML	0.002	0.001	−0.013	0.005	0.003	−0.030	0.003	0.002	−0.026
		(0.003)	(0.001)	(0.262)	(0.011)	(0.003)	(0.267)	(0.004)	(0.002)	(0.201)

Table 4. Bias and MSE for Series A of Model II (MSE in parentheses): CLS, MQL and CML.

N	Method	$\alpha_1^{(1)}$	$\alpha_1^{(2)}$	λ_1	$\alpha_2^{(1)}$	$\alpha_2^{(2)}$	λ_2	$\alpha_3^{(1)}$	$\alpha_3^{(2)}$	λ_3
50	CLS	−0.013	−0.005	0.021	−0.016	−0.014	0.024	−0.025	−0.011	0.044
		(0.073)	(0.011)	(0.247)	(0.291)	(0.032)	(0.408)	(0.330)	(0.024)	(0.449)
	MQL	−0.011	−0.005	0.019	−0.012	−0.013	0.019	−0.020	−0.010	0.040
		(0.067)	(0.011)	(0.228)	(0.287)	(0.032)	(0.402)	(0.330)	(0.024)	(0.439)
	CML	0.014	0.005	−0.026	0.041	0.011	−0.050	0.004	0.008	−0.016
		(0.016)	(0.005)	(0.076)	(0.040)	(0.010)	(0.158)	(0.020)	(0.007)	(0.153)
100	CLS	0.003	0.003	−0.011	−0.013	−0.011	0.027	−0.002	0.001	−0.005
		(0.032)	(0.005)	(0.116)	(0.145)	(0.016)	(0.195)	(0.170)	(0.012)	(0.219)
	MQL	0.001	0.002	−0.006	−0.011	−0.010	0.024	−0.001	0.001	−0.006
		(0.030)	(0.005)	(0.104)	(0.143)	(0.016)	(0.194)	(0.169)	(0.011)	(0.215)
	CML	0.006	0.004	−0.014	0.020	0.006	−0.021	0.005	0.004	−0.019
		(0.007)	(0.002)	(0.039)	(0.022)	(0.005)	(0.080)	(0.011)	(0.003)	(0.072)

Table 4. Cont.

N	Method	$\alpha_1^{(1)}$	$\alpha_1^{(2)}$	λ_1	$\alpha_2^{(1)}$	$\alpha_2^{(2)}$	λ_2	$\alpha_3^{(1)}$	$\alpha_3^{(2)}$	λ_3
300	CLS	0.001	0.000	−0.005	−0.003	0.000	0.009	−0.001	0.000	−0.001
		(0.011)	(0.002)	(0.039)	(0.050)	(0.006)	(0.067)	(0.052)	(0.004)	(0.077)
	MQL	0.000	0.000	−0.005	−0.003	−0.001	0.010	0.000	0.000	−0.002
		(0.010)	(0.001)	(0.034)	(0.049)	(0.006)	(0.067)	(0.052)	(0.004)	(0.076)
	CML	0.005	0.001	−0.011	0.000	0.004	0.000	0.002	0.002	−0.007
		(0.003)	(0.001)	(0.013)	(0.008)	(0.002)	(0.026)	(0.004)	(0.001)	(0.026)

Table 5. Bias and MSE for Series B of Model II (MSE in parentheses): CLS, MQL and CML.

N	Method	$\alpha_1^{(1)}$	$\alpha_1^{(2)}$	λ_1	$\alpha_2^{(1)}$	$\alpha_2^{(2)}$	λ_2	$\alpha_3^{(1)}$	$\alpha_3^{(2)}$	λ_3
50	CLS	0.009	0.003	−0.019	0.005	−0.012	0.016	0.006	−0.003	−0.002
		(0.038)	(0.007)	(2.068)	(0.382)	(0.043)	(5.495)	(0.217)	(0.026)	(5.702)
	MQL	0.008	0.003	−0.017	−0.055	−0.022	0.070	0.009	−0.002	−0.008
		(0.037)	(0.006)	(1.995)	(0.378)	(0.043)	(5.461)	(0.220)	(0.026)	(5.718)
	CML	0.007	0.004	−0.017	0.015	0.004	−0.025	0.014	0.007	−0.031
		(0.005)	(0.002)	(0.590)	(0.025)	(0.006)	(1.380)	(0.008)	(0.004)	(1.326)
100	CLS	−0.001	−0.002	0.007	−0.006	−0.004	0.007	−0.006	−0.002	0.011
		(0.019)	(0.003)	(1.143)	(0.190)	(0.023)	(3.017)	(0.114)	(0.011)	(2.871)
	MQL	0.000	−0.002	0.004	−0.005	−0.004	0.007	−0.006	−0.003	0.012
		(0.018)	(0.003)	(1.091)	(0.189)	(0.023)	(3.001)	(0.115)	(0.012)	(2.882)
	CML	0.006	0.002	−0.012	0.008	0.004	−0.017	0.001	0.007	−0.017
		(0.002)	(0.001)	(0.238)	(0.012)	(0.003)	(0.691)	(0.004)	(0.002)	(0.660)
300	CLS	−0.003	−0.001	0.004	−0.004	−0.001	−0.006	0.003	−0.002	−0.006
		(0.006)	(0.001)	(0.361)	(0.062)	(0.007)	(0.889)	(0.033)	(0.004)	(0.848)
	MQL	−0.003	0.000	0.003	−0.002	0.000	−0.008	0.004	−0.002	−0.007
		(0.006)	(0.001)	(0.345)	(0.062)	(0.007)	(0.887)	(0.033)	(0.004)	(0.849)
	CML	0.000	0.001	−0.001	0.001	0.001	−0.011	0.004	0.002	−0.015
		(0.001)	(0.000)	(0.069)	(0.004)	(0.001)	(0.205)	(0.001)	(0.001)	(0.222)

Table 6. Bias and MSE for Series C of Model II (MSE in parentheses): CLS, MQL and CML.

N	Method	$\alpha_1^{(1)}$	$\alpha_1^{(2)}$	λ_1	$\alpha_2^{(1)}$	$\alpha_2^{(2)}$	λ_2	$\alpha_3^{(1)}$	$\alpha_3^{(2)}$	λ_3
50	CLS	−0.004	−0.002	0.069	−0.019	−0.008	0.061	−0.011	−0.008	0.131
		(0.038)	(0.007)	(2.068)	(0.382)	(0.043)	(5.495)	(0.217)	(0.026)	(5.702)
	MQL	−0.004	−0.002	0.067	−0.016	−0.007	0.051	−0.009	−0.007	0.122
		(0.037)	(0.006)	(1.995)	(0.378)	(0.043)	(5.461)	(0.220)	(0.026)	(5.718)
	CML	0.010	0.005	−0.019	0.037	0.014	−0.152	0.013	0.009	−0.038
		(0.005)	(0.002)	(0.590)	(0.025)	(0.006)	(1.380)	(0.008)	(0.004)	(1.326)
100	CLS	0.000	0.000	−0.005	−0.020	−0.004	0.054	0.001	−0.008	0.046
		(0.019)	(0.003)	(1.143)	(0.190)	(0.023)	(3.017)	(0.114)	(0.011)	(2.871)
	MQL	−0.002	−0.001	0.006	−0.020	−0.004	0.054	0.002	−0.008	0.045
		(0.018)	(0.003)	(1.091)	(0.189)	(0.023)	(3.001)	(0.115)	(0.012)	(2.882)
	CML	0.008	0.003	−0.059	0.016	0.005	−0.068	0.009	0.003	−0.047
		(0.002)	(0.001)	(0.238)	(0.012)	(0.003)	(0.691)	(0.004)	(0.002)	(0.660)
300	CLS	0.000	−0.001	−0.007	−0.005	−0.001	0.010	−0.014	−0.004	0.071
		(0.006)	(0.001)	(0.361)	(0.062)	(0.007)	(0.889)	(0.033)	(0.004)	(0.848)
	MQL	0.000	−0.001	−0.008	−0.005	−0.001	0.011	−0.014	−0.004	0.072
		(0.006)	(0.001)	(0.345)	(0.062)	(0.007)	(0.887)	(0.033)	(0.004)	(0.849)
	CML	0.000	0.000	−0.012	0.005	0.001	−0.020	0.004	0.002	−0.021
		(0.001)	(0.000)	(0.069)	(0.004)	(0.001)	(0.205)	(0.001)	(0.001)	(0.222)

To demonstrate the robustness of the MQL method, we consider the simulations about Model III with different series by using CLS, MQL and CML methods, and set $N = 300$, $\rho = (0.9, 0.9, 0.9)$, $(0.8, 0.8, 0.8)$, respectively. From Tables 7–9, we can see that when ρ varies from $(0.9, 0.9, 0.9)$ down to $(0.8, 0.8, 0.8)$, the effect on CLS and MQL estimators is slight. Most of the biases and MSE of MQL estimators are smaller than CLS. But due to incorrect distribution used, the biases and MSE of CML estimators increase. This indicates that the MQL method is more robust than CLS and CML methods.

Table 7. Bias and MSE for Series A of Model III with $N = 300$ (MSE in parentheses).

ρ	Method	$\alpha_1^{(1)}$	$\alpha_1^{(2)}$	λ_1	$\alpha_2^{(1)}$	$\alpha_2^{(2)}$	λ_2	$\alpha_3^{(1)}$	$\alpha_3^{(2)}$	λ_3
(0.9, 0.9, 0.9)	CLS	0.002	0.002	−0.004	0.009	0.004	−0.014	−0.007	0.000	0.002
		(0.010)	(0.002)	(0.049)	(0.022)	(0.004)	(0.041)	(0.026)	(0.004)	(0.055)
	MQL	0.002	0.002	−0.004	0.009	0.004	−0.014	−0.007	−0.001	0.003
		(0.009)	(0.002)	(0.042)	(0.021)	(0.004)	(0.040)	(0.026)	(0.004)	(0.053)
	CML	−0.021	−0.009	0.046	−0.043	−0.018	0.057	−0.055	−0.022	0.081
		(0.006)	(0.001)	(0.027)	(0.013)	(0.003)	(0.030)	(0.012)	(0.003)	(0.034)
(0.8, 0.8, 0.8)	CLS	−0.001	−0.001	0.000	0.005	−0.004	0.005	−0.005	−0.004	0.012
		(0.010)	(0.002)	(0.048)	(0.026)	(0.004)	(0.044)	(0.030)	(0.004)	(0.056)
	MQL	−0.001	−0.001	0.000	0.005	−0.004	0.006	−0.008	−0.005	0.016
		(0.009)	(0.002)	(0.042)	(0.026)	(0.004)	(0.043)	(0.030)	(0.004)	(0.054)
	CML	−0.042	−0.018	0.088	−0.080	−0.040	0.122	−0.121	−0.049	0.183
		(0.007)	(0.002)	(0.033)	(0.015)	(0.004)	(0.041)	(0.028)	(0.004)	(0.067)

Table 8. Bias and MSE for Series B of Model III with $N = 300$ (MSE in parentheses).

ρ	Method	$\alpha_1^{(1)}$	$\alpha_1^{(2)}$	λ_1	$\alpha_2^{(1)}$	$\alpha_2^{(2)}$	λ_2	$\alpha_3^{(1)}$	$\alpha_3^{(2)}$	λ_3
(0.9, 0.9, 0.9)	CLS	0.003	0.001	−0.001	0.001	0.000	0.000	0.004	0.000	−0.006
		(0.020)	(0.003)	(0.041)	(0.031)	(0.005)	(0.068)	(0.018)	(0.004)	(0.083)
	MQL	0.002	0.001	0.001	0.001	−0.001	0.001	0.003	−0.001	−0.003
		(0.018)	(0.002)	(0.036)	(0.030)	(0.005)	(0.065)	(0.017)	(0.004)	(0.080)
	CML	−0.023	−0.009	0.041	−0.080	−0.033	0.122	−0.065	−0.030	0.140
		(0.007)	(0.001)	(0.018)	(0.019)	(0.004)	(0.050)	(0.014)	(0.003)	(0.069)
(0.8, 0.8, 0.8)	CLS	0.001	0.001	−0.006	−0.005	−0.005	0.017	−0.002	−0.002	0.009
		(0.023)	(0.003)	(0.045)	(0.033)	(0.006)	(0.070)	(0.018)	(0.004)	(0.083)
	MQL	0.002	0.002	−0.008	−0.004	−0.005	0.016	−0.002	−0.002	0.010
		(0.021)	(0.002)	(0.039)	(0.032)	(0.005)	(0.067)	(0.017)	(0.004)	(0.078)
	CML	−0.043	−0.015	0.064	−0.156	−0.065	0.240	−0.122	−0.054	0.263
		(0.009)	(0.001)	(0.021)	(0.040)	(0.008)	(0.104)	(0.023)	(0.005)	(0.119)

Table 9. Bias and MSE for Series C of Model III with $N = 300$ (MSE in parentheses).

ρ	Method	$\alpha_1^{(1)}$	$\alpha_1^{(2)}$	λ_1	$\alpha_2^{(1)}$	$\alpha_2^{(2)}$	λ_2	$\alpha_3^{(1)}$	$\alpha_3^{(2)}$	λ_3
(0.9, 0.9, 0.9)	CLS	0.003	0.001	−0.024	−0.014	−0.007	0.078	−0.002	−0.002	0.021
		(0.006)	(0.002)	(0.534)	(0.020)	(0.005)	(0.485)	(0.014)	(0.004)	(0.643)
	MQL	0.003	0.001	−0.020	−0.013	−0.007	0.077	−0.001	−0.002	0.018
		(0.005)	(0.002)	(0.472)	(0.020)	(0.005)	(0.484)	(0.014)	(0.004)	(0.630)
	CML	−0.044	−0.028	0.432	−0.122	−0.064	0.631	−0.186	−0.097	1.250
		(0.005)	(0.002)	(0.470)	(0.020)	(0.006)	(0.595)	(0.044)	(0.013)	(2.143)
(0.8, 0.8, 0.8)	CLS	0.001	−0.001	−0.003	−0.008	−0.003	0.036	0.000	−0.001	0.005
		(0.005)	(0.002)	(0.448)	(0.023)	(0.005)	(0.494)	(0.016)	(0.004)	(0.668)
	MQL	0.000	−0.001	0.002	−0.008	−0.002	0.034	−0.001	−0.002	0.007
		(0.005)	(0.002)	(0.407)	(0.023)	(0.005)	(0.490)	(0.015)	(0.004)	(0.661)
	CML	−0.074	−0.045	0.706	−0.158	−0.085	0.811	−0.296	−0.144	1.907
		(0.008)	(0.003)	(0.754)	(0.027)	(0.009)	(0.800)	(0.098)	(0.024)	(4.230)

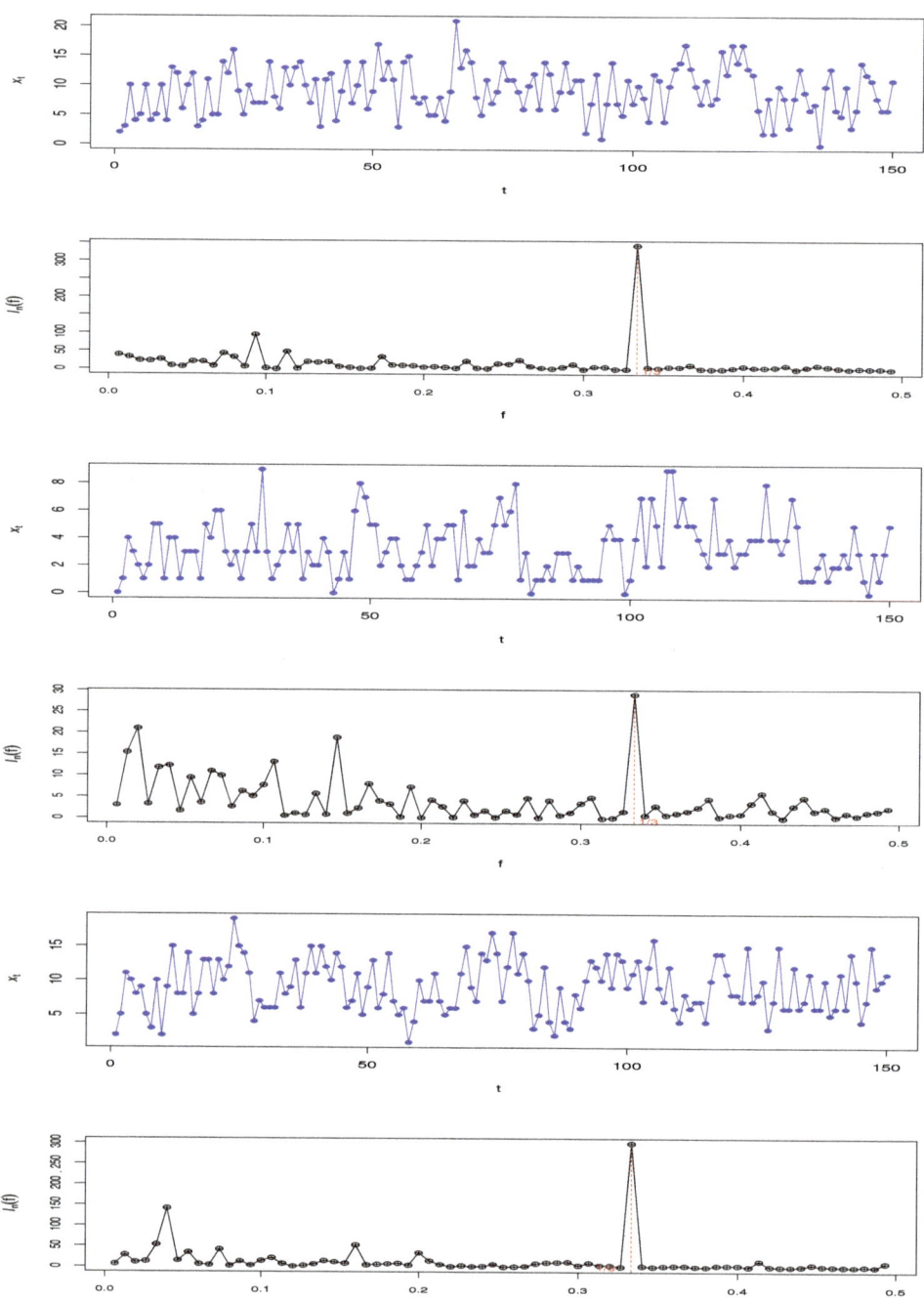

Figure 3. The sample path and periodogram of Series A(top), B(middle) and C(bottom) in Model I.

4.2. Performances of $\hat{r}_{MQL}$ and $\hat{r}_{CLS}$

As discussed in Section 3.2, we estimate the thresholds vector by using conditional least squares and modified quasi-likelihood methods. The performances of $\hat{r}_{MQL}$ and $\hat{r}_{CLS}$ are compared in this subsection through simulation studies. From the simulation results in Section 4.1, we find that the contaminated data generated from Model III has little influence on least squares and quasi-likelihood estimators, so we only simulate thresholds estimation for different series under Model I and Model II. We assess the performance of r by the bias, MSE and bias median, where the bias median is defined by:

$$\text{Bias median} = \underset{i \in \{1,2,\ldots,K\}}{\text{median}} (\hat{r}_{ij} - r_{0j}), j = 1, 2, \ldots, T,$$

where $\hat{r}_{ij}$ is the estimator of r_{0j}, r_{0j} is the true value with $j = 1, 2, \ldots, T$, and K is the number of replications. The simulation results are summarized in Tables 10–15.

Table 10. Bias, bias median and MSE for Series A of Model I.

N	Para.	MQL			CLS		
		Bias	Median	MSE	Bias	Median	MSE
50	r_1	−0.167	0	0.447	0.042	0	0.550
	r_2	0.422	0	1.986	0.723	0	2.841
	r_3	0.457	0	1.975	0.947	0	3.779
100	r_1	−0.107	0	0.151	−0.003	0	0.137
	r_2	0.224	0	1.378	0.570	0	2.428
	r_3	0.245	0	0.861	0.505	0	1.903
300	r_1	−0.007	0	0.007	0.000	0	0.002
	r_2	0.027	0	0.283	0.117	0	0.477
	r_3	0.021	0	0.035	0.066	0	0.200

Table 11. Bias, bias median and MSE for Series B of Model I.

N	Para.	MQL			CLS		
		Bias	Median	MSE	Bias	Median	MSE
50	r_1	0.499	0	2.129	1.294	1	4.176
	r_2	0.538	0	2.320	0.868	0	3.142
	r_3	0.139	0	2.687	0.634	0	3.610
100	r_1	0.555	1	1.933	1.301	1	3.597
	r_2	0.283	0	1.437	0.643	0	2.473
	r_3	0.107	0	2.537	0.599	0	3.431
300	r_1	0.480	1	1.518	1.215	1	2.485
	r_2	0.021	0	0.213	0.141	0	0.489
	r_3	−0.095	0	1.191	0.261	0	1.825

Table 12. Bias, bias median and MSE for Series C of Model I.

N	Para.	MQL			CLS		
		Bias	Median	MSE	Bias	Median	MSE
50	r_1	−0.012	0	0.588	0.023	0	0.661
	r_2	0.268	0	5.378	0.541	0	5.909
	r_3	0.155	0	1.433	0.216	0	1.750
100	r_1	0.015	0	0.079	0.023	0	0.081
	r_2	0.072	0	2.332	0.254	0	2.972
	r_3	0.041	0	0.325	0.050	0	0.330
300	r_1	0.000	0	0.000	0.000	0	0.000
	r_2	−0.015	0	0.317	0.027	0	0.457
	r_3	0.002	0	0.004	0.002	0	0.004

Table 13. Bias, bias median and MSE for Series A of Model II.

N	Para.	MQL			CLS		
		Bias	Median	MSE	Bias	Median	MSE
50	r_1	0.027	0	1.231	0.407	0	2.227
	r_2	1.025	0	4.897	1.293	1	6.051
	r_3	1.582	1	7.600	2.003	1	9.905
100	r_1	−0.013	0	0.489	0.185	0	0.723
	r_2	0.944	0	4.808	1.271	0	6.215
	r_3	1.539	0	8.391	2.005	1	11.269
300	r_1	−0.042	0	0.066	0.022	0	0.070
	r_2	0.652	0	3.560	0.940	0	5.088
	r_3	0.605	0	3.243	1.062	0	6.540

Table 14. Bias, bias median and MSE for Series B of Model II.

N	Para.	MQL			CLS		
		Bias	Median	MSE	Bias	Median	MSE
50	r_1	1.231	1	5.527	2.134	2	9.638
	r_2	1.307	1	6.063	1.633	1	7.439
	r_3	0.840	0	6.658	1.237	1	8.211
100	r_1	1.070	1	3.954	1.972	2	8.050
	r_2	1.208	0	5.772	1.561	1	7.375
	r_3	0.998	0	7.652	1.488	1	9.644
300	r_1	1.059	1	3.143	1.829	2	5.611
	r_2	0.717	0	3.465	1.031	0	4.961
	r_3	0.617	0	5.925	1.153	0	8.549

Table 15. Bias, bias median and MSE for Series C of Model II.

N	Para.	MQL			CLS		
		Bias	Median	MSE	Bias	Median	MSE
50	r_1	−1.066	0	11.494	−0.859	0	12.671
	r_2	0.006	0	18.430	0.149	0	19.137
	r_3	−0.337	−1	27.211	−0.206	−1	27.764
100	r_1	−0.130	0	4.220	0.078	0	5.250
	r_2	0.538	0	22.610	0.696	0	23.536
	r_3	0.241	0	26.911	0.386	0	28.340
300	r_1	−0.040	0	0.236	−0.016	0	0.262
	r_2	1.213	0	26.909	1.389	0	28.515
	r_3	0.794	0	19.586	0.961	0	21.521

From Tables 10–15, we can see that all the simulation results perform better as sample size N increases, which implies that the estimators are consistent. The results in Tables 10–12 have smaller biases, bias medians and MSE than in Tables 13–15. This might be because the variance of Model II is larger than Model I for each series. Moreover, almost all the biases, bias medians and MSE of MQL estimators are smaller than CLS estimators, and the MSE of some MQL estimators are even half of the CLS. Because the thresholds are integer values, when we assess the accuracy of the estimators, the bias medians estimated can be more reasonable. It is concluded that it is much better to estimate the thresholds with the MQL method than CLS.

In the process of simulation, we generate the data with $X_0 = 0$; however, 0 is not the mean of the process, so we generate a set of data, discard some data generated first, and use the remaining data for inference, namely, "burn in" samples. Here, we generate a set of data with a length of 1800. We do the simulations for Series A of Model I, Model II and Model III ($\rho = (0.8, 0.8, 0.8)$). Other simulation settings are the same as before. The simulation results are listed in Tables 16–20. From these tables, we can see that under the "burn in" samples, the estimated results are similar to that when the initial value is 0, which indicates that the initial value will not affect our estimated results.

Table 16. Bias and MSE for Series A of Model I with "burn in" samples (MSE in parentheses): CLS, MQL and CML.

N	Method	$\alpha_1^{(1)}$	$\alpha_1^{(2)}$	λ_1	$\alpha_2^{(1)}$	$\alpha_2^{(2)}$	λ_2	$\alpha_3^{(1)}$	$\alpha_3^{(2)}$	λ_3
50	CLS	−0.002	−0.008	0.012	−0.018	−0.012	0.029	0.001	0.004	−0.008
		(0.067)	(0.017)	(0.338)	(0.132)	(0.024)	(0.241)	(0.168)	(0.025)	(0.351)
	MQL	0.001	−0.006	0.006	−0.016	−0.012	0.027	0.002	0.004	−0.007
		(0.066)	(0.017)	(0.331)	(0.134)	(0.024)	(0.240)	(0.174)	(0.024)	(0.347)
	CML	0.032	0.008	−0.061	0.043	0.007	−0.044	0.000	0.004	−0.005
		(0.027)	(0.008)	(0.124)	(0.056)	(0.015)	(0.125)	(0.046)	(0.012)	(0.125)
100	CLS	−0.005	−0.006	0.014	−0.011	−0.005	0.012	−0.001	−0.002	0.011
		(0.030)	(0.009)	(0.153)	(0.063)	(0.011)	(0.106)	(0.081)	(0.012)	(0.166)
	MQL	−0.006	−0.006	0.017	−0.012	−0.006	0.013	0.000	−0.002	0.008
		(0.028)	(0.008)	(0.138)	(0.061)	(0.010)	(0.103)	(0.078)	(0.012)	(0.158)
	CML	0.006	−0.001	−0.010	0.025	0.006	−0.031	0.000	0.001	0.003
		(0.015)	(0.004)	(0.069)	(0.035)	(0.007)	(0.069)	(0.021)	(0.006)	(0.061)
300	CLS	−0.001	−0.002	0.002	−0.003	−0.001	0.009	0.002	0.001	0.000
		(0.010)	(0.003)	(0.052)	(0.019)	(0.004)	(0.034)	(0.024)	(0.004)	(0.050)
	MQL	0.000	−0.001	0.000	−0.003	−0.001	0.008	0.001	0.001	0.002
		(0.009)	(0.002)	(0.047)	(0.019)	(0.003)	(0.033)	(0.024)	(0.003)	(0.049)
	CML	0.001	−0.001	−0.003	0.003	0.001	0.001	0.001	0.001	0.000
		(0.006)	(0.002)	(0.027)	(0.015)	(0.003)	(0.025)	(0.007)	(0.002)	(0.021)

Table 17. Bias and MSE for Series A of Model II with "burn in" samples (MSE in parentheses): CLS, MQL and CML.

N	Method	$\alpha_1^{(1)}$	$\alpha_1^{(2)}$	λ_1	$\alpha_2^{(1)}$	$\alpha_2^{(2)}$	λ_2	$\alpha_3^{(1)}$	$\alpha_3^{(2)}$	λ_3
50	CLS	0.009	0.000	−0.017	0.023	−0.011	0.005	−0.011	−0.004	0.005
		(0.067)	(0.011)	(0.242)	(0.306)	(0.035)	(0.424)	(0.303)	(0.026)	(0.479)
	MQL	0.010	0.000	−0.017	0.028	−0.010	0.000	−0.018	−0.005	0.012
		(0.065)	(0.010)	(0.227)	(0.302)	(0.035)	(0.420)	(0.355)	(0.026)	(0.505)
	CML	0.022	0.006	−0.042	0.053	0.013	−0.048	0.007	0.007	−0.032
		(0.015)	(0.004)	(0.075)	(0.045)	(0.010)	(0.151)	(0.022)	(0.007)	(0.148)
100	CLS	−0.011	−0.001	0.026	−0.013	0.000	0.015	−0.024	−0.007	0.022
		(0.034)	(0.005)	(0.123)	(0.151)	(0.016)	(0.210)	(0.157)	(0.012)	(0.223)
	MQL	−0.011	−0.001	0.026	−0.013	−0.001	0.016	−0.022	−0.006	0.019
		(0.033)	(0.005)	(0.117)	(0.148)	(0.016)	(0.208)	(0.156)	(0.012)	(0.219)
	CML	0.006	0.005	−0.004	0.018	0.007	−0.013	0.009	0.003	−0.015
		(0.008)	(0.002)	(0.039)	(0.025)	(0.005)	(0.075)	(0.010)	(0.003)	(0.076)
300	CLS	−0.004	−0.001	0.005	−0.001	−0.002	0.008	0.000	−0.005	0.006
		(0.010)	(0.002)	(0.037)	(0.050)	(0.005)	(0.069)	(0.052)	(0.003)	(0.074)
	MQL	−0.003	−0.001	0.004	−0.001	−0.002	0.009	0.001	−0.005	0.004
		(0.010)	(0.001)	(0.034)	(0.049)	(0.005)	(0.068)	(0.051)	(0.003)	(0.073)
	CML	0.001	0.001	−0.005	0.007	0.001	−0.001	0.000	−0.001	−0.002
		(0.003)	(0.001)	(0.013)	(0.008)	(0.002)	(0.028)	(0.003)	(0.001)	(0.028)

Table 18. Bias and MSE for Series A of Model III ($\rho = (0.8, 0.8, 0.8)$) with "burn in" samples (MSE in parentheses): CLS, MQL and CML.

N	Method	$\alpha_1^{(1)}$	$\alpha_1^{(2)}$	λ_1	$\alpha_2^{(1)}$	$\alpha_2^{(2)}$	λ_2	$\alpha_3^{(1)}$	$\alpha_3^{(2)}$	λ_3
50	CLS	−0.087	−0.040	0.214	0.018	−0.004	−0.007	0.019	0.002	−0.030
		(0.068)	(0.016)	(0.339)	(0.153)	(0.025)	(0.248)	(0.203)	(0.026)	(0.381)
	MQL	−0.011	−0.007	0.026	0.019	−0.003	−0.008	0.019	0.002	−0.031
		(0.065)	(0.014)	(0.292)	(0.155)	(0.024)	(0.244)	(0.203)	(0.026)	(0.376)
	CML	−0.016	−0.009	0.039	−0.012	−0.024	0.047	−0.109	−0.045	0.153
		(0.022)	(0.007)	(0.118)	(0.042)	(0.015)	(0.140)	(0.091)	(0.017)	(0.245)
100	CLS	−0.044	−0.017	0.103	−0.015	−0.006	0.020	−0.005	−0.003	0.008
		(0.033)	(0.008)	(0.162)	(0.075)	(0.012)	(0.132)	(0.100)	(0.013)	(0.199)
	MQL	−0.008	−0.002	0.013	−0.015	−0.006	0.020	−0.004	−0.003	0.008
		(0.030)	(0.007)	(0.137)	(0.074)	(0.012)	(0.129)	(0.099)	(0.012)	(0.197)
	CML	−0.043	−0.017	0.088	−0.057	−0.033	0.093	−0.129	−0.048	0.186
		(0.014)	(0.004)	(0.073)	(0.027)	(0.008)	(0.083)	(0.062)	(0.010)	(0.156)
300	CLS	−0.016	−0.006	0.036	0.000	−0.002	0.000	0.004	−0.001	−0.003
		(0.010)	(0.002)	(0.048)	(0.026)	(0.004)	(0.043)	(0.030)	(0.004)	(0.057)
	MQL	−0.003	−0.001	0.003	−0.001	−0.002	0.003	0.004	−0.001	−0.003
		(0.009)	(0.002)	(0.043)	(0.025)	(0.004)	(0.042)	(0.029)	(0.004)	(0.054)
	CML	−0.047	−0.020	0.097	−0.081	−0.037	0.113	−0.112	−0.046	0.169
		(0.007)	(0.002)	(0.035)	(0.016)	(0.004)	(0.038)	(0.025)	(0.004)	(0.061)

Table 19. Bias, bias median and MSE for Series A of Model I with "burn in" samples.

N	Para.	MQL			CLS		
		Bias	Median	MSE	Bias	Median	MSE
50	r_1	−0.180	0	0.400	0.053	0	0.393
	r_2	0.390	0	1.960	0.720	0	2.894
	r_3	0.580	0	2.322	0.963	0	3.583
100	r_1	−0.099	0	0.143	−0.007	0	0.081
	r_2	0.198	0	1.142	0.491	0	1.975
	r_3	0.218	0	0.800	0.455	0	1.585
300	r_1	−0.015	0	0.015	−0.004	0	0.004
	r_2	0.018	0	0.268	0.098	0	0.416
	r_3	0.018	0	0.036	0.058	0	0.170

Table 20. Bias, bias median and MSE for Series A of Model II with "burn in" samples.

N	Para.	MQL			CLS		
		Bias	Median	MSE	Bias	Median	MSE
50	r_1	−0.071	0	0.835	0.252	0	1.394
	r_2	1.156	0	5.640	1.436	1	6.878
	r_3	1.616	1	7.974	2.046	1	10.284
100	r_1	−0.110	0	0.320	0.099	0	0.473
	r_2	1.172	0	5.662	1.477	1	7.063
	r_3	1.518	1	7.508	1.947	1	10.059
300	r_1	−0.041	0	0.055	0.027	0	0.055
	r_2	0.648	0	3.364	0.940	0	4.884
	r_3	0.574	0	3.532	0.854	0	5.338

5. Real Data Example

In this section, we use the PSETINAR$(2;1,1)_T$ process to fit the series of monthly counts of claimants collecting short-term disability benefits. In the dataset, all the claimants are male, have cuts, lacerations or punctures, and are between the ages of 35 and 54. In addition, they all work in the logging industry and collect benefits from the Workers' Compensation Board (WCB) of British Columbia. The dataset consists of 120 observations, from 1985 to 1994 (Freeland [36]). The simulations were performed on the $\mathcal{R}$3.6.2 software. The threshold vector was calculated by the algorithms (the three-step algorithm of NeSS combined with quasi-likelihood principle and the algorithm of NeSS combined with least squares principle) described in Section 3.2. We uses the command *constrOptim* to optimize the objective function of the maximum likelihood estimation. Figure 4 shows the sample path, ACF and PACF plots of the observations. It can be seen from Figure 4 that this dataset is a dependent counting time series with periodic characteristic.

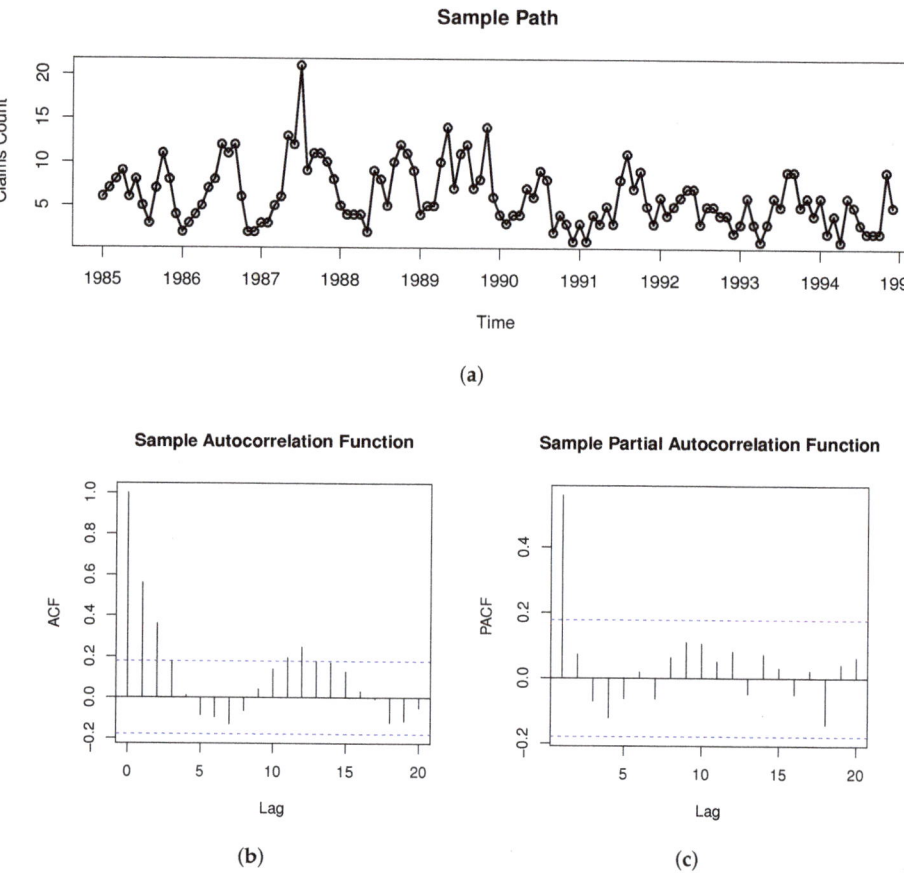

Figure 4. The sample path plot (**a**), ACF and PACF plots (**b**,**c**) for the counts of claimants.

We use the periodogram method to determine the period about this dataset and draw Figure 5, from which it can be seen that $I_n(f_k)$ reach maximum at $f_k = 1/12$, and concluded that $T = 12$. This displays the periodic characteristic of the data and exhibits a form of periodic change per year.

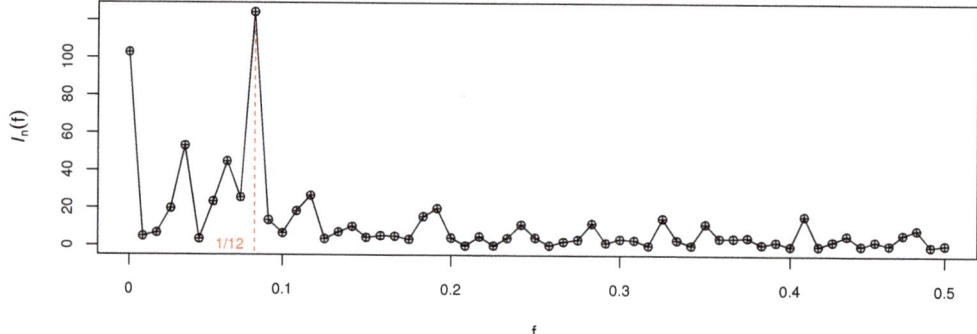

Figure 5. The periodogram plot for the monthly counts of claimants.

Table 21 displays the descriptive statistics for the monthly counts of claimants collecting short-term disability benefits from WCB. From Table 21, we can see that the mean and variance are approximately equal in some months. We can assume that the distribution of the innovations is a periodic Poisson. However, some months and the total data indicate overdispersion. We find that the dataset has no zero and the minimum value is one. This leads us to consider the periodic Poisson, periodic Geometric, zero-truncated periodic Poisson and zero-truncated periodic Geometric distributions for the innovations to fit the model, respectively. Before the model fitting, we first estimate the threshold vector. The $\hat{r}_{CLS}$ is calculated by (9) and the $\hat{r}_{MQL}$ is calculated through (10) by using the three-step algorithm. Table 22 summarizes the fitting results of $\hat{r}_{CLS}$ and $\hat{r}_{MQL}$. Due to the lesser data, to fit the model better, when the number of data in each regime is relatively smaller than two or the threshold is the maximum or minimum value of the boundary, we think that these monthly data do not have a piecewise phenomenon, that is, March, July, and August do not have piecewise phenomena.

Table 21. Summary statistics for the monthly counts of claimants.

	Whole Dataset	Jan.	Feb.	Mar.	Apr.	May	Jun.	Jul.	Aug.	Sep.	Oct.	Nov.	Dec.
Mean	6.1	4.2	3.8	4.6	4.9	7.0	7.1	8.5	7.5	7.2	7.2	7.2	4.4
Variance	11.8	2.2	3.3	1.8	9.0	14.7	5.9	28.9	12.5	12.0	12.2	14.8	6.9
Maximum	21	6	7	8	10	14	12	21	12	12	12	14	19
Minimum	1	2	1	3	1	2	3	3	2	2	2	2	1

Table 22. Threshold estimators for the monthly counts of claimants.

	Jan.	Feb.	Mar.	Apr.	May	Jun.	Jul.	Aug.	Sep.	Oct.	Nov.	Dec.
$\hat{r}_{CLS}$	3	4	7	5	5	6	10	4	9	6	7	6
$\hat{r}_{MQL}$	3	4	7	5	5	6	10	4	9	6	7	5

To capture the piecewise phenomenon of this time series dataset, we use PINAR(1)$_T$ and PSETINAR(2;1,1)$_T$ models with period $T = 12$ to fit the dataset, respectively. The PINAR(1) process proposed by Monteiro et al. [22] with the following recursive equation

$$X_t = \alpha_t \circ X_{t-1} + Z_t,\qquad (12)$$

with $\alpha_t = \alpha_j \in (0,1)$ for $t = j + sT (j = 1, \ldots, T, s \in \mathbb{N}_0)$, the definition of thinning operator "∘" and innovation process $\{Z_t\}$ is the same as the PSETINAR(2;1,1)$_T$ process.

It is worth mentioning that for this dataset, the conditional least squares and quasi-likelihood methods produce non-admissible estimators for some months, so we use the conditional maximum likelihood approach to estimate the parameters. Next, we use PSETINAR(2;1,1)$_{12}$ and PINAR(1)$_{12}$ models to fit the dataset in combination with the four innovation distributions mentioned before. Here, the threshold vectors are based on $\hat{r}_{MQL}$. The AIC and BIC are listed in Table 23. When we fit the dataset, we hope to get smaller AIC and BIC values. From the results in Table 23, we can conclude that the PSETINAR(2;1,1)$_{12}$ model with zero-truncated periodic Poisson distribution is more suitable. Then, we do the conditional maximum likelihood estimation, and the results are listed in Table 24. Some estimators of the parameters in Table 24, for example, the $\alpha^{(2)}$ of January, May, June, September, October and November, are not statistically significant, suggesting that on those months, the number of claims is mainly modeled through the innovation process.

Table 23. The AIC and BIC of the claims data.

PSETINAR(2;1,1)$_{12}$	AIC	BIC	PINAR(1)$_{12}$	AIC	BIC
Pois.	586.63	596.61	Pois.	592.12	599.38
Zero-truncated Pois.	581.65	591.64	Zero-truncated Pois.	594.44	601.71
Geom.	610.45	620.43	Geom.	605.56	612.82
Zero-truncated Geom.	586.36	596.34	Zero-truncated Geom.	595.15	602.42

Table 24. CML estimators in the dataset.

Month	$\alpha^{(1)}$	$\alpha^{(2)}$	λ
Jan.	0.112	8.907×10^{-08}	3.819
Feb.	0.227	0.032	3.060
Mar.	0.692	-	1.969
Apr.	0.999	0.240	2.048
May	0.586	8.521×10^{-09}	4.889
Jun.	0.265	4.316×10^{-08}	5.507
Jul.	0.360	-	5.942
Aug.	0.390	-	4.186
Sep.	0.380	3.366×10^{-07}	5.218
Oct.	0.502	1.027×10^{-07}	4.044
Nov.	0.433	2.776×10^{-08}	4.990
Dec.	0.508	0.222	1.000

Remark: "-" stand for not available.

To check the predictability of the PSETINAR(2;1,1)$_T$ model, we carry out the h-step-ahead forecasting for varying h of the PSETINAR(2;1,1)$_T$ model. The h-step-ahead conditional expectation point predictor of the PSETINAR(2;1,1)$_T$ model is given by

$$\hat{X}_{j+sT+h} = E\left[X_{j+sT+h}|X_{j+sT}\right], h = 1, 2, \ldots.$$

Specifically, the one-step-ahead conditional expectation point predictor is given by

$$\hat{X}_{j+sT+1} = E\left[X_{j+sT+1}|X_{j+sT}\right] = \alpha_{j+1}^{(1)} X_{j+sT} I_{j+sT}^{(1)} + \alpha_{j+1}^{(2)} X_{j+sT} I_{j+sT}^{(2)} + \lambda_{j+1}.$$

However, the conditional expectation will seldom produce integer-valued forecasts. Recently, coherent forecasting techniques have been recommended, which only produce forecasts in $\mathbb{N}_0$. This is achieved by computing the h-step-ahead forecasting conditional distribution. As pointed out by Möller et al. [37], this approach leads to forecasts themselves being easily obtained from the median or the mode of the forecasting distribution. In addition, Li et al. [38] and Kang et al. [8] have applied this method to forecast the integer-valued processes. Homburg et al. [39] discussed the prediction methods based on conditional distributions and Gaussian approximations and applied them to some integer-valued processes and compared them. For the PSETINAR(2;1,1)$_T$ process, the one-step-ahead conditional distribution of X_{j+sT+1} given X_{j+sT} is given by

$$P(X_{j+sT+1} = x_{j+sT+1}|X_{j+sT} = x_{j+sT})$$
$$= \sum_{i=1}^{\min\{x_{j+sT},x_{j+sT+1}\}} \sum_{k=1}^{2} \binom{x_{j+sT}}{i} \alpha_{j+1}^{(k)i} \left(1 - \alpha_{j+1}^{(k)}\right)^{x_{j+sT}-i} I_{j+sT}^{(k)} P(Z_{j+sT+1} = x_{j+sT+1} - i).$$

Due to the existence of the threshold, while we use the conditional expectation method to predict $X_{j+sT+h}, h > 1$, we have to predict the previous moment of $X_{j+sT+h-1}$ first and compare it with the corresponding threshold before we do the next prediction. We do the same for the conditional distribution method. (To prevent confusion, we call this method a point-wise conditional distribution forecast. The predictors completely based on h-step-ahead conditional distribution without intermediate step prediction will be discussed later.) The mode of h-step-ahead point-wise conditional distribution can be viewed as the point prediction. Here we compare the two forecasting methods, a standard descriptive measure of forecasting accuracy, namely, h-step-ahead predicted root mean squared error (PRMSE) is adopted. This measure can be given by

$$PRMSE = \sqrt{\frac{1}{K-K_0} \sum_{t=K_0+1}^{K} \left(X_{t+h} - \hat{X}_{t+h}\right)^2}, h = 1, 2, \ldots,$$

where K is the full sample size, we split the data into two parts, and the last $K - K_0$ observations as a forecasting evaluation sample. We forecast the value of the last year when $h = 1, 2, 3, 12$.

The PRMSEs of the h-step-ahead point predictors are list in Table 25. For conditional expectation point predictors, the PRMSEs of PSETINAR$(2; 1, 1)_{12}$ with zero-truncated periodic Poisson distribution are smaller than the PINAR$(1)_{12}$ with periodic Poisson and zero-truncated periodic Poisson distributions. This further shows the superiority of our model. The PRMSEs of the one-step-ahead point predictors are smaller than others. This is very natural because we use the value of the previous moment as the explanatory variable. For PSETINAR$(2; 1, 1)_{12}$ with zero-truncated periodic Poisson distribution, the PRMSEs of twelve-step-ahead predictors are smaller than other h-step-ahead predictors except for one-step-ahead. This may be because our period is 12. The PRMSE of one-step-ahead conditional expectation point predictors is smaller than point-wise conditional distribution point predictors. Thus, the former method is better for this dataset.

The PRMSEs of the one-step-ahead fitted series calculated by conditional expectation and conditional distribution are 2.434 and 3.565, respectively. This further illustrates that for our dataset, one-step-ahead forecasting conditional expectation is better than conditional distribution. The original data and the fitted series (calculated by the one-step-ahead conditional expectation based on the observations of the previous moments) by the PSETINAR$(2; 1, 1)_{12}$ model with zero-truncated periodic Poisson distribution are plotted in Figure 6. It is observed that the trend is similar to the original data. Except for the points with large value (the unexpected prediction may be due to the wrong judgement of regime), this model fits the data well.

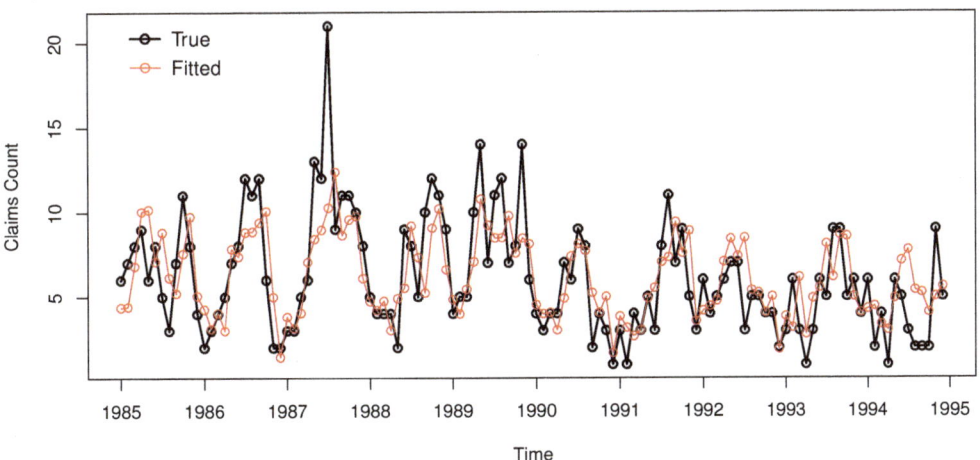

Figure 6. Plot of fitted curves of the claims data.

Table 25. PRMSE of the h-step-ahead point predictors.

		h	1	2	3	12
Conditional expectation	PSETINAR$(2;1,1)_{12}$ (Zero-truncated Pois.)		2.641	3.019	3.433	2.929
	PINAR$(1)_{12}$ (Zero-truncated Pois.)		2.753	3.377	3.567	3.788
	PINAR$(1)_{12}$ (Pois.)		2.724	3.407	3.704	4.008
Conditional distribution	PSETINAR$(2;1,1)_{12}$ (Zero-truncated Pois.)		2.814	3.000	3.109	2.930

Actually, we can get the h-step-ahead conditional distribution; here, we list the two-step-ahead and three-step-ahead conditional distributions as an example,

$$P(X_{j+sT+2} = x_{j+sT+2} | X_{j+sT} = x_{j+sT})$$
$$= \sum_{m=0}^{n} P(X_{j+sT+1} = m | X_{j+sT} = x_{j+sT}) P(X_{j+sT+2} = x_{j+sT+2} | X_{j+sT+1} = m),$$

and

$$P(X_{j+sT+3} = x_{j+sT+3} | X_{j+sT} = x_{j+sT})$$
$$= \sum_{m=0}^{n} P(X_{j+sT+2} = m | X_{j+sT} = x_{j+sT}) P(X_{j+sT+3} = x_{j+sT+3} | X_{j+sT+2} = m),$$

where $m \in \{0, 1, \ldots, n\}$ is the possible domain of X_{j+sT}, $j = 1, \ldots, T$, and $s \in \mathbb{N}_0$. When $h = 1, 2, 3$, we show the plots of the h-step-ahead conditional distribution in Figure 7, where x_{j+sT} represents the count of claimants in December 1993 and February 1994, respectively. The mode of h-step-ahead conditional distribution can be viewed as the point prediction. The PRMSEs of the two-step-ahead and three-step-ahead point predictors for the last year are 3.227 and 3.215, respectively, which is larger than the point-wise conditional distribution method described before. Maybe for other datasets or models, the h-step-ahead forecasting conditional distribution will show some advantages. We will not go into details here.

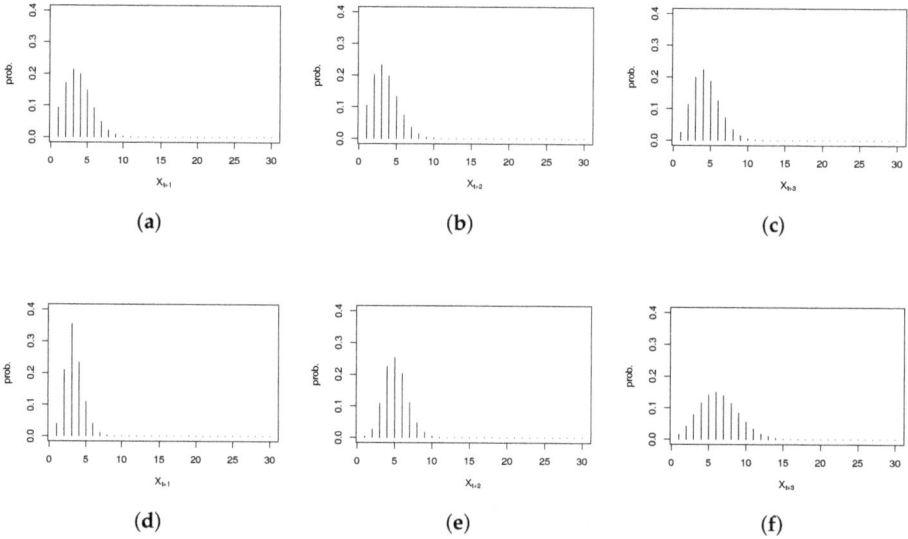

Figure 7. The h-step-ahead forecasting conditional distribution for the counts of claimants: (**a**–**c**) conditional on the count of claimants in December 1993; (**d**–**f**) conditional on the count of claimants in February 1994.

6. Conclusions

This paper extended the PSETINAR$(2;1,1)_T$ process proposed by Pereira et al. [25], by removing the assumption of Poisson distribution of $\{Z_t\}$ and considered the PSETINAR $(2;1,1)_T$ process under weak conditions that the second moment of $\{Z_t\}$ is finite. The ergodicity of the process is established. MQL-estimators of the model parameters vector β, MQL-estimators and CLS-estimators of the thresholds vector r are obtained. Moreover, through simulation, we can see the advantages of the quasi-likelihood method by comparing with the conditional maximum likelihood and conditional least square methods. An application to a real dataset is presented. In addition, the forecasting problem of this dataset is addressed.

In this paper, we only discuss the PSETINAR$(2;1,1)_T$ process for univariate time series. Hence, an extension for the multivariate PSETINAR$(2;1,1)_T$ process with a diagonal or cross-correlation autoregressive matrix is a topic for future investigation. Furthermore, it is also important to stress that beyond this extension, there are a number of interesting problems for future research in this area. For example, even a simple periodic model can have an inordinately large number of parameters. This is also true for PSETINAR$(2;1,1)_T$ models and even multi-period models. Therefore, the development of procedures of dimensionality reduction to overcome the computational difficulties is an impending problem. This remains a topic of future research.

Author Contributions: Conceptualization, C.L. and D.W.; methodology, C.L.; software, C.L.; validation, C.L., J.C. and D.W.; formal analysis, C.L.; investigation, C.L. and D.W.; resources, C.L. and D.W.; data curation, C.L.; writing—original draft preparation, C.L.; writing—review and editing, J.C. and D.W.; visualization, C.L.; supervision, J.C. and D.W.; project administration, D.W.; funding acquisition, D.W. All authors have read and agreed to the published version of the manuscript.

Funding: This work was supported by the National Natural Science Foundation of China (No. 11871028, 11731015, 11901053, 12001229), and the Natural Science Foundation of Jilin Province (No. 20180101216JC).

Institutional Review Board Statement: Not applicable.

Informed Consent Statement: Not applicable.

Data Availability Statement: The dataset is available in the book Freeland [36].

Conflicts of Interest: The authors declare no conflict of interest.

Appendix A

Proof of Theorem 1. According to Theorem 2 of Tweedie [40] (see also, Zheng and Basawa [41]), for the process defined by (2), and $\forall j = 1, 2, \ldots, T, s \in \mathbb{Z}$, we have

$$E(X_{j+sT}|X_{j+sT-1} = x) = \alpha_j^{(1)} x I_{j+sT-1}^{(1)} + \alpha_j^{(2)} x I_{j+sT-1}^{(2)} + \lambda_j \leq \alpha_{j,\max} x + \lambda_j,$$

where $\alpha_{j,\max} = \max\{\alpha_j^{(1)}, \alpha_j^{(2)}\} < 1$.

Let $K = \left[\frac{1+\lambda_j}{1-\alpha_{j,\max}}\right] + 1$, where $[\cdot]$ denotes the integer part of a number. Then for $x \geq K$, we have

$$E(X_{j+sT}|X_{j+sT-1} = x) \leq x - 1,$$

and for $x < K$,

$$E(X_{j+sT}|X_{j+sT-1} = x) \leq \alpha_{j,\max} x + \lambda_j \leq K + \lambda_j < \infty.$$

Therefore, the process $\{X_t\}$ for $t = j + sT$ defined in (2) is an ergodic Markov chain. □

Proof of Proposition 2. (i) From Proposition 1, we have

$$Var(X_{j+sT}|X_{j+sT-1}) = E\{X_{j+sT} - E(X_{j+sT}|X_{j+sT-1})\}^2$$
$$= E\left(X_{j+sT} - \alpha_j^{(1)} X_{j+sT-1} I_{j+sT-1}^{(1)} - \alpha_j^{(2)} X_{j+sT-1} I_{j+sT-1}^{(2)} - \lambda_j\right)^2,$$

and

$$Var(X_{j+sT}|X_{j+sT-1}) = \sum_{k=1}^{2} \alpha_j^{(k)}\left(1 - \alpha_j^{(k)}\right) X_{j+sT-1} I_{j+sT-1}^{(k)} + \sigma_{z,j}^2,$$

with $k = 1, 2, j = 1, \ldots, T, s \in \mathbb{N}_0$, so by substituting suitable consistent estimators of $\alpha_j^{(k)}$ and λ_j, we can get the consistent estimation of $\sigma_{z,j}^2$,

$$\hat{\sigma}_{1,z,j}^2 = \frac{1}{N} \sum_{s=0}^{N-1} \left(X_{j+sT} - \sum_{k=1}^{2} \hat{\alpha}_j^{(k)} X_{j+sT-1} I_{j+sT-1}^{(k)} - \hat{\lambda}_j\right)^2$$
$$- \frac{1}{N} \sum_{k=1}^{2} \sum_{s=0}^{N-1} \hat{\alpha}_j^{(k)}\left(1 - \hat{\alpha}_j^{(k)}\right) X_{j+sT-1} I_{j+sT-1}^{(k)}.$$

(ii) Moreover, from model (2), we have

$$Var(X_{j+sT}) = \sum_{k=1}^{2} Var\left\{\left(\alpha_j^{(k)} \circ X_{j+sT-1}\right) I_{j+sT-1}^{(k)}\right\} + Var(Z_{j+sT})$$
$$+ 2Cov\left\{\left(\alpha_j^{(1)} \circ X_{j+sT-1}\right) I_{j+sT-1}^{(1)}, \left(\alpha_j^{(2)} \circ X_{j+sT-1}\right) I_{j+sT-1}^{(2)}\right\},$$

where

$$Var\left\{\left(\alpha_j^{(k)} \circ X_{j+sT-1}\right) I_{j+sT-1}^{(k)}\right\}$$
$$= Var\left\{E\left[\left(\alpha_j^{(k)} \circ X_{j+sT-1}\right) I_{j+sT-1}^{(k)} | X_{j+sT-1}\right]\right\} + E\left\{Var\left[\left(\alpha_j^{(k)} \circ X_{j+sT-1}\right) I_{j+sT-1}^{(k)} | X_{j+sT-1}\right]\right\}$$
$$= \alpha_j^{(k)^2} Var\left(X_{j+sT-1} I_{j+sT-1}^{(k)}\right) + \alpha_j^{(k)}\left(1 - \alpha_j^{(k)}\right) E\left(X_{j+sT-1} I_{j+sT-1}^{(k)}\right),$$

and

$$2Cov\left\{\left(\alpha_j^{(1)} \circ X_{j+sT-1}\right) I_{j+sT-1}^{(1)}, \left(\alpha_j^{(2)} \circ X_{j+sT-1}\right) I_{j+sT-1}^{(2)}\right\}$$
$$= -2E\left\{\left(\alpha_j^{(1)} \circ X_{j+sT-1}\right) I_{j+sT-1}^{(1)}\right\} E\left\{\left(\alpha_j^{(2)} \circ X_{j+sT-1}\right) I_{j+sT-1}^{(2)}\right\}$$
$$= -2\alpha_j^{(1)} \alpha_j^{(2)} E\left\{X_{j+sT-1} I_{j+sT-1}^{(1)}\right\} E\left\{X_{j+sT-1} I_{j+sT-1}^{(2)}\right\}.$$

Note that

$$E\left\{X_{j+sT-1} I_{j+sT-1}^{(1)}\right\}$$
$$= E\left\{E\left[X_{j+sT-1} I_{j+sT-1}^{(1)} | I_{j+sT-1}^{(1)} = 1\right]\right\}$$
$$= E\left\{X_{j+sT-1} I_{j+sT-1}^{(1)} | I_{j+sT-1}^{(1)} = 1\right\} P\left(I_{j+sT-1}^{(1)} = 1\right)$$
$$= E\{X_{j+sT-1} | X_{j+sT-1} \leq r_j\} P\left(I_{j+sT-1}^{(1)} = 1\right).$$

Let $p_j = P\left(I^{(1)}_{j+sT-1} = 1\right)$ with $j = 1, \ldots, T, s \in \mathbb{N}_0$, we can estimate it with $\hat{p}_j = \frac{1}{N} \sum_{s=0}^{N-1} I^{(1)}_{j+sT-1}$. Therefore, by substituting a suitable consistent estimator of $\alpha_j^{(k)}$, based on moment estimation, we can get the estimator $\hat{\sigma}^2_{2,z,j}$ in Proposition 2. □

Proof of Theorem 2. Let $\mathcal{F}_{j+sT} = \sigma(X_0, X_1, \ldots, X_{j+sT})$ with $j = 1, \ldots, T, s \in \mathbb{N}_0$. First, we suppose θ is known, for the following estimation equations:

$$S^{(1)}_{N,j}(\theta_j, \beta_j) = \sum_{s=0}^{N-1} V_{\theta_j}^{-1} \left(X_{j+sT} - \alpha_j^{(1)} X_{j+sT-1} I^{(1)}_{j+sT-1} - \alpha_j^{(2)} X_{j+sT-1} I^{(2)}_{j+sT-1} - \lambda_j \right) X_{j+sT-1} I^{(1)}_{j+sT-1},$$

we have

$$E\left[V_{\theta_j}^{-1} \left(X_{j+sT} - \alpha_j^{(1)} X_{j+sT-1} I^{(1)}_{j+sT-1} - \alpha_j^{(2)} X_{j+sT-1} I^{(2)}_{j+sT-1} - \lambda_j \right) X_{j+sT-1} I^{(1)}_{j+sT-1} \big| \mathcal{F}_{j+sT-1} \right]$$
$$= V_{\theta_j}^{-1} X_{j+sT-1} I^{(1)}_{j+sT-1} E\left[(X_{j+sT} - E(X_{j+sT}|X_{j+sT-1})) | \mathcal{F}_{j+sT-1} \right]$$
$$= V_{\theta_j}^{-1} X_{j+sT-1} I^{(1)}_{j+sT-1} E\left[(X_{j+sT} - E(X_{j+sT}|X_{j+sT-1})) | X_{j+sT-1} \right]$$
$$= 0,$$

and

$$E\left[S^{(1)}_{s,j}(\theta_j, \beta_j) | \mathcal{F}_{j+sT-1} \right]$$
$$= E\left[V_{\theta_j}^{-1} \left(X_{j+sT} - \alpha_j^{(1)} X_{j+sT-1} I^{(1)}_{j+sT-1} - \alpha_j^{(2)} X_{j+sT-1} I^{(2)}_{j+sT-1} - \lambda_j \right) X_{j+sT-1} I^{(1)}_{j+sT-1} + S^{(1)}_{s-1,j}(\theta_j, \beta_j) | \mathcal{F}_{j+sT-1} \right]$$
$$= E\left[S^{(1)}_{s-1,j}(\theta_j, \beta_j) | \mathcal{F}_{j+sT-1} \right]$$
$$= S^{(1)}_{s-1,j}(\theta_j, \beta_j),$$

so $\{S^{(1)}_{s,j}(\theta_j, \beta_j), \mathcal{F}_{j+sT}, j = 1, 2, \ldots, T, s \in \mathbb{N}_0\}$ is a martingale. By Theorem 1.1 of Billingsley [42], we have

$$\frac{1}{N} \sum_{s=0}^{N-1} V_{\theta_j}^{-2}(X_{j+sT}|X_{j+sT-1}) \left(X_{j+sT} - \alpha_j^{(1)} X_{j+sT-1} I^{(1)}_{j+sT-1} - \alpha_j^{(2)} X_{j+sT-1} I^{(2)}_{j+sT-1} - \lambda_j \right)^2 X^2_{j+sT-1} I^{(1)}_{j+sT-1}$$
$$\xrightarrow{a.s.} E\left[V_{\theta_j}^{-2}(X_j|X_{j-1}) \left(X_j - \alpha_j^{(1)} X_{j-1} I^{(1)}_{j-1} - \alpha_j^{(2)} X_{j-1} I^{(2)}_{j-1} - \lambda_j \right)^2 X^2_{j-1} I^{(1)}_{j-1} \right]$$
$$= E\{ E\left[V_{\theta_j}^{-2}(X_j|X_{j-1}) \left(X_j - \alpha_j^{(1)} X_{j-1} I^{(1)}_{j-1} - \alpha_j^{(2)} X_{j-1} I^{(2)}_{j-1} - \lambda_j \right)^2 X^2_{j-1} I^{(1)}_{j-1} | X_{j-1} \right] \}$$
$$= E\{ V_{\theta_j}^{-2}(X_j|X_{j-1}) X^2_{j-1} I^{(1)}_{j-1} E\left[\left(X_j - \alpha_j^{(1)} X_{j-1} I^{(1)}_{j-1} - \alpha_j^{(2)} X_{j-1} I^{(2)}_{j-1} - \lambda_j \right)^2 | X_{j-1} \right] \}$$
$$= E\{ V_{\theta_j}^{-1}(X_j|X_{j-1}) X^2_{j-1} I^{(1)}_{j-1} \}$$
$$\triangleq H_{j,11}(\theta_j).$$

Thus, by the central limit theorem of martingale, we get

$$\frac{1}{\sqrt{N}} S^{(1)}_{N,j}(\theta_j, \beta_j) \xrightarrow{L} \mathcal{N}(0, H_{j,11}(\theta_j)).$$

Similarly,

$$\frac{1}{\sqrt{N}} S^{(2)}_{N,j}(\theta_j, \beta_j) \xrightarrow{L} \mathcal{N}(0, H_{j,22}(\theta_j)),$$

$$\frac{1}{\sqrt{N}} S_{N,j}^{(3)}(\boldsymbol{\theta}_j, \beta_j) \xrightarrow{L} \mathcal{N}(0, H_{j,33}(\boldsymbol{\theta}_j)),$$

where

$$S_{N,j}^{(2)}(\boldsymbol{\theta}_j, \beta_j) = \sum_{s=0}^{N-1} V_{\boldsymbol{\theta}_j}^{-1} \left(X_{j+sT} - \alpha_j^{(1)} X_{j+sT-1} I_{j+sT-1}^{(1)} - \alpha_j^{(2)} X_{j+sT-1} I_{j+sT-1}^{(2)} - \lambda_j \right) X_{j+sT-1} I_{j+sT-1}^{(2)},$$

and

$$S_{N,j}^{(3)}(\boldsymbol{\theta}_j, \beta_j) = \sum_{s=0}^{N-1} V_{\boldsymbol{\theta}_j}^{-1} \left(X_{j+sT} - \alpha_j^{(1)} X_{j+sT-1} I_{j+sT-1}^{(1)} - \alpha_j^{(2)} X_{j+sT-1} I_{j+sT-1}^{(2)} - \lambda_j \right).$$

For any $c = (c_1, c_2, \ldots, c_T)' \in \mathbb{R}^{3T} \setminus (0, 0, \ldots, 0)$, $c_j = \left(c_j^{(1)}, c_j^{(2)}, c_j^{(3)} \right)'$ with $j = 1, 2, \ldots, T$, to simplify, let

$$j + sT \neq i + kT, i, j = 1, 2, \ldots, T, s, k \in \mathbb{N}_0,$$

$$u_{j,s} = c_j^{(1)} X_{j+sT-1} I_{j+sT-1}^{(1)} + c_j^{(2)} X_{j+sT-1} I_{j+sT-1}^{(2)} + c_j^{(3)},$$

$$w_{j,s} = X_{j+sT} - E(X_{j+sT} | X_{j+sT-1}) = X_{j+sT} - \alpha_j^{(1)} X_{j+sT-1} I_{j+sT-1}^{(1)} - \alpha_j^{(2)} X_{j+sT-1} I_{j+sT-1}^{(2)} - \lambda_j,$$

and $n(T, N)$ is a constant associated with N and T, then

$$E\left[\sum_{j=1}^{T} \sum_{s=0}^{N-1} V_{\boldsymbol{\theta}_j}^{-1} \left(X_{j+sT} - E(X_{j+sT}|X_{j+sT-1}) \right) \left(c_j^{(1)} X_{j+sT-1} I_{j+sT-1}^{(1)} + c_j^{(2)} X_{j+sT-1} I_{j+sT-1}^{(2)} + c_j^{(3)} \right) \right]^2$$

$$= \sum_{j=1}^{T} \sum_{s=0}^{N-1} E\left[V_{\boldsymbol{\theta}_j}^{-2} w_{j,s}^2 u_{j,s}^2 \right] + n(T, N) E\left[V_{\boldsymbol{\theta}_j}^{-1} V_{\boldsymbol{\theta}_i}^{-1} w_{j,s} w_{i,k} u_{j,s} u_{i,k} \right], \tag{A1}$$

for the first item in the right side of Equation (A1), we have

$$E\left[V_{\boldsymbol{\theta}_j}^{-2} w_{j,s}^2 u_{j,s}^2 \right]$$
$$= E\{ E\left[V_{\boldsymbol{\theta}_j}^{-2} w_{j,s}^2 u_{j,s}^2 | X_{j+sT-1} \right] \}$$
$$= E\{ V_{\boldsymbol{\theta}_j}^{-2} u_{j,s}^2 E\left[w_{j,s}^2 | X_{j+sT-1} \right] \}$$
$$= E\left[V_{\boldsymbol{\theta}_j}^{-1} \left(c_j^{(1)} X_{j+sT-1} I_{j+sT-1}^{(1)} + c_j^{(2)} X_{j+sT-1} I_{j+sT-1}^{(2)} + c_j^{(3)} \right)^2 \right],$$

for the second item in the right side of Equation (A1), we have

$$E\left[V_{\boldsymbol{\theta}_j}^{-1} V_{\boldsymbol{\theta}_i}^{-1} w_{j,s} w_{i,k} u_{j,s} u_{i,k} \right]$$
$$= E\{ V_{\boldsymbol{\theta}_j}^{-1} V_{\boldsymbol{\theta}_i}^{-1} u_{j,s} u_{i,k} E\left[w_{j,s} w_{i,k} | X_{j+sT-1}, X_{i+kT-1} \right] \}$$
$$= 0,$$

which imply that $Cov(S_{N,j}, S_{N,i}) = 0$, where $S_{N,j} = \left(S_{N,j}^{(1)}(\boldsymbol{\theta}_j, \beta_j), S_{N,j}^{(2)}(\boldsymbol{\theta}_j, \beta_j), S_{N,j}^{(3)}(\boldsymbol{\theta}_j, \beta_j) \right)'$, $\forall i, j = 1, 2, \ldots, T, i \neq j$, then we have

$$\frac{c_j^T}{\sqrt{N}}\left(S_{N,j}^{(1)}(\theta_j,\beta_j), S_{N,j}^{(2)}(\theta_j,\beta_j), S_{N,j}^{(3)}(\theta_j,\beta_j)\right)'$$

$$= \frac{1}{\sqrt{N}}\sum_{i=1}^{3}c_j^{(i)}S_{N,j}^{(i)}(\theta_j,\beta_j)$$

$$= \frac{1}{\sqrt{N}}\sum_{s=0}^{N-1}V_{\theta_j}^{-1}[X_{j+sT}-E(X_{j+sT}|X_{j+sT-1})]\left(c_j^{(1)}X_{j+sT-1}I_{j+sT-1}^{(1)}+c_j^{(2)}X_{j+sT-1}I_{j+sT-1}^{(2)}+c_j^{(3)}\right)$$

$$\xrightarrow{L} \mathcal{N}\left(0, E\left[V_{\theta_j}^{-1}(X_j|X_{j-1})\left(c_j^{(1)}X_{j-1}I_{j-1}^{(1)}+c_j^{(2)}X_{j-1}I_{j-1}^{(2)}+c_j^{(3)}\right)^2\right]\right), \text{ as } N\to\infty,$$

therefore, the $\frac{c_j^T}{\sqrt{N}}\left(S_{N,j}^{(1)}(\theta_j,\beta_j), S_{N,j}^{(2)}(\theta_j,\beta_j), S_{N,j}^{(3)}(\theta_j,\beta_j)\right)'$ converges to a normal distribution with mean zero and variance $E\left[V_{\theta_j}^{-1}(X_j|X_{j-1})\left(c_j^{(1)}X_{j-1}I_{j-1}^{(1)}+c_j^{(2)}X_{j-1}I_{j-1}^{(2)}+c_j^{(3)}\right)^2\right]$.

Thus, by Cramer-wold device, it follows that

$$\frac{1}{\sqrt{N}}\begin{pmatrix}S_{N,1}^{(1)}(\theta_1,\beta_1)\\S_{N,1}^{(2)}(\theta_1,\beta_1)\\S_{N,1}^{(3)}(\theta_1,\beta_1)\\\vdots\\S_{N,T}^{(1)}(\theta_T,\beta_T)\\S_{N,T}^{(2)}(\theta_T,\beta_T)\\S_{N,T}^{(3)}(\theta_T,\beta_T)\end{pmatrix} \xrightarrow{L} \mathcal{N}\left(0, \begin{bmatrix}H_1(\theta) & 0 & \cdots & 0\\0 & H_2(\theta) & \cdots & 0\\\vdots & \vdots & \ddots & \vdots\\0 & 0 & \cdots & H_T(\theta)\end{bmatrix}\right),$$

the 0's are (3×3)-null matrices. Now, we replace $V_{\theta_j}(X_{j+sT}|X_{j+sT-1})$ by $V_{\hat{\theta}_j}(X_{j+sT}|X_{j+sT-1})$, where $\hat{\theta}_j$ is a consistent estimator of θ_j. We aim to get the result

$$\frac{1}{\sqrt{N}}\begin{pmatrix}S_{N,1}^{(1)}(\hat{\theta}_1,\beta_1)\\S_{N,1}^{(2)}(\hat{\theta}_1,\beta_1)\\S_{N,1}^{(3)}(\hat{\theta}_1,\beta_1)\\\vdots\\S_{N,T}^{(1)}(\hat{\theta}_T,\beta_T)\\S_{N,T}^{(2)}(\hat{\theta}_T,\beta_T)\\S_{N,T}^{(3)}(\hat{\theta}_T,\beta_T)\end{pmatrix} \xrightarrow{L} \mathcal{N}\left(0, \begin{bmatrix}H_1(\theta) & 0 & \cdots & 0\\0 & H_2(\theta) & \cdots & 0\\\vdots & \vdots & \ddots & \vdots\\0 & 0 & \cdots & H_T(\theta)\end{bmatrix}\right). \quad (A2)$$

To prove (A2), we need to check the following conclusion

$$\frac{1}{\sqrt{N}}S_{N,j}^{(i)}(\hat{\theta}_j,\beta_j) - \frac{1}{\sqrt{N}}S_{N,j}^{(i)}(\theta_j,\beta_j) \xrightarrow{P} 0, j = 1,2,\ldots,T, i = 1,2,3, N\to\infty. \quad (A3)$$

For $\forall \epsilon > 0, \exists \delta > 0$, we have

$$P\left(|\frac{1}{\sqrt{N}}S_N^{(1)}(\hat{\theta}_j,\beta_j)-\frac{1}{\sqrt{N}}S_N^{(1)}(\theta_j,\beta_j)|>\epsilon\right) \leq \sum_{k=1}^{2}P\left(|\hat{\theta}_{j_1}^{(k)}-\theta_j^{(k)}|>\delta\right) + P\left(|\hat{\sigma}_{z,j_1}^2-\sigma_{z,j}^2|>\delta\right)$$

$$+ P\left(\sup_D|\frac{1}{\sqrt{N}}S_N^{(1)}(\theta_{j_1},\beta_j)-\frac{1}{\sqrt{N}}S_N^{(1)}(\theta_j,\beta_j)|>\epsilon\right),$$

where $\boldsymbol{\theta}_{j_1} = \left(\theta_{j_1}^{(1)}, \theta_{j_1}^{(2)}, \sigma_{z,j_1}^2\right)'$, $D = \{\boldsymbol{\theta}_{j_1} : |\theta_{j_1}^{(1)} - \theta_j^{(1)}| < \delta, |\theta_{j_1}^{(2)} - \theta_j^{(2)}| < \delta, \sigma_{z,j_1}^2 - \sigma_{z,j}^2| < \delta\}$.
If $\hat{\boldsymbol{\theta}}_j$ is a consistent estimator of $\boldsymbol{\theta}_j$, then we just need to prove that

$$P\left(\sup_D \left| \frac{1}{\sqrt{N}} S_N^{(1)}(\boldsymbol{\theta}_{j_1}, \beta_j) - \frac{1}{\sqrt{N}} S_N^{(1)}(\boldsymbol{\theta}_j, \beta_j)\right| > \epsilon\right) \xrightarrow{P} 0, N \to \infty.$$

By the Markov inequality,

$$P\left(\sup_D \left| \frac{1}{\sqrt{N}} S_N^{(1)}(\boldsymbol{\theta}_{j_1}, \beta_j) - \frac{1}{\sqrt{N}} S_N^{(1)}(\boldsymbol{\theta}_j, \beta_j)\right| > \epsilon\right)$$

$$\leq \frac{1}{\epsilon^2} E\left(\sup_D \left(\frac{1}{\sqrt{N}} S_N^{(1)}(\boldsymbol{\theta}_{j_1}, \beta_j) - \frac{1}{\sqrt{N}} S_N^{(1)}(\boldsymbol{\theta}_j, \beta)\right)^2\right)$$

$$\leq \frac{1}{\epsilon^2} E\{\sup_D \frac{1}{N} \left[\sum_{s=0}^{N-1} \left(V_{\boldsymbol{\theta}_{j_1}}^{-1} - V_{\boldsymbol{\theta}_j}^{-1}\right)^2 (X_{j+sT} - E(X_{j+sT}|X_{j+sT-1}))^2 X_{j+sT-1}^2 I_{j+sT-1}^{(1)}\right]\}$$

$$= \frac{1}{\epsilon^2} E\left[\sup_D \left(V_{\boldsymbol{\theta}_{j_1}}^{-1}(X_j|X_{j-1}) - V_{\boldsymbol{\theta}_j}^{-1}(X_j|X_{j-1})\right)^2 \left(X_j - \alpha_j^{(1)} X_{j-1} I_{j-1}^{(1)} - \alpha_j^{(2)} X_{j-1} I_{j-1}^{(2)} - \lambda_j\right)^2 X_{j-1}^2 I_{j-1}^{(1)}\right]$$

$$= \frac{1}{\epsilon^2} E\{\sup_D \frac{\left[\left(\theta_j^{(1)} - \theta_{j_1}^{(1)}\right) X_{j-1} I_{j-1}^{(1)} + \left(\theta_j^{(1)} - \theta_{j_1}^{(2)}\right) X_{j-1} I_{j-1}^{(2)} + \left(\sigma_{z,j}^2 - \sigma_{z,j_1}^2\right)\right]^2}{V_{\boldsymbol{\theta}_{j_1}}^2(X_j|X_{j-1}) V_{\boldsymbol{\theta}_j}(X_j|X_{j-1})} X_{j-1}^2 I_{j-1}^{(1)}\}$$

$$\leq \frac{1}{\epsilon^2} \sup_D \{[\left(\theta_j^{(1)} - \theta_{j_1}^{(1)}\right)^2 m_1 + \left(\theta_j^{(2)} - \theta_{j_1}^{(2)}\right)^2 m_2 + \left(\sigma_{z,j}^2 - \sigma_{z,j_1}^2\right)^2 m_3 + 2m_4 |\theta_j^{(1)} - \theta_{j_1}^{(1)}||\theta_j^{(2)} - \theta_{j_1}^{(2)}|$$
$$+ 2m_5 |\theta_j^{(1)} - \theta_{j_1}^{(1)}||\sigma_{z,j}^2 - \sigma_{z,j_1}^2| + 2m_6 |\theta_j^{(2)} - \theta_{j_1}^{(2)}||\sigma_{z,j}^2 - \sigma_{z,j_1}^2|] X_{j-1}^2 I_{j-1}^{(1)}\}$$

$$\leq \frac{c\delta^2}{\epsilon^2},$$

where $m_1, m_2, \ldots, m_6$ are some finite moments of process $\{X_t\}$ under assumption (C2), and c is a positive constant. A similar argument can be used for $\frac{1}{\sqrt{N}} S_{N,j}^{(2)}(\boldsymbol{\theta}_j, \beta_j)$ and $\frac{1}{\sqrt{N}} S_{N,j}^{(3)}(\boldsymbol{\theta}_j, \beta_j), j = 1, \ldots, T$. Let $\delta \to 0$, we can get (A3).
By the ergodic theorem, we have

$$\frac{1}{N} Q_N \xrightarrow{P} H(\boldsymbol{\theta}).$$

After some calculation, we have

$$Q_N\left(\hat{\boldsymbol{\beta}}_{MQL} - \boldsymbol{\beta}\right)$$
$$= \left(S_{N,1}^{(1)}(\hat{\boldsymbol{\theta}}_1, \beta_1), S_{N,1}^{(2)}(\hat{\boldsymbol{\theta}}_1, \beta_1), S_{N,1}^{(3)}(\hat{\boldsymbol{\theta}}_1, \beta_1), \ldots, S_{N,T}^{(1)}(\hat{\boldsymbol{\theta}}_T, \beta_T), S_{N,T}^{(2)}(\hat{\boldsymbol{\theta}}_T, \beta_T), S_{N,T}^{(3)}(\hat{\boldsymbol{\theta}}_T, \beta_T)\right)',$$

Therefore,

$$\sqrt{N}\left(\hat{\boldsymbol{\beta}}_{MQL} - \boldsymbol{\beta}\right) \xrightarrow{L} \mathcal{N}\left(0, H^{-1}(\boldsymbol{\theta})\right).$$

This completes the proof. □

References

1. Al-Osh, M.A.; Alzaid, A.A. First-order integer-valued autoregressive (INAR(1)) process. *J. Time Ser. Anal.* **1987**, *8*, 261–275. [CrossRef]
2. Du, J.; Li, Y. The integer-valued autoregressive (INAR(p)) model. *J. Time Ser. Anal.* **1991**, *12*, 129–142.
3. Jung, R.C.; Ronning, G.; Tremayne, A.R. Estimation in conditional first order autoregression with discrete support. *Stat. Pap.* **2005**, *46*, 195–224. [CrossRef]
4. Weiß, C.H. Thinning operations for modeling time series of counts-a survey. *Asta-Adv. Stat. Anal.* **2008**, *92*, 319–341. [CrossRef]

5. Ristić, M.M.; Bakouch, H.S.; Nastić, A.S. A new geometric first-order integer-valued autoregressive (NGINAR(1)) process. *J. Stat. Plan. Infer.* **2009**, *139*, 2218–2226. [CrossRef]
6. Zhang, H.; Wang, D.; Zhu, F. Inference for INAR(p) processes with signed generalized power series thinning operator. *J. Stat. Plan. Infer.* **2010**, *140*, 667–683. [CrossRef]
7. Li, C.; Wang, D.; Zhang, H. First-order mixed integer-valued autoregressive processes with zero-inflated generalized power series innovations. *J. Korean Stat. Soc.* **2015**, *44*, 232–246. [CrossRef]
8. Kang, Y.; Wang, D.; Yang, K. A new INAR(1) process with bounded support for counts showing equidispersion, underdispersion and overdispersion. *Stat. Pap.* **2021**, *62*, 745–767. [CrossRef]
9. Yu, M.; Wang, D.; Yang, K.; Liu, Y. Bivariate first-order random coefficient integer-valued autoregressive processes. *J. Stat. Plan. Inference* **2020**, *204*, 153–176. [CrossRef]
10. Tong, H. On a threshold model. In *Pattern Recognition and Signal Processing*; Chen, C.H., Ed.; Sijthoff and Noordhoff: Amsterdam, The Netherlands, 1978; pp. 575–586.
11. Tong, H.; Lim, K.S. Threshold autoregression, limit cycles and cyclical data. *J. R. Stat. Soc. B* **1980**, *42*, 245–292. [CrossRef]
12. Monteiro, M.; Scotto, M.G.; Pereira, I. Integer-valued self-exciting threshold autoregressive processes. *Commun. Stat-Theory Methods* **2012**, *41*, 2717–2737. [CrossRef]
13. Wang, C.; Liu, H.; Yao, J.; Davis, R.A.; Li, W.K. Self-excited threshold poisson autoregression. *J. Am. Stat. Assoc.* **2014**, *109*, 777–787. [CrossRef]
14. Yang, K.; Li, H.; Wang, D. Estimation of parameters in the self-exciting threshold autoregressive processes for nonlinear time series of counts. *Appl. Math. Model.* **2018**, *57*, 226–247. [CrossRef]
15. Yang, K.; Wang, D.; Jia, B.; Li, H. An integer-valued threshold autoregressive process based on negative binomial thinning. *Stat. Pap.* **2018**, *59*, 1131–1160. [CrossRef]
16. Bennett, W.R. Statistics of regenerative digital transmission. *Bell Syst. Tech. J.* **1958**, *37*, 1501–1542. [CrossRef]
17. Gladyshev, E.G. Periodically and almost-periodically correlated random processes with a continuous time parameter. *Theory Probab. Appl.* **1963**, *8*, 173–177. [CrossRef]
18. Bentarzi, M.; Hallin, M. On the invertibility of periodic moving-average models. *J. Time Ser. Anal.* **1994**, *15*, 263–268. [CrossRef]
19. Lund, R.; Basawa, I.V. Recursive Prediction and Likelihood Evaluation for Periodic ARMA Models. *J. Time Ser. Anal.* **2000**, *21*, 75–93. [CrossRef]
20. Basawa, I.V.; Lund, R. Large sample properties of parameter estimates for periodic ARMA models. *J. Time Ser. Anal.* **2001**, *22*, 651–663. [CrossRef]
21. Shao, Q. Mixture periodic autoregressive time series models. *Stat. Probabil. Lett.* **2006**, *76*, 609–618. [CrossRef]
22. Monteiro, M.; Scotto, M.G.; Pereira, I. Integer-valued autoregressive processes with periodic structure. *J. Stat. Plan. Inference* **2010**, *140*, 1529–1541. [CrossRef]
23. Hall, A.; Scotto, M.; Cruz, J. Extremes of integer-valued moving average sequences. *Test* **2010**, *19*, 359–374. [CrossRef]
24. Santos, C.; Pereira, I.; Scotto, M.G. On the theory of periodic multivariate INAR processes. *Stat. Pap.* **2021**, *62*, 1291–1348. [CrossRef]
25. Pereira, I.; Scotto, M.G.; Nicolette, R. Integer-valued self-exciting periodic threshold autoregressive processes. In *Contributions in Statistics and Inference. Celebrating Nazaré Mendes Lopes' Birthday*; Gonçalves, E., Oliveira, P.E., Tenreiro, C., Eds.; Departamento de Matemática, Universidade de Coimbra/Mathematics Department of the University of Coimbra: Coimbra, Portugal, 2015; Volume 47, pp. 81–92.
26. Manaa, A.; Bentarzi, M. On a periodic SETINAR model. *Commun. Stat.-Simul. Comput.* **2021**. [CrossRef]
27. Li, D.; Tong, H. Nested sub-sample search algorithm for estimation of threshold models. *Stat. Sin.* **2016**, *26*, 1543–1554. [CrossRef]
28. Wedderburn, R.W.M. Quasi-likelihood functions, generalized linear models and the Gauss-Newton method. *Biometrika* **1974**, *61*, 439–447.
29. Azrak, R.; Mélard, G. The exact quasi-likelihood of time-dependent ARMA models. *J. Stat. Plan. Inference* **1998**, *68*, 31–45. [CrossRef]
30. Christou, V.; Fokianos, K. Quasi-likelihood inference for negative binomial time series models. *J. Time Ser. Anal.* **2014**, *35*, 55–78. [CrossRef]
31. Li, H.; Yang, K.; Wang, D. Quasi-likelihood inference for self-exciting threshold integer-valued autoregressive processes. *Comput. Stat.* **2017**, *32*, 1597–1620. [CrossRef]
32. Yang, K.; Kang, Y.; Wang, D.; Li, H.; Diao, Y. Modeling overdispersed or underdispersed count data with generalized Poisson integer-valued autoregressive processes. *Metrika* **2019**, *82*, 863–889. [CrossRef]
33. Zheng, H.; Basawa, I.V.; Datta, S. Inference for pth-order random coefficient integer-valued autoregressive processes. *J. Time Ser. Anal.* **2006**, *27*, 411–440. [CrossRef]
34. Schuster, A. On the investigation of hidden periodicities with application to a supposed 26 day period of meteorological phenomena. *Terr. Magn.* **1898**, *3*, 13–41. [CrossRef]
35. Fisher, R.A. Tests of significance in harmonic analysis. *Proc. Roy. Soc. A Math. Phys.* **1929**, *125*, 54–59.
36. Freeland, R.K. Statistical Analysis of Discrete Time Series with Application to the Analysis of Workers' Compensation Claims Data. Ph.D. Thesis, Management Science Division, Faculty of Commerce and Business Administration, University of British Columbia, Vancouver, BC, Canada, 1998.

37. Möller T.A.; Silva, M.E.; Weiß, C.H.; Scotto, M.G.; Pereira, I. Self-exciting threshold binomial autoregressive processes. *Asta-Adv. Stat. Anal.* **2016**, *100*, 369–400. [CrossRef]
38. Li, H.; Yang, K.; Zhao, S.; Wang, D. First-order random coefficients integer-valued threshold autoregressive processes. *Asta-Adv. Stat. Anal.* **2018**, *102*, 305–331. [CrossRef]
39. Homburg, A.; Weiß, C.H.; Alwan, L.C.; Frahm, G.; Göb, R. Evaluating Approximate Point Forecasting of Count Processes. *Econometrics* **2019**, *7*, 30. [CrossRef]
40. Tweedie, R.L. Sufficient conditions for regularity, recurrence and ergodicity of Markov processes. *Proc. Camb. Philos. Soc.* **1975**, *78*, 125–136. [CrossRef]
41. Zheng, H.; Basawa, I.V. First-order observation-driven integer-valued autoregressive processes. *Stat. Probabil. Lett.* **2008**, *78*, 1–9. [CrossRef]
42. Billingsley, P. *Statistical Inference for Markov Processes*; The University of Chicago Press: Chicago, IL, USA, 1961.

Article

Monitoring the Zero-Inflated Time Series Model of Counts with Random Coefficient

Cong Li [1], Shuai Cui [1] and Dehui Wang [1,2,*]

1. School of Mathematics, Jilin University, Changchun 130012, China; li_cong@jlu.edu.cn (C.L.); cuishuai@jlu.edu.cn (S.C.)
2. School of Economics, Liaoning University, Shenyang 110036, China
* Correspondence: wangdh@jlu.edu.cn

Abstract: In this research, we consider monitoring mean and correlation changes from zero-inflated autocorrelated count data based on the integer-valued time series model with random survival rate. A cumulative sum control chart is constructed due to its efficiency, the corresponding calculation methods of average run length and the standard deviation of the run length are given. Practical guidelines concerning the chart design are investigated. Extensive computations based on designs of experiments are conducted to illustrate the validity of the proposed method. Comparisons with the conventional control charting procedure are also provided. The analysis of the monthly number of drug crimes in the city of Pittsburgh is displayed to illustrate our current method of process monitoring.

Keywords: CUSUM control chart; INAR-type time series; statistical process monitoring; random survival rate; zero-inflation

Citation: Li, C.; Cui, S.; Wang, D. Monitoring the Zero-Inflated Time Series Model of Counts with Random Coefficient. *Entropy* **2021**, *23*, 372. https://doi.org/10.3390/e23030372

Academic Editor: Christian H. Weiss

Received: 19 January 2021
Accepted: 17 March 2021
Published: 20 March 2021

Publisher's Note: MDPI stays neutral with regard to jurisdictional claims in published maps and institutional affiliations.

Copyright: © 2021 by the authors. Licensee MDPI, Basel, Switzerland. This article is an open access article distributed under the terms and conditions of the Creative Commons Attribution (CC BY) license (https://creativecommons.org/licenses/by/4.0/).

1. Introduction

This work is motivated by an empirical analysis and process control of a monthly drug crime series, which contains excess zeros (over 40%) and shows clear serial dependence (see Section 5 for more details). To solve this problem, an appropriate integer-valued model is selected, further, control charts based on this model are developed. Among kinds of integer-valued models, a specific kind featured with first-order integer-valued autoregressive (INAR(1)) models plays an very important role and has been widely studied in the literature. In reality, serial dependence among the count data have been demonstrated to arise extensively in practice, typical examples are infectious disease counts, defect counts and unemployment counts, etc. These data are important indicators of the epidemic study, quality control and economics analysis, and the process monitoring is essential to detect the shifts in them.

The first INAR(1) model proposed by Al-Osh and Alzaid [1] is in the following form

$$X_t = \alpha \circ X_{t-1} + \varepsilon_t, \quad t = 1, 2, \cdots,$$

where the binomial thinning operator "$\circ$" is defined by Steutel and Van Harn [2], $\alpha \circ X_{t-1} = \sum_{i=1}^{X_{t-1}} Y_i$, $\alpha \in (0,1)$, $\{Y_i\}_N$ is a sequence of independent and identically distributed (i.i.d.) random variables with Bernoulli(α) distribution; and $\{\varepsilon_t\}_N$ is a sequence of i.i.d. random variables, independent of all $\{Y_i\}$. The INAR(1) model is currently applied in various kinds of real-world problems because of its good interpretability. As one example, we let X_t represent the number of patients of an infectious disease in a community at time t, ε_t the number of new patients entering the community at time t, and suppose each patient at time $t-1$ survives at time t with survival probability α. As for the crime data, $\alpha \circ X_{t-1}$ can be considered as the number of re-offendings provoked by X_{t-1} with probability α. Depending on the nature of this kind of observed data, the INAR(1) models have been modified and generalized with respect to their orders (Ristić and Nastić [3], Nastić, Laketa and Ristić [4]), dimensions (Pedeli and Karlis [5], Khan, Cekim and Ozel [6]), marginal distributions (Alzaid and Al-Osh [7], Alzaid and Al-Osh [8], Jazi, Jones and Lai [9], Ristić, Nastić and Bakouch [10], Barreto-Souza [11]), thinning operators (Ristić, Bakouch and Nastić[12], Liu and Zhu [13]), and mixed models (Ristić and Nastić [3], Li, Wang and Zhang [14], Orozco, Sales, Fernández and Pinho [15]). For more literature, we refer to review papers (Weiß [16], Scotto, Weiß and Gouveia [17]). Differing with the models that based on a fixed survival rate α, Zheng, Basawa and Datta [18] proposed the random coefficient INAR(1) model, supposing that the parameter α may be affected by various environmental factors, and could vary randomly over time. Some of the generalizations of random coefficient INAR models can also be found in Kang and Lee [19] and Zhang, Wang and Zhu [20]. In particular, considering both random survival probability and zero-inflation phenomenon, Bakouch, Mohammadpour and Shirozhan [21] purposed a zero-inflated geometric INAR(1) time series with random coefficient (short for the ZIGINAR$_{RC}$(1) process). The ZIGINAR$_{RC}$(1) model has simple structure and good properties, which turns out to be the best fit for the real data studied by us.

As the serial dependence shows big influence on the performance of the control chart, the traditional control charts under the assumption of independent observations are not appropriate in many cases. Therefore, the monitoring of INAR(1) models has received much attention. The related research includes but not limited to the control charts for the generally developed Poisson INAR(1) models (Weiß [22], Weiß and Testik [23], Weiß and Testik [24], Yontay, Weiß, Testik and Bayindir [25]), for zero-inflated or zero-deflated INAR(1) models (Rakitzis, Weiß and Castagliola [26], Li, Wang and Sun [27], Fernandes, Bourguignon and Ho [28]), for the mixed INAR(1) model (Sales, Pinho, Vivacqua and Ho [29]), etc. While, to the best of our knowledge, methods for monitoring the zero-inflated INAR(1) model with random coefficient have not been studied in the literature so far, which is exactly what we are going to explore. As cumulative sum (CUSUM) control charts are known to be sensitive in detecting small shifts, we study the performance of the CUSUM chart for monitoring ZIGINAR$_{RC}$(1) process. We investigate the practical guidelines for the statistical design and the methods for evaluating the chart performance. Besides monitoring mean shifts of the ZIGINAR$_{RC}$(1) model, our scope is also to monitor correlation shifts in the model. Meanwhile, we compare the performance of the CUSUM chart with the conventional Shewhart chart.

The rest of the article is outlined as follows. The ZIGINAR$_{RC}$(1) process and some properties of this process are introduced in Section 2. In Section 3, we present the monitoring procedure to detect the mean and correlation shifts of the process. Extensive computation results are discussed in Section 4. In Section 5, the applicability of the process monitoring is investigated using the monthly number of drug crimes in Pittsburgh. Finally conclusions and possible future lines of research are shown in Section 6.

2. The ZIGINAR$_{RC}$(1) Process

A randomized binomial thinning operation in Bakouch, Mohammadpour and Shirozhan [21] is defined by

$$\alpha_t \circ X = \begin{cases} \alpha \circ X, & \text{with probability } 1 - \beta, \\ 0, & \text{with probability } \beta, \end{cases}$$

where $\alpha, \beta \in (0,1)$, α_t is a binary random variable independent of discrete random variable X, $P(\alpha_t = 0) = \beta = 1 - P(\alpha_t = \alpha)$.

Based on the definition of the randomized binomial thinning operation, the ZIGINAR$_{RC}$(1) model $\{X_t\}$ presented by Bakouch, Mohammadpour and Shirozhan [21], is given by

$$X_t = \alpha_t \circ X_{t-1} + \varepsilon_t, \quad t = 1, 2, \cdots,$$

where the marginal distribution is a zero-inflated Geometric distribution (denoted as ZIG(p, θ)), $P(X_t = 0) = p + (1-p)/(1+\theta)$, $P(X_t = j) = (1-p)\theta^j/(1+\theta)^{j+1}$, $j = 1, 2, \cdots$, and $p \in (0,1)$, $\theta > 0$. $\{\varepsilon_t\}_N$ is independent of the past of the solution $\{X_s; s < t\}$ and the binary sequence $\{\alpha_t\}$, parameters are also constrained by the condition $p/(\beta + p(1-\beta)) < \alpha < 1$.

The ZIGINAR$_{RC}$(1) process is quite suitable for modelling some real-life phenomena in which counted events may survive or vanish with the random survival probability α_t. Such series are studied in Section 5 with an example of the counts of the drug crimes, where the re-offending rate may be affected by public security situation and financial situation. The mean, variance, and first-order autocorrelation function of the process are, respectively,

$$\mu_X \triangleq E(X_t) = (1-p)\theta, \quad \sigma_X^2 \triangleq \text{Var}(X_t) = (1-p)\theta[(1+p)\theta + 1],$$

$$\rho_X \triangleq \text{Corr}(X_t, X_{t+1}) = \alpha(1-\beta).$$

Obviously the process is characterized by the property of overdispersion, i.e., the variance greater than the expectation. Figure 1 shows some sample paths of simulated ZIGINAR$_{RC}$(1) processes for $\theta = 1, 3, 5$; $p = 0.2, 0.5$; $\alpha = 0.5, 0.8$ and $\beta = 0.3, 0.8$. As we can see, the model has larger process mean with larger θ, and larger percentage of zeros with larger p.

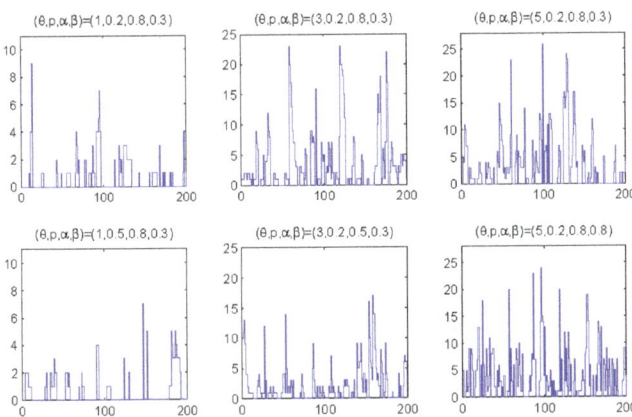

Figure 1. Sample path of ZIGINAR$_{RC}$(1) processes for $\theta = 1, 3, 5$, $p = 0.2, 0.5$, $\alpha = 0.5, 0.8$ and $\beta = 0.3, 0.8$.

Following Theorem 2.1 in Bakouch, Mohammadpour and Shirozhan [21], the ZIGINAR$_{RC}$(1) model has a unique, strictly stationary solution given by

$$\sum_{i=1}^{\infty} \left(\prod_{l=0}^{i-1} \alpha_{t-l} \right) \circ \varepsilon_{t-i} + \varepsilon_t.$$

Furthermore, the probability mass function of $\{\varepsilon_t\}_N$ is

$$P(\varepsilon_t = j) = \frac{p}{\beta + p(1-\beta)} I_{(j)} + \frac{(1-p)(1-\alpha)}{1 - \alpha[\beta + p(1-\beta)]} \frac{\theta^j}{(1+\theta)^{j+1}}$$
$$+ \frac{(1-p)(1-\beta)[\alpha(\beta + p(1-\beta)) - p]}{(1 - \alpha[\beta + p(1-\beta)])(\beta + p(1-\beta))}$$
$$\times \frac{[\alpha\theta(\beta + p(1-\beta))]^j}{[1 + \alpha\theta(\beta + p(1-\beta))]^{j+1}}, \quad j = 0, 1, 2, \cdots.$$

where $I_{(j)} = 1$ for $j = 0$ and 0 else. It can be deduced that the innovation series $\{\varepsilon_t\}_N$ is a mixture of three random variables, a degenerate distribution at 0, Geometric($\theta/(1+\theta)$) and Geometric($\alpha\theta(\beta + p(1-\beta))/(1 + \alpha\theta(\beta + p(1-\beta)))$) distributions with three different mixing portions. The following form is the transition probability of the process $\{X_t\}_{N_0}$

$$P(X_t = j | X_{t-1} = i) = \beta P(\varepsilon_t = j) + (1-\beta) \sum_{l=0}^{min(i,j)} \binom{i}{l} \alpha^l (1-\alpha)^{i-l} P(\varepsilon_t = j - l),$$

$$i, j = 0, 1, \cdots.$$

Some other important probabilistic properties of the process, like spectral density, multi-step conditional mean and variance, extreme order statistics, distributional properties of length of run of zeros, have also been discussed in Bakouch, Mohammadpour and Shirozhan [21]. Furthermore, the unknown parameters of the model could be estimated by conditional least squares or maximum likelihood methods.

3. Monitoring Procedure

In this section, we present a CUSUM chart for monitoring the ZIGINAR$_{RC}$(1) process. As this process is used to fit the number of crimes, an increase in the process mean usually means a deteriorating public security environment, and an increase in the process correlation usually means more re-offendings. Thus, our purpose is to detect the increasing of both mean shifts and correlation shifts in the ZIGINAR$_{RC}$(1) process. According to the model properties, the process mean is affected by the parameters θ and p, the correlation is affected by the parameters α and β. Let θ_0, p_0, α_0 and β_0 (θ_1, p_1, α_1 and β_1) denote the in-control (out-of-control) parameters of the processes, and μ_0, σ_0, ρ_0 (μ_1, σ_1, ρ_1) be the corresponding in-control (out-of-control) process mean, standard deviation and first-order correlation.

The CUSUM charts are commonly used charts in statistical process control, which were first proposed by Page [30]. The essential assumption underlying the design of CUSUM charts is that the process observations are independent (Montgomery [31], Alencar, Ho and Albarracin [32], Bourguignon, Medeiros, Fernandes and Ho [33]). While the violation of this major assumption seriously affects the monitoring performance of the charts (Harris and Ross [34], Triantafyllopoulos and Bersimis [35], Albarracin, Alencar and Ho [36]). Some authors have studied the performance of CUSUM charts for some integer-valued models (Weiß and Testik [23], Weiß and Testik [24], Yontay, Weiß, Testik and Bayindir [25], Rakitzis, Weiß and Castagliola [26], Li, Wang and Sun [27], Lee and Kim [37], Lee, Kim and Kim [38]).

Scheme (The ZIGINAR$_{RC}$(1) CUSUM chart). Let $\{X_t\}_{N_0}$ be a stationary ZIGINAR$_{RC}$(1) process, the CUSUM statistics C_t is defined as:

$$C_t = \max(0, X_t - k + C_{t-1}), \ t = 1, 2, \cdots,$$

where k is a positive integer constant representing the reference ($k \geqslant \mu_0$). This chart is said to be out-of-control when C_t falls outside the control limit h ($h \in N$), that is, $C_t > h$.

The initial value of the CUSUM statistics is set equal to the integer constant c_0, i.e., $C_0 = c_0$ with $c_0 < h$. The performance evaluation of this chart is accomplished based on the average run length (ARL) measures, which is defined as the average number of points to be plotted on the chart until the first out-of-control signal triggers. As $\{X_t, C_t\}_{t \in N_0}$ of the ZIGINAR$_{RC}$(1) process is a bivariate Markov chain, the Markov chain approach proposed by Brook and Evans [39] is adapted to evaluate the exact ARLs. Though this method has been described in detail in the relevant literature by Weiß [22], Weiß and Testik [23] and Weiß and Testik [24], we briefly introduce this method here for completeness. The reachable control region ($\mathcal{CR}$) of $\{X_t, C_t\}_{N_0}$ is given by

$$\mathcal{CR} \triangleq \{(n,i) \in \mathbb{N}_0 \times \{0,\cdots,h\}| \max(0, n-k+i) \in \{0,\cdots,h\}\}$$
$$= \{(n,i)|i \in \{0,\cdots,h\},\ n \in \{\max(0, i+k-h),\cdots,i+k\}\}.$$

Obviously $\mathcal{CR}$ has a finite number of elements and could be ordered in a certain manner. The transition probability matrix of $\{X_t, C_t\}_{\mathbb{N}_0}$ is $\mathbf{Q}^\top \triangleq (p(n,j|m,i))_{(n,j),(m,i) \in \mathcal{CR}}$,

$$p(n,j|m,i) \triangleq P(X_t = n, C_t = j | X_{t-1} = m, C_{t-1} = i) = I_{(j-\max(0,n-k+i))} P(X_t = n | X_{t-1} = m).$$

The initial probabilities are

$$p(n,j|c_0) \triangleq P(X_1 = n, C_1 = j | C_0 = c_0) = I_{(j-\max(0, n-k+c_0))} P(X_1 = n).$$

The conditional probability that the run length of $\{X_t, C_t\}_{\mathbb{N}_0}$ equals r is defined by

$$p_{m,i}(r) \triangleq P((X_{r+1}, C_{r+1}) \notin \mathcal{CR}, (X_r, C_r), \cdots, (X_2, C_2) \in \mathcal{CR} | (X_1, C_1) = (m, i)),$$

where $(m,i) \in \mathcal{CR}$. Let the vector $\boldsymbol{\mu}_{(k)}$ denote the k-th factorial moments that $(\boldsymbol{u}_{(k)})_{m,i} \triangleq \sum_{r=1}^{\infty} r_{(k)} p_{m,i}(r)$ where $k \geqslant 1$ and $r_{(k)} \triangleq r \times (r-1) \times \cdots \times (r-k+1)$. Then

$$p_{m,i}(r) = \sum_{(n,j) \in \mathcal{CR}} p_{n,j}(r-1) \times p(n,j|m,i),$$

$$(\boldsymbol{u}_{(1)})_{m,i} = 1 + \sum_{(n,j) \in \mathcal{CR}} p(n,j|m,i) \times (\boldsymbol{u}_{(1)})_{n,j},\ \text{i.e.,}\ (\boldsymbol{I} - \boldsymbol{Q})\boldsymbol{u}_{(1)} = \boldsymbol{1}.$$

The ARL is obtained as

$$\text{ARL} = \sum_{(m,i) \in \mathcal{CR}} (\boldsymbol{u}_{(1)})_{m,i} \times p(m,i|c_0).$$

For simplicity we do not repeat the proof methods, see Weiß [22], Weiß and Testik [23] and Weiß and Testik [24] for more details. It is expected that an efficient chart possesses a large in-control ARL (denoted as ARL_0) and a small out-of-control ARL. Along with the ARL, we also assess the performance of the charts through the standard deviation of run length (SDRL) suggested by Weiß [22]. The SDRL of the ZIGINAR$_{\text{RC}}$(1) CUSUM chart could again be computed efficiently by applying the Markov chain method. The second order factorial moments $\boldsymbol{u}_{(2)}$ can be determined recursively from the relation $(\boldsymbol{I} - \boldsymbol{Q})\boldsymbol{u}_{(2)} = 2\boldsymbol{Q}\boldsymbol{u}_{(1)}$. Then the SDRL is

$$\text{SDRL} = \sqrt{\sum_{(m,i) \in \mathcal{CR}} ((\boldsymbol{u}_{(2)})_{m,i} + (\boldsymbol{u}_{(1)})_{m,i}) \times p(m,i|c_0) - \text{ARL}^2}.$$

To implement the proposed monitoring scheme, the chart design pairs (h, k) need to be designed in advance. Generally, a fixed ARL_0 value is set to be the target value, and (h, k) is set accordingly. Some guidelines for the choices of them will be given in the next section.

4. Computation Results

In this section, we evaluate the ZIGINAR$_{\text{RC}}$(1) CUSUM chart performance basing on extensive numerical experiments and presume that the parameters in this model have already been known. In practice, the in-control parameters need to be estimated from the data, as shown in the next section. We search for possible chart designs (integer (h, k) pairs) in order to adjust the ARL_0 close to the target value. Here the target ARL_0 value is set to be 370, which is commonly used in the statistical process monitoring domain. Meanwhile, the values of ARL and SDRL are calculated accurately by the Markov chain method, and we only show the results with two decimal places for simplicity. We first compute ARL_0

and $SDRL_0$ of the CUSUM chart for different in-control process parameters and initial values in Table 1. The process parameters are: $\theta_0 = \{1, 5\}$; $p_0 = \{0.1, 0.3\}$; $\alpha_0 = \{0.5, 0.8\}$; $\beta_0 = \{0.5, 0.8\}$. Furthermore, initial values are $c_0 = \{0, 3, 6\}$. These chosen parameters could cover a broad range of different scenarios. Based on the results in Table 1, three important conclusions can be derived. First, it can be observed that when c_0 takes smaller value, the deviation of ARL_0 and $SDRL_0$ is small. When c_0 takes larger value, there might be a situation where the value of $SDRL_0$ is significantly greater than the value of ARL_0 (for example, $ARL_0 = 409.42$, $SDRL_0 = 442.13$ under $(\theta_0, p_0, \alpha_0, \beta_0, c_0) = (1, 0.3, 0.5, 0.8, 6)$). Thus, we assume that $c_0 = 0$ in the following studies to get better robust. Second, as the differences of the values between ARL and SDRL are small when $c_0 = 0$, we only use ARL as the measure in the following computations to save space. Last, the parameter θ_0 shows a great influence on the selection of control designs (h, k), with a larger θ_0 comes a larger pair of (h, k).

Table 1. ARL_0 and $SDRL_0$ of the CUSUM chart for various $\theta_0, p_0, \alpha_0, \beta_0$ and c_0.

$(\theta_0, p_0, \alpha_0, \beta_0)$	(h, k)	$(c_0, ARL_0, SDRL_0)$	$(c_0, ARL_0, SDRL_0)$	$(c_0, ARL_0, SDRL_0)$
(1, 0.1, 0.5, 0.5)	(9, 2)	(0, 340.55, 339)	(3, 336.84, 338.98)	(6, 322.88, 338.52)
(1, 0.1, 0.5, 0.8)	(8, 2)	(0, 428.55, 427.38)	(3, 423.5, 427.34)	(6, 398.83, 426.33)
(1, 0.1, 0.8, 0.5)	(12, 2)	(0, 368.36, 366.45)	(3, 365.76, 366.43)	(6, 358.76, 366.3)
(1, 0.1, 0.8, 0.8)	(9, 2)	(0, 428.69, 427.42)	(3, 424.79, 427.4)	(6, 408.35, 426.91)
(1, 0.3, 0.5, 0.5)	(8, 2)	(0, 385.69, 384.65)	(3, 381.9, 384.62)	(6, 365.66, 384.11)
(1, 0.3, 0.5, 0.8)	(7, 2)	(0, 444.16, 443.51)	(3, 438.89, 443.47)	(6, 409.42, 442.13)
(1, 0.3, 0.8, 0.5)	(10, 2)	(0, 359.91, 358.61)	(3, 357.26, 358.6)	(6, 349.32, 358.44)
(1, 0.3, 0.8, 0.8)	(8, 2)	(0, 469.37, 468.53)	(3, 465.3, 468.51)	(6, 446.23, 467.94)
(5, 0.1, 0.5, 0.5)	(60, 6)	(0, 379.61, 371.51)	(3, 379.07, 371.51)	(6, 378.25, 371.51)
(5, 0.1, 0.5, 0.8)	(49, 6)	(0, 376.02, 369.17)	(3, 375.4, 369.17)	(6, 374.38, 369.16)
(5, 0.1, 0.8, 0.5)	(75, 6)	(0, 371.37, 363.76)	(3, 370.91, 363.76)	(6, 370.28, 363.75)
(5, 0.1, 0.8, 0.8)	(54, 6)	(0, 378.57, 372.12)	(3, 378.01, 372.12)	(6, 377.14, 372.11)
(5, 0.3, 0.5, 0.5)	(46, 6)	(0, 383.15, 379.48)	(3, 382.64, 379.47)	(6, 381.86, 379.47)
(5, 0.3, 0.5, 0.8)	(38, 6)	(0, 386.29, 383.42)	(3, 385.68, 383.42)	(6, 384.7, 383.42)
(5, 0.3, 0.8, 0.5)	(59, 6)	(0, 378.46, 374.67)	(3, 378.04, 374.67)	(6, 377.45, 374.67)
(5, 0.3, 0.8, 0.8)	(42, 6)	(0, 379.79, 377.04)	(3, 379.26, 377.04)	(6, 378.45, 377.03)

Due to its simplicity, the conventional Shewhart chart is very popular in monitoring the process shifts. The upper limit for the Shewhart chart is denoted as UCL. For observations, when the value of the process $\{X_t\}_{N_0}$ exceeds the threshold value UCL ($X_t > UCL$), a fault is declared. Figures 2 and 3 investigate the CUSUM method preliminarily by comparing it with the Shewhart method. In both of these figures, we assume that the in-control parameters are $\theta_0 = 2$, $p_0 = 0.2$, $\alpha_0 = 0.5$ and $\beta_0 = 0.5$, which are selected based on the real drug crime data in Section 5. According to these parameters, the CUSUM chart designs can be determined, respectively, as $h = 31, k = 2$ (corresponding $ARL_0 = 383.74$); $h = 19, k = 3$ ($ARL_0 = 396.12$); $h = 14, k = 4$ ($ARL_0 = 373.27$); $h = 11, k = 5$ ($ARL_0 = 370.77$); $h = 9, k = 6$ ($ARL_0 = 394.03$). Furthermore, the Shewhart chart limit $UCL = 13$ ($ARL_0 = 381.31$) can be used. It should be noted that two types of changes are considered in Figure 2, which both lead to the upward mean shifts. The first type of changes occurs only in the parameter θ, with other parameters invariant, the results are listed in Figure 2a. Similarly, the second type of changes occurs only in the parameter p, with other parameters invariant, the results are in Figure 2b. From Figure 2, we can conclude that the CUSUM chart with the design $h = 31, k = 2$ outperforms the other

CUSUM charts under most shifts, while the Shewhart chart performs worst among them. For the upward correlation shift scenarios, ARL values under two types of the parameter changes are displayed in Figure 3. The first one considers changes only in the parameter α, and the second one considers changes only in the parameter β. For each scenario, the Shewhart chart performs increase ratio of ARL with the increase of first-order correlation ρ_X, and the CUSUM chart has the better behaviour in the figure. In a comprehensive view, the conventional Shewhart chart is insensitive for upward mean shifts caused by changes in parameter p, and fails to detect shifts in the correlation. While the proposed CUSUM chart could overcome these limitations and has superiorities in various coefficient shifts compared with the Shewhart chart. From the figures, we can also conclude that the smaller the value of k, the more sensitive the CUSUM chart is. As the constraint $k \geqslant \mu_0$ is required to make the chart reasonable, it is natural to recommend $k = \lceil \mu_0 \rceil$ (the smallest integer no less than μ_0), then we aim to select the value of h such that ARL_0 is close to 370. Now the computations of the CUSUM chart are extended to general cases with designs of experiments as follow.

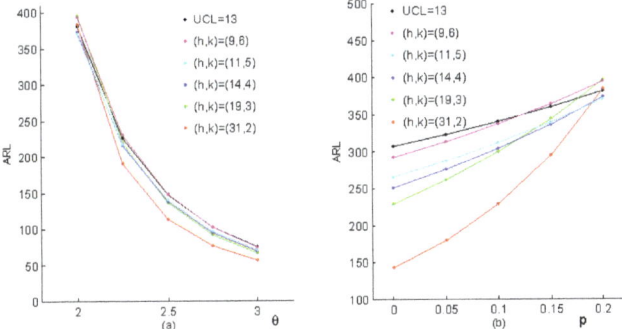

Figure 2. The performance of the Shewhart chart and the CUSUM chart to detect an increase in process mean, (**a**) changes only in θ, (**b**) changes only in p.

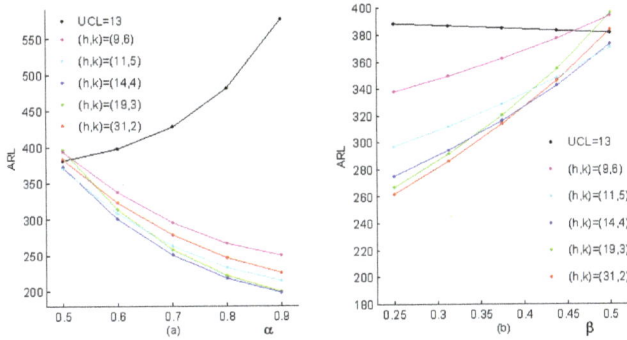

Figure 3. The performance of the Shewhart chart and the CUSUM chart to detect an increase in process first-order correlation, (**a**) changes only in α, (**b**) changes only in β.

In Tables 2–4, we focus on situations that there are increasing shifts in process mean, and the correlation remains the same. Each in-control parameter has three levels: $\theta_0 = \{1, 3, 5\}$; $p_0 = \{0.1, 0.2, 0.3\}$; $\alpha_0 = \{0.5, 0.6, 0.7\}$; $\beta_0 = \{0.5, 0.6, 0.7\}$. We consider the case when the changes only occur in θ, this is the most common case. The out-of-control process mean is $\mu_1 = \mu_0 + \delta\sigma_0$, the shift size δ considers potential values in set $\{0.5, 1, 1.5, 6\}$. The usual relative deviation (in %) in ARL is defined as $dev^{(\%)} = 100\% \times$

(ARL-ARL$_0$)/ARL$_0$ (Weiß and Testik [24]). From Tables 2–4, we can conclude that the ZIGINAR$_{RC}$(1) CUSUM chart performs quite well in detecting upward mean shifts for all scenarios. For the small shift size of $\delta = 0.5$, the CUSUM chart is efficient with the minimum 86.38% drop of ARL and the maximum 91.23% drop of ARL. Take larger δ for another illustration ($\delta = 1$), the drop of ARL is at least 92.81%, and up to 96.03% at most. It can also be obtained that δ has to be at least 6 to get an immediate signal with the out-of-control ARL closer to 1. In addition, extensive computation results show that the in-control parameters ($\theta_0, p_0, \alpha_0, \beta_0$) have little effect on the better performance of the CUSUM chart to detect the mean shifts.

Table 2. ARL profiles of the CUSUM chart versus mean shifts under $\theta_0 = 1$.

Process Parameters						ARL (dev$^{(\%)}$)				
p_0	α_0	β_0	μ_0	h	k	$\delta = 0$	$\delta = 0.5$	$\delta = 1$	$\delta = 1.5$	$\delta = 6$
0.1	0.5	0.5	0.9	22	1	348.22	38.62 (−88.91%)	19.31 (−94.45%)	12.94 (−96.28%)	3.44 (−99.01%)
0.1	0.5	0.6	0.9	21	1	357.74	36.77 (−89.72%)	18.28 (−94.89%)	12.21 (−96.59%)	3.21 (−99.1%)
0.1	0.5	0.7	0.9	20	1	365.71	34.91 (−90.45%)	17.27 (−95.28%)	11.51 (−96.85%)	3 (−99.18%)
0.1	0.6	0.5	0.9	24	1	370.79	42.46 (−88.55%)	21.23 (−94.27%)	14.23 (−96.16%)	3.78 (−98.98%)
0.1	0.6	0.6	0.9	22	1	358.58	38.82 (−89.17%)	19.32 (−94.61%)	12.91 (−96.4%)	3.39 (−99.05%)
0.1	0.6	0.7	0.9	21	1	378.05	36.86 (−90.25%)	18.23 (−95.18%)	12.15 (−96.79%)	3.16 (−99.16%)
0.1	0.7	0.5	0.9	25	1	358.79	44.75 (−87.53%)	22.43 (−93.75%)	15.04 (−95.81%)	3.99 (−98.89%)
0.1	0.7	0.6	0.9	23	1	359.01	40.95 (−88.59%)	20.39 (−94.32%)	13.63 (−96.2%)	3.57 (−99.01%)
0.1	0.7	0.7	0.9	21	1	351.73	37.15 (−89.44%)	18.41 (−94.77%)	12.28 (−96.51%)	3.19 (−99.09%)
0.2	0.5	0.5	0.8	18	1	371.21	37.8 (−89.82%)	18.14 (−95.11%)	12.01 (−96.76%)	3.26 (−99.12%)
0.2	0.5	0.6	0.8	17	1	378.49	35.72 (−90.56%)	17 (−95.51%)	11.22 (−97.04%)	3.01 (−99.2%)
0.2	0.5	0.7	0.8	16	1	382.96	33.57 (−91.23%)	15.88 (−95.85%)	10.45 (−97.27%)	2.77 (−99.28%)
0.2	0.6	0.5	0.8	19	1	360.14	40.29 (−88.81%)	19.42 (−94.61%)	12.87 (−96.43%)	3.49 (−99.03%)
0.2	0.6	0.6	0.8	18	1	382.06	38.15 (−90.01%)	18.17 (−95.24%)	11.99 (−96.86%)	3.2 (−99.16%)
0.2	0.6	0.7	0.8	16	1	346.1	33.8 (−90.23%)	16.06 (−95.36%)	10.58 (−96.94%)	2.81 (−99.19%)
0.2	0.7	0.5	0.8	20	1	349.23	42.95 (−87.7%)	20.77 (−94.05%)	13.77 (−96.06%)	3.71 (−98.94%)
0.2	0.7	0.6	0.8	18	1	338.16	38.56 (−88.6%)	18.49 (−94.53%)	12.21 (−96.39%)	3.25 (−99.04%)
0.2	0.7	0.7	0.8	17	1	365.85	36.24 (−90.09%)	17.17 (−95.31%)	11.29 (−96.91%)	2.98 (−99.19%)
0.3	0.5	0.5	0.7	15	1	376.37	37.77 (−89.96%)	17.57 (−95.33%)	11.54 (−96.93%)	3.25 (−99.14%)
0.3	0.5	0.6	0.7	14	1	374.56	35.37 (−90.56%)	16.31 (−95.65%)	10.67 (−97.15%)	2.97 (−99.21%)
0.3	0.5	0.7	0.7	13	1	368.03	32.86 (−91.07%)	15.04 (−95.91%)	9.81 (−97.33%)	2.71 (−99.26%)
0.3	0.6	0.5	0.7	16	1	369.03	40.69 (−88.97%)	19.04 (−94.84%)	12.51 (−96.61%)	3.5 (−99.05%)
0.3	0.6	0.6	0.7	14	1	323.29	35.58 (−88.99%)	16.61 (−94.86%)	10.89 (−96.63%)	3.03 (−99.06%)
0.3	0.6	0.7	0.7	13	1	328.19	33.03 (−89.94%)	15.25 (−95.35%)	9.96 (−96.97%)	2.75 (−99.16%)
0.3	0.7	0.5	0.7	17	1	360.19	43.81 (−87.84%)	20.59 (−94.28%)	13.53 (−96.24%)	3.75 (−98.96%)
0.3	0.7	0.6	0.7	15	1	335.26	38.66 (−88.47%)	18 (−94.63%)	11.79 (−96.48%)	3.24 (−99.03%)
0.3	0.7	0.7	0.7	14	1	357.18	36.06 (−89.9%)	16.52 (−95.37%)	10.76 (−96.99%)	2.93 (−99.18%)

Table 3. ARL profiles of the CUSUM chart versus mean shifts under $\theta_0 = 3$.

Process Parameters						ARL (dev$^{(\%)}$)				
p_0	α_0	β_0	μ_0	h	k	$\delta = 0$	$\delta = 0.5$	$\delta = 1$	$\delta = 1.5$	$\delta = 6$
0.1	0.5	0.5	2.7	54	3	364.48	38.52 (-89.43%)	19.06 (-94.77%)	12.72 (-96.51%)	3.37 (-99.08%)
0.1	0.5	0.6	2.7	51	3	366.54	36.32 (-90.09%)	17.87 (-95.12%)	11.9 (-96.75%)	3.11 (-99.15%)
0.1	0.5	0.7	2.7	48	3	365.43	34.07 (-90.68%)	16.69 (-95.43%)	11.09 (-96.97%)	2.87 (-99.21%)
0.1	0.6	0.5	2.7	58	3	373.32	41.8 (-88.8%)	20.71 (-94.45%)	13.83 (-96.3%)	3.66 (-99.02%)
0.1	0.6	0.6	2.7	54	3	374.45	38.8 (-89.64%)	19.09 (-94.9%)	12.71 (-96.61%)	3.32 (-99.11%)
0.1	0.6	0.7	2.7	50	3	369.98	35.75 (-90.34%)	17.51 (-95.27%)	11.63 (-96.86%)	3.01 (-99.19%)
0.1	0.7	0.5	2.7	61	3	365.9	44.55 (-87.82%)	22.1 (-93.96%)	14.76 (-95.97%)	3.89 (-98.94%)
0.1	0.7	0.6	2.7	56	3	366.07	40.68 (-88.89%)	20.03 (-94.53%)	13.33 (-96.36%)	3.47 (-99.05%)
0.1	0.7	0.7	2.7	52	3	374.26	37.49 (-89.98%)	18.34 (-95.1%)	12.17 (-96.75%)	3.14 (-99.16%)
0.2	0.5	0.5	2.4	43	3	371.94	36.87 (-90.09%)	17.43 (-95.31%)	11.48 (-96.91%)	3.1 (-99.17%)
0.2	0.5	0.6	2.4	40	3	363.7	34.31 (-90.57%)	16.11 (-95.57%)	10.58 (-97.09%)	2.82 (-99.22%)
0.2	0.5	0.7	2.4	38	3	377.02	32.59 (-91.36%)	15.16 (-95.98%)	9.92 (-97.37%)	2.62 (-99.31%)
0.2	0.6	0.5	2.4	46	3	368.96	39.91 (-89.18%)	18.93 (-94.87%)	12.47 (-96.62%)	3.35 (-99.09%)
0.2	0.6	0.6	2.4	43	3	379.62	37.34 (-90.16%)	17.5 (-95.39%)	11.48 (-96.98%)	3.04 (-99.2%)
0.2	0.6	0.7	2.4	39	3	360.73	33.72 (-90.65%)	15.72 (-95.64%)	10.29 (-97.15%)	2.7 (-99.25%)
0.2	0.7	0.5	2.4	49	3	363.38	43.17 (-88.12%)	20.52 (-94.35%)	13.51 (-96.28%)	3.6 (-99.01%)
0.2	0.7	0.6	2.4	45	3	370.51	39.59 (-89.31%)	18.57 (-94.99%)	12.17 (-96.72%)	3.2 (-99.14%)
0.2	0.7	0.7	2.4	41	3	369.61	35.87 (-90.3%)	16.68 (-95.49%)	10.89 (-97.05%)	2.84 (-99.23%)
0.3	0.5	0.5	2.1	36	3	380.79	36.83 (-90.33%)	16.87 (-95.57%)	11.01 (-97.11%)	3.08 (-99.19%)
0.3	0.5	0.6	2.1	33	3	357.51	33.81 (-90.54%)	15.39 (-95.7%)	10.01 (-97.2%)	2.78 (-99.22%)
0.3	0.5	0.7	2.1	31	3	360.6	31.82 (-91.18%)	14.33 (-96.03%)	9.28 (-97.43%)	2.55 (-99.29%)
0.3	0.6	0.5	2.1	38	3	358.5	39.26 (-89.05%)	18.15 (-94.94%)	11.86 (-96.69%)	3.3 (-99.08%)
0.3	0.6	0.6	2.1	35	3	355.89	36.32 (-89.79%)	16.56 (-95.35%)	10.76 (-96.98%)	2.96 (-99.17%)
0.3	0.6	0.7	2.1	33	3	378.11	34.37 (-90.91%)	15.41 (-95.92%)	9.96 (-97.37%)	2.71 (-99.28%)
0.3	0.7	0.5	2.1	41	3	359.33	43.13 (-88%)	19.97 (-94.44%)	13.03 (-96.37%)	3.58 (-99%)
0.3	0.7	0.6	2.1	38	3	377.93	40.22 (-89.36%)	18.23 (-95.18%)	11.81 (-96.88%)	3.2 (-99.15%)
0.3	0.7	0.7	2.1	34	3	362.28	35.83 (-90.11%)	16.08 (-95.56%)	10.38 (-97.13%)	2.8 (-99.23%)

Table 4. ARL profiles of the CUSUM chart versus mean shifts under $\theta_0 = 5$.

Process Parameters						ARL (dev$^{(\%)}$)				
p_0	α_0	β_0	μ_0	h	k	$\delta = 0$	$\delta = 0.5$	$\delta = 1$	$\delta = 1.5$	$\delta = 6$
0.1	0.5	0.5	4.5	85	5	365.12	38.31 (−89.51%)	18.91 (−94.82%)	12.61 (−96.55%)	3.33 (−99.09%)
0.1	0.5	0.6	4.5	80	5	364.27	36 (−90.12%)	17.67 (−95.15%)	11.76 (−96.77%)	3.07 (−99.16%)
0.1	0.5	0.7	4.5	76	5	371.67	34.09 (−90.83%)	16.64 (−95.52%)	11.04 (−97.03%)	2.86 (−99.23%)
0.1	0.6	0.5	4.5	91	5	370.55	41.46 (−88.81%)	20.5 (−94.47%)	13.68 (−96.31%)	3.61 (−99.03%)
0.1	0.6	0.6	4.5	85	5	374.72	38.61 (−89.7%)	18.95 (−94.94%)	12.6 (−96.64%)	3.29 (−99.12%)
0.1	0.6	0.7	4.5	79	5	373.87	35.71 (−90.45%)	17.43 (−95.34%)	11.56 (−96.91%)	2.99 (−99.2%)
0.1	0.7	0.5	4.5	97	5	373.34	44.81 (−88%)	22.15 (−94.07%)	14.77 (−96.04%)	3.89 (−98.96%)
0.1	0.7	0.6	4.5	89	5	373.74	40.89 (−89.06%)	20.06 (−94.63%)	13.34 (−96.43%)	3.47 (−99.07%)
0.1	0.7	0.7	4.5	81	5	365.37	36.94 (−89.89%)	18.03 (−95.07%)	11.96 (−96.73%)	3.08 (−99.16%)
0.2	0.5	0.5	4	118	4	373.44	45.69 (−87.77%)	24.01 (−93.57%)	16.42 (−95.6%)	4.62 (−98.76%)
0.2	0.5	0.6	4	112	4	369.47	43.25 (−88.29%)	22.65 (−93.87%)	15.45 (−95.82%)	4.3 (−98.84%)
0.2	0.5	0.7	4	107	4	369.02	41.18 (−88.84%)	21.48 (−94.18%)	14.63 (−96.04%)	4.03 (−98.91%)
0.2	0.6	0.5	4	122	4	366.95	47.66 (−87.01%)	25.12 (−93.15%)	17.21 (−95.31%)	4.86 (−98.68%)
0.2	0.6	0.6	4	115	4	365.11	44.74 (−87.75%)	23.47 (−93.57%)	16.03 (−95.61%)	4.47 (−98.78%)
0.2	0.6	0.7	4	110	4	371.48	42.56 (−88.54%)	22.22 (−94.02%)	15.14 (−95.92%)	4.17 (−98.88%)
0.2	0.7	0.5	4	128	4	371.06	50.55 (−86.38%)	26.69 (−92.81%)	18.29 (−95.07%)	5.17 (−98.61%)
0.2	0.7	0.6	4	120	4	372.35	47.07 (−87.36%)	24.71 (−93.36%)	16.88 (−95.47%)	4.71 (−98.74%)
0.2	0.7	0.7	4	112	4	368.48	43.62 (−88.16%)	22.79 (−93.82%)	15.53 (−95.79%)	4.28 (−98.84%)
0.3	0.5	0.5	3.5	80	4	367.77	38.09 (−89.64%)	19.11 (−94.8%)	12.93 (−96.48%)	3.77 (−98.97%)
0.3	0.5	0.6	3.5	75	4	363.52	35.68 (−90.18%)	17.79 (−95.11%)	12 (−96.7%)	3.45 (−99.05%)
0.3	0.5	0.7	3.5	71	4	367.7	33.69 (−90.84%)	16.7 (−95.46%)	11.23 (−96.95%)	3.2 (−99.13%)
0.3	0.6	0.5	3.5	85	4	366.71	40.93 (−88.84%)	20.6 (−94.38%)	13.95 (−96.2%)	4.06 (−98.89%)
0.3	0.6	0.6	3.5	79	4	366.41	37.96 (−89.64%)	18.95 (−94.83%)	12.78 (−96.51%)	3.67 (−99%)
0.3	0.6	0.7	3.5	74	4	372.31	35.4 (−90.49%)	17.55 (−95.29%)	11.8 (−96.83%)	3.35 (−99.1%)
0.3	0.7	0.5	3.5	90	4	364.09	43.96 (−87.93%)	22.17 (−93.91%)	15.02 (−95.87%)	4.35 (−98.81%)
0.3	0.7	0.6	3.5	83	4	368.41	40.36 (−89.04%)	20.16 (−94.53%)	13.59 (−96.31%)	3.88 (−98.95%)
0.3	0.7	0.7	3.5	76	4	365.16	36.7 (−89.95%)	18.2 (−95.02%)	12.23 (−96.65%)	3.46 (−99.05%)

The computation study in Tables 5 and 6 concerns the upward shifts in the process correlation. Two levels are accepted for each in-control parameter: $\theta_0 = \{1, 5\}$; $p_0 = \{0.1, 0.2\}$; $\alpha_0 = \{0.5, 0.6\}$; $\beta_0 = \{0.7, 0.8\}$. Two types of out-of-control pattern are considered here for comprehensive investigation, the first type is that the upward changes only exist in α (shown in Table 5), and the second type is that the downward changes only exist in β (shown Table 6). In Table 5, the shifts in the magnitude $\delta_\alpha = \alpha_1 - \alpha_0$ are from the set $\{0.1, 0.2, 0.3\}$. The results imply that the performance of CUSUM chart fluctuates greatly in detecting correlation shifts caused only by α. To be specific, when δ_α is 0.3, dev$^{(\%)}$ ranges from -6.81% to -23.68%. Another finding based on the design of experiments in Table 5 is that both a smaller θ_0 and a smaller β_0 could slightly improve detection efficiency, while p_0 and α_0 could not. In Table 6, the shifts magnitude $\delta_\beta = \beta_1 - \beta_0$ are from the set $\{-0.1, -0.2, -0.3\}$. From Table 6, we can see the CUSUM chart is more efficient in detecting the correlation shifts caused by β. As the absolute value of δ_β gets bigger, the decreasing proportion of ARL gradually increased. When $\delta_\beta = -0.3$, dev$^{(\%)}$ ranges from -21.3% to -40.39%. Meanwhile, we can further conclude that a smaller θ_0 and a larger α_0 often lead to better chart performance, and p_0, β_0 have little influence. Based on all the analysis above in this paragraph, we can further conclude that a smaller θ_0 and a larger value of initial correlation ρ_0 are helpful to detect the correlation shifts. Furthermore, that we cannot get an immediate signal when only correlation shifts occur.

Table 5. ARL profiles of the CUSUM chart versus correlation shifts caused by α.

Process Parameters							ARL (dev$^{(\%)}$)			
θ_0	p_0	α_0	β_0	ρ_0	h	k	$\delta_\alpha = 0$	$\delta_\alpha = 0.1$	$\delta_\alpha = 0.2$	$\delta_\alpha = 0.3$
1	0.1	0.5	0.7	0.15	20	1	365.71	339.16 (-7.26%)	316.72 (-13.4%)	298.07 (-18.5%)
1	0.1	0.5	0.8	0.1	19	1	371.95	353.06 (-5.08%)	336.45 (-9.54%)	322.04 (-13.42%)
1	0.1	0.6	0.7	0.18	21	1	378.05	351.73 (-6.96%)	329.83 (-12.75%)	311.98 (-17.48%)
1	0.1	0.6	0.8	0.12	19	1	353.06	336.45 (-4.7%)	322.04 (-8.79%)	309.72 (-12.28%)
1	0.2	0.5	0.7	0.15	16	1	382.96	346.1 (-9.63%)	316.72 (-17.44%)	292.28 (-23.68%)
1	0.2	0.5	0.8	0.1	15	1	384.25	357.92 (-6.85%)	335.41 (-12.71%)	316.44 (-17.65%)
1	0.2	0.6	0.7	0.18	16	1	346.1	316.19 (-8.64%)	292.28 (-15.55%)	273.5 (-20.98%)
1	0.2	0.6	0.8	0.12	15	1	357.92	335.41 (-6.29%)	316.44 (-11.59%)	300.62 (-16.01%)
5	0.1	0.5	0.7	0.15	76	5	371.67	340.88 (-8.28%)	315.5 (-15.11%)	295.25 (-20.56%)
5	0.1	0.5	0.8	0.1	72	5	377.08	355.15 (-5.82%)	336.23 (-10.83%)	320.32 (-15.05%)
5	0.1	0.6	0.7	0.18	79	5	373.87	344.75 (-7.79%)	321.44 (-14.02%)	303.45 (-18.84%)
5	0.1	0.6	0.8	0.12	73	5	367.42	347.51 (-5.42%)	330.75 (-9.98%)	316.97 (-13.73%)
5	0.2	0.5	0.7	0.15	107	4	369.02	353.52 (-4.2%)	339.82 (-7.91%)	328.1 (-11.09%)
5	0.2	0.5	0.8	0.1	103	4	372.99	362.65 (-2.77%)	353.34 (-5.27%)	345.14 (-7.47%)
5	0.2	0.6	0.7	0.18	110	4	371.48	356.88 (-3.93%)	344.32 (-7.31%)	333.97 (-10.1%)
5	0.2	0.6	0.8	0.12	104	4	369.05	359.52 (-2.58%)	351.13 (-4.86%)	343.92 (-6.81%)

Table 6. ARL profiles of the CUSUM chart versus correlation shifts caused by β.

Process Parameters							ARL (dev$^{(\%)}$)			
θ_0	p_0	α_0	β_0	ρ_0	h	k	$\delta_\alpha = 0$	$\delta_\alpha = 0.1$	$\delta_\alpha = 0.2$	$\delta_\alpha = 0.3$
1	0.1	0.5	0.7	0.15	20	1	365.71	321.34 (-12.13%)	284.33 (-22.25%)	252.99 (-30.82%)
1	0.1	0.5	0.8	0.1	19	1	371.95	325.82 (-12.4%)	287.74 (-22.64%)	255.77 (-31.24%)
1	0.1	0.6	0.7	0.18	21	1	378.05	324.46 (-14.18%)	281.01 (-25.67%)	245.09 (-35.17%)
1	0.1	0.6	0.8	0.12	19	1	353.06	303.31 (-14.09%)	263.34 (-25.41%)	230.49 (-34.72%)
1	0.2	0.5	0.7	0.15	16	1	382.96	325.3 (-15.06%)	280.11 (-26.86%)	243.87 (-36.32%)
1	0.2	0.5	0.8	0.1	15	1	384.25	324.54 (-15.54%)	278.48 (-27.53%)	241.96 (-37.03%)
1	0.2	0.6	0.7	0.18	16	1	346.1	288.47 (-16.65%)	244.5 (-29.36%)	209.96 (-39.34%)
1	0.2	0.6	0.8	0.12	15	1	357.92	295.66 (-17.39%)	249.23 (-30.37%)	213.36 (-40.39%)
5	0.1	0.5	0.7	0.15	76	5	371.67	322.61 (-13.2%)	282.59 (-23.97%)	249.42 (-32.89%)
5	0.1	0.5	0.8	0.1	72	5	377.08	325.96 (-13.56%)	284.72 (-24.49%)	250.82 (-33.48%)
5	0.1	0.6	0.7	0.18	79	5	373.87	316.52 (-15.34%)	271.1 (-27.49%)	234.34 (-37.32%)
5	0.1	0.6	0.8	0.12	73	5	367.42	310.19 (-15.58%)	265.46 (-27.75%)	229.57 (-37.52%)
5	0.2	0.5	0.7	0.15	107	4	369.02	340.45 (-7.74%)	313.89 (-14.94%)	289.18 (-21.64%)
5	0.2	0.5	0.8	0.1	103	4	372.99	344.5 (-7.64%)	318.09 (-14.72%)	293.53 (-21.3%)
5	0.2	0.6	0.7	0.18	110	4	371.48	337.45 (-9.16%)	306.12 (-17.59%)	277.2 (-25.38%)
5	0.2	0.6	0.8	0.12	104	4	369.05	336.01 (-8.95%)	305.7 (-17.17%)	277.78 (-24.73%)

5. Analyses of Drug Crime Count Time Series

In this section, we present a case study of crime count data in Pittsburgh. The data set contains multiple crime types, such as arson, drink-driving, robbery and so on. Monitoring of crime data is needed not only for early warnings of the organised crime, but also for assessments of the social security environment. For the crime data, the readers can download it from the Forecasting Principles site (http://www.forecastingprinciples.com, accessed on 20 March 2021), or email to the corresponding author to access. The subset we analyse is a monthly drug use count data collected from the 56th police car beat, which contains 144 observations from January 1990 to December 2001. There are 67 zeros in this drug use data (the proportion up to 46.53%), which have the greatest proportion among the other values for the data series. The sample mean, variance and first-order autocorrelation of the data are 1.7153, 6.4289 and 0.3886, respectively, which show strong overdispersion and autocorrelation. The sample path and the histogram of the series are in Figure 4. The histograms of estimated ZIG distribution, estimated Geometric distribution and estimated Poisson distribution are also given in Figure 4b, which indicate that the ZIG marginal is the most appropriate to describe the data. The sample autocorrelation function (ACF) and the sample partial autocorrelation function (PACF) in Figure 5 reveal that the series most likely comes from an AR-type process of order 3. While our intention is to illustrate the implementation of the proposed control chart, we will employ the first-order INAR models that are widely studied and applied in the literature. The consideration of more complex models will be left for future study.

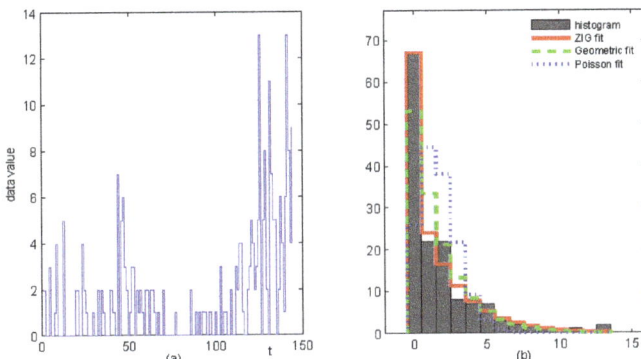

Figure 4. The plots about the zero-inflated drug crime series, (**a**) the sample path, (**b**) the histogram with ZIG fit, Geometric fit and Poisson fit.

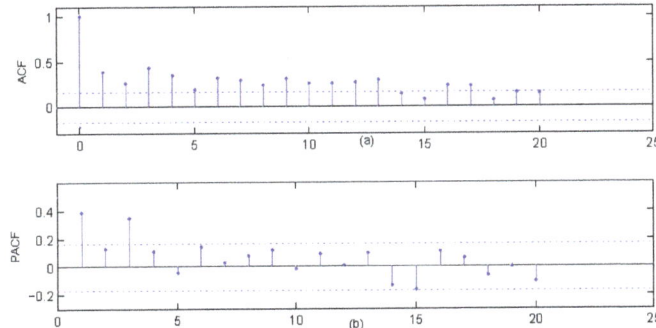

Figure 5. The plots about the zero-inflated drug crime series, (**a**) the ACF plot, (**b**) the PACF plot.

Except for the ZIGINAR$_{RC}$(1) model, some competitive models are also applied to the time series, such as Poisson INAR(1) (Al-Osh and Alzaid [1]), GINAR(1) (Alzaid and Al-Osh [7]), ZINAR(1) (Jazi, Jones and Lai [9]), ZMGINAR(1) (Barreto-Souza [11]), NGINAR(1) (Ristić, Bakouch and Nastić [12]), ZIMINAR(1) (Li, Wang and Zhang [14]). The Akaike Information Criterion (AIC) and Bayesian Information Criterion (BIC) are suggested to evaluate these models. Numerical results in Table 7 show that the ZIGINAR$_{RC}$(1) process has the best overall performance, compared with its competitors. Therefore, we assume that the drug crime data is from the ZIGINAR$_{RC}$(1) model and the estimated parameters in Table 7 are used in the process control procedures. Based on the computation results in Section 4, the CUSUM chart with designs $h = 34$, $k = 2$ (corresponding $ARL_0 = 364.44$) is the best choice, which is shown in Figure 6a. For comparison, we also present CUSUM control charts with designs h=15, $k = 4$ ($ARL_0 = 358.40$) in Figure 6b, designs $h = 12$, $k = 5$ ($ARL_0 = 372.28$) in Figure 6c, and the Shewhart chart with control limit $UCL = 13$ ($ARL_0 = 340.25$) in Figure 6d. We observe that all the CUSUM charts give out-of-control signals, while there is no outliers in the Shewhart chart. Because the Shewhart chart has been proved to be less effective than the CUSUM chart, the drug crime data set seems to be out-of-control with increasing mean shifts or increasing correlation shifts, and some investigation should be done for further explanation. The CUSUM control charts with three designs also display different detection efficiencies. The CUSUM chart with $k = 2$ first signals at $t = 131$ following with continuous alarms as t increases. The signals of the CUSUM chart with $k = 4$ are first given at $t = 133$, then go back below the control limit over a period of time, and come again at $t = 141$. While the outlier of the CUSUM chart with $k=5$ occurs at $t=144$. The analysis above proves again that the CUSUM chart design $k=\lceil\mu_0\rceil$ is the most effective in practice.

Table 7. The estimated parameters, AIC and BIC of candidate models.

Model	Estimated Parameters		AIC	BIC
Poisson INAR(1)	$\hat{\lambda}=1.2806$	$\hat{\alpha}=0.2614$	625.787	631.7266
GINAR(1)	$\hat{p}=0.3581$	$\hat{\alpha}=0.2005$	507.0551	512.9947
NGINAR(1)	$\hat{\mu}=1.7925$	$\hat{\alpha}=0.3028$	502.6322	508.5718
ZINAR(1)	$\hat{\alpha}=0.2157$ $\hat{p}=0.5351$	$\hat{\lambda}=2.9131$	527.1981	536.1075
ZMGINAR(1)	$\hat{\mu}=1.9712$ $\hat{\pi}=0.1197$	$\hat{\alpha}=0.29$	501.7477	510.6571
ZIMINAR(1)	$\hat{\alpha}=0.001$ $\hat{p}=0.4101$ $\hat{\rho}=0.548$	$\hat{\beta}=0.6731$ $\hat{\lambda}=2.2564$	503.457	518.3061
ZIGINAR$_{RC}$(1)	$\hat{\alpha}=0.547$ $\hat{p}=0.185$	$\hat{\theta}=2.0495$ $\hat{\beta}=0.5188$	494.906	506.7852

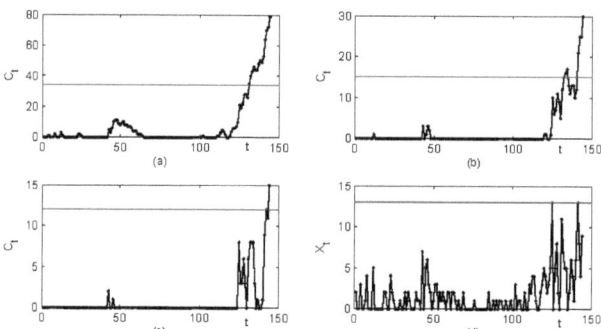

Figure 6. The control charts for the zero-inflated drug crime series, the CUSUM charts with designs (a) $h=34$, $k=2$, (b) $h=15$, $k=4$, (c) $h=12$, $k=5$, and (d) the Shewhart chart with $UCL=13$.

6. Conclusions

In this paper, we have made contributions on monitoring the zero-inflated autocorrelated count data which can be described by an INAR(1) process with random coefficient. ARL and SDRL are adopted to be the measures and calculated by the Markov chain approach. The design parameter k in the CUSUM chart is proved to have great influence on monitoring efficiency, and the smallest integer no less than μ_0 is the recommended value of k. The proposed ZIGINAR$_{RC}$(1)CUSUM chart is proved to have superiorities in various coefficient shifts compared with the conventional Shewhart chart. Computation results also show that the CUSUM chart performs quite well in detecting upward mean shifts, and shows fluctuation in detecting upward correlation shifts. Based on the design of experiment, we also find that a larger value of initial correlation ρ_0 is helpful to detect the correlation shifts. An immediate signal occurs after large upward mean shifts, while does not occur when only correlation shifts exist.

There are some possible topics for our future research. First, we can consider a different monitoring scheme for the ZIGINAR$_{RC}$(1) process and conduct a comparison study. Second, we can explore the monitoring of p-th random coefficient INAR model, which is suitable for the count data with high order dependence. Third, we can study a multivariate INAR(1) process with random coefficient to continuously monitor the serial correlated counts that we are interested in.

Author Contributions: Conceptualization, C.L. and D.W.; methodology, C.L.; software, S.C.; validation, C.L., S.C. and D.W.; formal analysis, D.W.; investigation, C.L.; resources, C.L. and D.W.; data curation, S.C.; writing—original draft preparation, S.C.; writing—review and editing, C.L.; visualization, S.C.; supervision, D.W.; project administration, C.L.; funding acquisition, C.L. and D.W. All authors have read and agreed to the published version of the manuscript.

Funding: This work is supported by National Natural Science Foundation of China grant numbers 12001229, 11871028 and 11731015, and Natural Science Foundation of Jilin Province (No. 20180101216JC).

Conflicts of Interest: The authors declare no conflict of interest.

References

1. Al-Osh, M.A.; Alzaid, A.A. First-order integer-valued autoregressive (INAR(1)) process. *J. Time. Ser. Anal.* **1987**, *8*, 261–275. [CrossRef]
2. Steutel, F.W.; Van Harn, K. Discrete analogues of self-decomposability and stability. *Ann. Probab.* **1979**, *7*, 893–899. [CrossRef]
3. Ristić, M.M.; Nastić, A.S. A mixed INAR(p) model. *J. Time. Ser. Anal.* **2012**, *33*, 903–915. [CrossRef]
4. Nastić, A.S.; Laketa, P.N.; Ristić, M.M. Random environment INAR models of higher order. *Revstat-Stat. J.* **2019**, *17*, 35–65.
5. Pedeli, X.; Karlis, D. A bivariate INAR(1) process with application. *Stat. Model.* **2011**, *11*, 325–349. [CrossRef]
6. Khan, N.M.; Cekim, H.O.; Ozel, G. The family of the bivariate integer-valued autoregressive process (BINAR(1)) with Poisson-Lindley (PL) innovations. *J. Stat. Comput. Sim.* **2020**, *90*, 624–637. [CrossRef]
7. Alzaid, A.A.; Al-Osh, M.A. First-order integer-valued autoregressive (INAR(1)) process: Distributional and regression properties. *Stat. Neerl.* **1988**, *42*, 53–61. [CrossRef]
8. Alzaid, A.A.; Al-Osh, M.A. Some autoregressive moving average processes with generalized Poisson marginal distributions. *Ann. I. Stat. Math.* **1993**, *45*, 223–232. [CrossRef]
9. Jazi, M.A.; Jones, G.; Lai, C.D. First-order integer valued AR processes with zero inflated Poisson innovations. *J. Time. Ser. Anal.* **2012**, *33*, 954–963. [CrossRef]
10. Ristić, M.M.; Nastić, A.S.; Bakouch, H.S. Estimation in an integer-valued autoregressive process with negative binomial marginals (NBINAR(1)). *Commun. Stat-Theor. M.* **2012**, *41*, 606–618. [CrossRef]
11. Barreto-Souza, W. Zero-modified Geometric INAR(1) process for modelling count time series with deflation or inflation of zeros. *J. Time. Ser. Anal.* **2015**, *36*, 839–852. [CrossRef]
12. Ristić, M.M.; Bakouch, H.S.; Nastić, A.S. A new geometric first-order integer-valued autoregressive (NGINAR(1)) process. *J. Stat. Plan. Infer.* **2009**, *139*, 2218–2226. [CrossRef]
13. Liu, Z.; Zhu, F. A new extension of thinning-based integer-valued autoregressive models for count data. *Entropy* **2021**, *23*, 62. [CrossRef] [PubMed]
14. Li, C.; Wang, D.; Zhang, H. First-order mixed integer-valued autoregressive processes with zero-inflated generalized power series innovations. *J. Korean Stat. Soc.* **2015**, *44*, 232–246. [CrossRef]
15. Orozco, D.L.; Sales, L.O.; Fernández, L.M.; Pinho, A.L. A new mixed first-order integer-valued autoregressive process with Poisson innovations. *ASTA-Adv. Stat. Anal.* **2020**, 1–22. [CrossRef]
16. Weiß, C.H. Thinning operations for modeling time series of counts-a survey. *Asta-Adv. Stat. Anal.* **2008**, *92*, 319–341. [CrossRef]
17. Scotto, M.; Weiß, C.H.; Gouveia, S. Thinning-based models in the analysis of integer-valued time series: A review. *Stat. Model.* **2015**, *15*, 590–618. [CrossRef]
18. Zheng, H.; Basawa, I.V.; Datta, S. Inference for pth-order random coefficient integer-valued autoregressive processes. *J. Time. Ser. Anal.* **2006**, *27*, 411–440. [CrossRef]
19. Kang, J.; Lee, S. Parameter change test for random coefficient integer-valued autoregressive processes with application to polio data analysis. *J. Time. Ser. Anal.* **2009**, *30*, 239–258. [CrossRef]
20. Zhang, H.; Wang, D.; Zhu, F. Empirical likelihood inference for random coefficient INAR(p) process. *J. Time. Ser. Anal.* **2011**, *32*, 195–203. [CrossRef]
21. Bakouch, H.S.; Mohammadpour, M.; Shirozhan, M. A zero-inflated geometric INAR(1) process with random coefficient. *Appl. Math.* **2018**, *63*, 79–105. [CrossRef]
22. Weiß, C.H. EWMA monitoring of correlated processes of Poisson counts. *Qual. Technol. Quant. M.* **2009**, *6*, 137–153. [CrossRef]
23. Weiß, C.H.; Testik, M.C. CUSUM monitoring of first-order integer-valued autoregressive processes of Poisson counts. *J. Qual. Technol.* **2009**, *41*, 389–400. [CrossRef]
24. Weiß, C.H.; Testik, M.C. The Poisson INAR(1) CUSUM chart under overdispersion and estimation error. *IIE Trans.* **2011**, *43*, 805–818. [CrossRef]
25. Yontay, P.; Weiß, C.H.; Testik, M.C.; Bayindir, Z.P. A two-sided cumulative sum chart for first-order integer-valued autoregressive processes of poisson counts. *Qual. Reliab. Eng. Int.* **2013**, *29*, 33–42. [CrossRef]

26. Rakitzis, A.C.; Weiß, C.H.; Castagliola, P. Control charts for monitoring correlated Poisson counts with an excessive number of zeros. *Qual. Reliab. Eng. Int.* **2017**, *33*, 413–430. [CrossRef]
27. Li, C.; Wang, D.; Sun, J. Control charts based on dependent count data with deflation or inflation of zeros. *J. Stat. Comput. Sim.* **2019**, *89*, 3273–3289. [CrossRef]
28. Fernandes, F.H.; Bourguignon, M.; Ho, L.L. Control charts to monitor integer valued autoregressive process with inflation or deflation of zeros. *J. Stat. Manag. Syst.* **2020**, *23*, 1463–1484. [CrossRef]
29. Sales, L.O.; Pinho, A.L.; Vivacqua, C.A.; Ho, L.L. Shewhart control chart for monitoring the mean of Poisson mixed integer autoregressive processes via Monte Carlo simulation. *Comput. Ind. Eng.* **2020**, *140*, 106245. [CrossRef]
30. Page, E.S. Cumulative sum charts. *Technometrics* **1961**, *3*, 1–9. [CrossRef]
31. Montgomery, D.C. *Introduction to Statistical Quality Control*, 6th ed.; Wiley: New York, NY, USA, 2009.
32. Alencar, A.P.; Ho, L.L.; Albarracin, O.Y.E. CUSUM control charts to monitor series of negative binomial count data. *Stat. Methods. Med. Res.* **2017**, *26*, 1925–1935. [CrossRef]
33. Bourguignon, M.; Medeiros, R.M.; Fernandes, F.H.; Ho, L.L. Simple and useful statistical control charts for monitoring count data. *Qual. Reliab. Eng. Int.* **2021**, *37*, 541–566. [CrossRef]
34. Harris, T.J.; Ross, W.H. Statistical process control procedures for correlated observations. *Can. J. Chem. Eng.* **1991**, *69*, 48–57. [CrossRef]
35. Triantafyllopoulos, K.; Bersimis, S. Phase II control charts for autocorrelated processes. *Qual. Technol. Quant. M.* **2016**, *13*, 88–108. [CrossRef]
36. Albarracin, O.Y.E.; Alencar, A.P.; Ho, L.L. Effect of neglecting autocorrelation in regression EWMA charts for monitoring count time series. *Qual. Reliab. Eng. Int.* **2018**, *34*, 1752–1762. [CrossRef]
37. Lee, S.; Kim, D. Monitoring parameter change for time series models of counts based on minimum density power divergence estimator. *Entropy* **2020**, *22*, 1304. [CrossRef] [PubMed]
38. Lee, S.; Kim, C.K.; Kim, D. Monitoring volatility change for time series based on support vector regression. *Entropy* **2020**, *22*, 1312. [CrossRef]
39. Brook, D.A.; Evans, D. An approach to the probability distribution of CUSUM run length. *Biometrika* **1972**, *59*, 539–549. [CrossRef]

Article
Robust Estimation for Bivariate Poisson INGARCH Models

Byungsoo Kim [1,*], Sangyeol Lee [2] and Dongwon Kim [2]

1. Department of Statistics, Yeungnam University, Gyeongsan 38541, Korea
2. Department of Statistics, Seoul National University, Seoul 08826, Korea; sylee@stats.snu.ac.kr (S.L.); dongwon.k@snu.ac.kr (D.K.)
* Correspondence: bkim@yu.ac.kr

Abstract: In the integer-valued generalized autoregressive conditional heteroscedastic (INGARCH) models, parameter estimation is conventionally based on the conditional maximum likelihood estimator (CMLE). However, because the CMLE is sensitive to outliers, we consider a robust estimation method for bivariate Poisson INGARCH models while using the minimum density power divergence estimator. We demonstrate the proposed estimator is consistent and asymptotically normal under certain regularity conditions. Monte Carlo simulations are conducted to evaluate the performance of the estimator in the presence of outliers. Finally, a real data analysis using monthly count series of crimes in New South Wales and an artificial data example are provided as an illustration.

Keywords: integer-valued time series; bivariate Poisson INGARCH model; outliers; robust estimation; minimum density power divergence estimator

1. Introduction

Integer-valued time series models have received widespread attention from researchers and practitioners, due to their versatile applications in many scientific areas, including finance, insurance, marketing, and quality control. Numerous studies focus on integer-valued autoregressive (INAR) models to analyze the time series of counts, see Weiß [1] and Scotto et al. [2] for general reviews. Taking a different approach, Ferland et al. [3] proposed using Poisson integer-valued generalized autoregressive conditional heteroscedastic (INGARCH) models and Fokianos et al. [4] developed Poisson AR models to generalize the linear assumption on INGARCH models. The Poisson assumption on INGARCH models has been extended to negative binomial INGARCH models (Davis and Wu [5] and Christou and Fokianos [6]), zero-inflated generalized Poisson INGARCH models (Zhu [7,8] and Lee et al. [9]), and one-parameter exponential family AR models (Davis and Liu [10]). We refer to the review papers by Fokianos [11,12] and Tjøstheim [13,14] for more details.

Researchers invested considerable efforts to extend the univariate integer-valued time series models to bivariate (multivariate) models. For INAR type models, Quoreshi [15] proposed bivariate integer-valued moving average models and Pedeli and Karlis [16] introduced bivariate INAR models with Poisson and negative binomial innovations. Liu [17] proposed bivariate Poisson INGARCH models with a bivariate Poisson distribution that was constructed via the trivariate reduction method and established the stationarity and ergodicity of the model. Andreassen [18] later verified the consistency of the conditional maximum likelihood estimator (CMLE) and Lee et al. [19] studied the asymptotic normality of the CMLE and developed the CMLE- and residual-based change point tests. However, this model has the drawback that it can only accommodate positive correlation between two time series of counts. To cope with this issue, Cui and Zhu [20] recently introduced a new bivariate Poisson INGARCH model based on Lakshminarayana et al.'s [21] bivariate Poisson distribution. Their model can deal with positive or negative correlation, depending on the multiplicative factor parameter. They employed the CMLE for parameter estimation. However, because the CMLE is unduly influenced by outliers, the robust estimation in bivariate Poisson INGARCH models is crucial and deserves thorough investigation.

As such, here we develop a robust estimator for Cui and Zhu's [20] bivariate Poisson INGARCH models. Among the robust estimation methods, we employ the minimum density power divergence estimator (MDPDE) approach that was originally proposed by Basu et al. [22], because it is well known to consistently provide robust estimators in various situations. For previous works in the context of time series of counts, see Kang and Lee [23], Kim and Lee [24,25], Diop and Kengne [26], Kim and Lee [27], and Lee and Kim [28], who studied the MDPDE for Poisson AR models, zero-inflated Poisson AR models, one-parameter exponential family AR models, and change point tests. For another robust estimation approach in INGARCH models, see Xiong and Zhu [29] and Li et al. [30], who studied Mallows' quasi-likelihood method. To the best of our knowledge, the robust estimation method for bivariate Poisson INGARCH models has not been previously studied. In earlier studies, the MDPDE was proven to possess strong robust properties against outliers with little loss in asymptotic efficiency relative to the CMLE. This study confirms the same conclusion for bivariate Poisson INGARCH models.

The rest of this paper is organized, as follows. Section 2 constructs the MDPDE for bivariate Poisson INGARCH models. Section 3 shows the asymptotic properties of the MDPDE. Section 4 conducts empirical studies to evaluate the performance of the MDPDE. Section 5 provides concluding remarks. Appendix A provides the proof.

2. MDPDE for Bivariate Poisson Ingarch Models

Basu et al. [22] defined the density power divergence d_α between two densities f and g, with a tuning parameter α, as

$$d_\alpha(g,f) = \begin{cases} \int \{f^{1+\alpha}(y) - (1+\frac{1}{\alpha})g(y)f^\alpha(y) + \frac{1}{\alpha}g^{1+\alpha}(y)\}dy, & \alpha > 0, \\ \int g(y)(\log g(y) - \log f(y))dy, & \alpha = 0. \end{cases}$$

For a parametric family $\{F_\theta; \theta \in \Theta\}$ having densities $\{f_\theta\}$ and a distribution G with density g, they defined the minimum density power divergence functional $T_\alpha(G)$ by $d_\alpha(g, f_{T_\alpha(G)}) = \min_{\theta \in \Theta} d_\alpha(g, f_\theta)$. If G belongs to $\{F_\theta\}$, which is, $G = F_{\theta^0}$ for some $\theta^0 \in \Theta$, then $T_\alpha(F_{\theta^0}) = \theta^0$. Let g be the density function of a random sample $Y_1, \ldots, Y_n$. Using the empirical distribution G_n to approximate G, Basu et al. [22] defined the MDPDE by

$$\hat{\theta}_{\alpha,n} = \operatorname*{argmin}_{\theta \in \Theta} H_{\alpha,n}(\theta),$$

where $H_{\alpha,n}(\theta) = \frac{1}{n}\sum_{t=1}^n h_{\alpha,t}(\theta)$ and

$$h_{\alpha,t}(\theta) = \begin{cases} \int f_\theta^{1+\alpha}(y)dy - \left(1+\frac{1}{\alpha}\right)f_\theta^\alpha(Y_t), & \alpha > 0, \\ -\log f_\theta(Y_t), & \alpha = 0. \end{cases}$$

The tuning parameter α controls the trade-off between the robustness and asymptotic efficiency of the MDPDE. Namely, relatively large α values improve the robustness but the estimator's efficiency decreases. The MDPDE with $\alpha = 0$ and 1 leads to the MLE and L_2-distance estimator, respectively. Basu et al. [22] showed the consistency and asymptotic normality of the MDPDE and demonstrated that the estimator is robust against outliers, but it still retains high efficiency when the true distribution belongs to a parametric family $\{F_\theta\}$ and α is close to zero.

We need to define the conditional version of the MDPDE in order to apply the above procedure to bivariate Poisson INGARCH models. Let $\{f_\theta(\cdot|\mathcal{F}_{t-1})\}$ denote the parametric family of autoregressive models, being indexed by the parameter θ, and let $f_{\theta^0}(\cdot|\mathcal{F}_{t-1})$ be the true conditional density of the time series Y_t given $\mathcal{F}_{t-1}$, where $\mathcal{F}_{t-1}$ is a σ-field generated by $Y_{t-1}, Y_{t-2}, \ldots$. Subsequently, the MDPDE of θ^0 is given by

$$\hat{\theta}_{\alpha,n} = \operatorname*{argmin}_{\theta \in \Theta} H_{\alpha,n}(\theta),$$

where $H_{\alpha,n}(\theta) = \frac{1}{n}\sum_{t=1}^{n} h_{\alpha,t}(\theta)$ and

$$h_{\alpha,t}(\theta) = \begin{cases} \int f_\theta^{1+\alpha}(y|\mathcal{F}_{t-1})dy - \left(1+\frac{1}{\alpha}\right)f_\theta^\alpha(Y_t|\mathcal{F}_{t-1}), & \alpha > 0, \\ -\log f_\theta(Y_t|\mathcal{F}_{t-1}), & \alpha = 0 \end{cases} \quad (1)$$

(cf. Section 2 of Kang and Lee [23]).

Let $\mathbf{Y}_t = (Y_{t,1}, Y_{t,2})^T$ be a two-dimensional vector of counts at time t, namely, $\{Y_{t,1}, t \geq 1\}$ and $\{Y_{t,2}, t \geq 1\}$ are the two time series of counts under consideration. Liu [17] proposed the bivariate Poisson INGARCH model, as follows

$$\mathbf{Y}_t|\mathcal{F}_{t-1} \sim BP^*(\lambda_{t,1}, \lambda_{t,2}, \phi), \quad \boldsymbol{\lambda}_t = (\lambda_{t,1}, \lambda_{t,2})^T = \boldsymbol{\omega} + \mathbf{A}\boldsymbol{\lambda}_{t-1} + \mathbf{B}\mathbf{Y}_{t-1},$$

where $\mathcal{F}_t$ is the σ-field generated by $\mathbf{Y}_t, \mathbf{Y}_{t-1}, \ldots$, $\phi \geq 0$, $\boldsymbol{\omega} = (\omega_1, \omega_2)^T \in \mathbb{R}_+^2$, $\mathbf{A} = \{a_{ij}\}_{i,j=1,2}$ and $\mathbf{B} = \{b_{ij}\}_{i,j=1,2}$ are 2×2 matrices with non-negative entries. $BP^*(\lambda_{t,1}, \lambda_{t,2}, \phi)$ denotes the bivariate Poisson distribution constructed via the trivariate reduction method, whose probability mass function (PMF) is

$$P(Y_{t,1} = y_1, Y_{t,2} = y_2|\mathcal{F}_{t-1})$$
$$= e^{-(\lambda_{t,1}+\lambda_{t,2}-\phi)} \frac{(\lambda_{t,1}-\phi)^{y_1}}{y_1!} \frac{(\lambda_{t,2}-\phi)^{y_2}}{y_2!} \sum_{s=0}^{\min(y_1,y_2)} \binom{y_1}{s}\binom{y_2}{s} s! \left\{\frac{\phi}{(\lambda_{t,1}-\phi)(\lambda_{t,2}-\phi)}\right\}^s.$$

In this model, $Cov(Y_{t,1}, Y_{t,2}|\mathcal{F}_{t-1}) = \phi \in [0, \min(\lambda_{t,1}, \lambda_{t,2}))$, so that the model has a drawback that it can only deal with positive correlation between two components.

To overcome this defect, Cui and Zhu [20] proposed a new bivariate Poisson IN-GARCH model using the distribution that was proposed by Lakshminarayana et al. [21]. They considered the model:

$$\mathbf{Y}_t|\mathcal{F}_{t-1} \sim BP(\lambda_{t,1}, \lambda_{t,2}, \delta), \quad \boldsymbol{\lambda}_t = (\lambda_{t,1}, \lambda_{t,2})^T = \boldsymbol{\omega} + \mathbf{A}\boldsymbol{\lambda}_{t-1} + \mathbf{B}\mathbf{Y}_{t-1} \quad (2)$$

and $BP(\lambda_{t,1}, \lambda_{t,2}, \delta)$ is the bivariate Poisson distribution constructed as a product of Poisson marginals with a multiplicative factor, whose PMF is given by

$$P(Y_{t,1} = y_1, Y_{t,2} = y_2|\mathcal{F}_{t-1})$$
$$= \frac{\lambda_{t,1}^{y_1}\lambda_{t,2}^{y_2}}{y_1!y_2!} e^{-(\lambda_{t,1}+\lambda_{t,2})} \left\{1 + \delta(e^{-y_1} - e^{-c\lambda_{t,1}})(e^{-y_2} - e^{-c\lambda_{t,2}})\right\}, \quad (3)$$

where $c = 1 - e^{-1}$. The marginal conditional distribution of $Y_{t,1}$ and $Y_{t,2}$ are Poisson with parameters $\lambda_{t,1}$ and $\lambda_{t,2}$, respectively, and $Cov(Y_{t,1}, Y_{t,2}|\mathcal{F}_{t-1}) = \delta c^2 \lambda_{t,1}\lambda_{t,2}e^{-c(\lambda_{t,1}+\lambda_{t,2})}$. Hence, this model supports positive or negative correlation, depending on the multiplicative factor parameter δ. Cui and Zhu [20] established the stationarity and ergodicity of the model under certain conditions and showed the consistency and asymptotic normality of the CMLE.

In this study, we apply the MDPDE to the model (2). We focus on the case that $\mathbf{A}$ is a diagonal matrix, because this simplification can reduce the number of model parameters and makes it easy to use in practice, as Heinen and Rengifo [31] suggested. Further, the diagonal setup of $\mathbf{A}$ eases the verification of the asymptotic properties of the MDPDE. Similar approaches can be found in Liu [17], Lee et al. [19], and Cui et al. [32]. Let $\mathbf{A} = diag(a_1, a_2)$. Subsequently, we set $\theta = (\theta_1^T, \theta_2^T, \delta)^T$, where $\theta_1 = (\omega_1, a_1, b_{11}, b_{12})^T$ and $\theta_2 = (\omega_2, a_2, b_{21}, b_{22})^T$, and write the true parameter as $\theta^0 = (\theta_1^{0T}, \theta_2^{0T}, \delta^0)^T$, where $\theta_1^0 = (\omega_1^0, a_1^0, b_{11}^0, b_{12}^0)^T$ and $\theta_2^0 = (\omega_2^0, a_2^0, b_{21}^0, b_{22}^0)^T$.

Given $\mathbf{Y}_1, \ldots, \mathbf{Y}_n$ that is generated from (2), from (1), we obtain the MDPDE of θ^0 by

$$\hat{\theta}_{\alpha,n} = \operatorname*{argmin}_{\theta \in \Theta} \tilde{H}_{\alpha,n}(\theta) = \operatorname*{argmin}_{\theta \in \Theta} \frac{1}{n}\sum_{t=1}^{n} \tilde{h}_{\alpha,t}(\theta),$$

where

$$\tilde{h}_{\alpha,t}(\theta) = \begin{cases} \sum_{y_1=0}^{\infty}\sum_{y_2=0}^{\infty} f_\theta^{1+\alpha}(y|\tilde{\lambda}_t) - \left(1+\frac{1}{\alpha}\right) f_\theta^\alpha(Y_t|\tilde{\lambda}_t), & \alpha > 0, \\ -\log f_\theta(Y_t|\tilde{\lambda}_t), & \alpha = 0, \end{cases} \quad (4)$$

$f_\theta(y|\lambda_t)$ for $y = (y_1, y_2)^T$ is the conditional PMF in (3), and $\tilde{\lambda}_t$ is recursively defined by

$$\tilde{\lambda}_t = (\tilde{\lambda}_{t,1}, \tilde{\lambda}_{t,2})^T = \omega + A\tilde{\lambda}_{t-1} + BY_{t-1}, \ t \geq 2$$

with an arbitrarily chosen initial value $\tilde{\lambda}_1$. We also use notations $\lambda_t(\theta)$ and $\tilde{\lambda}_t(\theta)$ to denote λ_t and $\tilde{\lambda}_t$, respectively, in order to emphasize the role of θ.

3. Asymptotic Properties of the MDPDE

In this section, we establish the consistency and asymptotic normality of the MDPDE. Throughout this study, $\|A\|_p$ denotes the p-induced norm of matrix A for $1 \leq p \leq \infty$ and $\|x\|_p$ is the p-norm of vector x. When $p = 1$ and ∞, $\|A\|_1 = \max_{1 \leq j \leq n} \sum_{i=1}^{m} |a_{ij}|$ and $\|A\|_\infty = \max_{1 \leq i \leq m} \sum_{j=1}^{n} |a_{ij}|$ for $A = \{a_{ij}\}_{1 \leq i \leq m, 1 \leq j \leq n}$, respectively. $E(\cdot)$ is taken under θ^0. We assume that the following conditions hold in order to verify the asymptotic properties of the MDPDE.

(A1) θ_1^0, θ_2^0, and δ^0 are interior points in the compact parameter spaces Θ_1, Θ_2, and Θ_3, respectively, and $\Theta = \Theta_1 \times \Theta_2 \times \Theta_3$. In addition, there exist positive constants ω_L, ω_U, a_L, a_U, b_L, b_U, and δ_U, such that for $i, j = 1, 2$,

$$0 < \omega_L \leq \omega_i \leq \omega_U, \ 0 < a_L \leq a_i \leq a_U, \ 0 < b_L \leq b_{ij} \leq b_U, \text{ and } |\delta| \leq \delta_U.$$

(A2) There exist positive constants φ_L and φ_U such that for $y = (y_1, y_2)^T \in \mathbb{N}_0^2$, $\lambda = (\lambda_1, \lambda_2)^T \in (0, \infty)^2$, and $\delta \in \Theta_3$,

$$0 < \varphi_L \leq \varphi(y, \lambda, \delta) \leq \varphi_U, \text{ where } \varphi(y, \lambda, \delta) = 1 + \delta(e^{-y_1} - e^{-c\lambda_1})(e^{-y_2} - e^{-c\lambda_2}).$$

(A3) There exists a $p \in [1, \infty]$ such that $\|A\|_p + 2^{(1-1/p)}\|B\|_p < 1$.

Remark 1. *These conditions can be found in Cui and Zhu [20]. According to Theorem 1 in their study, $\{(Y_t, \lambda_t)\}$ is stationary and ergodic under (A1) and (A3).*

Subsequently, we obtain the following results; the proofs are provided in the Appendix A.

Theorem 1. *Under the conditions (A1)–(A3),*

$$\hat{\theta}_{\alpha,n} \xrightarrow{a.s.} \theta^0 \text{ as } n \to \infty.$$

Theorem 2. *Under the conditions (A1)–(A3),*

$$\sqrt{n}(\hat{\theta}_{\alpha,n} - \theta^0) \xrightarrow{d} N(0, J_\alpha^{-1} K_\alpha J_\alpha^{-1}) \text{ as } n \to \infty,$$

where

$$J_\alpha = -E\left(\frac{\partial^2 h_{\alpha,t}(\theta^0)}{\partial \theta \partial \theta^T}\right), \quad K_\alpha = E\left(\frac{\partial h_{\alpha,t}(\theta^0)}{\partial \theta} \frac{\partial h_{\alpha,t}(\theta^0)}{\partial \theta^T}\right),$$

and $h_{\alpha,t}(\theta)$ is defined by replacing $\tilde{\lambda}_t(\theta)$ with $\lambda_t(\theta)$ in (4).

Remark 2. *Because the tuning parameter α controls the trade-off between the robustness and asymptotic efficiency, choosing the optimal α is an important issue in practice. Several researchers investigated the selection criterion of optimal α; see Fujisawa and Eguchi [33], Durio and Isaia [34],* and Toma and Broniatowski [35]. Among them, we adopt the method of Warwick [36] to choose α that minimizes the trace of the estimated asymptotic mean squared error ($\widehat{AMSE}$) defined by

$$\widehat{AMSE} = (\hat{\theta}_{\alpha,n} - \hat{\theta}_{1,n})(\hat{\theta}_{\alpha,n} - \hat{\theta}_{1,n})^T + \widehat{As.var}(\hat{\theta}_{\alpha,n}),$$

where $\hat{\theta}_{1,n}$ is the MDPDE with $\alpha = 1$ and $\widehat{As.var}(\hat{\theta}_{\alpha,n})$ is an estimate of the asymptotic variance of $\hat{\theta}_{\alpha,n}$, which is computed as

$$\widehat{As.var}(\hat{\theta}_{\alpha,n}) = \left(\sum_{t=1}^n \frac{\partial^2 \tilde{h}_{\alpha,t}(\hat{\theta}_{\alpha,n})}{\partial \theta \partial \theta^T}\right)^{-1} \left(\sum_{t=1}^n \frac{\partial \tilde{h}_{\alpha,t}(\hat{\theta}_{\alpha,n})}{\partial \theta} \frac{\partial \tilde{h}_{\alpha,t}(\hat{\theta}_{\alpha,n})}{\partial \theta^T}\right) \left(\sum_{t=1}^n \frac{\partial^2 \tilde{h}_{\alpha,t}(\hat{\theta}_{\alpha,n})}{\partial \theta \partial \theta^T}\right)^{-1}.$$

This criterion is applied to our empirical study in Section 4.2.

4. Empirical Studies

4.1. Simulation

In this section, we report the simulation results to evaluate the performance of the MDPDE. The simulation settings are described, as follows. Using the inverse transformation sampling method (cf. Section 2.3 of Verges [37]), we generate $Y_1, \ldots, Y_n$ from (2) with the initial value $\lambda_1 = (0,0)^T$. For the estimation, $\tilde{\lambda}_1$ is set to be the sample mean of the data. We first consider $\theta = (\omega_1, a_1, b_{11}, b_{12}, \omega_2, a_2, b_{21}, b_{22}, \delta)^T = (1, 0.2, 0.1, 0.2, 0.5, 0.3, 0.4, 0.2, 0.5)^T$, which satisfies (A3) with $p = 1$. In this simulation, we compare the performance of the MDPDE with $\alpha > 0$ with that of the CMLE ($\alpha = 0$). We examine the sample mean, variance, and mean squared error (MSE) of the estimators. The sample size under consideration is $n = 1000$ and the number of repetitions for each simulation is 1000. In Tables 1–16, the symbol * represents the minimal MSEs for each parameter.

Table 1 indicates that, when the data are not contaminated by outliers, the CMLE exhibits minimal MSEs for all parameters, and the MSEs of the MDPDE with small α are close to those of the CMLE. The MSE of the MDPDE shows an increasing tendency as α increases. Hence, we can conclude that the CMLE outperforms the MDPDE in the absence of outliers.

Table 1. Sample mean, variance, and mean squared error (MSE) of estimators when $\theta = (1, 0.2, 0.1, 0.2, 0.5, 0.3, 0.4, 0.2, 0.5)^T$, $n = 1000$, and no outliers exist.

α		$\hat{\omega}_1$	$\hat{a}_1$	$\hat{b}_{11}$	$\hat{b}_{12}$	$\hat{\omega}_2$	$\hat{a}_2$	$\hat{b}_{21}$	$\hat{b}_{22}$	$\hat{\delta}$
0(CMLE)	Mean	1.010	0.198	0.099	0.199	0.510	0.298	0.401	0.198	0.583
	Var × 10²	3.421	1.062	0.105	0.083	1.344	0.366	0.119	0.100	15.54
	MSE × 10²	3.429 *	1.061 *	0.105 *	0.083 *	1.352 *	0.366 *	0.119 *	0.100 *	16.22 *
0.1	Mean	1.012	0.198	0.099	0.199	0.510	0.297	0.401	0.199	0.577
	Var × 10²	3.527	1.091	0.108	0.083	1.379	0.372	0.121	0.103	15.83
	MSE × 10²	3.537	1.091	0.108	0.084	1.387	0.372	0.121	0.103	16.41
0.2	Mean	1.013	0.197	0.099	0.199	0.510	0.297	0.401	0.199	0.572
	Var × 10²	3.671	1.134	0.113	0.086	1.453	0.387	0.126	0.108	16.42
	MSE × 10²	3.684	1.134	0.113	0.086	1.463	0.388	0.126	0.108	16.92
0.3	Mean	1.013	0.197	0.099	0.199	0.511	0.296	0.401	0.199	0.568
	Var × 10²	3.870	1.195	0.120	0.090	1.555	0.410	0.133	0.114	17.22
	MSE × 10²	3.883	1.195	0.120	0.090	1.565	0.411	0.133	0.114	17.66
0.5	Mean	1.012	0.197	0.100	0.199	0.511	0.294	0.402	0.200	0.559
	Var × 10²	4.336	1.340	0.137	0.101	1.817	0.469	0.151	0.130	19.51
	MSE × 10²	4.347	1.340	0.137	0.101	1.828	0.472	0.152	0.130	19.84
1	Mean	1.007	0.198	0.101	0.200	0.513	0.289	0.405	0.203	0.544
	Var × 10²	6.094	1.864	0.198	0.148	2.805	0.690	0.222	0.189	29.18
	MSE × 10²	6.094	1.863	0.198	0.148	2.818	0.701	0.224	0.190	29.35

Now, we consider the situation that the data are contaminated by outliers. To this end, we generate contaminated data $Y_{c,t} = (Y_{c,t,1}, Y_{c,t,2})^T$ when considering

$$Y_{c,t,i} = Y_{t,i} + P_{t,i} Y_{o,t,i}, \quad i = 1, 2,$$

where $Y_{t,i}$ are generated from (2), $P_{t,i}$ are i.i.d. Bernoulli random variables with success probability p, and $Y_{o,t,i}$ are i.i.d. Poisson random variables with mean γ. We consider three cases: $(p, \gamma) = (0.03, 5)$, $(0.03, 10)$, and $(0.05, 10)$. Tables 2–4 report the results. In the tables, the MDPDE appears to have smaller MSEs than the CMLE for all cases, except for the case of $\alpha = 1$ when $(p, \gamma) = (0.03, 5)$. As p or γ increases, the MSEs of the CMLE increase faster than those of the MDPDE, which indicates that the MDPDE outperforms the CMLE, as the data are more contaminated by outliers. Moreover, as p or γ increases, the symbol * tends to move downward. This indicates that, when the data are severely contaminated by outliers, the MDPDE with large α performs better.

Table 2. Sample mean, variance, and MSE of estimators when $\theta = (1, 0.2, 0.1, 0.2, 0.5, 0.3, 0.4, 0.2, 0.5)^T$, $n = 1000$, and $(p, \gamma) = (0.03, 5)$.

α		$\hat{\omega}_1$	$\hat{a}_1$	$\hat{b}_{11}$	$\hat{b}_{12}$	$\hat{\omega}_2$	$\hat{a}_2$	$\hat{b}_{21}$	$\hat{b}_{22}$	$\hat{\delta}$
0(CMLE)	Mean	1.073	0.266	0.077	0.167	0.650	0.339	0.325	0.176	0.728
	Var × 10^2	6.363	1.707	0.109	0.105	2.333	0.553	0.168	0.115	17.16
	MSE × 10^2	6.897	2.140	0.160	0.213	4.577	0.704	0.736	0.170	22.36
0.1	Mean	1.028	0.264	0.080	0.170	0.607	0.335	0.331	0.179	0.697
	Var × 10^2	5.299	1.510	0.098	0.097	2.040	0.512	0.160	0.108	17.23
	MSE × 10^2	5.375	1.915	0.139	0.188	3.185	0.635	0.636	0.151	21.09
0.2	Mean	1.008	0.261	0.081	0.171	0.587	0.331	0.335	0.181	0.679
	Var × 10^2	5.114	1.491	0.098	0.097	2.031	0.526	0.165	0.110	17.70
	MSE × 10^2	5.116 *	1.855	0.133	0.179	2.789	0.621 *	0.583	0.147 *	20.87 *
0.3	Mean	1.000	0.257	0.083	0.172	0.578	0.327	0.339	0.182	0.662
	Var × 10^2	5.182	1.526	0.101	0.100	2.099	0.558	0.177	0.115	18.34
	MSE × 10^2	5.177	1.846 *	0.131 *	0.177 *	2.701 *	0.628	0.548	0.148	20.95
0.5	Mean	0.997	0.248	0.086	0.174	0.572	0.317	0.346	0.184	0.633
	Var × 10^2	5.729	1.682	0.114	0.116	2.381	0.658	0.220	0.136	20.02
	MSE × 10^2	5.724	1.910	0.134	0.183	2.899	0.686	0.516	0.162	21.77
1	Mean	1.007	0.230	0.094	0.179	0.578	0.296	0.363	0.191	0.587
	Var × 10^2	7.297	2.213	0.166	0.168	3.435	0.965	0.315	0.205	29.90
	MSE × 10^2	7.294	2.301	0.170	0.210	4.039	0.966	0.449 *	0.214	30.62

Table 3. Sample mean, variance, and MSE of estimators when $\theta = (1, 0.2, 0.1, 0.2, 0.5, 0.3, 0.4, 0.2, 0.5)^T$, $n = 1000$, and $(p, \gamma) = (0.03, 10)$.

α		$\hat{\omega}_1$	$\hat{a}_1$	$\hat{b}_{11}$	$\hat{b}_{12}$	$\hat{\omega}_2$	$\hat{a}_2$	$\hat{b}_{21}$	$\hat{b}_{22}$	$\hat{\delta}$
0(CMLE)	Mean	1.141	0.349	0.052	0.123	0.846	0.398	0.230	0.141	1.113
	Var × 10^2	16.43	3.478	0.101	0.140	5.886	1.087	0.265	0.138	21.91
	MSE × 10^2	18.39	5.702	0.335	0.736	17.88	2.051	3.138	0.487	59.51
0.1	Mean	1.015	0.329	0.057	0.131	0.706	0.382	0.248	0.150	0.865
	Var × 10^2	7.844	2.031	0.069	0.095	3.087	0.672	0.224	0.100	19.42
	MSE × 10^2	7.860	3.703	0.250	0.566	7.329	1.348	2.523	0.355	32.72
0.2	Mean	0.995	0.314	0.060	0.134	0.680	0.365	0.259	0.153	0.802
	Var × 10^2	7.073	1.948	0.068	0.095	2.912	0.677	0.244	0.104	19.42
	MSE × 10^2	7.068	3.252	0.225	0.529	6.156 *	1.105	2.245	0.321	28.54
0.3	Mean	1.002	0.298	0.064	0.137	0.681	0.349	0.269	0.157	0.765
	Var × 10^2	6.995	1.972	0.075	0.102	3.030	0.742	0.280	0.114	19.94
	MSE × 10^2	6.989 *	2.936	0.207	0.499	6.287	0.977	2.005	0.301	26.92
0.5	Mean	1.034	0.264	0.072	0.145	0.695	0.314	0.293	0.165	0.706
	Var × 10^2	7.365	2.137	0.097	0.125	3.415	0.913	0.382	0.146	21.81
	MSE × 10^2	7.475	2.545	0.176	0.430	7.223	0.932 *	1.536	0.266 *	26.01 *
1	Mean	1.088	0.198	0.095	0.167	0.719	0.242	0.353	0.191	0.604
	Var × 10^2	7.825	2.377	0.171	0.203	4.553	1.273	0.601	0.258	30.55
	MSE × 10^2	8.592	2.375 *	0.173 *	0.309 *	9.328	1.611	0.818 *	0.267	31.61

Table 4. Sample mean, variance, and MSE of estimators when $\theta = (1, 0.2, 0.1, 0.2, 0.5, 0.3, 0.4, 0.2, 0.5)^T$, $n = 1000$, and $(p, \gamma) = (0.05, 10)$.

α		$\hat{\omega}_1$	$\hat{a}_1$	$\hat{b}_{11}$	$\hat{b}_{12}$	$\hat{\omega}_2$	$\hat{a}_2$	$\hat{b}_{21}$	$\hat{b}_{22}$	$\hat{\delta}$
0(CMLE)	Mean	1.223	0.404	0.040	0.093	0.990	0.449	0.167	0.114	1.635
	Var × 10^2	28.47	4.763	0.086	0.128	11.74	1.691	0.229	0.131	29.21
	MSE × 10^2	33.40	8.909	0.442	1.281	35.70	3.897	5.645	0.867	158.1
0.1	Mean	1.012	0.390	0.046	0.103	0.772	0.437	0.185	0.125	1.057
	Var × 10^2	11.78	2.695	0.056	0.083	4.883	0.952	0.188	0.095	21.44
	MSE × 10^2	11.78	6.291	0.349	1.031	12.27	2.820	4.823	0.661	52.48
0.2	Mean	0.967	0.377	0.048	0.105	0.724	0.421	0.192	0.128	0.935
	Var × 10^2	9.531	2.414	0.052	0.080	4.163	0.896	0.203	0.093	20.74
	MSE × 10^2	9.633	5.529	0.324	0.986	9.168	2.359	4.525	0.608	39.63
0.3	Mean	0.971	0.361	0.050	0.107	0.720	0.405	0.199	0.131	0.879
	Var × 10^2	9.450	2.465	0.055	0.086	4.189	0.962	0.236	0.101	20.90
	MSE × 10^2	9.526 *	5.040	0.308	0.953	9.029 *	2.068	4.296	0.578	35.21
0.5	Mean	1.004	0.327	0.056	0.113	0.741	0.369	0.217	0.138	0.801
	Var × 10^2	9.878	2.724	0.071	0.112	4.689	1.209	0.363	0.132	22.32
	MSE × 10^2	9.870	4.336	0.269	0.861	10.51	1.687 *	3.700	0.511	31.33 *
1	Mean	1.102	0.229	0.084	0.142	0.807	0.257	0.300	0.170	0.651
	Var × 10^2	10.28	3.134	0.183	0.238	5.959	1.804	0.946	0.304	30.79
	MSE × 10^2	11.32	3.214 *	0.208 *	0.574 *	15.35	1.990	1.936 *	0.392 *	33.03

We also consider smaller sample size $n = 200$. The results are presented in Tables 5–8 and they show results similar to those in Tables 1–4. The variances and MSEs of both the CMLE and MDPDE are larger than those in Tables 1–4.

Table 5. Sample mean, variance, and MSE of estimators when $\theta = (1, 0.2, 0.1, 0.2, 0.5, 0.3, 0.4, 0.2, 0.5)^T$, $n = 200$, and no outliers exist.

α		$\hat{\omega}_1$	$\hat{a}_1$	$\hat{b}_{11}$	$\hat{b}_{12}$	$\hat{\omega}_2$	$\hat{a}_2$	$\hat{b}_{21}$	$\hat{b}_{22}$	$\hat{\delta}$
0(CMLE)	Mean	1.005	0.208	0.089	0.199	0.541	0.281	0.411	0.195	0.893
	Var × 10²	12.41	3.866	0.426	0.394	7.816	2.078	0.651	0.553	71.23
	MSE × 10²	12.40 *	3.869 *	0.437 *	0.394 *	7.973 *	2.112 *	0.663 *	0.555 *	86.57
0.1	Mean	0.975	0.203	0.087	0.193	0.529	0.271	0.400	0.191	0.786
	Var × 10²	14.98	3.919	0.439	0.498	8.317	2.212	1.097	0.649	59.99
	MSE × 10²	15.03	3.916	0.455	0.502	8.392	2.296	1.096	0.658	68.12
0.2	Mean	0.970	0.203	0.087	0.192	0.527	0.267	0.400	0.191	0.756
	Var × 10²	15.48	3.965	0.458	0.520	8.672	2.292	1.176	0.687	60.78
	MSE × 10²	15.55	3.962	0.473	0.526	8.734	2.396	1.174	0.695	67.27 *
0.3	Mean	0.962	0.204	0.088	0.191	0.525	0.263	0.400	0.191	0.730
	Var × 10²	16.41	4.166	0.477	0.555	9.040	2.366	1.274	0.734	63.50
	MSE × 10²	16.54	4.163	0.492	0.563	9.096	2.497	1.273	0.741	68.71
0.5	Mean	0.945	0.202	0.088	0.188	0.521	0.254	0.398	0.192	0.685
	Var × 10²	18.64	4.513	0.527	0.653	10.34	2.653	1.561	0.873	70.39
	MSE × 10²	18.93	4.509	0.540	0.666	10.38	2.863	1.560	0.879	73.75
1	Mean	0.968	0.209	0.102	0.204	0.537	0.249	0.433	0.213	0.684
	Var × 10²	18.37	5.307	0.757	0.817	11.99	3.117	1.327	1.159	135.3
	MSE × 10²	18.45	5.310	0.757	0.817	12.12	3.374	1.437	1.175	138.5

Table 6. Sample mean, variance and MSE of estimators when $\theta = (1, 0.2, 0.1, 0.2, 0.5, 0.3, 0.4, 0.2, 0.5)^T$, $n = 200$, and $(p, \gamma) = (0.03, 5)$.

α		$\hat{\omega}_1$	$\hat{a}_1$	$\hat{b}_{11}$	$\hat{b}_{12}$	$\hat{\omega}_2$	$\hat{a}_2$	$\hat{b}_{21}$	$\hat{b}_{22}$	$\hat{\delta}$
0(CMLE)	Mean	1.056	0.276	0.077	0.164	0.662	0.324	0.339	0.173	1.054
	Var × 10²	20.83	5.501	0.419	0.489	12.34	2.785	0.918	0.619	80.49
	MSE × 10²	21.13	6.078	0.471	0.616 *	14.94	2.839	1.292 *	0.690 *	111.2
0.1	Mean	0.992	0.262	0.077	0.163	0.605	0.311	0.334	0.171	0.925
	Var × 10²	20.38	5.118	0.411	0.510	11.65	2.801	1.153	0.641	67.19
	MSE × 10²	20.37	5.496	0.463 *	0.648	12.75	2.810 *	1.581	0.724	85.15
0.2	Mean	0.973	0.253	0.079	0.165	0.585	0.305	0.338	0.172	0.882
	Var × 10²	19.71	4.993	0.422	0.525	11.55	2.817	1.207	0.652	68.88
	MSE × 10²	19.76	5.265	0.465	0.645	12.26 *	2.816	1.594	0.730	83.37
0.3	Mean	0.958	0.247	0.081	0.165	0.577	0.296	0.340	0.172	0.840
	Var × 10²	19.93	5.028	0.445	0.563	12.33	2.962	1.321	0.690	70.67
	MSE × 10²	20.09	5.244	0.483	0.682	12.90	2.961	1.681	0.766	82.17 *
0.5	Mean	0.944	0.234	0.084	0.167	0.572	0.281	0.344	0.174	0.774
	Var × 10²	20.94	5.080	0.503	0.647	13.53	3.241	1.574	0.806	78.15
	MSE × 10²	21.23	5.193 *	0.528	0.756	14.04	3.273	1.885	0.873	85.55
1	Mean	0.960	0.236	0.101	0.187	0.592	0.266	0.388	0.198	0.770
	Var × 10²	19.00	5.571	0.755	0.859	15.57	3.851	1.689	1.119	147.0
	MSE × 10²	19.14 *	5.696	0.754	0.876	16.40	3.962	1.702	1.119	154.2

Table 7. Sample mean, variance, and MSE of estimators when $\theta = (1, 0.2, 0.1, 0.2, 0.5, 0.3, 0.4, 0.2, 0.5)^T$, $n = 200$, and $(p, \gamma) = (0.03, 10)$.

α		$\hat{\omega}_1$	$\hat{a}_1$	$\hat{b}_{11}$	$\hat{b}_{12}$	$\hat{\omega}_2$	$\hat{a}_2$	$\hat{b}_{21}$	$\hat{b}_{22}$	$\hat{\delta}$
0(CMLE)	Mean	1.128	0.349	0.052	0.126	0.860	0.388	0.241	0.135	1.365
	Var × 10²	38.79	8.126	0.345	0.618	26.05	4.690	1.145	0.746	96.45
	MSE × 10²	40.38	10.33	0.574	1.161	38.95	5.467	3.659	1.174	171.2
0.1	Mean	1.003	0.314	0.054	0.128	0.715	0.355	0.250	0.141	1.050
	Var × 10²	28.42	6.643	0.269	0.507	16.99	3.644	1.158	0.616	69.06
	MSE × 10²	28.39	7.925	0.480	1.021	21.59	3.938	3.403	0.961	99.19
0.2	Mean	0.980	0.296	0.057	0.130	0.679	0.337	0.258	0.146	0.953
	Var × 10²	26.04	6.348	0.270	0.505	15.82	3.612	1.262	0.628	67.71
	MSE × 10²	26.05	7.268	0.455 *	0.991	19.02	3.749	3.268	0.914	88.19
0.3	Mean	0.972	0.289	0.060	0.133	0.678	0.320	0.270	0.151	0.893
	Var × 10²	25.20	6.357	0.299	0.535	15.69	3.649	1.407	0.660	69.05
	MSE × 10²	25.26	7.142	0.457	0.987 *	18.84 *	3.683 *	3.096	0.894 *	84.43
0.5	Mean	0.974	0.264	0.070	0.139	0.673	0.287	0.294	0.160	0.783
	Var × 10²	24.72	6.143	0.399	0.643	16.00	3.836	1.847	0.794	75.64
	MSE × 10²	24.76	6.548	0.490	1.019	18.96	3.848	2.963	0.953	83.56 *
1	Mean	1.007	0.232	0.100	0.171	0.677	0.235	0.374	0.200	0.657
	Var × 10²	21.91	6.221	0.778	1.007	16.89	3.717	2.460	1.238	130.0
	MSE × 10²	21.89 *	6.319 *	0.777	1.088	20.01	4.133	2.526 *	1.237	132.3

Table 8. Sample mean, variance, and MSE of estimators when $\theta = (1, 0.2, 0.1, 0.2, 0.5, 0.3, 0.4, 0.2, 0.5)^T$, $n = 200$, and $(p, \gamma) = (0.05, 10)$.

α		$\hat{\omega}_1$	$\hat{a}_1$	$\hat{b}_{11}$	$\hat{b}_{12}$	$\hat{\omega}_2$	$\hat{a}_2$	$\hat{b}_{21}$	$\hat{b}_{22}$	$\hat{\delta}$
0(CMLE)	Mean	1.171	0.406	0.046	0.097	1.041	0.420	0.183	0.108	1.814
	Var $\times 10^2$	53.26	9.255	0.326	0.521	45.18	6.131	1.054	0.654	133.0
	MSE $\times 10^2$	56.14	13.48	0.619	1.572	74.38	7.569	5.761	1.504	305.4
0.1	Mean	1.037	0.347	0.047	0.102	0.821	0.389	0.192	0.117	1.203
	Var $\times 10^2$	36.17	7.578	0.227	0.430	26.89	4.810	1.034	0.549	80.49
	MSE $\times 10^2$	36.28	9.719	0.509	1.388	37.17	5.600	5.372	1.244	129.8
0.2	Mean	0.989	0.334	0.049	0.104	0.772	0.370	0.199	0.122	1.064
	Var $\times 10^2$	31.43	7.373	0.218	0.421	23.29	4.607	1.106	0.554	77.26
	MSE $\times 10^2$	31.41	9.171	0.477	1.344	30.69	5.097	5.144	1.156	108.9
0.3	Mean	0.989	0.320	0.051	0.106	0.762	0.355	0.207	0.126	0.984
	Var $\times 10^2$	30.35	7.338	0.234	0.443	22.64	4.685	1.247	0.602	76.99
	MSE $\times 10^2$	30.33	8.773	0.472 *	1.327	29.47	4.985	4.985	1.149 *	100.4
0.5	Mean	0.984	0.293	0.058	0.112	0.764	0.314	0.229	0.135	0.855
	Var $\times 10^2$	30.12	7.263	0.332	0.558	22.81	4.884	1.791	0.781	80.40
	MSE $\times 10^2$	30.12	8.122	0.505	1.331	29.73	4.897	4.726	1.206	92.95 *
1	Mean	1.046	0.239	0.097	0.151	0.774	0.243	0.333	0.178	0.696
	Var $\times 10^2$	23.99	6.497	0.805	1.059	21.95	4.517	3.261	1.366	136.2
	MSE $\times 10^2$	24.17 *	6.645 *	0.805	1.302 *	29.46 *	4.839 *	3.708 *	1.413	139.9

In order to evaluate the performance of the MDPDE for negatively cross-correlated data, we consider $\theta = (\omega_1, a_1, b_{11}, b_{12}, \omega_2, a_2, b_{21}, b_{22}, \delta)^T = (0.5, 0.1, 0.2, 0.4, 0.3, 0.3, 0.2, 0.1, -0.4)^T$ with the same p and γ, as above. The results are reported in Tables 9–16 for $n = 1000$ and 200, respectively. These tables exhibit results that are similar to those in Tables 1–8. Overall, our findings strongly support the assertion that the MDPDE is a functional tool for yielding a robust estimator for bivariate Poisson INGARCH models in the presence of outliers.

Table 9. Sample mean, variance, and MSE of estimators when $\theta = (0.5, 0.1, 0.2, 0.4, 0.3, 0.3, 0.2, 0.1, -0.4)^T$, $n = 1000$, and no outliers exist.

α		$\hat{\omega}_1$	$\hat{a}_1$	$\hat{b}_{11}$	$\hat{b}_{12}$	$\hat{\omega}_2$	$\hat{a}_2$	$\hat{b}_{21}$	$\hat{b}_{22}$	$\hat{\delta}$
0(CMLE)	Mean	0.501	0.103	0.199	0.397	0.306	0.296	0.200	0.098	−0.385
	Var $\times 10^2$	0.570	0.508	0.105	0.159	0.518	1.029	0.080	0.108	6.231
	MSE $\times 10^2$	0.569 *	0.508 *	0.105 *	0.160 *	0.522 *	1.030 *	0.080 *	0.109 *	6.247 *
0.1	Mean	0.501	0.103	0.199	0.397	0.306	0.295	0.200	0.098	−0.384
	Var $\times 10^2$	0.578	0.515	0.107	0.160	0.530	1.040	0.082	0.111	6.347
	MSE $\times 10^2$	0.578	0.515	0.108	0.161	0.534	1.041	0.082	0.112	6.367
0.2	Mean	0.501	0.103	0.199	0.397	0.307	0.295	0.200	0.098	−0.383
	Var $\times 10^2$	0.600	0.532	0.113	0.166	0.556	1.082	0.086	0.117	6.564
	MSE $\times 10^2$	0.600	0.533	0.113	0.167	0.560	1.083	0.086	0.117	6.588
0.3	Mean	0.501	0.104	0.199	0.397	0.307	0.294	0.200	0.098	−0.381
	Var $\times 10^2$	0.627	0.554	0.119	0.175	0.591	1.145	0.092	0.124	6.848
	MSE $\times 10^2$	0.627	0.555	0.119	0.176	0.595	1.147	0.092	0.125	6.876
0.5	Mean	0.500	0.105	0.198	0.398	0.308	0.292	0.201	0.099	−0.380
	Var $\times 10^2$	0.702	0.615	0.137	0.199	0.685	1.320	0.106	0.142	7.577
	MSE $\times 10^2$	0.701	0.617	0.137	0.200	0.690	1.325	0.106	0.142	7.610
1	Mean	0.495	0.110	0.198	0.399	0.310	0.287	0.203	0.100	−0.382
	Var $\times 10^2$	0.972	0.839	0.201	0.290	0.942	1.864	0.155	0.195	10.09
	MSE $\times 10^2$	0.974	0.848	0.201	0.290	0.951	1.878	0.156	0.194	10.12

Table 10. Sample mean, variance, and MSE of estimators when $\theta = (0.5, 0.1, 0.2, 0.4, 0.3, 0.3, 0.2, 0.1, -0.4)^T$, $n = 1000$, and $(p, \gamma) = (0.03, 5)$.

α		$\hat{\omega}_1$	$\hat{a}_1$	$\hat{b}_{11}$	$\hat{b}_{12}$	$\hat{\omega}_2$	$\hat{a}_2$	$\hat{b}_{21}$	$\hat{b}_{22}$	$\hat{\delta}$
0(CMLE)	Mean	0.633	0.194	0.143	0.269	0.399	0.368	0.147	0.064	−0.097
	Var $\times 10^2$	1.794	1.343	0.152	0.263	1.741	2.315	0.125	0.123	5.603
	MSE $\times 10^2$	3.560	2.219	0.474	1.974	2.728	2.769	0.409	0.255	14.79
0.1	Mean	0.572	0.186	0.149	0.280	0.350	0.360	0.153	0.067	−0.143
	Var $\times 10^2$	1.191	1.013	0.126	0.235	1.047	1.659	0.100	0.094	5.787
	MSE $\times 10^2$	1.711	1.743	0.390	1.676	1.297	2.016	0.325	0.205	12.38
0.2	Mean	0.550	0.177	0.151	0.286	0.335	0.350	0.155	0.068	−0.169
	Var $\times 10^2$	1.082	0.958	0.124	0.240	0.950	1.608	0.100	0.090	6.076
	MSE $\times 10^2$	1.335	1.543	0.361	1.536	1.074 *	1.861 *	0.305	0.191	11.43
0.3	Mean	0.543	0.167	0.154	0.292	0.333	0.340	0.156	0.070	−0.187
	Var $\times 10^2$	1.055	0.950	0.129	0.254	0.976	1.706	0.107	0.095	6.375
	MSE $\times 10^2$	1.241	1.401	0.344	1.427	1.083	1.868	0.297	0.184	10.89
0.5	Mean	0.542	0.148	0.159	0.304	0.339	0.318	0.161	0.075	−0.214
	Var $\times 10^2$	1.050	0.951	0.147	0.290	1.118	2.017	0.125	0.113	7.038
	MSE $\times 10^2$	1.229 *	1.185	0.315	1.203	1.270	2.049	0.279	0.175 *	10.49 *
1	Mean	0.548	0.112	0.176	0.340	0.360	0.268	0.176	0.090	−0.247
	Var $\times 10^2$	1.136	0.953	0.214	0.399	1.425	2.636	0.188	0.184	9.324
	MSE $\times 10^2$	1.363	0.966 *	0.271 *	0.756 *	1.783	2.733	0.244 *	0.194	11.64

Table 11. Sample mean, variance, and MSE of estimators when $\theta = (0.5, 0.1, 0.2, 0.4, 0.3, 0.3, 0.2, 0.1, -0.4)^T$, $n = 1000$, and $(p, \gamma) = (0.03, 10)$.

α		$\hat{\omega}_1$	$\hat{a}_1$	$\hat{b}_{11}$	$\hat{b}_{12}$	$\hat{\omega}_2$	$\hat{a}_2$	$\hat{b}_{21}$	$\hat{b}_{22}$	$\hat{\delta}$
0(CMLE)	Mean	0.774	0.295	0.087	0.149	0.525	0.415	0.094	0.037	0.336
	Var × 10²	7.976	4.279	0.176	0.346	7.305	6.070	0.187	0.103	7.338
	MSE × 10²	15.50	8.069	1.442	6.644	12.37	7.392	1.303	0.501	61.46
0.1	Mean	0.612	0.254	0.100	0.173	0.373	0.399	0.106	0.040	0.019
	Var × 10²	2.368	2.008	0.104	0.287	1.694	2.548	0.105	0.049	6.426
	MSE × 10²	3.628	4.369	1.109	5.441	2.231	3.521	0.982	0.406	23.94
0.2	Mean	0.596	0.227	0.103	0.182	0.364	0.378	0.108	0.042	−0.035
	Var × 10²	2.029	1.866	0.106	0.336	1.567	2.548	0.107	0.050	6.677
	MSE × 10²	2.944	3.490	1.048	5.081	1.971 *	3.160	0.949	0.384	20.01
0.3	Mean	0.600	0.203	0.108	0.195	0.372	0.355	0.112	0.046	−0.069
	Var × 10²	1.894	1.823	0.122	0.425	1.601	2.697	0.123	0.060	6.909
	MSE × 10²	2.884	2.873	0.973	4.637	2.111	2.997	0.889	0.353	17.86
0.5	Mean	0.611	0.145	0.125	0.240	0.401	0.287	0.130	0.061	−0.135
	Var × 10²	1.518	1.521	0.177	0.691	1.608	2.883	0.181	0.107	7.489
	MSE × 10²	2.744	1.725	0.742	3.259	2.619	2.898 *	0.674	0.259	14.50
1	Mean	0.594	0.059	0.178	0.360	0.440	0.155	0.185	0.107	−0.249
	Var × 10²	0.941	0.570	0.268	0.634	1.291	2.214	0.278	0.216	9.962
	MSE × 10²	1.828 *	0.737 *	0.316 *	0.794 *	3.262	4.327	0.301 *	0.220 *	12.22 *

Table 12. Sample mean, variance, and MSE of estimators when $\theta = (0.5, 0.1, 0.2, 0.4, 0.3, 0.3, 0.2, 0.1, -0.4)^T$, $n = 1000$, and $(p, \gamma) = (0.05, 10)$.

α		$\hat{\omega}_1$	$\hat{a}_1$	$\hat{b}_{11}$	$\hat{b}_{12}$	$\hat{\omega}_2$	$\hat{a}_2$	$\hat{b}_{21}$	$\hat{b}_{22}$	$\hat{\delta}$
0(CMLE)	Mean	0.870	0.382	0.059	0.086	0.645	0.451	0.062	0.027	0.829
	Var × 10²	17.32	6.408	0.129	0.206	14.69	8.150	0.135	0.073	8.777
	MSE × 10²	30.96	14.37	2.130	10.09	26.58	10.41	2.026	0.604	159.9
0.1	Mean	0.621	0.346	0.070	0.103	0.396	0.446	0.074	0.029	0.170
	Var × 10²	4.575	3.255	0.073	0.158	2.948	3.708	0.076	0.036	6.635
	MSE × 10²	6.034	9.327	1.762	8.971	3.862	5.837	1.659	0.545	39.12
0.2	Mean	0.585	0.327	0.070	0.104	0.370	0.431	0.073	0.029	0.089
	Var × 10²	3.641	2.988	0.065	0.164	2.294	3.311	0.068	0.031	6.911
	MSE × 10²	4.360	8.156	1.749	8.898	2.788 *	5.022	1.670	0.540	30.79
0.3	Mean	0.586	0.311	0.071	0.107	0.374	0.417	0.074	0.029	0.058
	Var × 10²	3.517	3.054	0.072	0.193	2.353	3.531	0.074	0.033	7.089
	MSE × 10²	4.249	7.500	1.727	8.805	2.893	4.895	1.661	0.532	28.06
0.5	Mean	0.608	0.265	0.080	0.124	0.399	0.371	0.083	0.035	0.016
	Var × 10²	3.465	3.335	0.114	0.398	2.559	4.119	0.120	0.055	7.515
	MSE × 10²	4.628	6.044	1.555	8.030	3.537	4.613 *	1.492	0.482	24.83
1	Mean	0.637	0.087	0.148	0.296	0.481	0.161	0.153	0.096	−0.144
	Var × 10²	1.536	1.591	0.410	1.613	1.732	3.105	0.408	0.306	9.724
	MSE × 10²	3.424 *	1.606 *	0.682 *	2.695 *	4.999	5.042	0.626 *	0.308 *	16.28 *

Table 13. Sample mean, variance, and MSE of estimators when $\theta = (0.5, 0.1, 0.2, 0.4, 0.3, 0.3, 0.2, 0.1, -0.4)^T$, $n = 200$, and no outliers exist.

α		$\hat{\omega}_1$	$\hat{a}_1$	$\hat{b}_{11}$	$\hat{b}_{12}$	$\hat{\omega}_2$	$\hat{a}_2$	$\hat{b}_{21}$	$\hat{b}_{22}$	$\hat{\delta}$
0(CMLE)	Mean	0.487	0.131	0.182	0.394	0.316	0.287	0.203	0.092	−0.313
	Var × 10²	2.173	2.095	0.526	0.806	2.213	4.245	0.411	0.475	33.35
	MSE × 10²	2.187 *	2.186 *	0.558 *	0.809 *	2.237 *	4.257 *	0.412 *	0.481 *	34.07
0.1	Mean	0.483	0.129	0.181	0.390	0.314	0.284	0.202	0.091	−0.294
	Var × 10²	2.391	2.104	0.571	0.958	2.337	4.358	0.452	0.487	31.43
	MSE × 10²	2.416	2.188	0.609	0.967	2.353	4.379	0.452	0.495	32.53 *
0.2	Mean	0.481	0.131	0.180	0.388	0.312	0.283	0.202	0.090	−0.285
	Var × 10²	2.542	2.158	0.608	1.040	2.427	4.490	0.484	0.512	31.25
	MSE × 10²	2.577	2.250	0.649	1.054	2.439	4.513	0.483	0.520	32.55
0.3	Mean	0.479	0.134	0.180	0.388	0.311	0.284	0.202	0.091	−0.284
	Var × 10²	2.612	2.225	0.636	1.069	2.531	4.657	0.503	0.537	32.24
	MSE × 10²	2.653	2.337	0.677	1.082	2.541	4.679	0.503	0.545	33.55
0.5	Mean	0.477	0.137	0.180	0.389	0.313	0.280	0.204	0.092	−0.285
	Var × 10²	2.860	2.423	0.726	1.183	2.681	4.733	0.554	0.601	35.16
	MSE × 10²	2.909	2.555	0.766	1.194	2.695	4.769	0.555	0.606	36.43
1	Mean	0.473	0.145	0.185	0.399	0.314	0.276	0.212	0.100	−0.324
	Var × 10²	3.364	3.057	1.016	1.549	3.003	5.321	0.746	0.858	48.77
	MSE × 10²	3.434	3.252	1.038	1.548	3.021	5.375	0.760	0.857	49.31

Table 14. Sample mean, variance, and MSE of estimators when $\theta = (0.5, 0.1, 0.2, 0.4, 0.3, 0.3, 0.2, 0.1, -0.4)^T$, $n = 200$, and $(p, \gamma) = (0.03, 5)$.

α		$\hat{\omega}_1$	$\hat{a}_1$	$\hat{b}_{11}$	$\hat{b}_{12}$	$\hat{\omega}_2$	$\hat{a}_2$	$\hat{b}_{21}$	$\hat{b}_{22}$	$\hat{\delta}$
0(CMLE)	Mean	0.611	0.215	0.138	0.270	0.396	0.349	0.157	0.068	−0.056
	Var × 10²	6.324	4.595	0.661	1.242	5.415	6.896	0.600	0.472	28.97
	MSE × 10²	7.555	5.916	1.050	2.927	6.330	7.124	0.786	0.574	40.76
0.1	Mean	0.561	0.202	0.141	0.278	0.348	0.345	0.160	0.068	−0.086
	Var × 10²	4.635	3.815	0.593	1.110	3.799	5.860	0.497	0.380	28.62
	MSE × 10²	5.004	4.853	0.942	2.597	4.023	6.053	0.653	0.483	38.44
0.2	Mean	0.537	0.192	0.142	0.282	0.329	0.338	0.161	0.068	−0.099
	Var × 10²	4.374	3.640	0.598	1.175	3.512	5.797	0.497	0.367	28.89
	MSE × 10²	4.504	4.487	0.930 *	2.562	3.592	5.933 *	0.649 *	0.468 *	37.90 *
0.3	Mean	0.526	0.187	0.144	0.287	0.325	0.330	0.162	0.070	−0.115
	Var × 10²	4.313	3.636	0.619	1.264	3.494	5.913	0.517	0.383	29.94
	MSE × 10²	4.377	4.383	0.932	2.529 *	3.553 *	5.998	0.660	0.472	38.02
0.5	Mean	0.516	0.177	0.149	0.300	0.329	0.312	0.166	0.075	−0.141
	Var × 10²	4.305	3.602	0.689	1.529	3.657	6.121	0.572	0.454	33.05
	MSE × 10²	4.327 *	4.188	0.950	2.532	3.739	6.128	0.689	0.514	39.75
1	Mean	0.503	0.162	0.167	0.340	0.346	0.272	0.183	0.092	−0.194
	Var × 10²	4.432	3.750	0.989	2.266	3.911	6.436	0.772	0.726	47.97
	MSE × 10²	4.428	4.130 *	1.100	2.619	4.119	6.507	0.799	0.731	52.14

Table 15. Sample mean, variance, and MSE of estimators when $\theta = (0.5, 0.1, 0.2, 0.4, 0.3, 0.3, 0.2, 0.1, -0.4)^T$, $n = 200$, and $(p, \gamma) = (0.03, 10)$.

α		$\hat{\omega}_1$	$\hat{a}_1$	$\hat{b}_{11}$	$\hat{b}_{12}$	$\hat{\omega}_2$	$\hat{a}_2$	$\hat{b}_{21}$	$\hat{b}_{22}$	$\hat{\delta}$
0(CMLE)	Mean	0.736	0.312	0.084	0.165	0.497	0.416	0.103	0.047	0.291
	Var × 10²	16.70	8.406	0.635	1.549	12.16	10.20	0.674	0.401	33.37
	MSE × 10²	22.23	12.91	1.971	7.078	16.04	11.54	1.610	0.680	81.11
0.1	Mean	0.600	0.264	0.092	0.181	0.381	0.368	0.110	0.047	0.032
	Var × 10²	8.133	5.973	0.445	1.314	5.185	7.196	0.442	0.233	27.95
	MSE × 10²	9.120	8.669	1.613	6.104	5.843	7.657	1.255	0.515	46.63
0.2	Mean	0.570	0.252	0.095	0.188	0.367	0.353	0.110	0.050	−0.012
	Var × 10²	7.112	5.817	0.451	1.441	4.576	6.859	0.446	0.244	28.91
	MSE × 10²	7.592	8.112	1.544	5.920	5.016	7.136	1.250	0.495 *	43.96
0.3	Mean	0.563	0.235	0.100	0.200	0.366	0.338	0.113	0.054	−0.046
	Var × 10²	6.572	5.489	0.513	1.736	4.477	6.853	0.502	0.294	29.07
	MSE × 10²	6.965	7.299	1.513	5.748	4.910	6.988	1.263	0.503	41.58
0.5	Mean	0.553	0.193	0.113	0.235	0.369	0.294	0.124	0.066	−0.101
	Var × 10²	6.273	4.985	0.736	2.594	4.586	7.030	0.689	0.441	29.86
	MSE × 10²	6.548	5.840	1.484	5.318	5.061	7.027	1.263	0.555	38.79 *
1	Mean	0.552	0.126	0.166	0.340	0.383	0.227	0.176	0.104	−0.262
	Var × 10²	4.097	3.239	1.193	3.141	3.787	6.138	1.084	0.816	46.47
	MSE × 10²	4.360 *	3.304 *	1.307 *	3.502 *	4.479 *	6.658 *	1.141 *	0.817	48.32

Table 16. Sample mean, variance, and MSE of estimators when $\theta = (0.5, 0.1, 0.2, 0.4, 0.3, 0.3, 0.2, 0.1, -0.4)^T$, $n = 200$, and $(p, \gamma) = (0.05, 10)$.

α		$\hat{\omega}_1$	$\hat{a}_1$	$\hat{b}_{11}$	$\hat{b}_{12}$	$\hat{\omega}_2$	$\hat{a}_2$	$\hat{b}_{21}$	$\hat{b}_{22}$	$\hat{\delta}$
0(CMLE)	Mean	0.829	0.378	0.064	0.102	0.613	0.435	0.081	0.037	0.749
	Var × 10²	27.18	10.13	0.521	0.889	18.60	10.72	0.611	0.307	39.05
	MSE × 10²	37.97	17.82	2.375	9.795	28.35	12.52	2.018	0.702	171.0
0.1	Mean	0.652	0.313	0.070	0.114	0.441	0.374	0.082	0.034	0.172
	Var × 10²	12.51	7.707	0.348	0.815	7.550	8.532	0.365	0.162	29.09
	MSE × 10²	14.81	12.21	2.040	9.010	9.521	9.078	1.748	0.592	61.80
0.2	Mean	0.612	0.294	0.071	0.117	0.417	0.354	0.080	0.035	0.097
	Var × 10²	9.876	7.005	0.317	0.871	6.395	8.221	0.332	0.147	30.13
	MSE × 10²	11.13	10.74	1.979	8.885	7.751	8.510	1.768	0.571	54.75
0.3	Mean	0.604	0.283	0.073	0.121	0.414	0.343	0.081	0.037	0.063
	Var × 10²	9.469	7.048	0.348	1.031	6.125	8.167	0.358	0.167	30.74
	MSE × 10²	10.54	10.38	1.970	8.819	7.414	8.347	1.771	0.567 *	52.11
0.5	Mean	0.607	0.241	0.085	0.151	0.420	0.310	0.091	0.048	−0.006
	Var × 10²	8.559	6.604	0.565	1.957	5.915	8.036	0.536	0.337	32.13
	MSE × 10²	9.688	8.590	1.881	8.142	7.350	8.038	1.713	0.608	47.63 *
1	Mean	0.600	0.135	0.146	0.292	0.425	0.220	0.152	0.097	−0.195
	Var × 10²	5.457	4.147	1.395	4.172	4.697	6.676	1.279	0.911	46.82
	MSE × 10²	6.453 *	4.268 *	1.690 *	5.343 *	6.263 *	7.302 *	1.508 *	0.910	50.98

4.2. Illustrative Examples

First, we illustrate the proposed method by examining the monthly count series of crimes provided by the New South Wales Police Force in Australia. The data set is classified by local government area and offence type. This data set has been studied in many literatures, including Lee et al. [9], Chen and Lee [38,39], Kim and Lee [24], and Lee et al. [40]. To investigate the behavior of the MDPDE in the presence of outliers, we consider the data series of liquor offences (LO) and transport regulatory offences (TRO) in Botany Bay from January 1995 to December 2012, which has 216 observations in each series. Figure 1 plots the monthly count series of LO and TRO and it shows the presence of some deviant observations in each series. The sample mean and variance are 1.912 and 13.14 for LO, and 2.463 and 20.41 for TRO. A large value of the variance of each series is expected to be influenced by outliers. The autocorrelation function (ACF) and partial autocorrelation function (PACF) of LO and TRO, as well as cross-correlation function (CCF) between two series, are given in Figure 2, indicating that the data are both serially and crossly correlated. The cross-correlation coefficient between two series is 0.141.

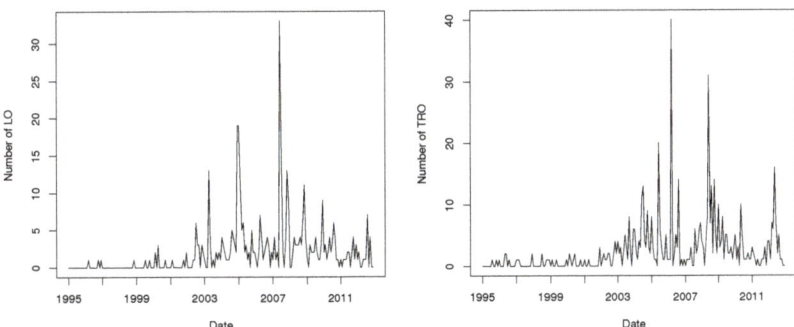

Figure 1. Monthly count series of liquor offences (LO) (**left**) and transport regulatory offences (TRO) (**right**) in Botany Bay.

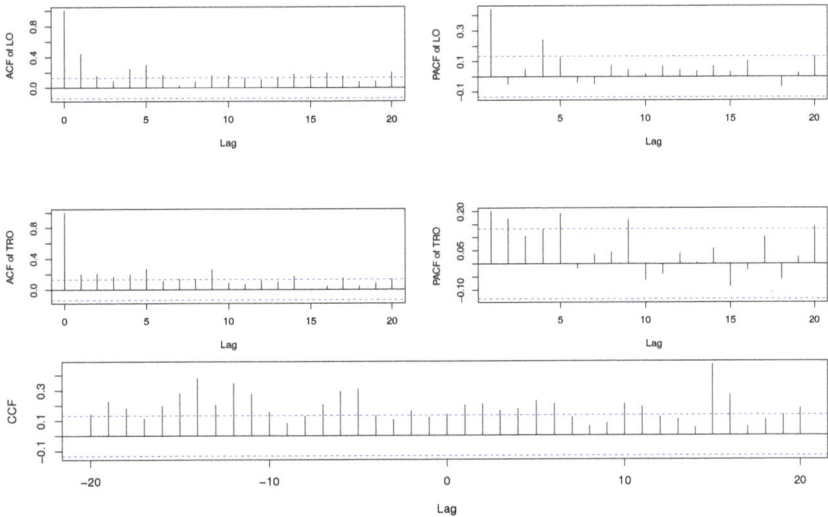

Figure 2. Autocorrelation function (ACF) and partial autocorrelation function (PACF) of LO (**top**) and TRO (**middle**), and cross-correlation function (CCF) (**bottom**) between two series.

We fit the model (2) to the data using both the CMLE and the MDPDE. $\tilde{\lambda}_1$ is set to be the sample mean of the data. Table 17 reports the estimated parameters with various α. The standard errors are given in parentheses and the symbol • represents the minimal $\widehat{AMSE}$ provided in Remark 2. In the table, we can observe that the MDPDE has smaller $\widehat{AMSE}$ than the CMLE and the optimal α is chosen to be 0.1. The MDPDE with optimal α is quite different from the CMLE, in particular, $\hat{\delta}$ is about half of the CMLE. This result indicates that outliers can seriously affect the parameter estimation and, thus, the robust estimation method is required when the data are contaminated by outliers.

Table 17. Parameter estimates for bivariate Poisson integer-valued generalized autoregressive conditional heteroscedastic (INGARCH) model for crime data; the symbol • represents the minimal $\widehat{\text{AMSE}}$.

α	$\hat{\omega}_1$	$\hat{a}_1$	$\hat{b}_{11}$	$\hat{b}_{12}$	$\hat{\omega}_2$	$\hat{a}_2$	$\hat{b}_{21}$	$\hat{b}_{22}$	$\hat{\delta}$	$\widehat{\text{AMSE}}$
0(CMLE)	0.019	0.779	0.125	0.073	0.032	0.865	0.090	0.057	1.312	1.578
	(0.054)	(0.290)	(0.166)	(0.075)	(0.028)	(0.090)	(0.032)	(0.086)	(1.096)	
0.1	0.034	0.609	0.172	0.094	0.097	0.654	0.095	0.156	0.685	0.699 •
	(0.034)	(0.149)	(0.104)	(0.026)	(0.047)	(0.091)	(0.043)	(0.069)	(0.678)	
0.2	0.026	0.643	0.134	0.087	0.117	0.575	0.124	0.159	0.509	0.858
	(0.032)	(0.163)	(0.109)	(0.026)	(0.060)	(0.121)	(0.052)	(0.069)	(0.692)	
0.3	0.021	0.666	0.113	0.085	0.129	0.523	0.154	0.155	0.401	0.991
	(0.029)	(0.149)	(0.096)	(0.027)	(0.067)	(0.133)	(0.053)	(0.068)	(0.710)	
0.4	0.019	0.673	0.107	0.085	0.130	0.508	0.176	0.145	0.356	1.081
	(0.029)	(0.143)	(0.093)	(0.029)	(0.067)	(0.135)	(0.055)	(0.069)	(0.736)	
0.5	0.018	0.675	0.105	0.086	0.125	0.514	0.196	0.131	0.365	1.108
	(0.029)	(0.138)	(0.093)	(0.032)	(0.065)	(0.136)	(0.059)	(0.071)	(0.768)	
0.6	0.017	0.676	0.104	0.088	0.119	0.527	0.216	0.115	0.418	1.094
	(0.029)	(0.133)	(0.091)	(0.036)	(0.062)	(0.135)	(0.065)	(0.073)	(0.807)	
0.7	0.017	0.675	0.104	0.089	0.114	0.540	0.238	0.100	0.509	1.073
	(0.029)	(0.130)	(0.092)	(0.041)	(0.059)	(0.133)	(0.075)	(0.075)	(0.859)	
0.8	0.018	0.674	0.104	0.090	0.111	0.551	0.261	0.087	0.638	1.079
	(0.031)	(0.130)	(0.094)	(0.045)	(0.057)	(0.133)	(0.089)	(0.076)	(0.929)	
0.9	0.018	0.672	0.104	0.091	0.109	0.560	0.285	0.076	0.808	1.158
	(0.033)	(0.133)	(0.098)	(0.050)	(0.056)	(0.134)	(0.105)	(0.077)	(1.021)	
1	0.019	0.668	0.104	0.092	0.108	0.568	0.312	0.066	1.025	1.383
	(0.035)	(0.138)	(0.103)	(0.054)	(0.057)	(0.136)	(0.122)	(0.079)	(1.143)	

We clean the data by using the approach that was introduced by Fokianos and Fried [41] and apply the CMLE and the MDPDE to this data in order to illustrate the behavior of the estimators in the absence of outliers. Table 18 reports the results. The standard errors and $\widehat{\text{AMSE}}$ tend to decrease compared to Table 17. The CMLE has minimal $\widehat{\text{AMSE}}$ and the MDPDE with small α appears to be similar to the CMLE.

Table 18. Parameter estimates for bivariate Poisson INGARCH model for cleaned data; the symbol • represents the minimal $\widehat{\text{AMSE}}$.

α	$\hat{\omega}_1$	$\hat{a}_1$	$\hat{b}_{11}$	$\hat{b}_{12}$	$\hat{\omega}_2$	$\hat{a}_2$	$\hat{b}_{21}$	$\hat{b}_{22}$	$\hat{\delta}$	$\widehat{\text{AMSE}}$
0(CMLE)	0.0018	0.943	0.025	0.021	0.092	0.682	0.067	0.184	0.118	0.430 •
	(0.007)	(0.118)	(0.084)	(0.017)	(0.050)	(0.115)	(0.069)	(0.069)	(0.609)	
0.1	0.0002	0.942	0.026	0.022	0.074	0.680	0.066	0.183	0.159	0.445
	(0.006)	(0.097)	(0.076)	(0.013)	(0.035)	(0.084)	(0.063)	(0.053)	(0.626)	
0.2	0.0001	0.940	0.025	0.023	0.066	0.679	0.066	0.182	0.199	0.497
	(0.009)	(0.100)	(0.079)	(0.011)	(0.032)	(0.075)	(0.063)	(0.049)	(0.657)	
0.3	0.0001	0.939	0.024	0.023	0.060	0.678	0.066	0.182	0.220	0.549
	(0.010)	(0.102)	(0.082)	(0.011)	(0.032)	(0.073)	(0.064)	(0.049)	(0.688)	
0.4	0.0001	0.939	0.024	0.023	0.057	0.678	0.066	0.182	0.228	0.591
	(0.010)	(0.104)	(0.086)	(0.011)	(0.032)	(0.074)	(0.066)	(0.049)	(0.715)	
0.5	0.0001	0.938	0.025	0.023	0.056	0.676	0.066	0.182	0.293	0.665
	(0.010)	(0.104)	(0.087)	(0.011)	(0.034)	(0.076)	(0.068)	(0.051)	(0.742)	
0.6	0.0001	0.938	0.024	0.023	0.054	0.677	0.066	0.182	0.263	0.688
	(0.009)	(0.102)	(0.088)	(0.011)	(0.035)	(0.077)	(0.071)	(0.053)	(0.769)	
0.7	0.0002	0.939	0.024	0.023	0.051	0.678	0.066	0.182	0.237	0.719
	(0.008)	(0.106)	(0.093)	(0.012)	(0.035)	(0.078)	(0.074)	(0.055)	(0.795)	
0.8	0.0002	0.940	0.015	0.027	0.053	0.678	0.067	0.179	0.100	0.812
	(0.011)	(0.126)	(0.093)	(0.013)	(0.038)	(0.081)	(0.076)	(0.057)	(0.873)	
0.9	0.0001	0.944	0.011	0.028	0.050	0.679	0.068	0.176	−0.029	0.924
	(0.012)	(0.150)	(0.108)	(0.015)	(0.039)	(0.083)	(0.080)	(0.058)	(0.933)	
1	0.0002	0.944	0.010	0.028	0.054	0.677	0.070	0.173	0.010	1.079
	(0.012)	(0.151)	(0.107)	(0.015)	(0.044)	(0.087)	(0.084)	(0.061)	(1.012)	

Now, we consider an artificial example that has negative cross-correlation coefficient. Following Cui and Zhu [20], we generate 1000 samples from univariate Poisson INGARCH model, i.e.,

$$X_t | \mathcal{F}_{t-1} \sim P(\lambda_t), \quad \lambda_t = 1 + 0.35\lambda_{t-1} + 0.45 X_{t-1},$$

where $P(\lambda_t)$ denotes the Poisson distribution with mean λ_t. Further, we observe the contaminated data $X_{c,t}$ as follows

$$X_{c,t} = X_t + P_t X_{o,t},$$

where P_t are i.i.d. Bernoulli random variables with a success probability of 0.03 and $X_{o,t}$ are i.i.d. Poisson random variables with mean 5. Let $Y_t = (Y_{t,1}, Y_{t,2})^T$, where $Y_{t,1} = X_{c,t}$ and $Y_{t,2} = X_{c,t+500}$ for $t = 1, \ldots, 500$. The sample mean and variance are 5.196 and 7.380 for $Y_{t,1}$, and 4.538 and 8.129 for $Y_{t,2}$. The cross-correlation coefficient between $Y_{t,1}$ and $Y_{t,2}$ is -0.161. We fit the model (2) to Y_t and the results are presented in Table 19. Similar to Table 17, the MDPDE has smaller $\widehat{AMSE}$ than the CMLE. The optimal α is chosen to be 0.3 and the corresponding $\hat{\delta}$ is -0.329, whereas the CMLE is 0.772.

Table 19. Parameter estimates for bivariate Poisson INGARCH model for artificial data; the symbol • represents the minimal $\widehat{AMSE}$.

α	$\hat{\omega}_1$	$\hat{a}_1$	$\hat{b}_{11}$	$\hat{b}_{12}$	$\hat{\omega}_2$	$\hat{a}_2$	$\hat{b}_{21}$	$\hat{b}_{22}$	$\hat{\delta}$	$\widehat{AMSE}$
0(CMLE)	1.507	0.274	0.438	0.000	0.976	0.410	0.000	0.375	0.772	9.468
	(0.442)	(0.102)	(0.053)	(0.031)	(0.241)	(0.069)	(0.029)	(0.048)	(2.939)	
0.1	1.442	0.274	0.449	0.000	0.952	0.412	0.000	0.372	0.308	7.647
	(0.432)	(0.100)	(0.053)	(0.031)	(0.236)	(0.066)	(0.030)	(0.048)	(2.688)	
0.2	1.402	0.273	0.457	0.000	0.918	0.417	0.000	0.371	−0.064	6.611
	(0.443)	(0.102)	(0.054)	(0.033)	(0.237)	(0.065)	(0.031)	(0.049)	(2.487)	
0.3	1.373	0.271	0.464	0.000	0.883	0.422	0.000	0.372	−0.329	6.216 •
	(0.465)	(0.105)	(0.056)	(0.034)	(0.242)	(0.065)	(0.033)	(0.050)	(2.367)	
0.4	1.349	0.269	0.471	0.000	0.849	0.425	0.000	0.375	−0.485	6.276
	(0.494)	(0.111)	(0.058)	(0.036)	(0.250)	(0.064)	(0.034)	(0.052)	(2.339)	
0.5	1.326	0.268	0.476	0.000	0.817	0.427	0.000	0.380	−0.540	6.607
	(0.528)	(0.118)	(0.060)	(0.039)	(0.259)	(0.064)	(0.037)	(0.054)	(2.388)	
0.6	1.302	0.267	0.482	0.000	0.786	0.428	0.000	0.386	−0.509	7.031
	(0.567)	(0.126)	(0.063)	(0.041)	(0.271)	(0.064)	(0.039)	(0.056)	(2.476)	
0.7	1.277	0.267	0.487	0.000	0.758	0.428	0.000	0.394	−0.407	7.412
	(0.610)	(0.135)	(0.066)	(0.044)	(0.285)	(0.064)	(0.042)	(0.058)	(2.566)	
0.8	1.250	0.267	0.491	0.000	0.732	0.427	0.000	0.401	−0.250	7.698
	(0.657)	(0.145)	(0.069)	(0.047)	(0.299)	(0.064)	(0.045)	(0.060)	(2.639)	
0.9	1.223	0.267	0.496	0.000	0.708	0.425	0.000	0.410	−0.055	7.916
	(0.707)	(0.156)	(0.072)	(0.050)	(0.314)	(0.065)	(0.048)	(0.062)	(2.688)	
1	1.196	0.268	0.500	0.000	0.686	0.423	0.000	0.418	0.165	8.131
	(0.761)	(0.168)	(0.076)	(0.053)	(0.330)	(0.065)	(0.051)	(0.064)	(2.719)	

5. Concluding Remarks

In this study, we developed the robust estimator for bivariate Poisson INGARCH models based on the MDPDE. In practice, this is important, because outliers can strongly affect the CMLE, which is commonly employed for parameter estimation in INGARCH models. We proved that the MDPDE is consistent and asymptotically normal under regularity conditions. Our simulation study compared the performances of the MDPDE and the CMLE, and confirmed the superiority of the proposed estimator in the presence of outliers. The real data analysis also confirmed the validity of our method as a robust estimator in practice. Although we focused on Cui and Zhu's [20] bivariate Poisson INGARCH models here, the MDPDE method can be extended to other bivariate INGARCH models. For example, one can consider the copula-based bivariate INGARCH models that were studied by Heinen and Rengifo [42], Andreassen [18], and Fokianos et al. [43]. We leave this issue of extension as our future research.

Author Contributions: Conceptualization, B.K. and S.L.; methodology, B.K., S.L. and D.K.; software, B.K. and D.K.; formal analysis, B.K., S.L. and D.K.; data curation, B.K.; writing—original draft preparation, B.K. and S.L.; writing—review and editing, B.K. and S.L.; funding acquisition, B.K. and S.L. All authors have read and agreed to the published version of the manuscript.

Funding: This work was supported by the National Research Foundation of Korea (NRF) grant funded by the Korea government (MSIT), grant no. NRF-2019R1C1C1004662 (B.K.) and 2021R1A2C1004009 (S.L.).

Data Availability Statement: Publicly available datasets were analyzed in this study. This data can be found here: data.gov.au (accessed on 19 March 2021).

Conflicts of Interest: The authors declare no conflict of interest.

Appendix A

In this Appendix, we provide the proofs for Theorems 1 and 2 in Section 3 when $\alpha > 0$. We refer to Cui and Zhu [20] for the case of $\alpha = 0$. In what follows, we denote

V and $\rho \in (0,1)$ as a generic positive integrable random variable and a generic constant, respectively, and $H_{\alpha,n}(\theta) = n^{-1}\sum_{t=1}^{n} h_{\alpha,t}(\theta)$. Furthermore, we employ the notation $\lambda_t = \lambda_t(\theta)$, $\tilde{\lambda}_t = \tilde{\lambda}_t(\theta)$, $\lambda_t^0 = \lambda_t(\theta^0)$, $\lambda_{t,i} = \lambda_{t,i}(\theta_i)$, $\tilde{\lambda}_{t,i} = \tilde{\lambda}_{t,i}(\theta_i)$, $\lambda_{t,i}^0 = \lambda_{t,i}(\theta_i^0)$ for $i=1,2$, and $u(y,\lambda) = e^{-y} - e^{-c\lambda}$ for brevity.

Lemma A1. *Under (A1)–(A3), we have for $i = 1, 2$,*

(i) $E(\sup_{\theta_i \in \Theta_i} \lambda_{t,i}) < \infty$.

(ii) $\lambda_{t,i} = \lambda_{t,i}^0$ a.s. implies $\theta_i = \theta_i^0$.

(iii) $\lambda_{t,i}$ is twice continuously differentiable with respect to θ_i and satisfies

$$E\left(\sup_{\theta_i \in \Theta_i}\left\|\frac{\partial \lambda_{t,i}}{\partial \theta_i}\right\|_1\right)^4 < \infty \text{ and } E\left(\sup_{\theta_i \in \Theta_i}\left\|\frac{\partial^2 \lambda_{t,i}}{\partial \theta_i \partial \theta_i^T}\right\|_1\right)^2 < \infty.$$

(iv) $\sup_{\theta_i \in \Theta_i} \left\|\frac{\partial \lambda_{t,i}}{\partial \theta_i} - \frac{\partial \tilde{\lambda}_{t,i}}{\partial \theta_i}\right\|_1 \leq V\rho^t$ a.s.

(v) $v^T \frac{\partial \lambda_{t,i}^0}{\partial \theta_i} = 0$ a.s. implies $v = 0$.

(vi) $\sup_{\theta_i \in \Theta_i} |\lambda_{t,i} - \tilde{\lambda}_{t,i}| \leq V\rho^t$ a.s.

Proof. By recursion of (2), we have

$$\lambda_t = (I_2 - A)^{-1}\omega + \sum_{k=0}^{\infty} A^k BY_{t-k-1},$$

$$\tilde{\lambda}_t = (I_2 + A + \cdots + A^{t-2})\omega + A^{t-1}\tilde{\lambda}_1 + \sum_{k=0}^{t-2} A^k BY_{t-k-1}$$

and thus, for $i = 1, 2$,

$$\lambda_{t,i} = \frac{\omega_i}{1 - a_i} + \sum_{k=0}^{\infty} a_i^k (b_{i1}Y_{t-k-1,1} + b_{i2}Y_{t-k-1,2}),$$

$$\tilde{\lambda}_{t,i} = \frac{\omega_i}{1 - a_i} + \sum_{k=0}^{t-2} a_i^k (b_{i1}Y_{t-k-1,1} + b_{i2}Y_{t-k-1,2}),$$

where I_2 denotes 2×2 identity matrix and the initial value $\tilde{\lambda}_{1,i}$ is taken as $\omega_i/(1-a_i)$ for simplicity. Subsequently, $(i) - (v)$ can be shown following the arguments in the proof of Theorem 3 in Kang and Lee [44]. For (vi), let $\rho = \sup_{\theta_i \in \Theta_i} a_i < 1$. Afterwards, from (2), it holds that

$$\sup_{\theta_i \in \Theta_i} |\lambda_{t,i} - \tilde{\lambda}_{t,i}| = \sup_{\theta_i \in \Theta_i} |a_i(\lambda_{t-1,i} - \tilde{\lambda}_{t-1,i})| = \cdots = \sup_{\theta_i \in \Theta_i} |a_i^{t-1}(\lambda_{1,i} - \tilde{\lambda}_{1,i})| \leq V\rho^t$$

with $V = \sup_{\theta_i \in \Theta_i} |\lambda_{1,i} - \tilde{\lambda}_{1,i}|/\rho$. Therefore, the lemma is established. □

Lemma A2. *Under (A1)–(A3), we have*

$$\sup_{\theta \in \Theta} |H_{\alpha,n}(\theta) - \tilde{H}_{\alpha,n}(\theta)| \xrightarrow{a.s.} 0 \text{ as } n \to \infty.$$

Proof. It is sufficient to show that

$$\sup_{\theta \in \Theta} |h_{\alpha,t}(\theta) - \tilde{h}_{\alpha,t}(\theta)| \xrightarrow{a.s.} 0 \text{ as } t \to \infty.$$

We write

$$\begin{aligned}
&|h_{\alpha,t}(\theta) - \tilde{h}_{\alpha,t}(\theta)| \\
&\leq \left|\sum_{y_1=0}^{\infty}\sum_{y_2=0}^{\infty}\left\{f_\theta^{1+\alpha}(y|\lambda_t) - f_\theta^{1+\alpha}(y|\tilde{\lambda}_t)\right\}\right| + \left(1+\frac{1}{\alpha}\right)|f_\theta^\alpha(Y_t|\lambda_t) - f_\theta^\alpha(Y_t|\tilde{\lambda}_t)| \\
&:= I_t(\theta) + II_t(\theta).
\end{aligned}$$

From **(A1)**, **(A2)**, and the mean value theorem (MVT), we have

$$\begin{aligned}
I_t(\theta) &= (1+\alpha)\left|\sum_{y_1=0}^{\infty}\sum_{y_2=0}^{\infty} f_\theta^{1+\alpha}(y|\lambda_t^*)\left\{\frac{y_1}{\lambda_{t,1}^*} - 1 + \frac{c\delta e^{-c\lambda_{t,1}^*}u(y_2,\lambda_{t,2}^*)}{\varphi(y,\lambda_t^*,\delta)}\right\}(\lambda_{t,1} - \tilde{\lambda}_{t,1})\right. \\
&\quad + \left.\sum_{y_1=0}^{\infty}\sum_{y_2=0}^{\infty} f_\theta^{1+\alpha}(y|\lambda_t^*)\left\{\frac{y_2}{\lambda_{t,2}^*} - 1 + \frac{c\delta e^{-c\lambda_{t,2}^*}u(y_1,\lambda_{t,1}^*)}{\varphi(y,\lambda_t^*,\delta)}\right\}(\lambda_{t,2} - \tilde{\lambda}_{t,2})\right| \\
&\leq (1+\alpha)\left[\sum_{y_1=0}^{\infty}\sum_{y_2=0}^{\infty} f_\theta(y|\lambda_t^*)\left\{\frac{y_1}{\lambda_{t,1}^*} + 1 + \frac{c|\delta|e^{-c\lambda_{t,1}^*}|u(y_2,\lambda_{t,2}^*)|}{\varphi(y,\lambda_t^*,\delta)}\right\}|\lambda_{t,1} - \tilde{\lambda}_{t,1}|\right. \\
&\quad + \left.\sum_{y_1=0}^{\infty}\sum_{y_2=0}^{\infty} f_\theta(y|\lambda_t^*)\left\{\frac{y_2}{\lambda_{t,2}^*} + 1 + \frac{c|\delta|e^{-c\lambda_{t,2}^*}|u(y_1,\lambda_{t,1}^*)|}{\varphi(y,\lambda_t^*,\delta)}\right\}|\lambda_{t,2} - \tilde{\lambda}_{t,2}|\right] \\
&\leq (1+\alpha)\left\{\left(1+1+\frac{2c\delta_U}{\varphi_L}\right)|\lambda_{t,1} - \tilde{\lambda}_{t,1}| + \left(1+1+\frac{2c\delta_U}{\varphi_L}\right)|\lambda_{t,2} - \tilde{\lambda}_{t,2}|\right\} \\
&= 2(1+\alpha)\left(1+\frac{c\delta_U}{\varphi_L}\right)(|\lambda_{t,1} - \tilde{\lambda}_{t,1}| + |\lambda_{t,2} - \tilde{\lambda}_{t,2}|),
\end{aligned}$$

where $\lambda_t^* = (\lambda_{t,1}^*, \lambda_{t,2}^*)^T$ and $\lambda_{t,i}^*$ is an intermediate point between $\lambda_{t,i}$ and $\tilde{\lambda}_{t,i}$ for $i = 1, 2$. Hence, $\sup_{\theta \in \Theta} I_t(\theta)$ converges to 0 a.s. as $t \to \infty$ by (vi) of Lemma A1.

Because $\lambda_{t,i}^* = m\lambda_{t,i} + (1-m)\tilde{\lambda}_{t,i}$ for some $m \in (0,1)$, it holds that $(\lambda_{t,i}^*)^{-1} \leq (m\lambda_{t,i})^{-1} \leq (m\omega_L)^{-1}$. Thus, we obtain

$$\begin{aligned}
II_t(\theta) &= (1+\alpha)\left|f_\theta^\alpha(Y_t|\lambda_t^*)\left\{\frac{Y_{t,1}}{\lambda_{t,1}^*} - 1 + \frac{c\delta e^{-c\lambda_{t,1}^*}u(Y_{t,2},\lambda_{t,2}^*)}{\varphi(Y_t,\lambda_t^*,\delta)}\right\}(\lambda_{t,1} - \tilde{\lambda}_{t,1})\right. \\
&\quad + \left.f_\theta^\alpha(Y_t|\lambda_t^*)\left\{\frac{Y_{t,2}}{\lambda_{t,2}^*} - 1 + \frac{c\delta e^{-c\lambda_{t,2}^*}u(Y_{t,1},\lambda_{t,1}^*)}{\varphi(Y_t,\lambda_t^*,\delta)}\right\}(\lambda_{t,2} - \tilde{\lambda}_{t,2})\right| \\
&\leq (1+\alpha)\left\{\left(\frac{Y_{t,1}}{m\omega_L} + 1 + \frac{2c\delta_U}{\varphi_L}\right)|\lambda_{t,1} - \tilde{\lambda}_{t,1}| + \left(\frac{Y_{t,2}}{m\omega_L} + 1 + \frac{2c\delta_U}{\varphi_L}\right)|\lambda_{t,2} - \tilde{\lambda}_{t,2}|\right\}.
\end{aligned}$$

According to Lemma 2.1 in Straumann and Mikosch [45], together with (vi) of Lemma A1, $\sup_{\theta \in \Theta} II_t(\theta)$ converges to 0 a.s. as $t \to \infty$. Therefore, the lemma is verified. □

Lemma A3. *Under (A1)–(A3), we have*

$$E\left(\sup_{\theta \in \Theta}|h_{\alpha,t}(\theta)|\right) < \infty \text{ and if } \theta \neq \theta^0, \text{ then } E(h_{\alpha,t}(\theta)) > E(h_{\alpha,t}(\theta^0)).$$

Proof. Because

$$|h_{\alpha,t}(\theta)| \leq \sum_{y_1=0}^{\infty}\sum_{y_2=0}^{\infty} f_\theta^{1+\alpha}(y|\lambda_t) + \left(1+\frac{1}{\alpha}\right)f_\theta^\alpha(Y_t|\lambda_t) \leq 2 + \frac{1}{\alpha},$$

the first part of the lemma is validated. Note that

$$
\begin{aligned}
&E(h_{\alpha,t}(\theta)) - E(h_{\alpha,t}(\theta^0)) \\
&= E\left[E\{h_{\alpha,t}(\theta) - h_{\alpha,t}(\theta^0)|\mathcal{F}_{t-1}\}\right] \\
&= E\left[\sum_{y_1=0}^{\infty}\sum_{y_2=0}^{\infty}\left\{f_{\theta}^{1+\alpha}(y|\lambda_t) - \left(1+\frac{1}{\alpha}\right)f_{\theta}^{\alpha}(y|\lambda_t)f_{\theta^0}(y|\lambda_t) + \frac{1}{\alpha}f_{\theta^0}^{1+\alpha}(y|\lambda_t)\right\}\right] \\
&\geq 0,
\end{aligned}
$$

where the equality holds if and only if $\delta = \delta^0$ and $\lambda_t = \lambda_t^0$ a.s. Therefore, the second part of the lemma is established by (ii) of Lemma A1. $\square$

Proof of Theorem 1. We can write

$$
\sup_{\theta\in\Theta}\left|\frac{1}{n}\sum_{t=1}^{n}\tilde{h}_{\alpha,t}(\theta) - E(h_{\alpha,t}(\theta))\right|
$$

$$
\leq \sup_{\theta\in\Theta}\left|\frac{1}{n}\sum_{t=1}^{n}\tilde{h}_{\alpha,t}(\theta) - \frac{1}{n}\sum_{t=1}^{n}h_{\alpha,t}(\theta)\right| + \sup_{\theta\in\Theta}\left|\frac{1}{n}\sum_{t=1}^{n}h_{\alpha,t}(\theta) - E(h_{\alpha,t}(\theta))\right|.
$$

The first term on the RHS of the inequality converges to 0 a.s. from Lemma A2. Moreover, because $h_{\alpha,t}(\theta)$ is stationary and ergodic with $E(\sup_{\theta\in\Theta}|h_{\alpha,t}(\theta)|) < \infty$ by Lemma A3, the second term also converges to 0 a.s. Finally, as $E(h_{\alpha,t}(\theta))$ has a unique minimum at θ^0 from Lemma A3, the theorem is established. $\square$

Now, we derive the first and second derivatives of $h_{\alpha,t}(\theta)$. The first derivatives are obtained as

$$
\begin{aligned}
\frac{\partial h_{\alpha,t}(\theta)}{\partial \theta} &= (1+\alpha)\left(D_{t,1}(\theta)s_{t,1}(\theta_1)^T, D_{t,2}(\theta)s_{t,2}(\theta_2)^T, D_{t,3}(\theta)\right)^T \\
&= (1+\alpha)\begin{pmatrix} D_{t,1}(\theta)I_4 & 0_{4\times 4} & 0_{4\times 1} \\ 0_{4\times 4} & D_{t,2}(\theta)I_4 & 0_{4\times 1} \\ 0_{1\times 4} & 0_{1\times 4} & D_{t,3}(\theta) \end{pmatrix}\begin{pmatrix} s_{t,1}(\theta_1) \\ s_{t,2}(\theta_2) \\ 1 \end{pmatrix} \\
&:= (1+\alpha)D_t(\theta)\Lambda_t(\theta),
\end{aligned}
$$

where I_4 denotes the 4×4 identity matrix, $0_{m\times n}$ means the $m\times n$ matrix with zero elements, and

$$
s_{t,i}(\theta_i) = \frac{\partial \lambda_{t,i}}{\partial \theta_i} \text{ for } i = 1,2,
$$

$$
D_{t,i}(\theta) = \sum_{y_1=0}^{\infty}\sum_{y_2=0}^{\infty}f_{\theta}^{1+\alpha}(y|\lambda_t)\left\{\frac{y_i}{\lambda_{t,i}} - 1 + \frac{c\delta e^{-c\lambda_{t,i}}u(y_j,\lambda_{t,j})}{\varphi(y,\lambda_t,\delta)}\right\}
$$

$$
- f_{\theta}^{\alpha}(Y_t|\lambda_t)\left\{\frac{Y_{t,i}}{\lambda_{t,i}} - 1 + \frac{c\delta e^{-c\lambda_{t,i}}u(Y_{t,j},\lambda_{t,j})}{\varphi(Y_t,\lambda_t,\delta)}\right\} \text{ for } (i,j) = (1,2),(2,1),
$$

$$
D_{t,3}(\theta) = \sum_{y_1=0}^{\infty}\sum_{y_2=0}^{\infty}f_{\theta}^{1+\alpha}(y|\lambda_t)\frac{u(y_1,\lambda_{t,1})u(y_2,\lambda_{t,2})}{\varphi(y,\lambda_t,\delta)} - f_{\theta}^{\alpha}(Y_t|\lambda_t)\frac{u(Y_{t,1},\lambda_{t,1})u(Y_{t,2},\lambda_{t,2})}{\varphi(Y_t,\lambda_t,\delta)}.
$$

The second derivatives are expressed as

$$
\frac{\partial^2 h_{\alpha,t}(\theta)}{\partial\theta\partial\theta^T} = (1+\alpha)\begin{pmatrix} F_{t,11}(\theta)s_{t,1}(\theta_1)s_{t,1}(\theta_1)^T & F_{t,12}(\theta)s_{t,1}(\theta_1)s_{t,2}(\theta_2)^T & F_{t,13}(\theta)s_{t,1}(\theta_1) \\ F_{t,21}(\theta)s_{t,2}(\theta_2)s_{t,1}(\theta_1)^T & F_{t,22}(\theta)s_{t,2}(\theta_2)s_{t,2}(\theta_2)^T & F_{t,23}(\theta)s_{t,2}(\theta_2) \\ F_{t,31}(\theta)s_{t,1}(\theta_1)^T & F_{t,32}(\theta)s_{t,2}(\theta_2)^T & F_{t,33}(\theta) \end{pmatrix}
$$

$$
+ (1+\alpha)\begin{pmatrix} D_{t,1}(\theta)s_{t,11}(\theta_1) & 0_{4\times 4} & 0_{4\times 1} \\ 0_{4\times 4} & D_{t,2}(\theta)s_{t,22}(\theta_2) & 0_{4\times 1} \\ 0_{1\times 4} & 0_{1\times 4} & 0 \end{pmatrix}
$$

$$
:= (1+\alpha)\left\{F_t(\theta) + D_t(\theta)\frac{\partial\Lambda_t(\theta)}{\partial\theta^T}\right\},
$$

where

$$s_{t,ii}(\theta_i) = \frac{\partial^2 \lambda_{t,i}}{\partial \theta_i \partial \theta_i^T} \text{ for } i=1,2,$$

$$F_{t,ii}(\theta) = \sum_{y_1=0}^{\infty}\sum_{y_2=0}^{\infty} f_\theta^{1+\alpha}(y|\lambda_t)\left[(1+\alpha)\left\{\frac{y_i}{\lambda_{t,i}}-1+\frac{c\delta e^{-c\lambda_{t,i}}u(y_j,\lambda_{t,j})}{\varphi(y,\lambda_t,\delta)}\right\}^2\right.$$
$$\left. -\frac{y_i}{\lambda_{t,i}^2}-\frac{c^2\delta e^{-c\lambda_{t,i}}u(y_j,\lambda_{t,j})\{1+\delta e^{-y_i}u(y_j,\lambda_{t,j})\}}{\varphi(y,\lambda_t,\delta)^2}\right]$$
$$-f_\theta^\alpha(Y_t|\lambda_t)\left[\alpha\left\{\frac{Y_{t,i}}{\lambda_{t,i}}-1+\frac{c\delta e^{-c\lambda_{t,i}}u(Y_{t,j},\lambda_{t,j})}{\varphi(Y_t,\lambda_t,\delta)}\right\}^2\right.$$
$$\left.-\frac{Y_{t,i}}{\lambda_{t,i}^2}-\frac{c^2\delta e^{-c\lambda_{t,i}}u(Y_{t,j},\lambda_{t,j})\{1+\delta e^{-Y_{t,i}}u(Y_{t,j},\lambda_{t,j})\}}{\varphi(Y_t,\lambda_t,\delta)^2}\right]$$

for $(i,j)=(1,2),(2,1)$,

$$F_{t,33}(\theta) = \alpha\sum_{y_1=0}^{\infty}\sum_{y_2=0}^{\infty} f_\theta^{1+\alpha}(y|\lambda_t)\left\{\frac{u(y_1,\lambda_{t,1})u(y_2,\lambda_{t,2})}{\varphi(y,\lambda_t,\delta)}\right\}^2$$
$$-(\alpha-1)f_\theta^\alpha(Y_t|\lambda_t)\left\{\frac{u(Y_{t,1},\lambda_{t,1})u(Y_{t,2},\lambda_{t,2})}{\varphi(Y_t,\lambda_t,\delta)}\right\}^2,$$

$$F_{t,12}(\theta) = \sum_{y_1=0}^{\infty}\sum_{y_2=0}^{\infty} f_\theta^{1+\alpha}(y|\lambda_t)\left[(1+\alpha)\left\{\frac{y_1}{\lambda_{t,1}}-1+\frac{c\delta e^{-c\lambda_{t,1}}u(y_2,\lambda_{t,2})}{\varphi(y,\lambda_t,\delta)}\right\}\right.$$
$$\left.\times\left\{\frac{y_2}{\lambda_{t,2}}-1+\frac{c\delta e^{-c\lambda_{t,2}}u(y_1,\lambda_{t,1})}{\varphi(y,\lambda_t,\delta)}\right\}+\frac{c^2\delta e^{-c(\lambda_{t,1}+\lambda_{t,2})}}{\varphi(y,\lambda_t,\delta)^2}\right]$$
$$-f_\theta^\alpha(Y_t|\lambda_t)\left[\alpha\left\{\frac{Y_{t,1}}{\lambda_{t,1}}-1+\frac{c\delta e^{-c\lambda_{t,1}}u(Y_{t,2},\lambda_{t,2})}{\varphi(Y_t,\lambda_t,\delta)}\right\}\right.$$
$$\left.\times\left\{\frac{Y_{t,2}}{\lambda_{t,2}}-1+\frac{c\delta e^{-c\lambda_{t,2}}u(Y_{t,1},\lambda_{t,1})}{\varphi(Y_t,\lambda_t,\delta)}\right\}+\frac{c^2\delta e^{-c(\lambda_{t,1}+\lambda_{t,2})}}{\varphi(Y_t,\lambda_t,\delta)^2}\right],$$

$$F_{t,i3}(\theta) = \sum_{y_1=0}^{\infty}\sum_{y_2=0}^{\infty} f_\theta^{1+\alpha}(y|\lambda_t)\left[(1+\alpha)\left\{\frac{y_i}{\lambda_{t,i}}-1+\frac{c\delta e^{-c\lambda_{t,i}}u(y_j,\lambda_{t,j})}{\varphi(y,\lambda_t,\delta)}\right\}\right.$$
$$\left.\times\frac{u(y_1,\lambda_{t,1})u(y_2,\lambda_{t,2})}{\varphi(y,\lambda_t,\delta)}+\frac{ce^{-c\lambda_{t,i}}u(y_j,\lambda_{t,j})}{\varphi(y,\lambda_t,\delta)^2}\right]$$
$$-f_\theta^\alpha(Y_t|\lambda_t)\left[\alpha\left\{\frac{Y_{t,i}}{\lambda_{t,i}}-1+\frac{c\delta e^{-c\lambda_{t,i}}u(Y_{t,j},\lambda_{t,j})}{\varphi(Y_t,\lambda_t,\delta)}\right\}\frac{u(Y_{t,1},\lambda_{t,1})u(Y_{t,2},\lambda_{t,2})}{\varphi(Y_t,\lambda_t,\delta)}\right.$$
$$\left.+\frac{ce^{-c\lambda_{t,i}}u(Y_{t,j},\lambda_{t,j})}{\varphi(Y_t,\lambda_t,\delta)^2}\right] \text{ for } (i,j)=(1,2),(2,1).$$

The following four lemmas are helpful for proving Theorem 2.

Lemma A4. *Let $\widetilde{D}_{t,i}(\theta)$ denote the counterpart of $D_{t,i}(\theta)$ by substituting λ_t with $\tilde{\lambda}_t$ for $i=1,2,3$. Subsequently, under (A1)–(A3), we have that for $i=1,2$,*

$$|D_{t,i}(\theta)|\leq C(Y_{t,i}+1),\ |\widetilde{D}_{t,i}(\theta)|\leq C(Y_{t,i}+1),\ |D_{t,3}(\theta)|\leq C,\ |\widetilde{D}_{t,3}(\theta)|\leq C,$$
$$|F_{t,ii}(\theta)|\leq C(Y_{t,i}^2+Y_{t,i}+1),\ |F_{t,33}(\theta)|\leq C,\ |F_{t,12}(\theta)|\leq C(Y_{t,1}Y_{t,2}+Y_{t,1}+Y_{t,2}+1),$$
$$|F_{t,i3}(\theta)|\leq C(Y_{t,i}+1),$$

and for $(i,j) = (1,2), (2,1)$,

$$|D_{t,i}(\theta) - \widetilde{D}_{t,i}(\theta)| \leq C(Y_{t,i}^2 + Y_{t,i} + 1)|\lambda_{t,i} - \tilde{\lambda}_{t,i}|$$
$$+ C(Y_{t,1}Y_{t,2} + Y_{t,1} + Y_{t,2} + 1)|\lambda_{t,j} - \tilde{\lambda}_{t,j}|,$$
$$|D_{t,3}(\theta) - \widetilde{D}_{t,3}(\theta)| \leq C(Y_{t,1} + 1)|\lambda_{t,1} - \tilde{\lambda}_{t,1}| + C(Y_{t,2} + 1)|\lambda_{t,2} - \tilde{\lambda}_{t,2}|,$$

where C is some positive constant.

Proof. From **(A1)–(A3)** and the fact that $\lambda_{t,i}^{-1} \leq \omega_L^{-1}$, we obtain

$$
\begin{aligned}
|D_{t,i}(\theta)| &\leq \sum_{y_1=0}^{\infty}\sum_{y_2=0}^{\infty} f_\theta(y|\lambda_t)\left\{\frac{y_i}{\lambda_{t,i}} + 1 + \frac{c|\delta|e^{-c\lambda_{t,i}}|u(y_j,\lambda_{t,j})|}{\varphi(y,\lambda_t,\delta)}\right\} \\
&\quad + \left\{\frac{Y_{t,i}}{\lambda_{t,i}} + 1 + \frac{c|\delta|e^{-c\lambda_{t,i}}|u(Y_{t,j},\lambda_{t,j})|}{\varphi(Y_t,\lambda_t,\delta)}\right\} \\
&\leq 1 + 1 + \frac{2c\delta_U}{\varphi_L} + \frac{Y_{t,i}}{\omega_L} + 1 + \frac{2c\delta_U}{\varphi_L} \\
&= \frac{Y_{t,i}}{\omega_L} + 3 + \frac{4c\delta_U}{\varphi_L}
\end{aligned}
$$

for $(i,j) = (1,2), (2,1)$ and

$$
\begin{aligned}
|D_{t,3}(\theta)| &\leq \sum_{y_1=0}^{\infty}\sum_{y_2=0}^{\infty} f_\theta(y|\lambda_t) \frac{|u(y_1,\lambda_{t,1})||u(y_2,\lambda_{t,2})|}{\varphi(y,\lambda_t,\delta)} + \frac{|u(Y_{t,1},\lambda_{t,1})||u(Y_{t,2},\lambda_{t,2})|}{\varphi(Y_t,\lambda_t,\delta)} \\
&= \frac{8}{\varphi_L}.
\end{aligned}
$$

The second and fourth parts of the lemma also hold in the same manner. Furthermore, we can show that

$$
\begin{aligned}
|F_{t,ii}(\theta)| &\leq \sum_{y_1=0}^{\infty}\sum_{y_2=0}^{\infty} f_\theta(y|\lambda_t)\left[2(1+\alpha)\left\{\left(\frac{y_i-\lambda_{t,i}}{\lambda_{t,i}}\right)^2 + \left(\frac{c\delta e^{-c\lambda_{t,i}}u(y_j,\lambda_{t,j})}{\varphi(y,\lambda_t,\delta)}\right)^2\right\}\right. \\
&\quad \left. + \frac{y_i}{\lambda_{t,i}^2} + \frac{c^2|\delta|e^{-c\lambda_{t,i}}|u(y_j,\lambda_{t,j})|\{1+|\delta|e^{-y_i}|u(y_j,\lambda_{t,j})|\}}{\varphi(y,\lambda_t,\delta)^2}\right] \\
&\quad + 2\alpha\left\{\left(\frac{Y_{t,i}}{\lambda_{t,i}} - 1\right)^2 + \left(\frac{c\delta e^{-c\lambda_{t,i}}u(Y_{t,j},\lambda_{t,j})}{\varphi(Y_t,\lambda_t,\delta)}\right)^2\right\} \\
&\quad - \frac{Y_{t,i}}{\lambda_{t,i}^2} - \frac{c^2|\delta|e^{-c\lambda_{t,i}}|u(Y_{t,j},\lambda_{t,j})|\{1+|\delta|e^{-Y_{t,i}}|u(Y_{t,j},\lambda_{t,j})|\}}{\varphi(Y_t,\lambda_t,\delta)^2} \\
&\leq 2(1+\alpha)\frac{1}{\omega_L} + 2(1+\alpha)\frac{4c^2\delta_U^2}{\varphi_L^2} + \frac{1}{\omega_L} + \frac{2c^2\delta_U(1+2\delta_U)}{\varphi_L^2} \\
&\quad + 4\alpha\frac{Y_{t,i}^2}{\omega_L^2} + 4\alpha + 2\alpha\frac{4c^2\delta_U^2}{\varphi_L^2} + \frac{Y_{t,i}}{\omega_L^2} + \frac{2c^2\delta_U(1+2\delta_U)}{\varphi_L^2} \\
&= \frac{4\alpha}{\omega_L^2}Y_{t,i}^2 + \frac{1}{\omega_L^2}Y_{t,i} + \frac{3+2\alpha}{\omega_L} + \frac{4c^2\delta_U\{4(1+\alpha)\delta_U + 1\}}{\varphi_L^2} + 4\alpha
\end{aligned}
$$

for $(i,j) = (1,2), (2,1)$,

$$|F_{t,33}(\theta)| \leq \alpha \sum_{y_1=0}^{\infty}\sum_{y_2=0}^{\infty} f_\theta(y|\lambda_t)\left\{\frac{u(y_1,\lambda_{t,1})u(y_2,\lambda_{t,2})}{\varphi(y,\lambda_t,\delta)}\right\}^2$$

$$+(1+\alpha)\left\{\frac{u(Y_{t,1},\lambda_{t,1})u(Y_{t,2},\lambda_{t,2})}{\varphi(Y_t,\lambda_t,\delta)}\right\}^2$$

$$\leq \frac{16\alpha}{\varphi_L^2} + \frac{16(1+\alpha)}{\varphi_L^2}$$

$$= \frac{16(1+2\alpha)}{\varphi_L^2},$$

$$|F_{t,12}(\theta)| \leq \sum_{y_1=0}^{\infty}\sum_{y_2=0}^{\infty} f_\theta(y|\lambda_t)\left[(1+\alpha)\left\{\frac{y_1}{\lambda_{t,1}}+1+\frac{c|\delta|e^{-c\lambda_{t,1}}|u(y_2,\lambda_{t,2})|}{\varphi(y,\lambda_t,\delta)}\right\}\right.$$

$$\times \left\{\frac{y_2}{\lambda_{t,2}}+1+\frac{c|\delta|e^{-c\lambda_{t,2}}|u(y_1,\lambda_{t,1})|}{\varphi(y,\lambda_t,\delta)}\right\}+\frac{c^2|\delta|e^{-c(\lambda_{t,1}+\lambda_{t,2})}}{\varphi(y,\lambda_t,\delta)^2}\right]$$

$$+\alpha\left\{\frac{Y_{t,1}}{\lambda_{t,1}}+1+\frac{c|\delta|e^{-c\lambda_{t,1}}|u(Y_{t,2},\lambda_{t,2})|}{\varphi(Y_t,\lambda_t,\delta)}\right\}$$

$$\times \left\{\frac{Y_{t,2}}{\lambda_{t,2}}+1+\frac{c|\delta|e^{-c\lambda_{t,2}}|u(Y_{t,1},\lambda_{t,1})|}{\varphi(Y_t,\lambda_t,\delta)}\right\}+\frac{c^2|\delta|e^{-c(\lambda_{t,1}+\lambda_{t,2})}}{\varphi(Y_t,\lambda_t,\delta)^2}$$

$$\leq (1+\alpha)\left\{c^2\delta_U+1+\left(1+\frac{2c\delta_U}{\varphi_L}\right)+\left(1+\frac{2c\delta_U}{\varphi_L}\right)+\frac{4c^2\delta_U^2}{\varphi_L^2}+\frac{4c\delta_U}{\varphi_L}+1\right\}$$

$$+\frac{c^2\delta_U}{\varphi_L^2}+\alpha\left\{\frac{Y_{t,1}Y_{t,2}}{\omega_L^2}+\frac{1}{\omega_L}\left(1+\frac{2c\delta_U}{\varphi_L}\right)(Y_{t,1}+Y_{t,2})+\frac{4c^2\delta_U^2}{\varphi_L^2}+\frac{4c\delta_U}{\varphi_L}+1\right\}$$

$$+\frac{c^2\delta_U}{\varphi_L^2}$$

$$= \frac{\alpha}{\omega_L^2}Y_{t,1}Y_{t,2}+\frac{\alpha}{\omega_L}\left(1+\frac{2c\delta_U}{\varphi_L}\right)(Y_{t,1}+Y_{t,2})+4(1+2\alpha)\frac{c^2\delta_U^2}{\varphi_L^2}+\frac{2c^2\delta_U}{\varphi_L^2}$$

$$+4(2+3\alpha)\frac{c\delta_U}{\varphi_L}+(1+\alpha)c^2\delta_U+4+5\alpha,$$

and

$$|F_{t,i3}(\theta)| \leq \sum_{y_1=0}^{\infty}\sum_{y_2=0}^{\infty} f_\theta(y|\lambda_t)\left[(1+\alpha)\left\{\frac{y_i}{\lambda_{t,i}}+1+\frac{c|\delta|e^{-c\lambda_{t,i}}|u(y_j,\lambda_{t,j})|}{\varphi(y,\lambda_t,\delta)}\right\}\right.$$

$$\times \frac{|u(y_1,\lambda_{t,1})||u(y_2,\lambda_{t,2})|}{\varphi(y,\lambda_t,\delta)}+\frac{ce^{-c\lambda_{t,i}}|u(y_j,\lambda_{t,j})|}{\varphi(y,\lambda_t,\delta)^2}\right]$$

$$+\alpha\left\{\frac{Y_{t,i}}{\lambda_{t,i}}+1+\frac{c|\delta|e^{-c\lambda_{t,i}}|u(Y_{t,j},\lambda_{t,j})|}{\varphi(Y_t,\lambda_t,\delta)}\right\}\frac{|u(Y_{t,1},\lambda_{t,1})||u(Y_{t,2},\lambda_{t,2})|}{\varphi(Y_t,\lambda_t,\delta)}$$

$$+\frac{ce^{-c\lambda_{t,i}}|u(Y_{t,j},\lambda_{t,j})|}{\varphi(Y_t,\lambda_t,\delta)^2}$$

$$\leq \frac{4(1+\alpha)}{\varphi_L}\left(1+1+\frac{2c\delta_U}{\varphi_L}\right)+\frac{2c}{\varphi_L^2}+\frac{4\alpha}{\varphi_L}\left(\frac{Y_{t,i}}{\omega_L}+1+\frac{2c\delta_U}{\varphi_L}\right)+\frac{2c}{\varphi_L^2}$$

$$= \frac{4\alpha}{\varphi_L\omega_L}Y_{t,i}+\frac{4c\{1+2(1+2\alpha)\delta_U\}}{\varphi_L^2}+\frac{4(2+3\alpha)}{\varphi_L}$$

for $(i,j) = (1,2), (2,1)$.

Now, we prove the last two parts of the lemma. Because $F_{t,ij}(\theta) = \partial D_{t,i}(\theta)/\partial \lambda_{t,j}$ for $i = 1,2,3$, $j = 1,2$, owing to MVT, it holds that for $i = 1,2,3$,

$$\begin{aligned}|D_{t,i}(\theta) - \widetilde{D}_{t,i}(\theta)| &\leq \left|\frac{\partial D_{t,i}(\theta)}{\partial \lambda_{t,1}}\right|_{\lambda_t = \lambda_t^*} |\lambda_{t,1} - \widetilde{\lambda}_{t,1}| + \left|\frac{\partial D_{t,i}(\theta)}{\partial \lambda_{t,2}}\right|_{\lambda_t = \lambda_t^*} |\lambda_{t,2} - \widetilde{\lambda}_{t,2}| \\ &= \left|F_{t,i1}(\theta)\right|_{\lambda_t = \lambda_t^*} |\lambda_{t,1} - \widetilde{\lambda}_{t,1}| + \left|F_{t,i2}(\theta)\right|_{\lambda_t = \lambda_t^*} |\lambda_{t,2} - \widetilde{\lambda}_{t,2}|,\end{aligned}$$

where $F_{t,ij}(\theta)|_{\lambda_t = \lambda_t^*}$ is the same as $F_{t,ij}(\theta)$ with λ_t replaced by λ_t^* for $j = 1,2$. Because $(\lambda_{t,i}^*)^{-1} \leq \omega_L^{-1}$, it can be easily shown that $\left|F_{t,ij}(\theta)\right|_{\lambda_t = \lambda_t^*}$ has the same upper bound as $|F_{t,ij}(\theta)|$ by following the aforementioned arguments. Therefore, the lemma is established. □

Lemma A5. *Under (A1)–(A3), we have*

$$E\left(\sup_{\theta \in \Theta} \left\|\frac{\partial^2 h_{\alpha,t}(\theta)}{\partial \theta \partial \theta^T}\right\|_1\right) < \infty \text{ and } E\left(\sup_{\theta \in \Theta} \left\|\frac{\partial h_{\alpha,t}(\theta)}{\partial \theta}\frac{\partial h_{\alpha,t}(\theta)}{\partial \theta^T}\right\|_1\right) < \infty.$$

Proof. We can write

$$E\left(\sup_{\theta \in \Theta} \left\|\frac{\partial^2 h_{\alpha,t}(\theta)}{\partial \theta \partial \theta^T}\right\|_1\right) \leq (1+\alpha)\left\{E\left(\sup_{\theta \in \Theta} \|F_t(\theta)\|_1\right) + E\left(\sup_{\theta \in \Theta} \left\|D_t(\theta)\frac{\partial \Lambda_t(\theta)}{\partial \theta^T}\right\|_1\right)\right\}.$$

Hence, to show the first part of the lemma, it is sufficient to show that, for $i,j = 1,2$,

$$E\left(\sup_{\theta \in \Theta}\left\|F_{t,ij}(\theta)\frac{\partial \lambda_{t,i}}{\partial \theta_i}\frac{\partial \lambda_{t,j}}{\partial \theta_j^T}\right\|_1\right) < \infty, \quad E\left(\sup_{\theta \in \Theta}\left\|F_{t,i3}(\theta)\frac{\partial \lambda_{t,i}}{\partial \theta_i}\right\|_1\right) < \infty,$$

$$E\left(\sup_{\theta \in \Theta} |F_{t,33}(\theta)|\right) < \infty, \text{ and } E\left(\sup_{\theta \in \Theta}\left\|D_{t,i}(\theta)\frac{\partial^2 \lambda_{t,i}}{\partial \theta_i \partial \theta_i^T}\right\|_1\right) < \infty,$$

which can be directly obtained from (iii) of Lemma A1, Lemma A4, and Cauchy–Schwarz inequality. For example,

$$\begin{aligned}&E\left(\sup_{\theta \in \Theta}\left\|F_{t,12}(\theta)\frac{\partial \lambda_{t,1}}{\partial \theta_1}\frac{\partial \lambda_{t,2}}{\partial \theta_2^T}\right\|_1\right) \\ &\leq \left\{E\left(\sup_{\theta \in \Theta}|F_{t,12}(\theta)|\right)^2\right\}^{1/2}\left\{E\left(\sup_{\theta \in \Theta}\left\|\frac{\partial \lambda_{t,1}}{\partial \theta_1}\frac{\partial \lambda_{t,2}}{\partial \theta_2^T}\right\|_1\right)^2\right\}^{1/2} \\ &\leq \left[E\{C(Y_{t,1}Y_{t,2} + Y_{t,1} + Y_{t,2} + 1)\}^2\right]^{1/2} \\ &\quad \times \left\{E\left(\sup_{\theta_1 \in \Theta_1}\left\|\frac{\partial \lambda_{t,1}}{\partial \theta_1}\right\|_1\right)^4\right\}^{1/4}\left\{E\left(\sup_{\theta_2 \in \Theta_2}\left\|\frac{\partial \lambda_{t,2}}{\partial \theta_2}\right\|_1\right)^4\right\}^{1/4} \\ &< \infty\end{aligned}$$

and

$$E\left(\sup_{\theta\in\Theta}\left\|D_{t,1}(\theta)\frac{\partial^2\lambda_{t,1}}{\partial\theta_1\partial\theta_1^T}\right\|_1\right)$$

$$\leq \left\{E\left(\sup_{\theta\in\Theta}|D_{t,1}(\theta)|\right)^2\right\}^{1/2}\left\{E\left(\sup_{\theta_1\in\Theta_1}\left\|\frac{\partial^2\lambda_{t,1}}{\partial\theta_1\partial\theta_1^T}\right\|_1\right)^2\right\}^{1/2}$$

$$\leq \left[E\{C(Y_{t,1}+1)\}^2\right]^{1/2}\left\{E\left(\sup_{\theta_1\in\Theta_1}\left\|\frac{\partial^2\lambda_{t,1}}{\partial\theta_1\partial\theta_1^T}\right\|_1\right)^2\right\}^{1/2}$$

$$< \infty.$$

The second part of the lemma can be shown in the same manner. □

Lemma A6. *Under (A1)–(A3), we have*

$$\frac{1}{\sqrt{n}}\sum_{t=1}^{n}\sup_{\theta\in\Theta}\left\|\frac{\partial h_{\alpha,t}(\theta)}{\partial\theta}-\frac{\partial\tilde{h}_{\alpha,t}(\theta)}{\partial\theta}\right\|_1 \xrightarrow{a.s.} 0 \text{ as } n\to\infty.$$

Proof. Owing to (*iv*) and (*vi*) of Lemma A1 and Lemma A4, we obtain a.s.,

$$\frac{1}{1+\alpha}\sup_{\theta\in\Theta}\left\|\frac{\partial h_{\alpha,t}(\theta)}{\partial\theta}-\frac{\partial\tilde{h}_{\alpha,t}(\theta)}{\partial\theta}\right\|_1$$

$$\leq \sup_{\theta\in\Theta}\left\|\tilde{D}_t(\theta)\right\|_1\sup_{\theta\in\Theta}\left\|\Lambda_t(\theta)-\tilde{\Lambda}_t(\theta)\right\|_1+\sup_{\theta\in\Theta}\left\|\Lambda_t(\theta)\right\|_1\sup_{\theta\in\Theta}\left\|D_t(\theta)-\tilde{D}_t(\theta)\right\|_1$$

$$\leq \left(\sum_{i=1}^{3}\sup_{\theta\in\Theta}|\tilde{D}_{t,i}(\theta)|\right)\left(\sum_{i=1}^{2}\sup_{\theta_i\in\Theta_i}\left\|\frac{\partial\lambda_{t,i}}{\partial\theta_i}-\frac{\partial\tilde{\lambda}_{t,i}}{\partial\theta_i}\right\|_1\right)$$

$$+\left(\sum_{i=1}^{2}\sup_{\theta_i\in\Theta_i}\left\|\frac{\partial\lambda_{t,i}}{\partial\theta_i}\right\|_1+1\right)\left(\sum_{i=1}^{3}\sup_{\theta\in\Theta}|D_{t,i}(\theta)-\tilde{D}_{t,i}(\theta)|\right)$$

$$\leq 2C(Y_{t,1}+Y_{t,2}+3)V\rho^t$$

$$+\left(\sum_{i=1}^{2}\sup_{\theta_i\in\Theta_i}\left\|\frac{\partial\lambda_{t,i}}{\partial\theta_i}\right\|_1+1\right)\times C\left\{Y_{t,1}^2+Y_{t,2}^2+2Y_{t,1}Y_{t,2}+4(Y_{t,1}+Y_{t,2})+6\right\}V\rho^t,$$

where $\tilde{D}_t(\theta)$ and $\tilde{\Lambda}_t(\theta)$ are the same as $D_t(\theta)$ and $\Lambda_t(\theta)$ with λ_t replaced by $\tilde{\lambda}_t$. Therefore, from Lemma 2.1 in Straumann and Mikosch [45], together with (*iii*) of Lemma A1, the RHS of the last inequality converges to 0 exponentially fast a.s. and, thus, the lemma is validated. We refer the reader to Straumann and Mikosch [45] and Cui and Zheng [46] for more details on exponentially fast a.s. convergence. □

Lemma A7. *Let $\hat{\theta}_{\alpha,n}^H = \operatorname{argmin}_{\theta\in\Theta} H_{\alpha,n}(\theta)$. Subsequently, under (A1)–(A3), we have*

$$\hat{\theta}_{\alpha,n}^H \xrightarrow{a.s.} \theta^0 \text{ and } \sqrt{n}(\hat{\theta}_{\alpha,n}^H-\theta^0) \xrightarrow{d} N(0, J_\alpha^{-1}K_\alpha J_\alpha^{-1}) \text{ as } n\to\infty.$$

Proof. As seen in the proof of Theorem 1, $\sup_{\theta\in\Theta}|n^{-1}\sum_{t=1}^{n}h_{\alpha,t}(\theta)-E(h_{\alpha,t}(\theta))|$ converges to 0 a.s. and $E(h_{\alpha,t}(\theta))$ has a unique minimum at θ^0. Hence, the first part of the lemma is verified.

Next, we handle the second part. Let $\theta(i), i=1,\ldots,9$ be the i-th element of θ. Using MVT, we have

$$0 = \frac{1}{\sqrt{n}}\sum_{t=1}^{n}\frac{\partial h_{\alpha,t}(\theta^0)}{\partial\theta(i)}+\sqrt{n}(\hat{\theta}_{\alpha,n}^H-\theta^0)^T\left(\frac{1}{n}\sum_{t=1}^{n}\frac{\partial^2 h_{\alpha,t}(\theta_{\alpha,n,i}^*)}{\partial\theta\partial\theta(i)}\right)$$

for some vector $\theta^*_{\alpha,n,i}$ between θ_0 and $\hat{\theta}^H_{\alpha,n}$, so that, eventually, we can write

$$0 = \frac{1}{\sqrt{n}} \sum_{t=1}^n \frac{\partial h_{\alpha,t}(\theta^0)}{\partial \theta} + \sqrt{n}(\hat{\theta}^H_{\alpha,n} - \theta^0)^T \left(\frac{1}{n} \sum_{t=1}^n \frac{\partial^2 h_{\alpha,t}(\theta^*_{\alpha,n})}{\partial \theta \partial \theta^T} \right),$$

where the term $\partial^2 h_{\alpha,t}(\theta^*_{\alpha,n})/\partial \theta \partial \theta^T$ actually represents a 9×9 matrix whose (i,j)-th entry is $\partial^2 h_{\alpha,t}(\theta^*_{\alpha,n,ij})/\partial \theta(i) \partial \theta(j)$ for some vector $\theta^*_{\alpha,n,ij}$ between θ_0 and $\hat{\theta}^H_{\alpha,n}$. We first show that

$$\frac{1}{\sqrt{n}} \sum_{t=1}^n \frac{\partial h_{\alpha,t}(\theta^0)}{\partial \theta} \xrightarrow{d} N(0, K_\alpha). \tag{A1}$$

For $v = (v_1^T, v_2^T, v_3)^T \in \mathbb{R}^4 \times \mathbb{R}^4 \times \mathbb{R}$, we obtain

$$\begin{aligned} E\left(v^T \frac{\partial h_{\alpha,t}(\theta^0)}{\partial \theta} \middle| \mathcal{F}_{t-1} \right) &= (1+\alpha) \left\{ v_1^T \frac{\partial \lambda^0_{t,1}}{\partial \theta_1} E\left(D_{t,1}(\theta^0) | \mathcal{F}_{t-1} \right) \right. \\ &\quad + v_2^T \frac{\partial \lambda^0_{t,2}}{\partial \theta_2} E\left(D_{t,2}(\theta^0) | \mathcal{F}_{t-1} \right) + v_3 E\left(D_{t,3}(\theta^0) | \mathcal{F}_{t-1} \right) \right\} \\ &= 0 \end{aligned}$$

and

$$E\left(v^T \frac{\partial h_{\alpha,t}(\theta^0)}{\partial \theta} \right)^2 = v^T E\left(\frac{\partial h_{\alpha,t}(\theta^0)}{\partial \theta} \frac{\partial h_{\alpha,t}(\theta^0)}{\partial \theta^T} \right) v < \infty$$

by Lemma A5. Hence, it follows from the central limit theorem in Billingsley [47] that

$$\frac{1}{\sqrt{n}} \sum_{t=1}^n v^T \frac{\partial h_{\alpha,t}(\theta^0)}{\partial \theta} \xrightarrow{d} N(0, v^T K_\alpha v),$$

which implies (A1).

Now, we claim that

$$-\frac{1}{n} \sum_{t=1}^n \frac{\partial^2 h_{\alpha,t}(\theta^*_{\alpha,n,ij})}{\partial \theta(i) \partial \theta(j)} \xrightarrow{a.s.} J^{ij}_\alpha, \tag{A2}$$

where J^{ij}_α denotes the (i,j)-th entry of J_α. From Lemma A5, J_α is finite. Further, after some algebras, we have

$$\begin{aligned} v^T(-J_\alpha) v &= (1+\alpha) E \left[\sum_{y_1=0}^\infty \sum_{y_2=0}^\infty f^{1+\alpha}_{\theta^0}(y|\lambda_t) \left\{ \left(v_1^T \frac{\partial \lambda^0_{t,1}}{\partial \theta_1} \right) \left(\frac{y_1}{\lambda^0_{t,1}} - 1 + \frac{c\delta^0 e^{-c\lambda^0_{t,1}} u(y_2, \lambda^0_{t,2})}{\varphi(y, \lambda^0_t, \delta^0)} \right) \right. \right. \\ &\quad + \left(v_2^T \frac{\partial \lambda^0_{t,2}}{\partial \theta_2} \right) \left(\frac{y_2}{\lambda^0_{t,2}} - 1 + \frac{c\delta^0 e^{-c\lambda^0_{t,2}} u(y_1, \lambda^0_{t,1})}{\varphi(y, \lambda^0_t, \delta^0)} \right) \\ &\quad \left. \left. + v_3 \frac{u(y_1, \lambda^0_{t,1}) u(y_2, \lambda^0_{t,2})}{\varphi(y, \lambda^0_t, \delta^0)} \right\}^2 \right] > 0 \end{aligned}$$

by (v) of Lemma A1, which implies that J_α is non-singular. Note that we can write

$$\left| \frac{1}{n} \sum_{t=1}^{n} \frac{\partial^2 h_{\alpha,t}(\theta^*_{\alpha,n,ij})}{\partial \theta(i) \partial \theta(j)} - E\left(\frac{\partial^2 h_{\alpha,t}(\theta^0)}{\partial \theta(i) \partial \theta(j)} \right) \right|$$
$$\leq \sup_{\theta \in \Theta} \left| \frac{1}{n} \sum_{t=1}^{n} \frac{\partial^2 h_{\alpha,t}(\theta)}{\partial \theta(i) \partial \theta(j)} - E\left(\frac{\partial^2 h_{\alpha,t}(\theta)}{\partial \theta(i) \partial \theta(j)} \right) \right| + \left| E\left(\frac{\partial^2 h_{\alpha,t}(\theta^*_{\alpha,n,ij})}{\partial \theta(i) \partial \theta(j)} \right) - E\left(\frac{\partial^2 h_{\alpha,t}(\theta^0)}{\partial \theta(i) \partial \theta(j)} \right) \right|.$$

Because $\partial^2 h_{\alpha,t}(\theta)/\partial \theta(i) \partial \theta(j)$ is stationary and ergodic, from Lemma A5, the first term on the RHS of the inequality converges to 0 a.s. Moreover, the second term converges to 0 by the dominated convergence theorem. Hence, (A2) is asserted. Therefore, from (A1) and (A2), the second part of the lemma is established. □

Proof of Theorem 2. From MVT, we get

$$\frac{1}{n} \sum_{t=1}^{n} \frac{\partial h_{\alpha,t}(\hat{\theta}^H_{\alpha,n})}{\partial \theta(i)} - \frac{1}{n} \sum_{t=1}^{n} \frac{\partial h_{\alpha,t}(\hat{\theta}_{\alpha,n})}{\partial \theta(i)} = (\hat{\theta}^H_{\alpha,n} - \hat{\theta}_{\alpha,n})^T \left(\frac{1}{n} \sum_{t=1}^{n} \frac{\partial^2 h_{\alpha,t}(\zeta_{\alpha,n,i})}{\partial \theta \partial \theta(i)} \right)$$

for some vector $\zeta_{\alpha,n,i}$ between $\hat{\theta}^H_{\alpha,n}$ and $\hat{\theta}_{\alpha,n}$. Thus, we can write

$$\frac{1}{n} \sum_{t=1}^{n} \frac{\partial h_{\alpha,t}(\hat{\theta}^H_{\alpha,n})}{\partial \theta} - \frac{1}{n} \sum_{t=1}^{n} \frac{\partial h_{\alpha,t}(\hat{\theta}_{\alpha,n})}{\partial \theta} = (\hat{\theta}^H_{\alpha,n} - \hat{\theta}_{\alpha,n})^T \left(\frac{1}{n} \sum_{t=1}^{n} \frac{\partial^2 h_{\alpha,t}(\zeta_{\alpha,n})}{\partial \theta \partial \theta^T} \right),$$

where the (i,j)-th entry of $\partial^2 h_{\alpha,t}(\zeta_{\alpha,n})/\partial \theta \partial \theta^T$ is $\partial^2 h_{\alpha,t}(\zeta_{\alpha,n,ij})/\partial \theta(i) \partial \theta(j)$ for some vector $\zeta_{\alpha,n,ij}$ between $\hat{\theta}^H_{\alpha,n}$ and $\hat{\theta}_{\alpha,n}$. Since $n^{-1} \sum_{t=1}^{n} \partial h_{\alpha,t}(\hat{\theta}^H_{\alpha,n})/\partial \theta = 0$ and $n^{-1} \sum_{t=1}^{n} \partial \tilde{h}_{\alpha,t}(\hat{\theta}_{\alpha,n})/\partial \theta = 0$, we have

$$\frac{1}{\sqrt{n}} \sum_{t=1}^{n} \frac{\partial \tilde{h}_{\alpha,t}(\hat{\theta}_{\alpha,n})}{\partial \theta} - \frac{1}{\sqrt{n}} \sum_{t=1}^{n} \frac{\partial h_{\alpha,t}(\hat{\theta}_{\alpha,n})}{\partial \theta} = \sqrt{n}(\hat{\theta}^H_{\alpha,n} - \hat{\theta}_{\alpha,n})^T \left(\frac{1}{n} \sum_{t=1}^{n} \frac{\partial^2 h_{\alpha,t}(\zeta_{\alpha,n})}{\partial \theta \partial \theta^T} \right).$$

The LHS of the above equation converges to 0 a.s. by Lemma A6, and $n^{-1} \sum_{t=1}^{n} \partial^2 h_{\alpha,t}(\zeta_{\alpha,n})/\partial \theta \partial \theta^T$ converges to $-J_\alpha$ a.s. in a similar way as in the proof of Lemma A7. Therefore, the theorem is established due to Lemma A7. □

References

1. Weiß, C.H. Thinning operations for modeling time series of counts-a survey. *AStA Adv. Stat. Anal.* **2008**, *92*, 319–341. [CrossRef]
2. Scotto, M.G.; Weiß, C.H.; Gouveia, S. Thinning-based models in the analysis of integer-valued time series: A review. *Stat. Model.* **2015**, *15*, 590–618. [CrossRef]
3. Ferl, ; R.; Latour, A.; Oraichi, D. Integer-valued GARCH processes. *J. Time Ser. Anal.* **2006**, *27*, 923–942. [CrossRef]
4. Fokianos, K.; Rahbek, A.; Tjøstheim, D. Poisson autoregression. *J. Am. Stat. Assoc.* **2009**, *104*, 1430–1439. [CrossRef]
5. Davis, R.A.; Wu, R. A negative binomial model for time series of counts. *Biometrika* **2009**, *96*, 735–749. [CrossRef]
6. Christou, V.; Fokianos, K. Quasi-likelihood inference for negative binomial time series models. *J. Time Ser. Anal.* **2014**, *35*, 55–78. [CrossRef]
7. Zhu, F. Modeling overdispersed or underdispersed count data with generalized poisson integer-valued garch models. *J. Math. Anal. Appl.* **2012**, *389*, 58–71. [CrossRef]
8. Zhu, F. Zero-inflated Poisson and negative binomial integer-valued GARCH models. *J. Stat. Plan. Infer.* **2012**, *142*, 826–839. [CrossRef]
9. Lee, S.; Lee, Y.; Chen, C.W.S. Parameter change test for zero-inflated generalized Poisson autoregressive models. *Statistics* **2016**, *50*, 540–557. [CrossRef]
10. Davis, R.A.; Liu, H. Theory and inference for a class of observation-driven models with application to time series of counts. *Stat. Sin.* **2016**, *26*, 1673–1707.
11. Fokianos, K. Count time series models. In *Handbook of Statistics: Time Series Analysis-Methods and Applications*; Rao, T.S., Rao, S.S., Rao, C.R., Eds.; Elsevier: Amsterdam, The Netherlands, 2012; Volume 30, pp. 315–347.
12. Fokianos, K. Statistical analysis of count time series models: A GLM perspective. In *Handbook of Discrete-Valued Time Series*; Davis, R.A., Holan, S.H., Lund,R., Ravishanker, N., Eds.; Chapman and Hall/CRC: Boca Raton, FL, USA, 2016; pp. 3–27.

13. Tjøstheim, D. Some recent theory for autoregressive count time series (with discussions). *Test* **2012**, *21*, 413–476. [CrossRef]
14. Tjøstheim, D. Count time series with observation-driven autoregressive parameter dynamics. In *Handbook of Discrete-Valued Time Series*; Davis, R.A., Holan, S.H., Lund, R., Ravishanker, N., Eds.; Chapman and Hall/CRC: Boca Raton, FL, USA, 2016; pp. 77–100.
15. Quoreshi, A.S. Bivariate time series modeling of financial count data. *Commun. Stat. Theory Methods* **2006**, *35*, 1343–1358. [CrossRef]
16. Pedeli, X.; Karlis, D. A bivariate INAR(1) process with application. *Stat. Model.* **2011**, *11*, 325–349. [CrossRef]
17. Liu, H. Some Models for Time Series of Counts. Ph.D. Thesis, Columbia University, New York, NY, USA, 2012.
18. Andreassen, C.M. Models and Inference for Correlated Count Data. Ph.D. Thesis, Aarhus University, Aarhus, Denmark, 2013.
19. Lee, Y.; Lee, S.; Tjøstheim, D. Asymptotic normality and parameter change test for bivariate Poisson INGARCH models. *Test* **2018**, *27*, 52–69. [CrossRef]
20. Cui, Y.; Zhu, F. A new bivariate integer-valued GARCH model allowing for negative cross-correlation. *Test* **2018**, *27*, 428–452. [CrossRef]
21. Lakshminarayana, J.; Pandit, S.N.N.; Rao, K.S. On a bivariate Poisson distribution. *Commun. Stat. Theory Methods* **1999**, *28*, 267–276. [CrossRef]
22. Basu, A.; Harris, I.R.; Hjort, N.L.; Jones, M.C. Robust and efficient estimation by minimizing a density power divergence. *Biometrika* **1998**, *85*, 549–559. [CrossRef]
23. Kang, J.; Lee, S. Minimum density power divergence estimator for Poisson autoregressive models. *Comput. Stat. Data Anal.* **2014**, *80*, 44–56. [CrossRef]
24. Kim, B.; Lee, S. Robust estimation for zero-inflated Poisson autoregressive models based on density power divergence. *J. Stat. Comput. Simul.* **2017**, *87*, 2981–2996. [CrossRef]
25. Kim, B.; Lee, S. Robust estimation for general integer-valued time series models. *Ann. Inst. Stat. Math.* **2020**, *72*, 1371–1396. [CrossRef]
26. Diop, M.L.; Kengne, W. Density power divergence for general integer-valued time series with multivariate exogenous covariate. *arXiv* **2020**, arXiv:2006.11948.
27. Kim, B.; Lee, S. Robust change point test for general integer-valued time series models based on density power divergence. *Entropy* **2020**, *22*, 493. [CrossRef]
28. Lee, S.; Kim, D. Monitoring parameter change for time series models of counts based on minimum density power divergence estimator. *Entropy* **2020**, *22*, 1304. [CrossRef]
29. Xiong, L.; Zhu, F. Robust quasi-likelihood estimation for the negative binomial integer-valued GARCH(1,1) model with an application to transaction counts. *J. Stat. Plan. Infer.* **2019**, *203*, 178–198. [CrossRef]
30. Li, Q.; Chen, H.; Zhu, F. Robust estimation for Poisson integer-valued GARCH models using a new hybrid loss. *J. Syst. Sci. Complex.* **2021**, in press. [CrossRef]
31. Heinen, A.; Rengifo, E. Multivariate modeling of time series count data: An AR conditional Poisson model. In *CORE Discussion Paper*; Université Catholique de Louvain, Center for Operations Research and Econometrics (CORE): Ottignies-Louvain-la-Neuve, Belgium, 2003.
32. Cui, Y.; Li, Q.; Zhu, F. Flexible bivariate Poisson integer-valued GARCH model. *Ann. Inst. Stat. Math.* **2020**, *72*, 1449–1477. [CrossRef]
33. Fujisawa, H.; Eguchi, S. Robust estimation in the normal mixture model. *J. Stat. Plan. Infer.* **2006**, *136*, 3989–4011. [CrossRef]
34. Durio, A.; Isaia, E.D. The minimum density power divergence approach in building robust regression models. *Informatica* **2011**, *22*, 43–56. [CrossRef]
35. Toma, A.; Broniatowski, M. Dual divergence estimators and tests: Robustness results. *J. Multivar. Anal.* **2011**, *102*, 20–36. [CrossRef]
36. Warwick, J. A data-based method for selecting tuning parameters in minimum distance estimators. *Comput. Stat. Data Anal.* **2005**, *48*, 571–585. [CrossRef]
37. Verges, Y. The Bivariate Integer-Valued GARCH Model: A Bayesian Estimation Framework. Master's Thesis, University of São Paulo, São Paulo, Brazil, 2019.
38. Chen, C.W.S.; Lee, S. Generalized Poisson autoregressive models for time series of counts. *Comput. Stat. Data Anal.* **2016**, *99*, 51–67. [CrossRef]
39. Chen, C.W.S.; Lee, S. Bayesian causality test for integer-valued time series models with applications to climate and crime data. *J. R. Stat. Soc. Ser. C Appl. Stat.* **2017**, *66*, 797–814. [CrossRef]
40. Lee, S.; Park, S.; Chen, C.W.S. On Fisher's dispersion test for integer-valued autoregressive Poisson models with applications. *Commun. Stat. Theory Methods* **2017**, *46*, 9985–9994. [CrossRef]
41. Fokianos, K.; Fried, R. Interventions in INGARCH processes. *J. Time Ser. Anal.* **2010**, *31*, 210–225. [CrossRef]
42. Heinen, A.; Rengifo, E. Multivariate autoregressive modeling of time series count data using copulas. *J. Empir. Financ.* **2007**, *14*, 564–583. [CrossRef]
43. Fokianos, K.; Støve, B.; Tjøstheim, D.; Doukhan, P. Multivariate count autoregression. *Bernoulli* **2020**, *26*, 471–499. [CrossRef]
44. Kang, J.; Lee, S. Parameter change test for Poisson autoregressive models. *Scand. J. Stat.* **2014**, *41*, 1136–1152. [CrossRef]
45. Straumann, D.; Mikosch, T. Quasi-maximum-likelihood estimation in conditionally heteroscedastic time series: A stochastic recurrence equations approach. *Ann. Stat.* **2006**, *34*, 2449–2495. [CrossRef]

46. Cui, Y.; Zheng, Q. Conditional maximum likelihood estimation for a class of observation-driven time series models for count data. *Stat. Probab. Lett.* **2017**, *123*, 193–201. [CrossRef]
47. Billingsley, P. The Lindeberg-Lévy theorem for martingales. *Proc. Am. Math. Soc.* **1961**, *12*, 788–792.

Article

Multivariate Count Data Models for Time Series Forecasting

Yuliya Shapovalova [1,*], Nalan Baştürk [2] and Michael Eichler [2]

[1] Institute for Computing and Information Sciences, Radboud University Nijmegen, Toernooiveld 212, 6525 EC Nijmegen, The Netherlands
[2] School of Business and Economics, Maastricht University, Tongersestraat 53, 6211 LM Maastricht, The Netherlands; n.basturk@maastrichtuniversity.nl (N.B.); m.eichler@maastrichtuniversity.nl (M.E.)
* Correspondence: yuliya.shapovalova@ru.nl

Abstract: Count data appears in many research fields and exhibits certain features that make modeling difficult. Most popular approaches to modeling count data can be classified into observation and parameter-driven models. In this paper, we review two models from these classes: the log-linear multivariate conditional intensity model (also referred to as an integer-valued generalized autoregressive conditional heteroskedastic model) and the non-linear state-space model for count data. We compare these models in terms of forecasting performance on simulated data and two real datasets. In simulations, we consider the case of model misspecification. We find that both models have advantages in different situations, and we discuss the pros and cons of inference for both models in detail.

Keywords: multivariate count data; INGACRCH; state-space model; bank failures; transactions

Citation: Shapovalova, Y.; Baştürk, N.; Eichler, M. Multivariate Count Data Models for Time Series Forecasting. *Entropy* **2021**, *23*, 718. https://doi.org/10.3390/e23060718

Academic Editor: Christian H. Weiss

Received: 30 April 2021
Accepted: 2 June 2021
Published: 5 June 2021

Publisher's Note: MDPI stays neutral with regard to jurisdictional claims in published maps and institutional affiliations.

Copyright: © 2020 by the authors. Licensee MDPI, Basel, Switzerland. This article is an open access article distributed under the terms and conditions of the Creative Commons Attribution (CC BY) license (https://creativecommons.org/licenses/by/4.0/).

1. Introduction

Modeling time series of counts is relevant in a range of application areas, including the dynamics of the number of infectious diseases, number of road accidents or number of bank failures. In many applications, such count data dynamics are correlated across several data series. Examples include from correlated number of bank failures [1], number of crimes [2] to COVID-19 contagion dynamics [3]. The analysis of such correlations provides detailed information about the overall connectedness of the series, as well as the dynamics of an individual series conditional on the others. Several multivariate count data models have been proposed to capture the overall connectedness of multivariate count data. Each one of these models has different underlying assumptions as well as computational challenges. We present a comparative study of two families of multivariate count data models, namely State Space Models (SSM) and log-linear multivariate autoregressive conditional intensity (MACI) models, based on simulation studies and two empirical applications.

We provide some examples of the count data and discuss particular properties that one desires to model when dealing with such data. In this paper, we assume that the counts are unbounded and we assume both models to be stationary. For discussion on the difference between bounded and unbounded count data and the difference of the modeling approaches for these data we refer to [4]. The top panels in Figure 1 present two conventional data sets that have been used for univariate illustrations, namely the monthly number of cases of poliomyelitis in the U.S. between 1970 and 1983, and asthma presentations at a Sydney hospital. The middle panel in Figure 1 presents the number of bank failures in the U.S. over time, a dataset that we also analyze in this paper, and the number of transactions for BMW in a 30 second interval. The bottom panel in Figure 1 presents a number of car crashes and a number of earthquakes. The former, number of car crashes over time is analyzed in Park and Lord [5] with a multivariate Poisson log-normal model with correlations for modeling the crash frequency by severity. The authors demonstrate that, accounting for the correlations in the multivariate model can improve the accuracy of the estimation. A common feature in all presented datasets is the autocorrelation present in count data over time that is visible in the time series plots. In multivariate count time series data, this correlation generalizes to a correlation between past and current values of a specific series as well as between different series.

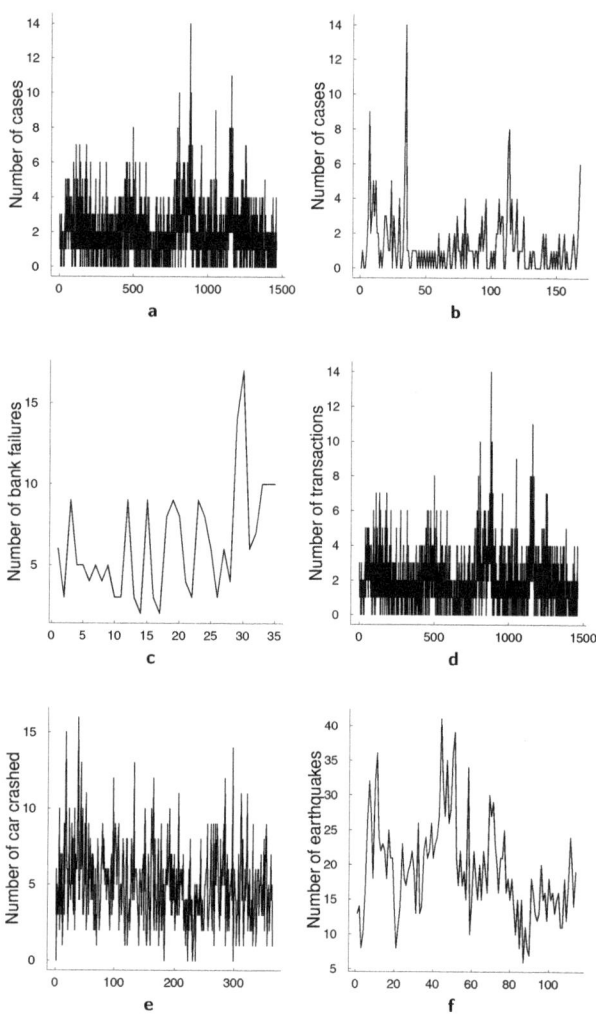

Figure 1. Typical examples of count data coming from applications in different scientific fields. (**a**) Monthly number of cases of poliomyelitis in the U.S. (1970–1983). (**b**) Asthma presentations at a Sydney hospital (**c**) Number of bank failures in US (**d**) Number of transactions for BMW on 30 s interval (**e**) Number of car crashes (**f**) Number of earthquakes.

Models for multivariate count time series typically rely on multivariate Poisson distributions, where time-variation is defined through one or more rate parameters [6]. In some cases, Gaussian approximations are used but, as has been shown in [7], this can lead to reduced performance in the risk forecasting assessment. In general, the quality of such approximations depends on a particular problem [8]. Estimation of these models is computationally demanding for high numbers of counts as the estimation relies on the sum over all counts. In addition, these models typically have positivity restrictions on the conditional intensity function that governs the Poisson process over time and the correlation between different time series. A few exceptions to the positive correlation assumption exist, see for example, [9,10].

An alternative model for the joint distribution of count data is the copula model. A number of papers proposed different copula models for multivariate count time series, see, for example, [11–14]. Copulas are generally used for modeling dependency in multiple series, which makes them attractive methods also for multiple count time series. However, several issues arise in their applications to count data such as unidentifiability and not accounting for potential overdispersion—a property that is common for count data. Genest and Nešlehová [15] provides a detailed overview of copula models for count data, and proposes and compares Bayesian and classical inferential approaches for multivariate Poisson regression. They show that computationally Bayesian and classical approaches are of a similar order.

Both approaches of modeling joint distribution of count data—multivariate Poisson distribution and copulas—can be incorporated in the autoregressive conditional intensity (ACI) framework, often also referred to as an integer-valued generalized autoregressive conditional heteroskedasticity model (INGARCH). This model belongs to the class of observation driven models as opposed to parameter driven models, a classification proposed by Cox et al. [16]. ACI models have been dominating the literature for quite a long time despite their restrictiveness: these models only allow only for positive coefficients in the equation for conditional intensity. These bounded coefficients lead to several problems besides potentially unrealistic dependence structure for some data. In particular, the problem of calculating confidence sets for the parameters that are close to or on the boundary rises and has not been yet solved in the literature. Another observation driven model that has been proposed as an alternative to ACI framework is log-linear model, see Fokianos and Tjøstheim [17], a multivariate extension of which has been considered in Doukhan et al. [10]. Even though the problem of modeling joint distribution remains, the advantage of this approach is that no restrictions on the parameter space are required due to the log-transform of the data.

Another class of models that can be considered for modeling count data, but is rarely used in the literature, is parameter driven models and, in particular, non-linear state-space models. In this framework, the observations are driven by an independent unobserved stochastic process which, for instance, can be a (vector) autoregressive process ($VAR(p)$). These models have been discussed extensively in the univariate case, see, for example Davis et al. [18]. However, these models are rarely used in multivariate applications due to the computationally demanding estimation methods that have to be used. To our knowledge, only one very recent study has considered them in a multivariate application Zhang et al. [19]. Non-linear state-space models are capable of modeling and inferring complex dependence structures in the data. They allow for both negative and positive contemporaneous correlation, as well as for both negative and positive Granger-causal feedback. Thereby, these models avoid the problem of modeling the joint distribution of time series of counts and provide a coherent inferential tool in the Bayesian framework. This is what distinguishes our approach from the approach discussed in Zhang et al. [19] who consider frequentist estimation of these models. We also compare SSM to log-linear models instead of MACI models since they allow for negative dependence between the intensities and hence appear to be more natural competitors of SSM models.

In this paper, we compare two classes of models, observation driven and parameter driven models, in terms of their forecasting performances. We estimate the observation driven models the quasi-maximum likelihood method. Parameter driven models, however, fit very well into the Bayesian paradigm and that is what we use for estimation. Certain advantages come together with this framework, such as those naturally obtained from the posterior distribution uncertainty about the parameters of the model and forecast of the multivariate time series [20]. We in particular use particle Markov chain Monte Carlo (pMCMC) [21] for the estimation of the parameter driven model. As is discussed in [22], pMCMC outperforms other methods (variational Bayes [23], integrated nested Laplace approximation [24] and Riemann manifold Hamiltonian Monte Carlo [25]) in terms of parameter estimation. There are other recent methods for the estimation of the state-space models such as auxiliary likelihood-based approximate Bayesian computation [26] and variational Sequential Monte Carlo [27], but their performance has to be investigated further, which is outside of the scope of this paper.

We present a set of simulation studies to show how these models perform when they are correctly specified and misspecified. The simulation results show that, as expected, the correctly specified models perform generally well, but there are exceptions. Particularly, parameter driven models have better forecast performances in some simulations even if they are misspecified. In addition to these simulation studies, we compare the performances of these models in two real data applications. The two data sets we analyze exhibit different sample sizes, standard deviation, dispersion and maximum counts. We show that the overall forecast performances of the models can be very different, depending on the applications. Furthermore, for the second data set we analyze, we find that observation driven models capture extreme data values better than parameter driven models.

The remainder of this paper is as follows: Sections 2 and 3 summarize observation and parameter driven models, respectively. Section 4 presents the model and forecast comparison tools we use for multivariate count data models. Section 5 presents simulation results. Section 6 presents results from applications to two data sets. Finally, Section 7 concludes the paper.

2. Observation Driven Models

In this section, we summarize two observation driven models: multivariate autoregressive conditional intensity model and log-linear analog of it. Both of these models are characterized by the dynamics that depend on the past of the process itself and some noise. Both models have been considered in Doukhan et al. [10], where the authors discussed some theoretical properties and proposed to use copula approach for modeling joint count distribution. Copulas are flexible tools for modeling dependence structure but their use in count time series models brings challenges. We first summarize the use of Poisson distribution for count data, analyze both models under an independence assumption in the Poisson random variables, and at the end of this section, we discuss the extension of modeling multiple count time series with multivariate Poisson distribution.

2.1. Poisson Distribution

Many of the count time series models take their origins in the idea of Poisson regression model, an extensive overview of these models is given in Fokianos [28]. Specifically, both models considered in this section as well as the parameter driven models in Section 3 rely on Poisson distributions. We therefore first provide some background on the Poisson distribution. Poisson distribution has played an important role in modeling count time series data as its interpretation is the number of independent events that occur in a time period. The Poisson distribution is defined for a random variable x takes integer values in $\{0, 1, \ldots\}$. The mean of the distribution, λ, describes the average occurrences per interval, the distribution has the equi-dispersion property since the variance its variance is also λ, and the probability mass function (pmf) of the distribution is

$$p(x) = \frac{\lambda^x e^{-\lambda}}{x!}, \quad x = 0, 1, 2, 3 \ldots, \tag{1}$$

with $\mathbb{E}(x) = Var(x) = \lambda$.

For the simple multivariate case, two Poisson random variables, say x_1 and x_2, the joint pmf reads

$$p(x_1, x_2) = \prod_{i=1}^{2} \frac{e^{-\lambda_i} \lambda_i^{x_i}}{x_i!}. \tag{2}$$

The multivariate extension in (2) is rather naive due to the underlying independence assumption between x_1 and x_2. Such a model would ignore potential dependency of the real world data, thus is potentially not suitable for the majority of applications. One way to use the Poisson distribution for modeling count data in multivariate case and incorporate correlation structure is through the so-called *trivariate reduction* [13,29]. The idea is that correlation can be modeled through the third Poisson variable. Assume we have three independent random variables $x_i \sim \text{Poisson}(\lambda_{t,i} - \varphi)$, where $0 \leq \varphi \leq \min\{\lambda_{t,1}, \lambda_{t,2}\}$. Define $Y_{t,1} = X_1 + X_2$ and $Y_{t,2} = X_2 + X_3$. In this way, the random variable X_2 is exploited to model the dependence between Y_1 and Y_2. The restriction of this approach is that the correlation is the same between all the series (in case one wants to model the systems beyond bivariate case) and the dependence can only be positive. We further discuss the trivariate reduction technique in the context of ACI/INGARCH models. In particular, the difficulties of extending this to higher dimensions is of interest and presents one with a challenging task.

2.2. MACI (INGARCH)

The Poisson integer-valued generalized autoregressive conditional heteroscedastic process (INGARCH) models [30]—also called multivariate autoregressive conditional intensity models (MACI) in the literature—are built upon GARCH framework and are capable of capturing time series properties of count data. As for GARCH-type models, it is assumed that the conditional mean of the process at time t depends on the value of the process at period $t-1$ and its conditional mean at time $t-1$. The time series of counts follow Poisson process with the conditional mean λ_t, that is,

$$X_{i,t} \mid \mathcal{F}_{t-1} \sim \text{Poisson}(\lambda_{i,t}), \quad i = 1, \ldots, n. \tag{3}$$

The corresponding joint pmf reads

$$P(X_{1t} = x_{1t}, \ldots, X_{nt} = x_{nt} \mid \mathcal{F}_{t-1}) = \prod_{i=1}^{n} \frac{e^{-\lambda_{it}} \lambda_{it}^{x_{it}}}{x_{it}!}. \tag{4}$$

The dynamics of the conditional intensity $\lambda_t = \mathbb{E}[X_t \mid \mathcal{F}_{t-1}]$ follows

$$\lambda_t = \omega + \sum_{i=1}^{n} A^i \lambda_{t-i} + \sum_{j=1}^{q} B^j X_{t-j}. \tag{5}$$

Note that the elements of ω, a^i, b^j are assumed to be positive to ensure the positivity of the intensity process λ. (Doukhan et al. [10] argue that the condition $\|A + B\|_2 < 1$ guarantees stationarity.) In addition, we assume no contemporaneous correlation in the counts. Consider the bivariate case for the conditional intensity process

$$\begin{bmatrix} \lambda_{1t} \\ \lambda_{2t} \end{bmatrix} = \begin{bmatrix} \omega_1 \\ \omega_2 \end{bmatrix} + \begin{bmatrix} a_{11} & a_{12} \\ a_{21} & a_{22} \end{bmatrix} \begin{bmatrix} \lambda_{1t-i} \\ \lambda_{2t-i} \end{bmatrix} + \begin{bmatrix} b_{11} & b_{12} \\ b_{21} & b_{22} \end{bmatrix} \begin{bmatrix} X_{1t-j} \\ X_{2t-j} \end{bmatrix}, \quad t = 0, \pm 1, \pm 2, \ldots. \tag{6}$$

From Equation (6) it is clear that when A and B are diagonal, there is no dependence structure between the intensities. Further, when $a_{12} = 0$ and $b_{12} = 0$ then the intensity of the first process, $\lambda_{1,t}$, depends only on its own past while the second process can depend on the dynamics of the first one. Finally, if we restrict A to be diagonal and B to be non-diagonal, every intensity process would depend on its past and possibly on the past of all of the observations. This constraint is relevant when one wants to apply graphical modeling to this problem.

2.3. Quasi-Maximum Likelihood for MACI Models

In this section, we discuss how the inference for MACI/INGARCH models can be executed. The details for the multivariate case have also been discussed in Doukhan et al. [10]. For these models, we make use of the classical estimation framework and in particular use quasi-maximum likelihood estimation. The conditional quasi-likelihood for this model and θ reads

$$L(\theta) = \prod_{t=1}^{T}\prod_{i=1}^{n}\left(\frac{\exp(-\lambda_{i,t}(\theta))\lambda_{i,t}^{x_{i,t}}(\theta)}{x_{i,t}!}\right), \quad (7)$$

where θ are the parameters of interest. It follows the the quasi log-likelihood function is

$$l(\theta) = \sum_{t=1}^{T}\sum_{i=1}^{n}(x_{i,t}\log\lambda_{i,t}(\theta) - \lambda_{i,t}(\theta)), \quad (8)$$

and the corresponding score function reads

$$S_T(\theta) = \sum_{t=1}^{T}\sum_{i=1}^{n}\left(\frac{x_{i,t}}{\lambda_{i,t}} - 1\right)\frac{\partial\lambda_{i,t}(\theta)}{\partial\theta}$$
$$= \sum_{t=1}^{T}\frac{\partial\lambda_t^T(\theta)}{\partial\theta}D_t^{-1}(\theta)(X_t - \lambda_t(\theta)) \equiv \sum_{t=1}^{T}s_t(\theta), \quad (9)$$

where $\partial\lambda_t/\partial\theta^T$ is $n \times d$ matrix with $d \equiv n(1+2n)$ being the dimension of the parameter vector θ, D_t is an $n \times n$ diagonal matrix and its diagonal elements are $\lambda_{i,t}(\theta)$, $i = 1, 2, \ldots, n$, and X_t consists of elements $x_{i,t}$, $i = 1, 2, \ldots, n$, $t = 1, 2, \ldots, T$. Thus the recursions for the quasi-maximum likelihood estimation follow

$$\frac{\partial\lambda_t}{\partial\omega^T} = I_n + A\frac{\partial\lambda_{t-1}}{\partial\omega^T}, \quad (10)$$

$$\frac{\partial\lambda_t}{\partial vec^T(A)} = (\lambda_{t-1}\otimes I_n)^T + A\frac{\partial\lambda_{t-1}}{\partial vec^T(A)}, \quad (11)$$

$$\frac{\partial\lambda_t}{\partial vec^T(B)} = (X_{t-1}\otimes I_n)^T + A\frac{\partial\lambda_{t-1}}{\partial vec^T(B)}. \quad (12)$$

Finally, the Hessian matrix and the conditional information matrix are correspondingly

$$H_T(\theta) = \sum_{t=1}^{T}\sum_{i=1}^{n}\frac{x_{i,t}}{\lambda_{i,t}^2(\theta)}\frac{\partial\lambda_{i,t}(\theta)}{\partial\theta}\frac{\partial\lambda_{i,t}(\theta)}{\partial\theta^T} - \sum_{t=1}^{T}\sum_{i=1}^{n}\left(\frac{x_{i,t}}{\lambda_{i,t}(\theta)} - 1\right)\frac{\partial^2\lambda_{i,t}(\theta)}{\partial\theta\partial\theta^T}, \quad (13)$$

$$G_T = \sum_{t=1}^{T}\frac{\partial\lambda_t^T(\theta)}{\partial\theta}D_t^{-1}(\theta)\Sigma_t D_t^{-1}(\theta)\frac{\lambda_t(\theta)}{\partial\theta^T}. \quad (14)$$

Further, one can show that $S_n(\theta) = 0$ has a unique solution, $\hat{\theta}$, which is strongly consistent and asymptotically normal. For further details of these properties, we refer the reader to Doukhan et al. [10]. However, that theoretical properties of $\hat{\theta}$ are proven under assumption that the true value θ_0 belongs to the interior of the parameter space Θ. The problems certainly arise when the true parameter is close or on the boundary of the parameter space. Dealing with the theoretical problems of the constrained optimization and parameters near or on the boundary of parameter space is out of the scope of this paper and generally establishing the theory for this case is a complicated task. One of the possible solutions is to exploit bootstrap methods for this task, see Hilmer et al. [31] for a

2.4. Log-Linear Autoregressive Model

Log-linear models have appeared in the count data literature in the recent years [10] and have good potential since they allow for both positive/negative correlation and avoid parameter boundary problems which MACI models suffer from.

$$X_{i,t} \mid \mathscr{F}_{t-1}^{X,\lambda} \sim \text{Poisson}(\lambda_{i,t}), \tag{15}$$

$$\nu_t = \omega + A\nu_{t-1} + B\log(X_{t-1} + \mathbf{1}_n), \quad t \geq 1, \tag{16}$$

where $\mathscr{F}_{t-1}^{Y,\lambda}$ is the σ-field generated by $\{X_0, \ldots, X_t, \lambda_0\}$, $\mathbf{1}_n$ is the n-dimensional vector of ones, $\nu_t \equiv \log \lambda_t$. Parameters of this model, ω, A, and B, do not have to be positive which makes this model more attractive than MACI.

2.5. Quasi-Maximum Likelihood for Log-Linear Models

The inference in log-linear models is very similar to the quasi-maximum likelihood approach derived for MACI models in Section 2.3. Only minor adjustments have to be made in corresponding recursions [10]. In particular, the score function for the log-linear model reads

$$S_T(\theta) = \sum_{t=1}^{T}\sum_{i=1}^{n}(x_{i,t} - \exp(\nu_{i,t}(\theta)))\frac{\partial \nu_{i,t}(\theta)}{\partial \theta} = \sum_{t=1}^{T}\frac{\partial \nu_t^T(\theta)}{\partial \theta}(X_t - \exp(\nu_t(\theta))), \tag{17}$$

the Hessian matrix is

$$H_T(\theta) = \sum_{t=1}^{T}\sum_{i=1}^{n}\exp(\nu_{i,t}(\theta))\frac{\partial \nu_{i,t}(\theta)}{\partial \theta}\frac{\partial \nu_{i,t}(\theta)}{\partial \theta^T} - \sum_{t=1}^{T}\sum_{i=1}^{n}(x_{i,t} - \exp(\nu_{i,t}(\theta)))\frac{\partial^2 \nu_{i,t}(\theta)}{\partial \theta \partial \theta^T}, \tag{18}$$

and the conditional information matrix for the log-linear model reads

$$G_T(\theta) = \sum_{t=1}^{T}\sum_{i=1}^{n}\exp(\nu_{i,t}(\theta))\frac{\partial \nu_{i,t}(\theta)}{\partial \theta}\frac{\partial \nu_{i,t}(\theta)}{\partial \theta^T}. \tag{19}$$

Doukhan et al. [10] prove theoretical properties of this model. In particular, they show that there exists a unique solution $\hat{\theta}$ which is strongly consistent and asymptotically normal. The authors also show that the condition $\sum_{j=0}^{\infty} \| A^j B \|_2 < 1$ guarantees both stationarity and weak dependence.

2.6. Multivariate Poisson Distribution

To allow for contemporaneous correlation, we need to use trivariate reduction technique discussed before. We consider the bivariate case to give an example, assume that there are three independent random variables Y_1, Y_2, Y_3 with positive means λ_1, λ_2, λ_3 respectively. Define random variables $X_1 = Y_1 + Y_3$ and $X_2 = Y_2 + Y_3$. The new random variables will have means $\lambda_1 + \lambda_3$ and $\lambda_2 + \lambda_3$, where λ_3 would also correspond to the covariance between X_1 and X_2. The covariance is clearly restricted to be positive, while correlation will lie between 0 and $\min\{\frac{\sqrt{\lambda_1+\lambda_3}}{\sqrt{\lambda_2+\lambda_3}}, \frac{\sqrt{\lambda_2+\lambda_3}}{\sqrt{\lambda_1+\lambda_3}}\}$. Thereby the joint pmf of interest, alternative to what we have in Equation (3), becomes

$$\begin{aligned}P(X_{1t} = x_{1t}, X_{2t} = x_{2t} \mid \mathscr{F}_{t-1}) =& e^{-(\lambda_1+\lambda_2+\lambda_3)}\frac{\lambda_1^{x_1}\lambda_2^{x_2}}{x_1!x_2!}\\ & \times \sum_{i=0}^{\min(x_1,x_2)}\binom{x_1}{i}\binom{x_2}{i}i!\left(\frac{\lambda_3}{\lambda_1\lambda_2}\right)^i.\end{aligned} \tag{20}$$

Extending this approach to contemporaneous correlation in higher dimensions is not trivial. Suppose that we would like to model n Poisson random variables, thus $X_i \sim \text{Poisson}(\lambda_i)$, $i = 1, \ldots, n$. By defining a random variable $X_0 \sim \text{Poisson}(\lambda_0)$ and extending the argument of the trivariate reduction we can define random variables $X_1 = Y_1 + Y_0, X_2 = Y_2 + Y_0, \ldots, X_n = Y_n + Y_0$. The joint pmf is

$$P(X_1 = x_1, X_2 = x_2, \ldots, X_n = x_n) = \exp(-\sum_{i=1}^{n} \lambda_i) \prod_{i=1}^{n} \frac{\lambda_i^{x_i}}{x_i!} \\ \times \sum_{i=0}^{m} \sum_{j=1}^{n} \binom{x_j}{i} i! \left(\frac{\lambda_0}{\prod_{i=1}^{n} \lambda_i}\right)^i, \quad (21)$$

where $m = \min(x_1, x_2, \ldots, x_n)$. This approach assumes that the covariance is the same for all the pairs of Poisson random variables which is very restrictive. Karlis and Meligkotsidou [9] consider the case with richer covariance structure which we discuss further. For simplicity, assume we want to model trivariate time series of counts Y_1, Y_2, Y_3. As before, let us specify through X_i and X_{ij} univariate Poisson random variables, i.e., $X_i \sim \text{Poisson}(\lambda_i)$ and $X_{ij} \sim \text{Poisson}(\lambda_{ij})$ with $i, j \in \{1, 2, 3\}$, $i < j$. Then the random variables Y_i with $i \in \{1, 2, 3\}$ are defined in the following way

$$\begin{aligned} Y_1 &= X_1 + X_{12} + X_{13}, \\ Y_2 &= X_2 + X_{12} + X_{23}, \\ Y_3 &= X_3 + X_{13} + X_{23}. \end{aligned} \quad (22)$$

Thus, $Y_i \sim \text{Poisson}(\lambda_i + \lambda_{ij} + \lambda_{ik})$, where $i, j, k \in \{1, 2, 3\}, i \neq j \neq k$. Finally, these random variables follow the Poisson distribution with $\boldsymbol{\lambda} = (\lambda_1, \lambda_2, \lambda_3, \lambda_{12}, \lambda_{13}, \lambda_{23})$, and hence with the mean vector $\mathbf{A}\boldsymbol{\lambda} = (\lambda_1 + \lambda_{12} + \lambda_{13}, \lambda_2 + \lambda_{12} + \lambda_{23}, \lambda_3 + \lambda_{13} + \lambda_{23})'$. The variance-covariance matrix for this distribution is given by

$$A\Sigma A' = \begin{pmatrix} \lambda_1 + \lambda_{12} + \lambda_{13} & \lambda_{12} & \lambda_{13} \\ \lambda_{12} & \lambda_2 + \lambda_{12} + \lambda_{23} & \lambda_{23} \\ \lambda_{13} & \lambda_{23} & \lambda_3 + \lambda_{13} + \lambda_{23} \end{pmatrix} \quad (23)$$

It is clear from the above examples that the modeling of the time series of counts with multivariate Poisson distribution in higher dimensions is restrictive and cumbersome. It is restrictive, since it allows only for positive dependency in the data, which can be unreasonable for real-world applications. It is cumbersome since the method is only computationally tractable when one has low counts data, see Equation (21) in which the number of terms in the sum depends on the number of observed counts. Methods such as expectation maximization can be applied in this case but they are not trivial and stable in case of high counts. Moreover, in this case, incorporation of the multivariate Poisson distribution into MACI or log-linear models also affects computational speed substantially, and these models lose their attractiveness over more complex models such as nonlinear state-space models in the next section.

3. Parameter Driven Model: Nonlinear State-Space Model

The advantage of parameter driven models is the clear interpretability of the model parameters and the high degree of flexibility. The model can easily incorporate different distributions and extends easily to the multivariate framework. Moreover, in the Bayesian framework, we have coherent inferential tools derived from the posterior distributions of the parameters, such as highest posterior density intervals. These models also take into account uncertainty about the parameters which is incorporated into predictions. The disadvantage of this approach is challenging estimation procedures that are computationally intensive. Hence, even though theoretically estimation methodologies are possible to extend to any dimension, in practice that is not feasible due to the time constraints. In this paper, we estimate the parameter-driven model for multivariate count data using Sequential Monte Carlo and particle Markov Chain Monte Carlo. These methods became popular with the availability of more computational power. They are restricted in some ways, and we will discuss these restrictions in the next subsections after introducing the nonlinear state space model (SSM), which we will compare to the observation driven models.

3.1. Multivariate SSM

A state-space model is usually presented by an observation equation and a state equation. The state equation represents a latent process, say h_t, which drives the dynamics of observations y_t. In the multivariate SSM for count data below, this dependence between the observations and the state is nonlinear

$$X_{it} \sim \text{Poisson}(\lambda_{it}), \quad i = 1, 2, \ldots, n \tag{24}$$

$$\lambda_t = \beta \exp(h_t) \tag{25}$$

$$h_t = \sum_{i=1}^{n} \Phi^i h_{t-i} + \eta_t, \quad \Sigma_\eta = \begin{pmatrix} \sigma_{\eta_1}^2 & \rho_{\eta_{12}} & \cdots & \rho_{\eta_{1n}} \\ \rho_{\eta_{21}} & \sigma_{\eta_2}^2 & \cdots & \rho_{\eta_{2n}} \\ \vdots & \vdots & \ddots & \vdots \\ \rho_{\eta_{n1}} & \cdots & \cdots & \sigma_{\eta_{n'}}^2 \end{pmatrix} \tag{26}$$

where $\eta_t \sim N(0, \Sigma_\eta)$. Equation (24) shows that the observations have Poisson distribution with mean λ_t defined through the Equation (25), and λ_t nonlinearly depends on the latent process h_t which is defined through Equation (26). Note that the latent process is defined through a VAR(p) process, and hence corresponding theory applies. In particular, the stationarity condition is that the roots of the Equation (27) must lie outside the unit circle

$$\mid \lambda^p I_n - \lambda^{p-1} \Phi_1 - \ldots \Phi_p \mid = 0. \tag{27}$$

The dependence structure between counts is modeled through the dependence in the latent process. Conditioned on the latent process $\{h_t\}_{t=1}^T$ the observations $\{X_t\}_{t=1}^T$ are independent. Furthermore, since the latent process of the model is a VAR(p), we can account for various dependence structures: positive and negative contemporaneous correlation, and positive and negative Granger-causal feedback.

These models are challenging to estimate, and an assumption of $p = 1$ can simplify the inference. (For extending the model to $p > 1$ we advise the reader to consider using sparse priors, such as Minnesota prior, spike and slab or horseshoe prior.) Bivariate specification of the nonlinear state-space model with the lag $p = 1$ reads

$$X_{it} \sim \text{Poisson}(\lambda_{it}), \quad i = 1, 2 \tag{28}$$

$$\lambda_{it} = \beta_i \exp(h_{it}), \quad i = 1, 2. \tag{29}$$

$$\begin{pmatrix} h_{1,t+1} \\ h_{2,t+1} \end{pmatrix} = \begin{pmatrix} \phi_{11} & \phi_{12} \\ \phi_{21} & \phi_{22} \end{pmatrix} \begin{pmatrix} h_{1,t} \\ h_{2,t} \end{pmatrix} + \begin{pmatrix} \eta_{1t+1} \\ \eta_{2t+1} \end{pmatrix}, \quad \Sigma_\eta = \begin{pmatrix} \sigma_{\eta_1}^2 & \rho_{\eta_{12}} \\ \rho_{\eta_{21}} & \sigma_{\eta_2}^2 \end{pmatrix} \tag{30}$$

The dependence structure between series is described by the Granger-causal relationship in the latent processes h_{it} and contemporaneous relations that are incorporated in Σ_η. For example, we say that $h_{2,t}$ does not Granger-cause $h_{1,t}$ if $\phi_{12} = 0$. Correlation coefficient ρ in this model allows us to model both positive and negative correlation between the counts.

3.2. Bayesian Inference in Multivariate SSM

The estimation of nonlinear state-space models naturally fits into the Bayesian framework. The presence of the unobservable process in the model and nonlinear dependence of the observations on this unobservable process leads to an intractable likelihood and posterior. For this reason, and due to the nonlinear SSM structure, we use particle Markov Chain Monte Carlo (PMCMC) for the estimation of the posterior distribution of the model parameters Andrieu et al. [21]. The method consists of two parts. First, the likelihood is estimated in a sequential manner through a *particle filter*. Second, this estimate is used within an MCMC sampler, in our case Metropolis-Hastings algorithm. An extensive introduction to nonlinear state-space models and particle filtering can be found, for example, in Särkkä [32].

Recall the Bayes rule on which the inference is based

$$p(h_{1:T}|x_{1:T}) = \frac{p(x_{1:T}|h_{1:T})\pi(h_{1:T})}{p(x_{1:T})}, \qquad (31)$$

where $\pi(h_{1:T})$ is the prior distribution on the parameters of the volatility process defined by the dynamic model, $p(x_{1:T}|h_{1:T})$ is the likelihood of the observations, $p(x_{1:T})$ is the normalization constant that is ignored during the inference. Thus, we use Bayes rule in proportionality terms

$$p(h_{1:T}|x_{1:T}) \propto p(x_{1:T}|h_{1:T})\pi(h_{1:T}). \qquad (32)$$

We use particle Metropolis-Hastings to estimate the posterior distribution of the parameters of the model since neither the likelihood nor the posterior are available in closed form. We use Metropolis-Hastings algorithm to sample from the posterior of the parameters. Algorithm 1 presents an iteration of the Metropolis-Hastings algorithm. At every iteration of the algorithm we make a new proposal θ_i^c for the parameter vector using a proposal mechanism $q(\cdot|\theta^{(i)})$. Then we accept the proposed candidate θ_i^c with probability α. The acceptance probability in Algorithm 1 depends on $p(\theta, h_{1:T}|x_{1:T})$—target distribution—and $q(\cdot)$—proposal distribution. How well we manage to explore the posterior distribution depends on the acceptance rate of the algorithm. When the acceptance rate is too high it is often related to too small proposal steps, and the other way around. Overall, either case slows down the convergence of the Markov Chain. General advice for the optimal performance of the algorithm is an acceptance rate that is around 0.234 [33].

Algorithm 1 Particle Metropolis-Hastings Algorithm

1: Given $\theta^{(i)}$,
2: Generate $\theta_i^c \sim q(\cdot|\theta^{(i)})$,
3: Take

$$\theta^{(i+1)} = \begin{cases} \theta_i^c, & \text{with probability } \rho(\theta^{(i)}, \theta_i^c) \\ \theta^{(i)} & \text{with probability } 1 - \rho(\theta^{(i)}, \theta_i^c), \end{cases}$$

where

$$\rho(\theta^{(i)}, \theta_t^c) = \min\left(\frac{p_{\theta_i^c}^c(x_{1:T})}{p_\theta^{(i)}(x_{1:T})} \frac{\pi(\theta_i^c)}{\pi(\theta_i)} \frac{q(\theta^{(i)}|\theta_i^c)}{q(\theta_i^c|\theta^{(i)})}, 1\right)$$

Using Algorithm 1 we obtain samples from otherwise intractable distribution $p(\theta, h_{1:T}|x_{1:T})$. Note, that $p_{\theta_i^c}^c(x_{1:T})$ and $p_\theta^{(i)}(x_{1:T})$ are also intractable. Thus, in practice we substitute them with the estimates $\hat{p}_{\theta_i^c}(x_{1:T})$ and $\hat{p}_{\theta^{(i)}}(x_{1:T})$ obtained with Sequential Monte Carlo.

We further discuss how $p_\theta(x_{1:T})$ can be estimated.

3.3. Estimation of the Likelihood with SMC

Sampling from the posterior distribution with algorithms such as Metropolis-Hastings requires evaluationg the likelihood. In case of non-linear state-space models, this likelihood evaluation is not straightforward since the likelihood is a high dimensional integral

$$L(x_{1:T}) = \int p(x_{1:T}, h_{1:T})dh_{1:T} = \int p(x_{1:T}|h_{1:T})p(h_{1:T})dh_{1:T}$$
$$= \int p(x_1|h_1)p(h_1)\prod_{t=2}^{T} p(x_t|h_t)p(h_t|h_{t-1})dh_1\ldots h_T, \quad (33)$$

and this likelihood is not analytically tractable. Instead of relying on an analytical result, the integral from Equation (33) can be approximated using Sequential Monte Carlo methods, also known as particle filters. This estimate of the likelihood is then used in Algoperithm 1 as $\hat{p}_\theta(x_{1:T})$. Algorithm 2 represents a simple version of a particle filter which we use in this paper. The algorithm consists of three main steps: prediction, updating and resampling. In the prediction step we sample N particles according to the assumed dynamics of the latent process $p(h_t|h_{t-1})$. Then we weight each particle according to the distribution of the data given the latent state $p(x_t|h_t)$. Finally, in the resampling step we resample the particles according to these weights. Resampling step is meant to solve the known problem of particle degeneracy: without resampling we end up only with a few particles with high weights over time.

Algorithm 2 Bootstrap Particle Filter with resampling

1: Draw a new point $h_t^{(i)}$ for each point in the sample set $\{h_{k-1}^{(i)} : i = 1, \ldots, N\}$ from the dynamic model:
$$h_t^{(i)} \sim p(h_t|h_{t-1}^{(i)}), \quad i = 1, \ldots, N.$$

2: Calculate the weights
$$\omega_t^{(i)} \sim p(x_t|h_t^{(i)}), \quad ,1, \ldots, N,$$
and normalize them to sum to unity.

3: Compute the estimate of $p(x_t|x_{1:t-1}, \theta)$ as
$$p(x_t|x_{1:t-1}, \theta) = \sum_i \log \omega_t^{(i)}.$$

Perform resampling:

4: Interpret each weight $\omega_t^{(i)}$ as the probability of obtaining the sample index i in the set $\{h_t^{(i)} : i = 1, \ldots, N\}$.

5: Draw N samples from that discrete distribution defined by the weights and replace the old samples set with this new one.

The particle filter provides us with the sequence of distributions $p(h_t|x_t)$, however due to particle degeneracy problem discussed previously, sampling from $p(h_{1:T}|x_{1:T})$ and approximating $p(h_k|x_{1:T}), k = 1, \ldots, T$, is inefficient. One of the possible solutions to this problem is using so called forward filtering - backward smoothing recursions [34]. The algorithms starts with sampling $h_T^* \sim \hat{p}(h_T|x_{1:T})$, and then backwards for $k = T-1, T-2, \ldots, 1$, we sample $h_k^* \sim \hat{p}(h_k|h_{k+1}^*, x_{1:n})$. Then we can approximate the distribution $\hat{p}(h_k|x_{1:T})$ as follows

$$\hat{p}(h_k|x_{1:T}) = \sum_{i=1}^{N} W_k^i \times \left[\sum_{j=1}^{N} W_{k+1|T}^j \frac{f(h_{k+1}^{*,j}|h_k^{*,i})}{[\sum_{l=1}^{N} W_k^l f(h_{k+1}^{*,j}|h_k^{*,l})]} \delta_{h_k^{*,i}}(h_k) \right]$$
$$= \sum_{i=1}^{N} W_{k|T}^i \delta_{h_k^{*,i}}(h_k). \quad (34)$$

The smoothing comes at cost of $O(NT)$ operations to sample a path from $p(h_{1:T}|x_{1:T})$ and $O(N^2T)$ operations to approximate $p(h_k|x_{1:T})$. This method works very well, in particular when dealing with large sample sizes. However, its performance comes at the price of a high computational costs. Thereby, it is generally recommended to use it when the sample size of the data is large and hence Sequential Monte Carlo is more likely to suffer from particle degeneracy. There exist other methods that are computationally less expensive [34]. However, in higher dimensions, they would be less reliable, and it would be recommendable to use more expensive methods.

3.4. Forecasting with SSM

One of the interests of statistical inference is the ability to perform forecasting exercises and thus handling the uncertainty about the future in the best possible way. In this section, we will discuss how forecasting task fits into the Bayesian framework, and in particular how it can be done for models of our interest.

Recall that we estimated multivariate SSM model for count data in the Bayesian framework. Observing the data $x = (x_1, \ldots, x_T)$ we estimated the posterior distribution of the parameters in our model using particle Markov Chain Monte Carlo methods $p(\theta|x)$. Suppose that we are interested in predicting the next s observations, that is, $x_{T+1}, \ldots, x_{T+s}$. First, note that the prediction equation for the next step reads

$$p(x_{t+1}|x_t) = \int p(x_{t+1}|h_{t+1})p(h_{t+1}|x_t)dh_t. \tag{35}$$

In the framework of particle Markov Chain Monte Carlo, it is natural to adopt a sequential nature of SMC and the fact that we obtain posterior draws in the MCMC part of the algorithm. Thereby, for every vector θ of the parameters drawn in the MCMC, we can propagate the particles obtained at time T and based on those make one-step ahead forecast. The similar idea extends to s-steps ahead forecasts. In this case the uncertainty about the parameters is included in the forecasts.

When forecasting, the most natural but cumbersome approach would be to update the posterior distribution every time we get a new observation. It would mean that we generate as many MCMC chains as we have steps for forecasting. This can be carried out in a straightforward way by re-estimating the posterior distribution every time or more efficiently by incorporating this into the SMC framework. However, for large enough samples, adding an extra estimation into the PMCMC framework should not change the results substantially. Ignoring this update also makes the forecasting performance of the frequentist and Bayesian approaches more comparable. Both frameworks are estimated in different paradigms. While SSM naturally fits into the Bayesian paradigm, the log-linear model is usually estimated using frequentist methods (quasi-maximum likelihood in this case). Since our goal is not to compare the two approaches to statistics, this design of forecasting exercise is more fair.

Figure 2 illustrates the forecasting approach we undertake with the state-space model in a graphical representation. In particular, one can see that we do not re-estimate the posterior distribution every time we receive a data point.

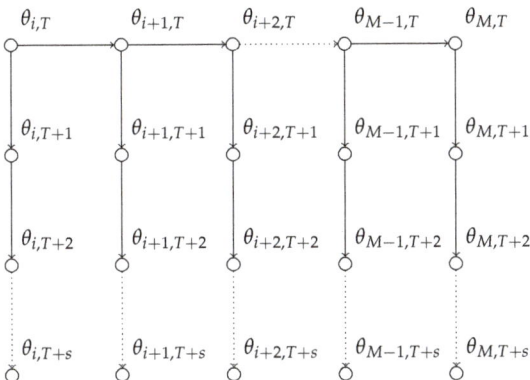

Figure 2. Visual representation of forecasting with the state-space model for count data.

4. Model Comparison and Prediction Assessment

We next summarize the measures using which we compare the models in Sections 2 and 3. Observation driven models in this comparison are represented by the log-linear autoregressive model. The log-linear autoregressive model is more flexible than the MACI model since it can account for a negative correlation and thus it is a fairer competitor. The parameter driven approach is the state-space model, where observations are generated from the Poisson distribution and dependency is modeled through a latent process. Note that, for the latter framework, we follow a fully Bayesian approach. Thereby, we compare these two classes of models by model fit and forecasting performance criteria. The standard measures to access the model fit and forecast accuracy would be Mean Squared Error (MSE) and Mean Absolute Error (MAE) defined in Equation (36) respectively.

$$MSE = \frac{1}{s}\sum_{i=1}^{s}(x_i - \hat{x}_i)^2, \qquad MAE = \frac{1}{s}\sum_{i=1}^{s}|x_i - \hat{x}_i|^2. \tag{36}$$

In Czado et al. [35] the authors propose comparing forecast performance using some scoring rules. To define scoring rules, let P be the predictive distribution and x be the counts; then the penalty is defined through $s(P, x)$. Table 1 presents some of the scoring rules one can use for comparing the performance of count data models.

Table 1. Scoring rules for assessment of the forecasts. The table summarizes scoring rules that we use to assess forecasting performance of the models under consideration, proposed in Czado et al. [35] for count data.

Logarithmic score	$\log(P, x) = -\log p_x$
Quadratic score	$qs(P, x) = -2p_x + \| p \|^2$
Spherical score	$sphs(P, x) = -\frac{p_i}{\|p\|}$, where $\| p \|^2 = \sum_{k=0}^{\infty} p_k^2$
Rank probability score	$rps(P, x) = \sum_{k=0}^{\infty}\{P_k - 1(x \geq k)\}^2$
Squared error score	$ses(P, x) = (x - \mu_p)^2$, where μ_p is the mean of P

Note, that in practice, one calculates the mean score

$$S = \frac{1}{n} \sum_{i=1}^{n} s(P^{(i)}, x^{(i)}). \qquad (37)$$

To compare our results with the conclusions in Zhang et al. [19] we also report Dawid-Sebastiani (DS) score which is defined in Equation (38)

$$DSS_{t,i}(X_{t,i}) = \frac{X_{t,i} - \mu_{t,i}}{\sigma_{t,i}} + 2\log(\sigma_{t,i}). \qquad (38)$$

5. Simulation Examples

In this section, we demonstrate the performances of the models based on simulated data. We generate data from various specifications of SSM and log-linear MACI models and compare the models on forecasting performance. We assess forecasting performance based on six different scoring measures discussed in the previous section. The design of the simulation study allows us to assess forecasting performance in the cases of both correct model specification and misspecified case. Table 2 summarizes three different specifications of the state-space approach for data generation and Table 3 summarizes specifications of the log-linear MACI for data generation. Figure 3 illustrates two examples of bivariate time series generated from these models. For each simulation setting, we generate ten datasets with different random seeds and report the average results from these ten datasets. State-space model was estimated using particle Metropolis-Hastings algorithm with $N = 5000$ particles and $M = 20000$ Metropolis-Hastings step with a warm-up period of 5000 steps. The acceptance rate was targeted to be between 25% and 40%.

Table 2. True parameters for the data sets generated from the state-space model in the simulation examples. All data generating processes include a one-directional Granger-causal feedback through a non-zero coefficient ϕ_{21} and different correlation structures: SSM_1 has a positive correlation coefficient ρ, SSM_2 has a negative correlation coefficient ρ and SSM_3 has no correlation.

Data Set	β_1	β_2	ϕ_{11}	ϕ_{21}	ϕ_{12}	ϕ_{22}	σ_{η_1}	σ_{η_2}	ρ
SSM_1	1.0	2.0	0.5	0.3	0.0	0.5	0.5	0.5	0.3
SSM_2	1.0	2.0	0.5	0.3	0.0	0.5	0.5	0.5	−0.3
SSM_3	1.0	2.0	0.5	0.3	0.0	0.5	0.5	0.5	0.0

Table 3. True parameters for the data sets generated from the log-linear MACI model in the simulated examples.

Data Set	ω_1	ω_2	a_{11}	a_{22}	b_{11}	b_{12}	b_{21}	b_{22}
LL_1	0.9	0.4	−0.5	0.2	0.5	0.2	0.0	0.4
LL_2	0.2	0.3	0.2	0.4	0.5	0.2	0.0	0.4

We assess the forecasting performance of two models for multivariate count data: state-space model and log-linear model. Tables 4 and 5 summarize the forecasting performances of the models according to various scoring rules. The rows of the tables correspond to a particular data generating process (see for details of specification Tables 2 and 3) and columns for a particular scoring rule (see scoring rules specification in Table 1). In particular, Table 4 shows performance of the state-space model and Table 5 the performance of the log-linear model. The state-space model outperforms the log-linear MACI model when the data are generated from SSM_1 (SSM with positive correlation) and LL_1 (log-linear model with a negative a_{11} coefficient). It is particularly interesting that when the data is simulated from LL_1, SSM performs best according to all measures despite being a misspecified model.

When the data are generated from SSM_2 and SSM_3, the state-space approach performs best based on most measures. This result is expected as SSM is the correct model specification for these simulated data. Finally, log-linear MACI model performs best in the case of data set LL_2—in the case when the model is correctly specified and all the coefficients are positive—according to most measures.

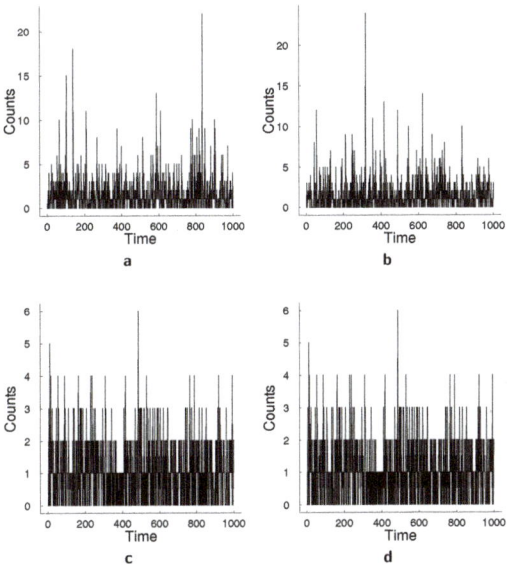

Figure 3. Examples of the data generated with the state-space and log-linear MACI models. (**a**) Dimension 1 of the bivariate time series generated from SSM2 in Table 2. (**b**) Dimension 2 of the bivariate time series generated from SSM2 in Table 2. (**c**) Dimension 1 of the bivariate time series generated from LL2 in Table 2. (**d**) Dimension 2 of the bivariate time series generated from LL2 in Table 2.

Table 4. Scores for the forecasting exercise with the state-space model, according to the definitions in Table 1. The smaller score indicates a better result. DGP column corresponds to the data generating processes in this simulation study, the true parameters are presented in Tables 2 and 3.

DGP	log	qs	sph	rps	ds	se
SSM_1	1.484	−0.229	−0.440	0.770	2.352	2.634
SSM_2	1.861	−0.235	−0.487	1.000	3.136	3.551
SSM_3	1.967	−0.224	−0.475	0.948	3.599	4.075
LL_1	1.959	−0.164	−0.405	0.974	2.176	3.214
LL_2	1.351	−0.293	−0.543	0.545	1.103	1.087

Table 5. Scores for the forecasting exercise with the log-linear MACI model. The smaller score indicates a better result. DGP column corresponds to the data generating processes in this simulation study, the true parameters are presented in Tables 2 and 3.

DGP	log	qs	sph	rps	ds	se
SSM_1	1.636	−0.321	−0.553	0.999	2.612	3.088
SSM_2	2.089	−0.164	−0.391	1.333	**2.614**	5.180
SSM_3	**1.929**	−0.220	−0.469	**0.948**	3.464	4.187
LL_1	1.985	−0.159	−0.400	0.996	2.238	3.357
LL_2	**1.320**	−0.309	−0.555	0.555	**1.036**	**1.023**

6. Empirical Applications

In this section, we compare the models in two empirical applications—bank failures and transactions data. These data sets exhibit different sample sizes, standard deviation, dispersion and maximum counts. In particular, bank failure time series reach a maximum of 10 and 24 counts while transaction data reach up to 67 and 60 counts with comparable mean counts.

6.1. Bank Failures

Bank failures have been analyzed using a univariate Poisson process [36]. A number of researches have investigated bank failure data of different time periods, see e.g., Schoenmaker [1] for an analysis of contagion risk in banking. Overall, it is reasonable to expect that bank failures in different countries are driven by similar economic phenomena, and possible contagion/spillover effects exist between economies of different countries.

For this application, we analyze count data using a bivariate data set of bank failures in the U.S. and Russia that has not been considered in the literature before. We use monthly number of bank failures for the period between January 2008 and December 2012 for both countries and apply the bi-variate specifications of the models in Sections 2 and 3. Especially due to the global financial crisis included in this period, it is important to allow for potential correlation between the number of bank failures in the U.S. and Russia using the multivariate count data models. Figure 4 illustrates these time series and Table 6 presents descriptive statistics for this data set.

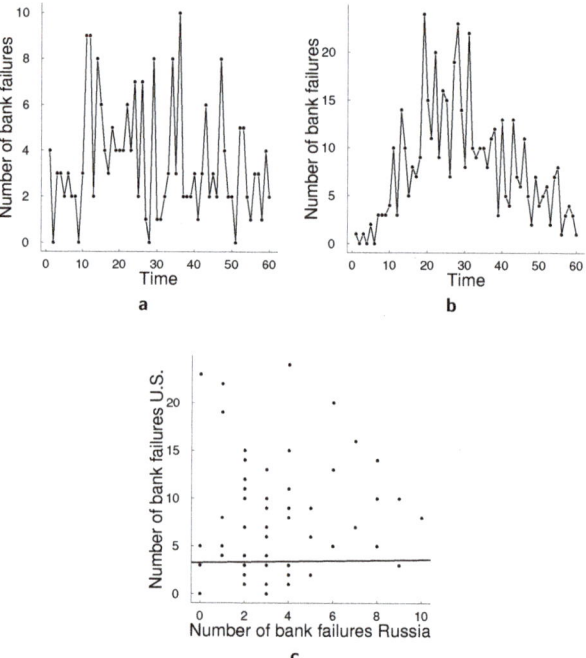

Figure 4. Data for bank failures empirical application. (**a**) Monthly bank failures in Russia January 2008–December 2012. (**b**) Monthly bank failures in the U.S. January 2008–December 2012. (**c**) Scatter plot of data bank failures in subplots (**a**,**b**).

Table 6. Descriptive statistics for the bank failures data for the period January 2008 until December 2012 for Russia and U.S.

	Russia	U.S.
mean	3.51	7.93
median	3	7
st.d.	2.46	5.93
minimum	0	0
maximum	10	24

The estimation results from both models are presented in Tables 7 and 8. Based on the state-space model, the correlation is estimated as being low negative and 0 is included in highest posterior density interval for this parameter. Despite that log-linear MACI model estimates correlation coefficient to be positive, it provides a large confidence interval for this parameter which also includes 0. Thus, for this relatively small data set, we do not find an indication of correlated bank failures using both models. We also note that some confidence intervals in Table 8 include point 0. As discussed in Section 2, applying observation driven models with positivity constraints would be problematic for these data especially in terms of the calculation of confidence intervals.

Table 7. Posterior moments of the parameters of the state-space model for bank failures.

	Mean	Median	Mode	HPD_l 95%	HPD_u 95%
β_1	1.1450	0.8867	0.0503	0.0503	2.9481
β_2	4.0414	3.7053	2.3414	1.8570	7.2843
ϕ_{11}	0.9569	0.9648	0.9757	0.8968	0.9991
ϕ_{21}	0.0101	0.0071	−0.0372	−0.0646	0.0856
ϕ_{12}	0.1193	0.0935	−0.1856	−0.2367	0.5438
ϕ_{22}	0.7387	0.7518	0.8862	0.4942	0.9733
σ_{η_1}	0.3335	0.3340	0.0122	0.1722	0.5400
σ_{η_2}	0.3302	0.3252	0.3189	0.1496	0.5176
ρ	−0.0845	−0.0848	−0.0703	−0.1879	0.0209

Table 8. Parameter estimates of the log-linear MACI model for bank failures.

	Estimate	CI_l 95%	CI_u 95%
w_1	0.1259	−0.7545	1.0064
w_2	−0.1307	−0.4348	0.1733
a_{11}	0.0732	−0.5380	0.6844
a_{22}	0.6923	0.5535	0.8312
b_{11}	0.0403	−0.2816	0.3621
b_{21}	0.1638	0.0190	0.3086
b_{12}	0.3879	0.0212	0.7546
b_{22}	0.2521	0.1069	0.3974
ρ	0.6513	−0.2131	1.5158

We next compare the models in terms of their forecast performances. For this comparison, we take a sample size of $T = 55$, and we make five steps ahead predictions using the log-linear model and the state-space approach. Table 9 presents scores for this forecasting exercise. Based on all scores, except for the rank probability score (rps), the state-space model outperforms the log-linear model in terms of forecasting. Based on the simulation results in Section 5, we conjecture the following: The better performance of the state-space model can be due to this model being close to the true data generating process, or due to its property of capturing data properties well even if it is misspecified.

Table 9. Scores for the forecasting exercise with bank failure data. This table shows scoring rules for the forecasting exercise in the bivariate model for bank failure data.

Model	log	qs	sph	rps	ds	se
SSM	**1.9026**	−0.1738	**−0.4189**	0.9755	**2.1619**	**3.0841**
Log-Linear	2.0244	−0.1623	−0.3996	**0.8862**	2.2934	4.2031

6.2. Transactions

In this empirical application, we analyze the number of transactions on 30 s intervals for Deutsche Bank AG and Bayer AG (the datawere obtained from FactSet, in the time period of 3 August 2015 09:05:30 until 3 August 2015 12:25:00 for the training data). We expect such transactions to be correlated due to their dependence on the time of the day and the market conditions. The sample size in this application is $T = 400$, which is significantly larger than the sample size in the bank failures application. The summary statistics for this data set are provided in Table 10 and Figure 5 illustrates these time series. Both time series have fat tails with a few very high values, concentrated around observation 100 and 1 for Deutsche Bank and Bayer AG, respectively.

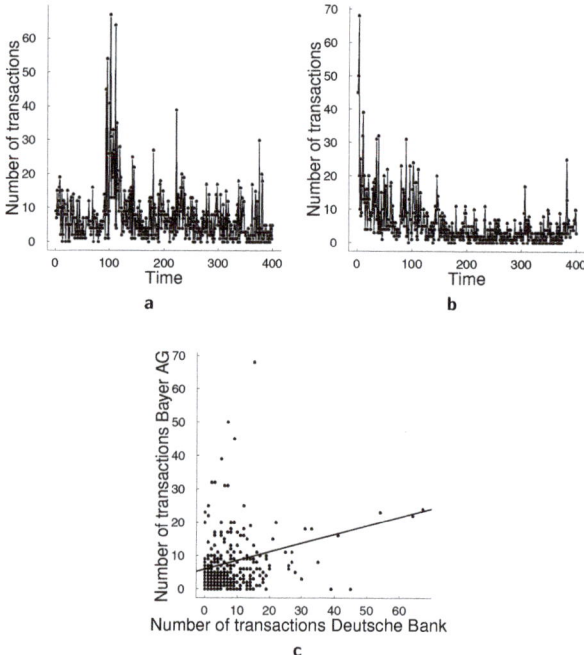

Figure 5. Data for transactions empirical application. (**a**) Transactions in 30 s. interval for Deutsche Bank AG. (**b**) Transactions in 30 sec. interval for Bayer AG. (**c**) Scatter plot of transactions in (**a**,**b**).

Table 10. Descriptive statistics for the transactions data.

	Deutsche Bank AG	**Bayer AG**
mean	6.95	7.716
median	5	5
st.d.	8.2462	8.227
minimum	0	0
maximum	67	60

We apply the bi-variate counterparts of the count data models in Sections 2 and 3 to these data and compare the model performances based on 100 steps ahead forecasts. In Tables 11 and 12 we present parameter estimates of both models. Both models estimate positive correlation between these time series. However, in the case of the log-linear MACI model the estimated correlation coefficient is much higher. In addition, the confidence intervals of parameter estimates such as b_{12} and b_{22} in Table 12 include point 0. Thus, true parameters being non-positive is a potential problem if other observation driven models, with positivity constraints, in Section 2 were applied to these data.

Table 13 presents the scores of each model in the forecast sample. In this application, the log-linear model performs best according to all scoring rules. Based on the simulation results in Section 5, we conjecture that the log-linear model is potentially closer to the true data generating process compared to the state space model. We further analyze the forecasting performances of the models in Figure 6. Particularly in Figure 6b, we observe that the log-linear MACI model captures better high spikes of the counts and then returns to the original level of the data. The forecast with the state-space model appears to be too smooth compared to the data points. Thus, the better forecast performance of the log-linear MACI is potentially due to its ability to capture these extreme data values successfully.

The under-performance and over-smoothing of the state-space approach can be mitigated by implementing a different particle filter. For example, one could take the direction of implementing look-ahead particle filters such as [37,38]. General idea of the look-ahead approaches is that in the particle filtering algorithm we make a proposal not just according to the dynamics of the model $p(h_t|h_{t-1})$, but taking the current observation into account $p(h_t|h_{t-1},y_t)$ or taking into account the complete time series $p(h_t|h_{t-1},y_{1:T})$ as in [38]. These methods, however, have not been developed for the distributional assumptions we are considering in this paper and further research is needed in this direction.

Table 11. Posterior moments of the parameters for the state space model for transactions.

	Mean	Median	Mode	HPD_l 95%	HPD_u 95%
β_1	4.5049	4.4860	4.3782	3.9274	5.1238
β_2	5.4475	5.4260	5.2161	4.7971	6.1541
ϕ_{11}	0.3058	0.3048	0.3137	0.1778	0.4316
ϕ_{12}	0.0180	0.0181	−0.0469	−0.1007	0.1342
ϕ_{21}	0.0518	0.0533	0.1520	−0.1118	0.1890
ϕ_{22}	0.3788	0.3795	0.5341	0.2414	0.5126
σ_{η_1}	0.8864	0.8853	0.8759	0.8059	0.9675
σ_{η_2}	0.7521	0.7519	0.7246	0.6835	0.8236
ρ	0.2400	0.2397	0.2499	0.1932	0.2875

Table 12. Parameter estimates of the log-linear MACI model for transactions.

	Estimate	CI_l 95%	CI_h 95%
w_1	0.1232	0.0304	0.2161
w_2	−0.0741	−0.1666	0.0184
a_{11}	0.7518	0.6826	0.8211
a_{22}	0.5333	0.4586	0.6080
b_{11}	0.1832	0.1471	0.2193
b_{21}	0.0315	−0.0028	0.0659
b_{12}	0.0024	−0.0256	0.0305
b_{22}	0.4591	0.3847	0.5334
ρ	0.7967	0.6053	0.9881

Table 13. Scores for transaction forecasts.

Model	log	qs	sph	rps	ds	se
SSM	5.0	−0.0171	−0.2059	3.4601	14.5471	49.97
Log-Linear	**4.4549**	**−0.0232**	**−0.2152**	**3.1621**	**11.3674**	**44.429**

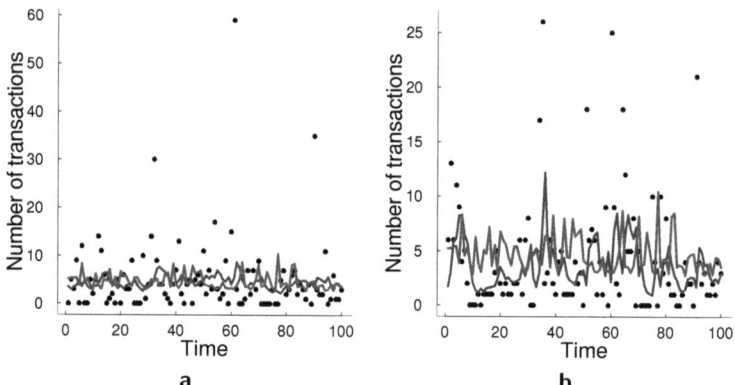

Figure 6. Forecasts and true data for the transactions empirical application. (**a**) Transactions in 30 s. interval for Deutsche Bank AG (black circles), forecast with SSM (green), forecast with log-linear MACI (red). (**b**) Transactions in 30 s. interval for Bayer AG (black circles), forecast with SSM (green), forecast with log-linear MACI (red).

7. Discussion

In this paper, we have reviewed and compared two approaches for modeling multivariate count time series data. One of the challenges that appears in the literature and have not been resolved is modeling the dependency between counts in a flexible way that would also allow for feasible estimation. We have discussed multivariate autoregressive conditional intensity models (MACI), their log-linear alternative which we refer to as the multivariate log-linear model and nonlinear state-space model. Both models have advantages and disadvantages. In particular, the nonlinear state-space framework allows for various interpretable dependencies that one cannot easily incorporate into MACI or log-linear approach. However, these models can be computationally expensive to estimate, in particular in higher-dimensions. Challenges in estimation arise from different sources. State-space models naturally fit into the Bayesian framework, however, since both the likelihood and the posterior of the model are analytically intractable this leads to computationally expensive procedure. MACI models, on the other hand, are quite restrictive: they restrict coefficients in the model to be positive as well as the correlation between time series. Both assumptions can be unrealistic in many real-world applications. Log-linear model avoids the problem of only positive coefficients in the model by logarithmic transformation of the data. However, estimation can be unstable, and good starting points need to be chosen for the estimation. When the dimension of the model grows, it becomes harder to choose good starting points for the optimization problem. The computational advantage of log-linear and MACI models decreases with the increase in either dimensionality of the model or the number of counts. This reduction in the computational advantage is due to the usage of the multivariate Poisson distribution as every pairwise correlation has to be modeled as a separate Poisson random variable. Moreover, the summation in the specification of the joint distribution runs through the number of counts. Generally, one could say that estimation of log-linear models much faster than of the state-space models. In low dimensions and with the small number of counts these models do not require much of computational power, however, once the number of counts increases and once we deal with higher dimensions, the computations become much more extensive due to large sums in the multivariate Poisson distribution. Moreover, while running the model on simulated and empirical data, we found that the estimation can be numerically unstable and can highly depend on the starting values in the estimation procedure. We follow the suggestion of Doukhan et al. [10], and the first estimate the model for univariate time series. These estimates we further use in multivariate estimation. However, the problem of numerical instability especially remains in small samples according to our experience. Nevertheless, in terms of flexibility, this model is the best competitor for the state-space approach.

We have compared log-linear models and state-space models for count data in terms of forecasting performance on multiple simulated data sets and real data applications. We found that on the simulated data state-space framework generally outperforms log-linear model, sometimes even under model misspecification. On the real data sets, the state-space model performs better in bank failures applications which consists of two time series of bank failures in Russia and U.S. and the counts remain relatively low and the data are relative smooth. The log-linear model performs better in the transactions applications in which we consider two time series of transactions counts in 30 seconds intervals. The challenge of transactions application is that there are spikes of counts which deviate a lot from the mean. In this case, we notice that the log-linear model approximates these spikes better. Thus, a possible direction for future research is adapting a multivariate state-space model for count data to capture such spikes better. A possible way to improve the model in this regard would be to adapt the particle filtering algorithm. We used bootstrap particle filter which does not take into account observations when making a proposal for particles, but taking current (or all) observation into account in the proposal mechanism for the particles can help approximating the spikes in the data. There have been proposed multiple look-ahead approaches for particle filters [37,38], but they have not been adapted to count data.

Finally, both approaches have their drawbacks. In particular, the log-linear model seems to have numerical stability issues and finding optimal starting values for optimization can be a challenge. In the state-space approach, the challenging part is the estimation of the likelihood, which is intractable and sampled from the posterior distribution. Additionally, the state-space model in its current implementation is challenged by possible spikes in the data to a larger degree than the log-linear model.

Author Contributions: Conceptualization, Y.S., N.B., and M.E.; methodology, Y.S. and M.E.; formal analysis, Y.S.; investigation, Y.S.; writing—original draft preparation, Y.S., N.B., and M.E.; writing—review and editing, Y.S. and N.B.; visualization, N.B. and Y.S.; supervision, M.E. All authors have read and agreed to the published version of the manuscript.

Funding: N.B. is partially supported by an NWO grant VI.Vidi. 195.187.

Data Availability Statement: The data for Polio and Asthma cases is publicly available from R library glarma [39]. Number of car crashes previously published in [40]. The earthquakes data is available from https://earthquake.usgs.gov/ (accessed on 14 June 2017). Transactions data can be downloaded from FactSet. The data for bank failures is publicly available at www.banki.ru (Russia) (accessed on 2 June 2021) and www.fdic.gov (U.S.) (accessed on 2 June 2021).

Conflicts of Interest: The authors declare no conflict of interest.

Abbreviations

The following abbreviations are used in this manuscript:

SSM	State-Space Model
MACI	Multivariate Autoregressive Conditional Intensity
INGARCH	Integer-Valued Generalized Autoregressive Conditional Heteroskedastic Model
MCMC	Markov Chain Monte Carlo
PMCMC	Particle Markov Chain Monte Carlo
SMC	Sequential Monte Carlo

References

1. Schoenmaker, D. *Contagion Risk in Banking*; LSE Financial Markets Group: London, UK, 1996.
2. Kim, B.; Lee, S.; Kim, D. Robust Estimation for Bivariate Poisson INGARCH Models. *Entropy* **2021**, *23*, 367. [CrossRef]
3. Agosto, A.; Giudici, P. A Poisson Autoregressive Model to Understand COVID-19 Contagion Dynamics. *Risks* **2020**, *8*, 77. [CrossRef]
4. Weiß, C.H. Stationary count time series models. *Wiley Interdiscip. Rev. Comput. Stat.* **2021**, *13*, e1502. [CrossRef]
5. Park, E.; Lord, D. Multivariate Poisson-lognormal models for jointly modeling crash frequency by severity. *Transp. Res. Rec. J. Transp. Res. Board* **2007**, *2019*, 1–6. [CrossRef]
6. Agosto, A.; Cavaliere, G.; Kristensen, D.; Rahbek, A. Modeling corporate defaults: Poisson autoregressions with exogenous covariates (PARX). *J. Empir. Financ.* **2016**, *38*, 640–663. [CrossRef]
7. Homburg, A.; Weiß, C.H.; Frahm, G.; Alwan, L.C.; Göb, R. Analysis and Forecasting of Risk in Count Processes. *J. Risk Financ. Manag.* **2021**, *14*, 182. [CrossRef]
8. Homburg, A. Criteria for evaluating approximations of count distributions. *Commun. Stat.-Simul. Comput.* **2020**, *49*, 3152–3170. [CrossRef]
9. Karlis, D.; Meligkotsidou, L. Multivariate Poisson regression with covariance structure. *Stat. Comput.* **2005**, *15*, 255–265. [CrossRef]
10. Doukhan, P.; Fokianos, K.; Støve, B.; Tjøstheim, D. Multivariate Count Autoregression. *arXiv* **2017**, arXiv:1704.02097.
11. Heinen, A.; Rengifo, E. Multivariate autoregressive modeling of time series count data using copulas. *J. Empir. Financ.* **2007**, *14*, 564–583. [CrossRef]
12. Nikoloulopoulos, A.K.; Karlis, D. Modeling multivariate count data using copulas. *Commun. Stat.-Simul. Comput.* **2009**, *39*, 172–187. [CrossRef]
13. Andreassen, C.M. Models and Inference for Correlated Count Data. Ph.D. Thesis, Department of Mathematics, Aarhus University, Aarhus, Denmark, 2013.
14. Lennon, H. Gaussian Copula Modelling for Integer-Valued Time Series. Ph.D. Thesis, University of Manchester, Manchester, UK, 2016.
15. Genest, C.; Nešlehová, J. A primer on copulas for count data. *ASTIN Bull. J. IAA* **2007**, *37*, 475–515. [CrossRef]
16. Cox, D.R.; Gudmundsson, G.; Lindgren, G.; Bondesson, L.; Harsaae, E.; Laake, P.; Juselius, K.; Lauritzen, S.L. Statistical analysis of time series: Some recent developments [with discussion and reply]. *Scand. J. Stat.* **1981**, *8*, 93–115.
17. Fokianos, K.; Tjøstheim, D. Log-linear Poisson autoregression. *J. Multivar. Anal.* **2011**, *102*, 563–578. [CrossRef]
18. Davis, R.A.; Dunsmuir, W.T.; Wang, Y. Modeling time series of count data. *Stat. Textb. Monogr.* **1999**, *158*, 63–114.
19. Zhang, C.; Chen, N.; Li, Z. State space modeling of autocorrelated multivariate Poisson counts. *IISE Trans.* **2017**, *49*, 518–531. [CrossRef]
20. Zellner, A. *An Introduction to Bayesian Inference in Econometrics*; Wiley: New York, NY, USA, 1971; Volume 156.
21. Andrieu, C.; Doucet, A.; Holenstein, R. Particle Markov chain Monte Carlo methods. *J. R. Stat. Soc. Ser. B Stat. Methodol.* **2010**, *72*, 269–342. [CrossRef]
22. Shapovalova, Y. "Exact" and Approximate Methods for Bayesian Inference: Stochastic Volatility Case Study. *Entropy* **2021**, *23*, 466. [CrossRef]
23. Salimans, T.; Knowles, D.A. Fixed-form variational posterior approximation through stochastic linear regression. *Bayesian Anal.* **2013**, *8*, 837–882. [CrossRef]
24. Rue, H.v.; Martino, S.; Chopin, N. Approximate Bayesian inference for latent Gaussian models by using integrated nested Laplace approximations. *J. R. Stat. Soc. Ser. B Stat. Methodol.* **2009**, *71*, 319–392. [CrossRef]
25. Girolami, M.; Calderhead, B. Riemann manifold Langevin and Hamiltonian Monte Carlo methods. *J. R. Stat. Soc. Ser. B Stat. Methodol.* **2011**, *73*, 123–214. [CrossRef]
26. Martin, G.M.; McCabe, B.P.; Frazier, D.T.; Maneesoonthorn, W.; Robert, C.P. Auxiliary likelihood-based approximate Bayesian computation in state space models. *J. Comput. Graph. Stat.* **2019**, *28*, 508–522. [CrossRef]
27. Naesseth, C.; Linderman, S.; Ranganath, R.; Blei, D. Variational sequential monte carlo. In Proceedings of the International Conference on Artificial Intelligence and Statistics, PMLR, Playa Blanca, Lanzarote, Spain, 9–11 April 2018; pp. 968–977.
28. Fokianos, K. Count time series models. *Time Ser. Appl. Handb. Stat.* **2012**, *30*, 315–347.
29. Liu, H. Some Models for Time Series of Counts. Ph.D. Thesis, Columbia University, New York, NY, USA, 2012.
30. Ferland, R.; Latour, A.; Oraichi, D. Integer-Valued GARCH Process. *J. Time Ser. Anal.* **2006**, *27*, 923–942. [CrossRef]
31. Hilmer, C.E.; Holt, M.T. *A Comparison of Resampling Techniques when Parameters Are on a Boundary: The Bootstrap, Subsample Bootstrap, and Subsample Jackknife*; North Carolina State University: Raleigh, NC, USA, 2000.
32. Särkkä, S. *Bayesian Filtering and Smoothing*; Cambridge University Press: Cambridge, UK, 2013; Volume 3.
33. Roberts, G.O.; Gelman, A.; Gilks, W.R. Weak convergence and optimal scaling of random walk Metropolis algorithms. *Ann. Appl. Probab.* **1997**, *7*, 110–120. [CrossRef]
34. Doucet, A.; Johansen, A.M. A tutorial on particle filtering and smoothing: Fifteen years later. *Handb. Nonlinear Filter.* **2009**, *12*, 3.
35. Czado, C.; Gneiting, T.; Held, L. Predictive model assessment for count data. *Biometrics* **2009**, *65*, 1254–1261. [CrossRef]
36. Davutyan, N. Bank failures as Poisson variates. *Econ. Lett.* **1989**, *29*, 333–338. [CrossRef]
37. Pitt, M.K.; Shephard, N. Filtering via simulation: Auxiliary particle filters. *J. Am. Stat. Assoc.* **1999**, *94*, 590–599. [CrossRef]

38. Guarniero, P.; Johansen, A.M.; Lee, A. The iterated auxiliary particle filter. *J. Am. Stat. Assoc.* **2017**, *112*, 1636–1647. [CrossRef]
39. Dunsmuir, W.T.M.; Scott, D.J. The glarma package for observation-driven time series regression of counts. *J. Stat. Softw.* **2015**, *67*, 1–36. [CrossRef]
40. Brijs, T.; Karlis, D.; Wets, G. Studying the effect of weather conditions on daily crash counts using a discrete time-series model. *Accid. Anal. Prev.* **2008**, *40*, 1180–1190. [CrossRef] [PubMed]

Article

Count Data Time Series Modelling in Julia—The CountTimeSeries.jl Package and Applications

Manuel Stapper

Institute of Econometrics and Economic Statistics, Westfälische Wilhelms-Universität Münster, 48149 Münster, Germany; manuel.stapper@wiwi.uni-muenster.de

Abstract: A new software package for the Julia language, CountTimeSeries.jl, is under review, which provides likelihood based methods for integer-valued time series. The package's functionalities are showcased in a simulation study on finite sample properties of Maximum Likelihood (ML) estimation and three real-life data applications. First, the number of newly infected COVID-19 patients is predicted. Then, previous findings on the need for overdispersion and zero inflation are reviewed in an application on animal submissions in New Zealand. Further, information criteria are used for model selection to investigate patterns in corporate insolvencies in Rhineland-Palatinate. Theoretical background and implementation details are described, and complete code for all applications is provided online. The CountTimeSeries package is available at the general Julia package registry.

Keywords: count data; time series analysis; Julia programming language

Citation: Stapper, M. Count Data Time Series Modelling in Julia—The CountTimeSeries.jl Package and Applications. *Entropy* **2021**, *23*, 666. https://doi.org/10.3390/e23060666

Academic Editor: Christian H. Weiss

Received: 15 April 2021
Accepted: 23 May 2021
Published: 25 May 2021

Publisher's Note: MDPI stays neutral with regard to jurisdictional claims in published maps and institutional affiliations.

Copyright: © 2021 by the author. Licensee MDPI, Basel, Switzerland. This article is an open access article distributed under the terms and conditions of the Creative Commons Attribution (CC BY) license (https://creativecommons.org/licenses/by/4.0/).

1. Introduction

The collection of count data dates back to at least 4000 years ago when the first census was documented. Nowadays, it still plays an important role in our everyday life as it occurs in various fields. For instance, it arises in insurance with the number of claims, in epidemiology where the number of infected patients is recorded, in urban planning as traffic volume, or in marketing with the number of certain items sold, just to name a few. During the past decades, researchers have developed methods to gain insight into the structure of the data and derive conclusions from it. Important frameworks for such analyses are integer counterparts of the well-known ARMA and GARCH models, the INARMA model introduced by Alzaid and Al-Osh [1], and the INGARCH model described, for example, by Ferland et al. [2]. Both models, the INARMA and the INGARCH model, have been further extended thereafter and are well established. In the following applications, a software package for the Julia Programming Language (Bezanzon et al. [3]) is used, which provides methods for these two models, CountTimeSeries.jl. After the Julia language was first published in 2012, the number of users grew remarkably during past years. On the one hand, it is as easy to learn as an interpreted language like R or Python, and on the other hand, it is a compiled language which makes its computation time comparable to C and Fortran. This combination makes Julia attractive for both researchers and practitioners. The CountTimeSeries package follows the idea of the language itself by providing a toolbox for count data time series that is simple to use and at the same time fast enough to allow analyzing large time series or carry out simulation studies. Covering some important generalization of INARMA and INGARCH models, the package allows to simulate from them, estimate parameters and conduct inference, assess the model choice, and compute forecasts.

There is no comparable package in R or Julia that covers the same functionalities. One important package for count data time series is R's tscount package, see Liboschik et al. [4]. In contrast to the CountTimeSeries package, it provides residual assessment, calibration methods, and intervention analysis. However, it does not cover the INARMA framework at all. Further, it also does not incorporate zero inflation, a model property frequently used in count data analysis. The thinning-based INARMA model can be formulated as Hidden Markov model, as for example discussed by Weiß et al. [5], so that the R packages HiddenMarkov (Harte [6]), and HMM (Himmelmann [7]) can be used to estimate its parameters. Both packages have been developed for general HMMs, thus the CountTimeSeries package offers the advantage of a more convenient usage and similar notation for both frameworks making it easy to switch between frameworks or compare results between those. As the name already suggests, count regression was not the goal of the CountTimeSeries package, but can be used for both a Poisson and a Negative Binomial regression. In R this is provided, for example, by the pscl package (Jackman [8]) with extensions by Zeileis et al. [9]. tscount can only be used for (Quasi-)Poisson regression, as it was designed only for Poisson and Quasi-Poisson INGARCH models.

In Julia, there is no package dealing with INGARCH and INARMA models, but like in R, there are packages for Hidden Markov models, like HMMBase, see Mouchet [10].

In the remainder of this paper, the CountTimeSeries package is first motivated and all its functionalities are explained. Then, four applications showcase how the package is used it practice. The first application deals with the spread of COVID-19 and focuses on the INGARCH framework and prediction. The second application analyzes the number of animals which were submitted to a New Zealand health laboratory and suffered from anorexia or skin lesions. Thinning-based INARMA models are used to shed more light on a discussion about the need of incorporating conditional overdispersion and zero inflation for this data set. During the analysis, inference is conducted on parameter estimates and information criteria used. Application 3 uses data first analyzed by Weiß and Feld [11] containing the number of corporate insolvencies in Rhineland-Palatinate. Thereby, the focus lies on the model choice, 54 models with different time trends and model specifications are fitted and information criteria used to select the best model. It is discussed whether there are spatial clustering effects in the characteristics of selected models and if there is a downwards trend in the number of insolvencies. Application 4 waives real data, a simulation study is carried out to assess the finite sample properties of Maximum Likelihood estimation for a Poisson INGARCH(1, 1). Different methods to initialize the conditional mean recursion, different parameter values, and different lengths of time series are compared. The simulation study is conducted in both Julia and R to compare the estimation time. Finally, Section 7 summarizes results, and gives an outlook and concludes.

2. The CountTimeSeries Package

When a new software package is developed, its rationale for existence should be addressed first. In other programming languages, there are many established packages to model count data time series. Three reasons spoke for the development of a new Julia package. Despite Julia's strong growth in popularity during recent years, no package for count data existed. Methods covered by other programming languages should be provided for Julia users. The goal was not only to translate functions, but also to create a package that can be easily extended for other models and alternative estimation methods. Julia's multiple dispatch feature allows adding model types and methods to existing functions for a unified notation. The third reason for the new package is its aforementioned advantages im terms of computing time. In the following, the general structure of the package is described in detail.

2.1. INGARCH Framework

By now, the package covers the INGARCH and INARMA framework and some important generalizations. The former was first described by Ferland et al. [2] as an integer process $\{Y_t\}_{t \in \mathbb{Z}}$ with

$$Y_t|\mathcal{F}_{t-1} \sim \text{Pois}(\lambda_t)$$
$$\lambda_t = \beta_0 + \sum_{i=1}^{p} \alpha_i Y_{t-i} + \sum_{i=1}^{q} \beta_i \lambda_{t-i} \tag{1}$$

where $\mathcal{F}_t$ denotes the information set available at time t. Alternatively, the conditional distribution can be replaced by a Negative Binomial distribution with overdispersion parameter $\phi > 0$ such that the conditional variance becomes $\text{Var}(Y_t|\mathcal{F}_{t-1}) = \lambda_t + \lambda_t^2/\phi$.

Different extensions have proven useful in practice. Currently covered by the Count-TimeSeries package are additional regressors, a log-linear link function, and zero inflation. In line with Liboschik et al. [4], regressors can be included in two different ways—as internal or as external regressors. By adding internal regressors $X_t^{(I)}$ in (1), the conditional mean becomes

$$\lambda_t = \beta_0 + \sum_{i=1}^{p} \alpha_i Y_{t-i} + \sum_{i=1}^{q} \beta_i \lambda_{t-i} + \eta' X_t^{(I)} \tag{2}$$

To further include external regressors $X_t^{(E)}$, the conditional mean is defined as

$$\lambda_t = \nu_t + \eta' X_t^{(E)}$$
$$\nu_t = \beta_0 + \sum_{i=1}^{p} \alpha_i Y_{t-i} + \sum_{i=1}^{q} \beta_i \nu_{t-i} + \zeta' X_t^{(I)} \tag{3}$$

In contrast to external regressors, internal regressors enter the recursive definition of conditional means. Often, it is not crucial how a regressor is declared by the user. The approach has been developed by Liboschik et al. [12] to model interventions that affect the observable process Y_t without affecting the underlying dynamics of the mean recursion, see Liboschik et al. [12].

If one wants to replace the linear link function in above models with a log-linear link, Equation (3) become

$$\log(\lambda_t) = \nu_t + \eta' X_t^{(E)}$$
$$\nu_t = \beta_0 + \sum_{i=1}^{p} \alpha_i \log(Y_{t-i} + 1) + \sum_{i=1}^{q} \beta_i \nu_{t-i} + \zeta' X_t^{(I)} \tag{4}$$

If zero inflation should be included, only the first line in (1) is changed. Then, given $\mathcal{F}_{t-1}$, Y_t is assumed to follow a singular distribution at zero with probability ω and to follow a Poisson (or Negative Binomial) distribution with probability $1 - \omega$. The conditional probability to observe a zero then is $P(Y_t = 0|\mathcal{F}_{t-1}) = \omega + (1-\omega)\exp\{-\lambda_t\}$. The CountTimeSeries package supports all combinations of these generalizations of the INGARCH model.

2.2. INARMA Framework

Besides the INGARCH framework, the CountTimeSeries package also covers another important model class widely used in count data analysis, INARMA models. Alzaid and Al-Osh [1] introduced an integer counterpart of AR(p) processes based on binomial thinning, which was then further developed and extended. The simple INAR(1) model is defined as $Y_t = R_t + \alpha \circ Y_{t-1}$, where R_t is a latent, non-negative, and integer-valued process often called innovation or immigration process. The thinning operator "$\circ$" is defined for a constant $\alpha \in [0,1]$ and a non-negative, integer random variable X as $\alpha \circ X = \sum_{i=1}^{X} Z_i$ for $X > 0$, where $Z_i \stackrel{iid}{\sim} \text{Bern}(\alpha)$ and zero if $X = 0$.

In line with the notation of Weiß et al. [5], a non-negative and integer-valued process $\{Y_t\}$ follows INARMA(p,q) process if

$$Y_t = R_t + \sum_{i=1}^{q} \beta_i \circ R_{t-i} + \sum_{i=1}^{p} \alpha_i \circ Y_{t-i} \tag{5}$$
$$R_t \sim \text{Pois}(\beta_0)$$

The distribution of R_t can also be chosen as a Negative Binomial distribution with mean β_0 and an overdispersion parameter ϕ, such that its variance is $\beta_0 + \beta_0^2/\phi$. One way to include regressors in above model would be to add another additive component Z_t to the first line, whose mean depends on regressors $X_t^{(E)}$. In the CountTimeSeries package, the distribution of Z_t can be chosen as Poisson or as Negative Binomial. The regressors included that way are referred to as external regressors, as they do not affect the innovation process R_t. Another possible inclusion of regressors is implemented as internal regressors $X_t^{(I)}$. The Poisson INARMA(p, q) with regressors then becomes

$$Y_t = R_t + \sum_{i=1}^{q} \beta_i \circ R_{t-i} + \sum_{i=1}^{p} \alpha_i \circ Y_{t-i} + Z_t$$
$$R_t \sim \text{Pois}\left(\beta_0 + \eta' X_t^{(I)}\right) \tag{6}$$
$$Z_t \sim \text{Pois}\left(\zeta' X_t^{(E)}\right)$$

Note that the mean of Z_t only depends on the regressors and no intercept parameter for identification. To include zero inflation, the distribution of R_t is altered to be zero with probability ω and to follow a Poisson with probability $1 - \omega$, see, for example, Aghababaei Jazi et al. [13]. This approach inflates the probability of Y_t to be zero indirectly, in contrast to the approach for the INGARCH framework. The reason for it is found when taking a closer look at how the likelihood is computed for an INARMA model. As Y_t is composed of multiple, non-observable parts, the distribution of Y_t given past observations is the result of a convolution. A convolution simplifies from a computational point of view if possible values for each component are bounded. For the autoregressive part, each result of a thinning operation is bounded from above by the corresponding past observation. Including zero inflation as described above preserves the property $R_t \leq Y_t$, which would not hold if zero inflation was included in the same way as for INGARCH models.

2.3. Package Structure

The general structure of the package can be divided into four parts: model specification, generating data from a model, fitting a model to data, and prediction.

For the specification of a model, object types have been implemented. Starting from the top, the type CountModel covers every possible model described in the previous section. Models in the INGARCH or INARMA framework are collected in the types INGARCH and INARMA, respectively. Subtypes of these two are finally INGARCHModel, INARCHModel and IIDModel, as well as INARMAModel, INARModel, and INMAModel. This definition of a type tree allows to implement methods for certain groups of models.

A wrapper function Model() is implemented to specify a model. The user provides the model framework, INGARCH or INARMA, the distribution, the link function, model orders p and q, regressors if wanted with an indicator whether they should be treated as internal or external and whether or not zero inflation should be considered. Default setting is a simple Poisson IID model. A Negative Binomial INARCH(1) with zero inflation is for example specified by

Model(distr = "NegativeBinomial", pastObs = 1, zi = true).

To generate artificial data from a model, the specification is paired with parameter values. This can generally be done in two ways: as a vector of values or as a newly defined data type parameter, with entries for β_0, α, β, η, ϕ, and ω. Note that in the implementation, no notational distinction is made between parameters for internal and external regressors. The two ways of providing parameter values are useful for example during the optimization of the likelihood, which uses the parameter vector, whereas estimation results are more convenient to handle as the parameter type.

Once a model and parameter values are specified, data are generated with the `simulate()` function by calling `simulate(T, model, parameter)` for a time series of length T.

Before fitting a model to a time series, settings for the maximization of the likelihood can be provided. The user can specify initial values and the method for the optimization routine with the `MLESettings()` function. The optimization routine can be `"NelderMead"`, `"BFGS"`, or `"LBFGS"`. If inference in terms of confidence intervals shall be conducted, the argument `ci` needs to be set to true. Standard errors are then computed from the numerical Hessian matrix.

Estimation of parameters is carried out by the `fit()` function, which takes the time series, the model, and, if chosen, the settings as input. The likelihood is maximized while considering constraints on the parameters. Constraints include positivity of conditional means at any time, proper thinning probabilities and constraints that ensure stability of the process.

The function `parametercheck` checks whether parameters are valid whenever calling the log-likelihood function. If invalid parameters are put in, the log-likelihood function simply returns negative infinity. This approach is usually unproblematic if starting values for the optimization are valid and not too close to being invalid.

The `fit()` function returns an object with estimates, standard errors, log-likelihood for estimates, and many more. This result object can be forwarded to functions for information criteria, `AIC()`, `BIC()`, and `HQIC()`, or to the function `pit()`. Then, the non-randomized probability integral transform histogram (see Czado et al. [14]) is plotted.

A forecast can be carried out by the `predict()` function. For models of the INGARCH framework, two options are available. Predictions can either be deterministic or simulation-based. In the deterministic approach, conditional means are used as prediction, and if an observation in the definition of the conditional mean is not observed, it is replaced by its corresponding prediction. In the simulation-based approach, the time series is continued many times with random realizations, or chains, following the process. This conveniently provides prediction intervals as the quantiles of the chains.

Assembling all parts together, this section is closed with an example of a Poisson INARCH(1). The code to specify the model, simulate a time series of length 500 with parameters $\beta_0 = 3$ and $\alpha_1 = 0.95$, fitting and predicting 100 steps into the future with prediction intervals from 10,000 chains would be

```
model = Model(pastObs = 1)
y = simulate(500, model, [3, 0.95])[1]
result = fit(y, model)
pred = predict(result, 100, 10000)
```

3. Application: COVID-19

For the first application, we take a closer look at the spread of COVID-19 in Germany. The global pandemic has moved disease spread into public focus. Being able to predict the number of newly infected patients is a cornerstone for policy making and resource planning in hospitals and medication supply. Many sophisticated models have been developed to predict the progression of infection rates. In the following, the CountTimeSeries package is utilized to conduct such predictions with a rather simple INGARCH model and daily data from German administrative districts. As an example, the application focusses on the number of new infections in Limburg-Weilburg, which is the district with a population density closest to the nationwide population density.

Data on infections of contagious, notifiable diseases are collected from the districts health authorities by the Robert–Koch Institut, see RKI [15]. The data set used in the course of the following analysis was prepared and published by NPGEO [16]. Besides general information on districts it contains the daily number of newly reported COVID-19 cases for all 401 German administrative districts on the NUTS 3 resolution. The observed time ranges from 7 January 2020 to 15 March 2021, totaling 434 days.

A first look at the data for the district of Limburg-Weilburg reveals that ~35% of observations are zero and 9% of observations exceed 50, which underlines the suitability and importance of a count data model. The excess of zeros in the data is partly explained by the beginning of the pandemic with only few cases and the rather low counts in general are due to the fine resolution in both, the time and space domain.

While advanced approaches for disease modeling incorporate many driving factors, this analysis relies purely on the number of infections in Limburg-Weilburg and its neighboring districts. Demographic factors, holidays and school closings, contact limitations, introduction of mandatory face masks, and occupancy of hospitals and care facilities are left out.

3.1. Model

To predict the number of new COVID-19 infections in Limburg-Weilburg, the widely used INGARCH(1, 1) model with log-linear link is chosen as a starting point. According to the World Health Organization [17], the incubation time of COVID-19 is on average five to six days. It is further stated that virus transmission occurred more often from patients with symptoms in empirical studies. Therefore, and to capture the weekly seasonality pattern visible in Figure 1, the lag order 7 is included in the model as well.

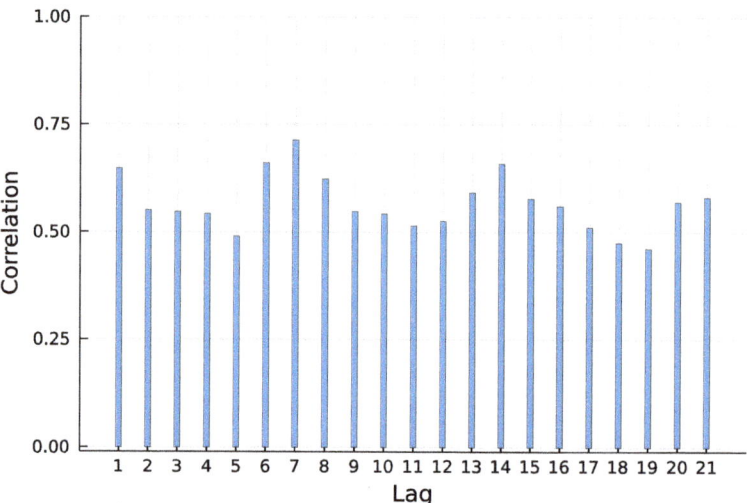

Figure 1. Autocorrelation function (ACF) of daily new infections in Limburg-Weilburg.

In the time series, the mean count is around 14 whereas the variance is 642. This marginal overdispersion does not necessarily indicate that a Negative Binomial must be considered as conditional distribution. The Poisson INGARCH processes with large dependency, meaning a sum of α_i and β_i parameters close to one, can exhibit such an overdispersion. In the following, three different approaches are considered and compared. Besides a Poisson and a Negative Binomial model, a retrospective estimation of the overdispersion parameter after fitting a Poisson model, as, for example, described by Christou and Fokianos [18], is also considered. Let the sequence $\hat{\lambda}_t$ be the conditional means on the basis of estimates from a Poisson fit, T the number of observations, and m the number of parameters. The retrospective estimate of the overdispersion parameter ϕ is then given by solving the moment equation

$$\sum_{t=1}^{T} \frac{(Y_t - \hat{\lambda}_t)^2}{\hat{\lambda}_t(1 + \hat{\lambda}_t/\phi)} = T - m \qquad (7)$$

for ϕ. The data-generating process is assumed to be a Negative Binomial process, whereas all parameters apart from ϕ are estimated by fitting a Poisson model. Therefore, this approach falls into the class of Quasi-Maximum Likelihood estimation with additional moment based estimation of ϕ. In the following, this approach is referred to as a Quasi-Poisson approach.

Infection counts of neighboring districts from seven days prior are included to account for potential spillover effects between districts. For each of Limburg-Weilburgs five neighbouring districts—Lahn-Dill-Kreis, Hochtaunuskreis, Rheingau-Taunus-Kries, Rhein-Lahn-Kreis, and Westerwaldkreis—and one internal regressor is added. Figure 2 displays the rolling means of infection counts for all districts under consideration with a window size of one week. Let $N_{i,t}$ denote the number of new infections in neighboring district i at time t. In the same fashion as for the autoregressive part, regressors are included as $\log(N_{i,t} + 1)$ and collected in X_t. Then, the model to be fitted can be summarized according to Equation (8) for the Poisson case.

$$Y_t \sim \text{Pois}(\lambda_t)$$
$$\log(\lambda_t) = \beta_0 + \alpha_1 \log(Y_{t-1} + 1) + \alpha_7 \log(Y_{t-7} + 1) + \beta_1 \log(\lambda_{t-1}) + \zeta' X_t \qquad (8)$$

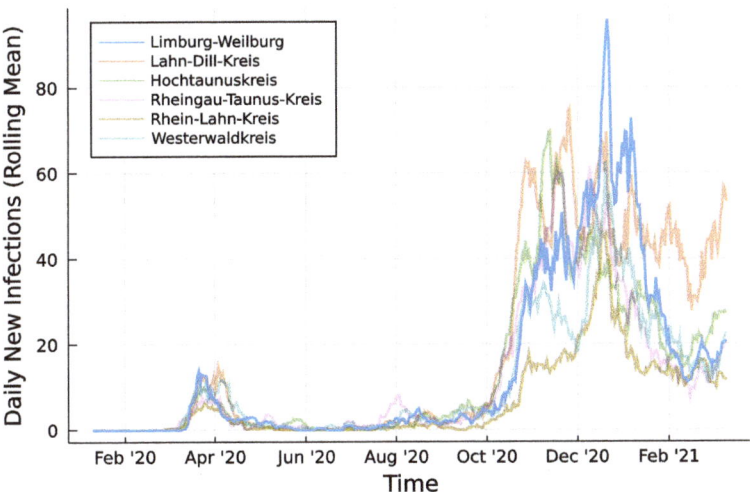

Figure 2. Daily infections in Limburg-Weilburg and Neighbouring Districts—7 day rolling mean.

3.2. Implementation

First step in the Julia implementation from above is to specify the models. With the regressors collected in matrix X, the Poisson model is created by running

```
modelPois = Model(model = "INGARCH",
pastObs = [1, 7],
pastMean = 1,
distr = "Poisson",
link = "Log",
external = fill(false, 5),
X = X)
```

For the alternative model, the distribution is simply replaced by "NegativeBinomial". To fit the models to the time series y, settings are chosen as the Nelder–Mead optimization routine and no inference. Then, the Poisson model is fitted by running

```
settingPois = MLESetting(y, modelPois, inits, optimizer = "NelderMead",
ci = false)
resultsPois = fit(y, modelPois, settingPois)
```

and for the Negative Binomial model accordingly. The retrospective estimation of the overdispersion parameter for the Quasi-Poisson approach is done by QPois(resultsPois). The function returns an updated results object.

In the last step, simulation-based prediction is performed using the function predict. For each of the three results, a matrix of new values for regressors is needed and saved as xNew. Then, the Julia code for a prediction with 10,000 chains is

```
predict(resultsPois, 7, 10000, xNew)
```

The application starts with using the first 322 observations until 30 November 2020 for estimation of parameters and predicting the following week. Successively, one further observation is used for estimation and another prediction performed until the first 420 observations are used for estimation.

3.3. Results

To assess the predictions, a first look reveals only minor visual differences between Poisson fit and the Quasi-Poisson approach as seen in Figure A1 in the Appendix A. Table 1 summarizes two different accuracy measures and the percentage of observed values which lie inside the prediction intervals. For every prediction horizon $h = 1, 2, \ldots, 7$, the root mean squared prediction error (RMSPE) is given. Further, the median absolute prediction error (MedAPE) is given to account for few large deviations that may influence the RMSPE. For both measures and all three approaches, one can see that a larger prediction horizon does not affect the accuracy in a negative way. The predictions even tend to be closer to the observed values for a larger horizon. The Quasi-Poisson approach is slightly more accurate compared to the Poisson approach. The absolute difference of prediction and observed counts is smaller than 10 in ~50% of cases for the two approaches regardless of the horizon. The Negative Binomial model appears to be less accurate compared to the other two. One reason for it might be the additional parameter during the maximization of the likelihood. This might cause a higher uncertainty in the estimation of all parameters and thus a less precise prediction. The difference between a Poisson model and the Quasi-Poisson can be seen by the percentage of observations inside the corresponding 95% prediction intervals. The Poisson case is far away from the desired 95%, prediction intervals are too narrow. The model simply can not account for the conditional overdispersion present in the data. The Quasi-Poisson and the Negative Binomial model are able to account for it.

To put these accuracy measures into context, the mean number of new infections being predicted is around 40. None of the three approaches can be considered very precise. However, this application showed that the Quasi-Poisson estimation can combine the advantages of a Poisson model and the Negative Binomial model.

In addition to the prediction results, Table 2 displays estimates, standard errors, and 95% confidence intervals when considering the complete data set in the estimation. On the left hand side, standard errors are computed from the Poisson model. The estimate of ϕ stems from the Quasi-Poisson approach. One can see that both autoregressive parameters α_1 and α_7 are significant, whereas the parameter β_1 is not. An INARCH model might be an alternative to the model considered here. The spillover effects are negative only for the first neighboring district, but not significant. The remaining four are significant and positive. The same can be found in the Negative Binomial case. In contrast to the Poisson model, neither the parameter β_1 nor both autoregressive parameters α_1 and α_7 are significant.

Table 1. Prediction Limburg–Weilburg: Root Mean Squared Prediction Error, Median Absolute Prediction Error and Percentage of Observations Inside the 95% Prediction Interval.

Criterion	Model	Prediction Horizon						
		1	2	3	4	5	6	7
RMSPE	Poisson	29.53	30.18	30.11	30.14	28.81	28.07	27.86
	Quasi-Poisson	29.53	30.07	30.04	29.98	28.59	27.89	27.71
	Negative Binomial	30.95	31.14	30.89	30.86	30.38	29.68	30.09
MedAPE	Poisson	10.14	10.39	10.19	10.27	10.38	10.20	10.03
	Quasi-Poisson	10.07	10.22	10.26	10.12	10.28	9.96	9.69
	Negative Binomial	12.55	12.43	12.19	12.05	11.71	11.97	11.75
Inside PI	Poisson	46.5	46.5	46.5	48.5	50.5	49.5	49.5
	Quasi-Poisson	97.0	97.0	97.0	97.0	97.0	97.0	97.0
	Negative Binomial	98.0	98.0	99.0	99.0	99.0	99.0	99.0

These findings hint at the existence and importance of including spillover effects in an epidemiological model. However, they do not prove any causal relationship, as the the models lack important driving factors of disease transmission. For more reliable results, a multivariate model or spatiotemporal model with all districts jointly enables to include spillover effects in both directions and is better suited for such epidemiological applications.

Table 2. Estimation Results, standard errors, and 95% confidence intervals.

	(Quasi-)Poisson			NegativeBinomial		
	Estimate	Std. Err.	Conf. Interval	Estimate	Std. Err.	Conf. Interval
β_0	−0.159	0.049	(−0.254, −0.063)	−0.540	0.107	(−0.749, −0.330)
α_1	0.041	0.018	(0.006, 0.077)	0.115	0.063	(−0.010, 0.239)
α_7	0.122	0.022	(0.079, 0.165)	0.088	0.080	(−0.068, 0.245)
β_1	0.053	0.030	(−0.006, 0.112)	0.005	0.100	(−0.190, 0.200)
ζ_1	−0.032	0.015	(−0.061, −0.003)	−0.034	0.057	(−0.146, 0.078)
ζ_2	0.221	0.027	(0.168, 0.274)	0.252	0.076	(0.103, 0.401)
ζ_3	0.314	0.029	(0.257, 0.370)	0.356	0.078	(0.203, 0.509)
ζ_4	0.149	0.024	(0.102, 0.196)	0.194	0.084	(0.030, 0.359)
ζ_5	0.265	0.024	(0.218, 0.312)	0.317	0.074	(0.172, 0.463)
ϕ	1.405			1.516	0.164	(1.195, 1.837)

4. Application: Animal Health in New Zealand

Aghababaei Jazi et al. [13] were the first to analyze submissions to animal health laboratories in New Zealand. The data contain the monthly number of submissions with anorexia and skin lesions from 2003 to 2009. The authors use a zero inflated Poisson distribution for the innovations, where the probability of $R_t = 0$ is inflated by a parameter ω. For the time series of animal submissions with skin lesions, this zero inflated INAR(1) process yields a significantly better fit compared to a simple INAR(1).

Mohammadpour et al. [19] argue that the empirical overdispersion in the data cannot be captured by the Poisson distribution. A test developed by Schweer and Weiß [20] was applied and reveals that this overdispersion is significant at the 5% level. To incorporate it, the authors suggest to use either a Negative Binomial distribution or the approach developed by them.

4.1. Model and Implementation

Figure 3 shows both time series. One can see that counts are rather low, a trend is not visible and especially for anorexia data, many observations are zero. The autocorrelation function of both time series in Figure 4 does not speak against a low-order INAR model. Eight different models are compared for the data, an INAR(1) and an INAR(2) model with both a Poisson and a Negative Binomial distribution, and for all of these four once with and once without zero inflation. Other regressors are left out, making the choice of a link function dispensable.

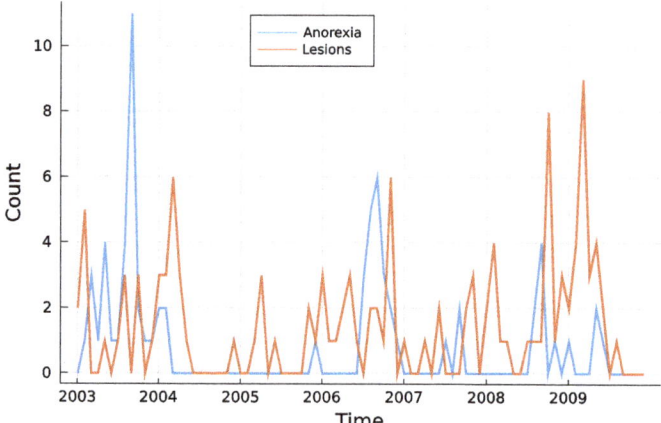

Figure 3. Monthly submissions with anorexia or skin lesions.

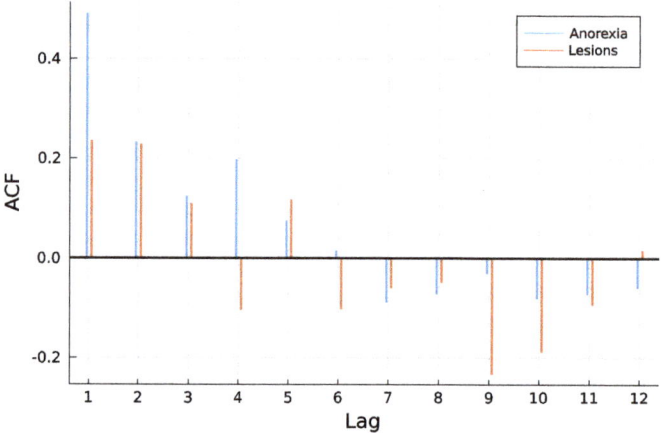

Figure 4. ACF of time series.

The first step in the analysis is to define the eight models, for example, a Poisson INAR(1) as

```
Model(model = "INARMA",
pastObs = 1,
zi = true)
```

making use of default arguments in the function. If a Negative Binomial shall be chosen, one simply needs to add distr = "NegativeBinomial". For a model order $p = 2$, pastObs = 1 is changed to pastObs = 1:2.

With starting values for optimization saved as a vector inits of type parameter, the function MLESettings is again used to specify estimation settings as

```
settings = MLESettings.(fill(dat.Anorexia, 8), models,
optimizer = "NelderMead", ci = true)
```

The dot behind a function name indicates that it shall be applied element-wise, in our case on the vector inits. The output then is a vector of estimation settings. Following that manner, the results for both time series can be computed all at once by

```
results = Array{INARMAResults, 2}(undef, (8, 2))
results[:, 1] = fit.(fill(dat.Anorexia, 8), models, settings)
results[:, 2] = fit.(fill(dat.Lesions, 8), models, settings)
```

If one is now interested in diagnostics, for example, an information criterion, functions can be applied on this array of results as well. In case of the AIC, the user runs

```
AIC.(results, 2)
```

At this point, one needs to keep in mind that information criteria are based on the likelihood. When computing the likelihood, the number of observations to condition on varies with the autoregression order. Especially for rather short time series, this property needs to be accounted for. In the above computation of the AIC, the second argument indicates that for each model, the contribution of the first two observations to the likelihood are ignored making information criteria comparable.

4.2. Results

Table 3 summarizes the results of the estimation including zero inflation. Besides estimates, standard errors from the Hessian and the AIC are shown. For parameters β_0, α_1, α_2, and ω, colors indicate the significance, where dark green refers to significance at the 0.1% level, medium green at 1%, and light green at 5%. The p-value of the overdispersion parameter ϕ is ignored, as it can not be zero. In addition, Table A1 in the Appendix A shows results for all models without zero inflation.

Mohammadpour et al. [19] analyze the data on lesions and find that a Poisson distribution does not capture the overdispersion. The results below confirm these findings in terms of the AIC. For an INAR(1) and an INAR(2), the Negative Binomial dominates the Poisson distribution. Moreover, including the lag order $p = 2$ lowers information criteria for both distributions and the estimates of α_1 are not significant at the 5% level in that case.

Aghababaei Jazi et al. [13] emphasize the benefit of incorporating zero inflation in the innovation process for the analysis of animal submissions with skin lesions. Only when considering the Poisson models, this is supported by the highly significant zero inflation parameter estimates and a lower AIC. However, when a Negative Binomial is used, the zero inflation parameters are not significant. In fact, dropping zero inflation for the NB-INAR(2) yields an even lower AIC. The results are similar for Anorexia data, although estimates of α_2 are not significant at 5%, both INAR(2) models have a lower AIC than the corresponding INAR(1).

In conclusion, this study has confirmed the findings of Mohammadpour et al. [19]. When only looking at first-order INAR models, there is a conditional overdispersion in the number of animal submissions with skin lesions that cannot be captured by a Poisson distribution. The findings of Aghababaei Jazi et al. [13] can also been confirmed. In case of a conditional Poisson distribution, it is useful to include zero inflation. The need for a zero inflation vanishes in the example when switching to the Negative Binomial. This is comprehensible, as the probability of a Negative Binomial being zero is larger than the probability of a Poisson with the same mean being zero. Therefore, switching to a Negative Binomial also inflates the zero probability. Although this application is only one specific example, we can draw general conclusions from it: The need to extend a given model during an application highly depends on the model you start with. When choosing a model, it can be useful to cover a broader range of potential models to uncover patterns. Especially for higher-order INARMA type models, the computer intensive evaluation of the likelihood makes such broad comparisons cumbersome. However, the CountTimeSeries package provides an efficient implementation of the likelihood and makes such broad comparisons feasible.

Table 3. Estimation results with zero inflation: significance highlighting for 5% level (light green), 1% level (medium green), 0.1% level (dark green), and not significant at the 5% level (grey).

	Model	$\hat{\beta}_0$	$\hat{\alpha}_1$	$\hat{\alpha}_2$	$\hat{\phi}$	$\hat{\omega}$	AIC
Anorexia	Pois-INAR(1)	2.215 (0.428)	0.338 (0.074)	-	-	0.669 (0.069)	183.00
	Pois-INAR(2)	2.692 (0.536)	0.249 (0.090)	0.118 (0.070)	-	0.752 (0.061)	179.85
	NB-INAR(1)	1.642 (0.831)	0.310 (0.077)	-	1.278 (1.766)	0.424 (0.283)	182.25
	NB-INAR(2)	2.363 (0.794)	0.225 (0.093)	0.120 (0.073)	2.668 (3.561)	0.630 (0.178)	179.94
Lesions	Pois-INAR(1)	2.042 (0.278)	0.175 (0.071)	-	-	0.372 (0.077)	276.26
	Pois-INAR(2)	2.055 (0.333)	0.110 (0.073)	0.176 (0.073)	-	0.464 (0.089)	271.32
	NB-INAR(1)	1.344 (0.599)	0.130 (0.077)	-	1.018 (1.081)	0.047 (0.242)	270.60
	NB-INAR(2)	1.252 (0.754)	0.084 (0.076)	0.166 (0.075)	0.920 (1.218)	0.103 (0.313)	267.76

5. Application: Corporate Insolvencies in Rhineland-Palatinate

After both model frameworks are introduced, the focus is now put on model choice and diagnostics. Weiß and Feld [11] were the first to analyze corporate insolvencies in Rhineland-Palatinate, one of the 16 federal states in Germany. The data were made available as supplementary material and are used in the scope of the following. Rhineland-Palatinate consists of 36 administrative districts at the NUTS-3 level: 24 rural districts and 12 district-free cities. For each of these districts, the monthly number of corporate insolvencies between 2008 and 2016 is observed. The data are used to investigate whether there is a downwards trend in insolvency numbers. Röhl and Vogt [21] state that this trend is especially visible after the financial crisis.

5.1. Models and Implementation

The question whether there was a trend in insolvencies is going to be answered for each district individually. This enables to descriptively check for regional differences in insolvency counts. Although insolvencies are presumably influenced by macroeconomic factors, the only external influence incorporated here is the potential trend. Besides that, the process is assumed to be self-driven. Looking at the total counts in Figure 5, one can see the downwards trend starting around 2010, as Röhl and Vogt [21] state. Further, there seems to be no seasonality pattern in the time series.

Weiß and Feld [11] investigated the model choice considering IID data with and without a linear trend and either a Poisson or Negative Binomial distribution. Here, we include IID data and both frameworks, INGARCH and INARMA. Potential orders are chosen to be (1, 0), (2, 0), (1, 1), and (2, 1) to capture serial correlation if present. Both distributions, Poisson and Negative Binomial, are incorporated. A trend component may be omitted, included for the complete range of time, or starting in January 2010. Thereby, all link functions are chosen to be a log-linear link. That way, negative coefficients of the trend component are no threat to the restriction of positive conditional means. This sums up to 54 potential models to be estimated for each of the 36 time series. To save computation time, inference is only conducted for chosen models and not for all 1944 combinations of models and time series. In the package, standard errors are computed via the numerical Hessian.

After the estimation, information criteria can be computed by running AIC(results), BIC(results), or HQIC(results). An additional argument dropfirst can be passed to these functions to suppress the likelihood contributions of first observations. This comes handy when comparing for example an INARCH(1) and an INARCH(2). Likelihood based estimation of these conditions on the first one or two observations respectively. For rather short time series, this difference might be crucial.

To check if a choice of model is suitable, the non-randomized probability integral transform (PIT) histogram is used. It was developed by Czado et al. [14] for count data time series and does not rely on random numbers. The packages function pit produces a histogram running pit(results, nbins = 10, level = 0.95) that is uniformly distributed if the model choice is correct. The argument nbins specifies the number of bins and the argument level can be used to test the uniform distribution at the $(1 - \text{level})$-level. If level is put in, lines are drawn in the histogram and if at least one bin exceeds the lines, the null hypothesis of a uniform distribution is rejected.

In the analysis, potential models are defined and fitted to all 36 time series. Then, the best model is chosen for each district and verified by the packages diagnostic tool.

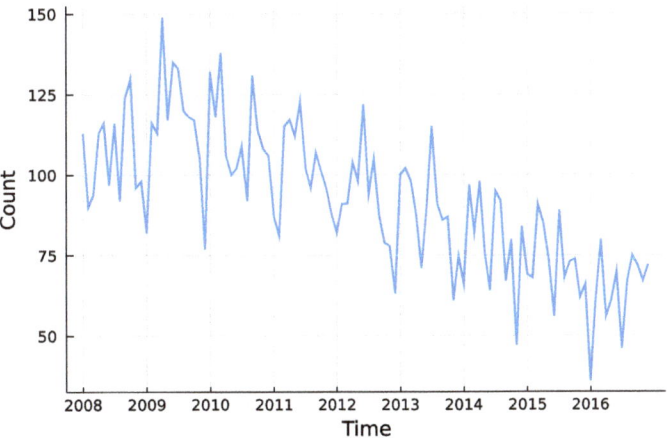

Figure 5. Total number of monthly insolvencies in Rhineland-Palatinate.

5.2. Results

In the fitting of 54 models to each of the 36 districts, all optimization procedures converged. For districts with a marginal underdispersion, all models with a Negative Binomial distribution have not been fitted. Inference has not been conducted so that results purely rely on the model selection criterion. Table A2 summarizes the model selection for all 36 districts together with mean and variance of time series. Many of the models selected exhibit no serial dependence. In these cases, and as is to be expected, the Negative Binomial was selected when the variance is clearly larger than the mean.

All selected models have been checked with the non-randomized PIT histogram. None of the selected models produce a PIT histogram for which the null hypothesis of a uniform distribution was rejected. Figure 6 shows an example PIT histogram for Trier, where the dashed lines represent critical values for the height on bins for which, if exceeded, the null hypothesis would be rejected. The confidence bank is rather wide due to few observations.

The (conditional) distribution is an indicator of how much the insolvency count varies compared to its mean. Figure A2 in the Appendix A displays the selected distributions where a clustering of districts with the same distribution is only vaguely recognizable.

A non-zero model order p or q implies a serial dependence besides a potential trend. In the case at hand, for districts with serial dependence, all estimates indicate that a larger/smaller number of insolvencies in one month tends to go along with a larger/smaller number in the month after. One reason for it can be an interconnected economy within the district, but not necessarily. Macroeconomic factors can also cause such dependence. Figure 7 displays the selected models orders. One can see that those districts with dependence seem to cluster.

The selected time trends and the sign of the corresponding coefficient are summarized in Figure 8. A grey district has no trend component in the selected model, light colors represent districts with a trend starting in 2010 being selected, and dark colors represent a selected trend along the complete observed time. Thereby, districts are colored green for a downwards trend in the insolvency counts and red for an upwards trend.

Only one district, the urban area of Trier, exhibits an upwards trend. Seven districts do not have a trend in the selected model and the remaining 28 show a downward trend. Districts in the center of Rhineland-Palatinate and the district far east tend to have a trend component starting in 2010. A complete trend was chosen more often in the southern region and the north.

This analysis did not incorporate any macroeconomic variables to explain the insolvencies and also no information about the companies being bankrupt, yet it showed that characteristics in terms of dispersion, autocorrelation and time trend cluster spatially. The results suggest that there could be an interconnection between districts and that it would be interesting to conduct the study in a multivariate setting instead of looking at districts individually.

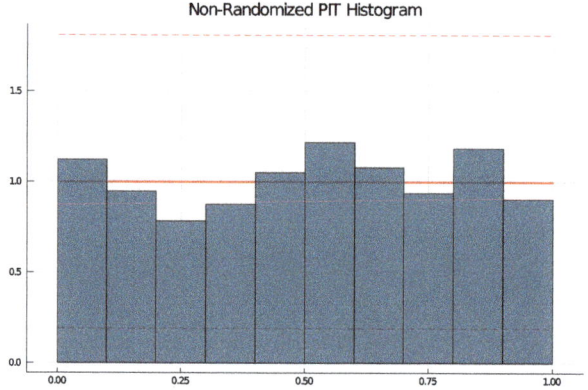

Figure 6. PIT histogram for Trier.

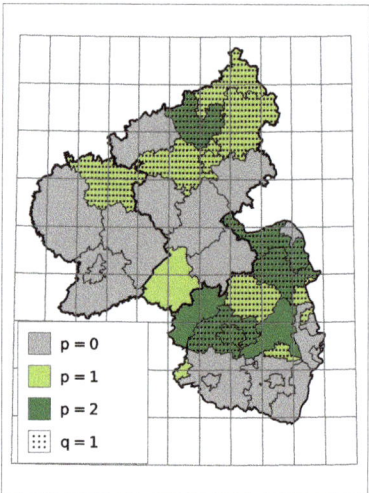

Figure 7. Selected Model Order.

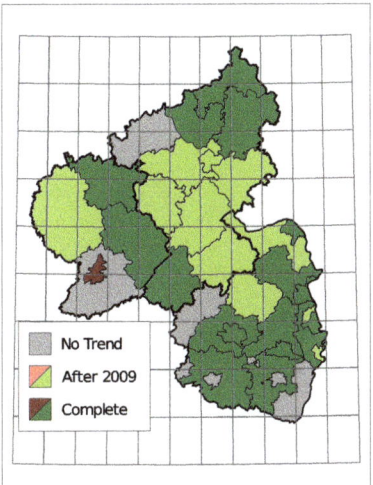

Figure 8. Selected Trend: upwards (red) or downwards (green).

6. Simulation Study: Finite Sample ML—Estimation

In many applications, the length of count data time series is rather short. Consistency and asymptotic normality are proven for Maximum Likelihood estimates of an INGARCH processes parameters, but a potential finite sample bias is often not addressed.

This last application pursues two goals. On the one hand, finite sample properties of a Maximum Likelihood estimation are investigated in a simulation study. On the other hand, the analysis is performed with both the CountTimeSeries package in Julia and the tscount package in R. Computation time for the estimation is compared between these two.

6.1. Study and Implementation

The general strategy for this study is to generate time series following an INGARCH process and then estimate its parameters by Maximum Likelihood. The study focusses on Poisson INGARCH(1, 1) processes with identity link, as it is commonly used in applications. The general framework covered by the CountTimeSeries package as described in Section 2 thus simplifies to

$$Y_t | \mathcal{F}_{t-1} \sim \text{Pois}(\lambda_t)$$
$$\lambda_t = \beta_0 + \alpha_1 Y_{t-1} + \beta_1 \lambda_{t-1}$$

Every combination of parameters $\alpha_1, \beta_1 \in \{0.1, 0.2, \ldots, 0.8\}$ with a sum smaller than 1 is incorporated in this study, which ensures stationarity. The intercept is chosen in such a way that the marginal mean $\beta_0/(1 - \alpha_1 - \beta_1)$ equals 15 for every choice of α_1 and β_1. Then, for each set of parameters, a total of 1000 time series of lengths $T \in \{50, 200, 1000\}$ are generated and its parameters estimated with four slightly different methods. The conditional mean recursion λ_t above can be initialized with either the marginal mean, the first observation or the intercept in the CountTimeSeries package. The fourth possibility, not covered by both packages, is to treat the initial value of the recursion as additional parameter and estimate it.

Time series can be generated with the CountTimeSeries package by running

```
simulate(T, model, truepar)
```

for a model specification `model` and parameters `truepar`. Then, with estimation settings collected in the object `setting`, parameters are estimated by

```
fit(y, model, setting, initiate = initiate, printResults = false)
```

The optional argument `initiate` specifies how the recursion λ_t is started and the argument `printResults` can be set to `false` to omit that results are printed in the console.

6.2. Results

Figures 9 and 10 display the mean bias of estimates α_1 and β_1, respectively, and negative values thereby indicate an estimate smaller than the true value. Note that for a true value of 0.1, there are eight different parameter combinations aggregated in the bias, but only one combination for a true value of 0.8 on the x axis due to the stationarity restriction. Initializing the recursion with the marginal mean and the first value appear to yield similar biases. This is a plausible result, considering that the first observation should equal the marginal mean on average. Treating the first conditional mean as parameter also gives similar biases in case of moderate ($T = 200$) and long ($T = 1000$) time series. In case of short time series $T = 50$, this approach exhibits the largest absolute bias for α_1 while at the same time being the least biased for β_1. Initializing the recursion with the intercept seems to perform poorly especially for the estimate of β_1 for larger true values of β_1. In general, the absolute bias should be larger for larger sums of α_1 and β_1, as the deviation between intercept and expectation of λ_1 is larger.

As to be expected by the consistency of Maximum Likelihood estimates, longer time series go along with smaller biases. In addition to Figures 9 and 10, Figures A3 and A4 in the Appendix A show relative biases. Thereby, biases are scaled by the true value of the parameter to be estimated. It reveals that the estimate α_1 exhibits larger relative biases for smaller true values.

Only considering the first three initialization methods, one single estimation took around 0.003 s for $T = 50$, 0.007 s for $T = 200$ and 0.02 s for $T = 1000$ on average in Julia. The initialization method does not affect the computation time while larger values of $\alpha_1 + \beta_1$ increased the computation time. Using the R package tscount, estimation took on average 0.11, 0.18 and 0.55 s for the three lengths of time series and thus around 25 times longer.

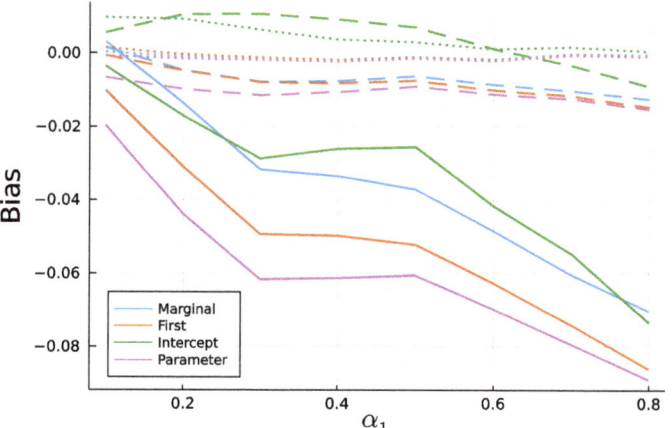

Figure 9. Mean Bias of $\hat{\alpha}_1$ for T = 50 (solid), T = 200 (dashed), and T = 1000 (dotted).

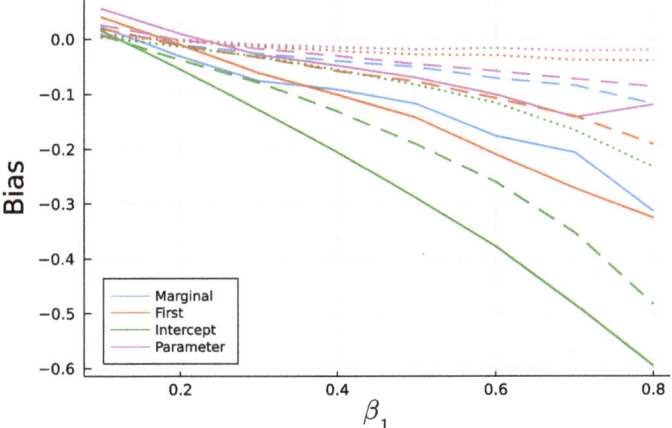

Figure 10. Mean Bias of $\hat{\beta}_1$ for for T = 50 (solid), T = 200 (dashed), and T=1000 (dotted).

Conclusively, this simulation study revealed the importance of a well chosen initialization method as well as systematic biases for Maximum Likelihood fitting of short INGARCH(1, 1) time series. Starting the mean recursion with the intercept should be avoided. Treating the first conditional mean as parameter might be the least restrictive among the four methods, but as an additional parameter, it might add noise to the estimation of the other parameters. It is recommended to initialize the mean recursion by the processes marginal mean or the first observed value in case of an INGARCH(1, 1), since the biases are comparable to the parameter initializer. However, for some processes in the INGARCH framework, for example an INGARCH(1, 1) with log-linear link, computation of the marginal mean is computer intensive or must be approximated. Therefore, starting the recursion with the first observation appears to be a good choice. It is also the default setting in the CountTimeSeries package.

7. Discussion and Outlook

Three real-life applications and the simulation study were chosen to give a broad overview on the CountTimeSeries package. All functions of the package have been shown: specifying a model, generating time series, estimating parameters, using information criteria, as well as a diagnostic tool and forecasting. In the spirit of the Julia programming language, the package is easy to learn, covers a broad range of models, and is fast compared with interpreted languages. It is the first software package that covers both widely used frameworks, INGARCH and INARMA processes.

Application 1 showed how the CountTimeSeries package can be used to investigate complex non-stationary time series from epidemiology and compute forecasts. Comparing prediction accuracy the prediction intervals, it revealed the advantage of the Quasi-Poisson approach. Application 2 and 3 showed how to fit multiple models to data simultaneously thanks to element-wise application of functions. In the second application, a contribution was made to an existing discussion about the need to incorporate zero inflation and overdispersion in a model for animals with skin lesions in New Zealand. Although the results of the previous literature have been confirmed, increasing the model order has put these findings in a different light.

Application 3 used corporate insolvency data to analyze if model choices based on information criteria cluster spatially. In line with previous literature, a downwards trend in the number of insolvencies was visible for most districts in Rhineland-Palatinate. Districts where this trend started in 2010 after the financial crisis appeared to come in clusters. Similar clustering effects were visible for dispersion characteristics and the order of dependency.

The simulation study in the end showed that there is a systematic bias when fitting an INGARCH(1, 1) model to a rather short time series. Thereby, it was revealed that one should not initialize the conditional mean recursion by the intercept. In comparison with R, the Julia package was faster, making broad simulation studies computationally more feasible.

Although the CountTimeSeries package already covers many important models, some extensions would be useful supplements. A function to compute the mean, variance, and ACF of a given model and parameters would first enable to compare empirical moments and theoretical moments from estimation results. In applications this might reveal weaknesses in the model choice. Further, such a function could be used to estimate parameters in a GMM approach. Especially for higher-order INARMA models, likelihood evaluation is computationally expensive. GMM might enable estimation in acceptable runtime. For models with no closed form moments, a simulation-based approximation of moments could be an alternative. If GMM estimation was integrated into the package, the computation of empirical moments could easily be replaced by robust moment estimators, like trimmed means, to provide robust estimation techniques.

When dealing with real-life data, extreme observations, not following the usual model, are commonly found. Other approaches to account for possible extreme observations include classical M-estimators or the recently published method by Li et al. [22] or approaches based on the density power divergence, see for example Xiong and Zhu [23].

Another family of models that would enrich the Julia package are models with bounded counts, see for example Weiß [24]. Examples for the usefulness of such models include time series or cross sectional data on product ratings or the number of rainy days in a month. An approach to include zero inflation for such models is discussed by Möller et al. [25]. As an alternative to zero inflation as presented here, would be hurdle models to account for a surplus of zeros.

Above extensions match the notation and the current infrastructure of the package and thus do not need a substantial reconstruction of the package. An inclusion of multivariate models and spatiotemporal models brings along new challenges for the inclusion in the Julia package, but would be rewarding for many applications. The COVID-19 application discussed above is only one example where a multivariate model is useful. Modeling multiple interacting count data processes jointly is also frequently used in finance, see, for example, Quoreshi [26].

Supplementary Materials: The following are available online at https://www.mdpi.com/article/10.3390/e23060666/s1, Julia and R code for all four applications.

Funding: The author acknowledges support from the Open Access Publication Fund of the University of Münster.

Data Availability Statement: Data used in application 1 are provided by RKI [15] and published by NPGEO [16] at https://npgeo-corona-npgeo-de.hub.arcgis.com/datasets/dd4580c810204019a7b8eb3e0b329dd6_0/data (accessed on 16 March 2021). Data on animal submissions in application 2 is provided in the article by Aghababaei Jazi et al. [13]. The data set on corporate insolvencies is found in the supplementary material of Weiß and Feld [11]. Shapefiles for Germany used to create Figures in application 3 are provided by eurostat [27].

Acknowledgments: The author further acknowledges the valuable advice from all reviewers.

Conflicts of Interest: The author declares no conflict of interest.

Appendix A

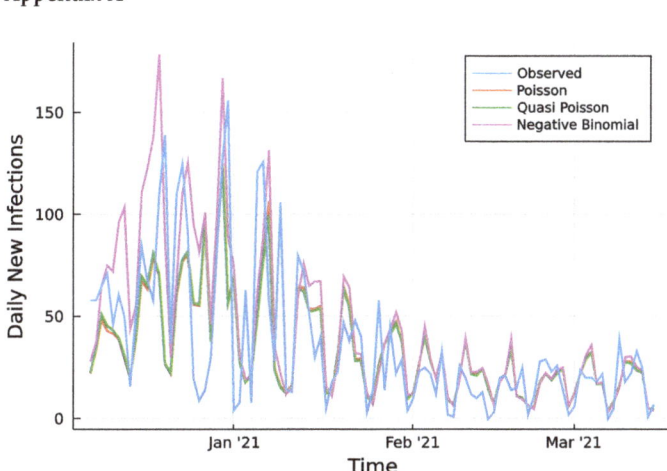

Figure A1. 7 days ahead prediction and observed counts.

Table A1. Estimation results without zero inflation—significance highlighting for 5% level (light green), 1% level (medium green), 0.1% level (dark green), and not significant at the 5% level (gray).

	Model	$\hat{\beta}_0$	$\hat{\alpha}_1$	$\hat{\alpha}_2$	$\hat{\phi}$	AIC
Anorexia	Pois-INAR(1)	0.511 (0.087)	0.385 (0.073)	-	-	225.05
	Pois-INAR(2)	0.450 (0.088)	0.353 (0.078)	0.098 (0.072)	-	224.70
	NB-INAR(1)	0.579 (0.173)	0.304 (0.078)	-	0.194 (0.011)	181.64
	NB-INAR(2)	0.545 (0.188)	0.220 (0.096)	0.118 (0.078)	0.139 (0.059)	180.37
Lesions	Pois-INAR(1)	1.172 (0.146)	0.173 (0.068)	-	-	294.77
	Pois-INAR(2)	0.976 (0.153)	0.145 (0.068)	0.132 (0.067)	-	292.17
	NB-INAR(1)	1.236 (0.217)	0.128 (0.076)	-	0.839 (0.006)	268.59
	NB-INAR(2)	1.017 (0.214)	0.084 (0.076)	0.164 (0.075)	0.608 (0.248)	265.85

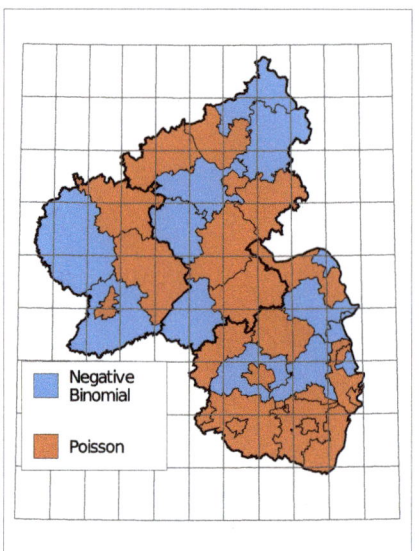

Figure A2. Selected distribution.

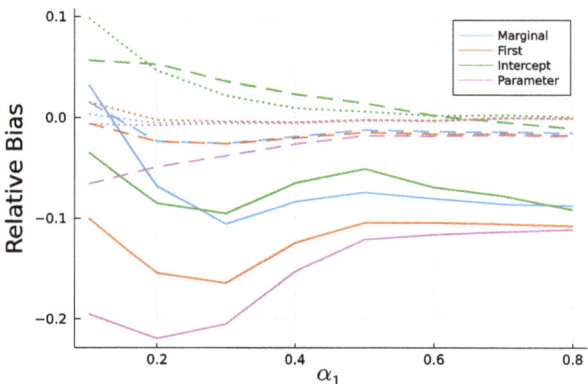

Figure A3. Mean relative bias of $\hat{\alpha}_1$ for T = 50 (solid), T = 200 (dashed), and T = 1000 (dotted).

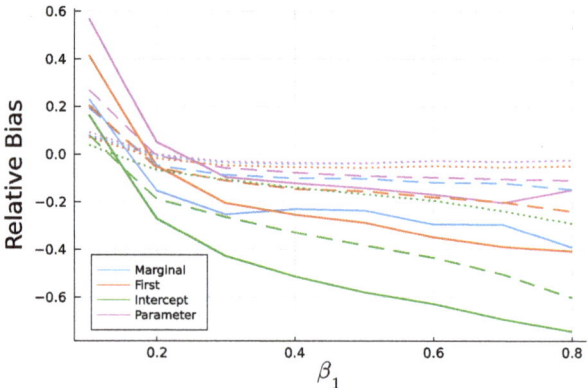

Figure A4. Mean relative bias of $\hat{\beta}_1$ for T = 50 (solid), T = 200 (dashed), and T = 1000 (dotted).

Table A2. Model choice for each district.

District	Mean	Var	Model	Trend
Ahrweiler	3.11	3.39	Pois IID	No
Altenkirchen	3.03	5.34	NB-INGARCH(1, 1)	Yes
Alzey-Worms	2.76	4.61	NB-INGARCH(2, 1)	Yes
Bad Dürckheim	1.88	2.57	NB INARCH(2)	Yes
Bad Kreuznach	4.94	5.90	Pois IID	Partly
Bernkastell-Wittlich	3.25	4.58	Pois IID	Yes
Birkenfeld	2.29	3.22	NB-INARCH(1)	Yes
Cochem-Zell	1.29	1.72	NB IID	Partly
Donnersbergkreis	1.38	1.86	P-INGARCH(1, 1)	Partly
Eifelkr. Bitburg-Prüm	2.03	3.28	NB-IID	Partly
Frankenthal, kfr. S.	0.96	1.29	P-INGARCH(1, 1)	Partly
Gemersheim	1.71	1.93	Pois IID	No
Kaiserslautern, kfr. S.	2.95	4.59	P-INGARCH(2, 1)	Yes
Kaiserslautern	2.73	6.03	NB-INGARCH(2, 1)	Yes
Koblenz	3.49	4.79	P-INGARCH(1, 1)	Partly
Kusel	1.27	1.53	P-INARCH(2)	No
Landau i.d.P	0.83	0.91	Pois IID	No
Ludwigshafen	3.06	4.73	NB IID	Yes
Mainz	4.90	10.62	NB IID	Yes
Mainz-Bingen	4.55	8.21	P-INGARCH(2, 1)	Partly
Mayen-Koblenz	5.24	6.67	NB-INGARCH(1, 1)	Partly
Neustadt a.d.W.	1.06	1.23	P-INGARCH(1, 1)	Yes
Neuwied	6.73	11.84	P-INGARCH(2, 1)	Yes
Pirmasens	0.94	0.91	Pois IID	No
Rhein-Hunsrück-Kreis	2.56	3.07	Pois IID	Partly
Rhein-Lahn-Kreis	2.88	3.83	Pois IID	Partly
Rhein-Pfalz-Kreis	2.40	2.82	Pois IID	Yes
Speyer	1.01	1.04	Pois IID	Partly
Südliche Weinstraße	1.76	1.98	Pois IID	Yes
Südwestpfalz	1.61	1.87	Pois IID	Yes
Trier, kfr. S.	1.87	1.85	Pois IID	Yes
Trier-Saarburg	1.46	1.90	NB IID	No
Vulkaneifel	1.41	1.70	P-INGARCH(1, 1)	Yes
Westerwald	5.30	8.66	NB-INGARCH(1, 1)	Yes
Worms	2.94	8.88	NB-INGARCH(1, 1)	Yes
Zweibrücken	0.88	1.13	P-INARCH(1)	No

References

1. Alzaid, A.; Al-Osh, M. First-Order Integer-Valued Autoregressive (INAR(1)) Process: Distributional and Regression Properties. *Stat. Neerl.* **1988**, *41*, 53–60. [CrossRef]
2. Ferland, R.; Latour, A.; Oraichi, D. Integer-Valued GARCH Process. *J. Time Ser. Anal.* **2006**, *27*, 923–942. [CrossRef]
3. Bezanson, J.; Karpinski, S.; Shah, V.B.; Edelman, A. Julia: A Fast Dynamic Language for Technical Computing. *arXiv* **2012**, arXiv:1209.5145.
4. Liboschik, T.; Fried, R.; Fokianos, K.; Probst, P. tscount: Analysis of Count Time Series; R Package Version 1.4.1. Available online: https://cran.r-project.org/web/packages/tscount/index.html (accessed on 16 March 2021).
5. Weiß, C.H.; Feld, M.H.J.M.; Mamode Khan, N.; Sunecher, Y. INARMA Modeling of Count Time Series. *Stats* **2019**, *2*, 284–320. [CrossRef]
6. Harte, D. *HiddenMarkov: Hidden Markov Models*; R Package Version 1.8-11; Statistics Research Associates: Wellington, New Zealand, 2017.
7. Himmelmann, L. *HMM: HMM—Hidden Markov Models*; R Package Version 1.0. Available online: https://cran.r-project.org/web/packages/HMM/index.html (accessed on 16 March 2021).
8. Jackman, S. pscl: Classes and Methods for R Developed in the Political Science Computational Laboratory; R Package Version 1.5.5. Available online: https://github.com/atahk/pscl/ (accessed on 16 March 2021).
9. Zeileis, A.; Kleiber, C.; Jackman, S. Regression Models for Count Data in R. *J. Stat. Softw.* **2008**, *27*, 1–25. [CrossRef]

10. Mouchet, M. *HMMBase—A Lightweight and Efficient Hidden Markov Model Abstraction*. 2020. Available online: https://github.com/maxmouchet/HMMBase.jl (accessed on 16 March 2021).
11. Weiß, C.H.; Feld, M. On the performance of information criteria for model identification of count time Series. *Stud. Nonlinear Dyn. Econom.* **2019**, *24*. [CrossRef]
12. Liboschik, T.; Kerschke, P.; Fokianos, K.; Fried, R. Modelling interventions in INGARCH processes. *Int. J. Comput. Math.* **2016**, *93*, 640–657. [CrossRef]
13. Aghababaei Jazi, M.; Jones, G.; Lai, C.D. First-order integer valued AR processes with zero inflated poisson innovations. *J. Time Ser. Anal.* **2012**, *33*, 954–963. [CrossRef]
14. Czado, C.; Gneiting, T.; Held, L. Predictive Model Assessment for Count Data. *Biometrics* **2009**, *65*, 1254–1261. [CrossRef] [PubMed]
15. RKI. Robert-Koch-Institut: SurvStat@RKI 2.0. 2021. Available online: https://survstat.rki.de/ (accessed on 16 March 2021).
16. NPGEO. RKI COVID19. 2021. Available online: https://npgeo-corona-npgeo-de.hub.arcgis.com/datasets/dd4580c810204019a7b8eb3e0b329dd6_0 (accessed on 16 March 2021).
17. World Health Organization. Transmission of SARS-CoV-2: Implications for Infection Prevention Precautions. 2020. Available online: https://www.who.int/news-room/commentaries/detail/transmission-of-sars-cov-2-implications-for-infection-prevention-precautions (accessed on 7 May 2021).
18. Christou, V.; Fokianos, K. Quasi-Likelihood Inference for Negative Binomial Time Series Models. *J. Time Ser. Anal.* **2014**, *35*, 55–78. [CrossRef]
19. Mohammadpour, M.; Bakouch, H.; Shirozhan, M. Poisson-Lindley INAR(1) model with applications. *Braz. J. Probab. Stat.* **2018**, *32*, 262–280. [CrossRef]
20. Schweer, S.; Weiß, C.H. Compound Poisson INAR(1) processes: Stochastic properties and testing for overdispersion. *Comput. Stat. Data Anal.* **2014**, *77*, 267–284. [CrossRef]
21. Röhl, K.H.; Vogt, G. Unternehmensinsolvenzen in Deutschland. 2019. Available online: https://www.iwkoeln.de/studien/iw-trends/beitrag/klaus-heiner-roehl-unternehmensinsolvenzen-in-deutschland-trendwende-voraus-449151.html (accessed on 16 March 2021)
22. Li, Q.; Chen, H.; Zhu, F. Robust Estimation for Poisson Integer-Valued GARCH Models Using a New Hybrid Loss. *J. Syst. Sci. Complex.* **2021**. [CrossRef]
23. Xiong, L.; Zhu, F. Minimum Density Power Divergence Estimator for Negative Binomial Integer-Valued GARCH Models. *Commun. Math. Stat.* **2021**. [CrossRef]
24. Weiß, C.H. Stationary count time series models. *WIREs Comput. Stat.* **2021**, *13*. [CrossRef]
25. Möller, T.; Weiß, C.; Kim, H.Y.; Sirchenko, A. Modeling Zero Inflation in Count Data Time Series with Bounded Support. *Methodol. Comput. Appl. Probab.* **2018**, *20*. [CrossRef]
26. Quoreshi, A.M.M.S. Bivariate Time Series Modeling of Financial Count Data. *Commun. Stat. Theory Methods* **2006**, *35*, 1343–1358. [CrossRef]
27. Eurostat. GISCO: Geographische Informationen und Karten. 2021. Available online: https://ec.europa.eu/eurostat/de/web/gisco/geodata/reference-data/administrative-units-statistical-units/nuts (accessed on 16 March 2021).

MDPI
St. Alban-Anlage 66
4052 Basel
Switzerland
Tel. +41 61 683 77 34
Fax +41 61 302 89 18
www.mdpi.com

Entropy Editorial Office
E-mail: entropy@mdpi.com
www.mdpi.com/journal/entropy

www.ingramcontent.com/pod-product-compliance
Lightning Source LLC
LaVergne TN
LVHW070240100526
838202LV00015B/2156